The Science and Engineering of Materials

The Science and Engineering of Materials

Editor: Heather Dale

NY RESEARCH PRESS

New York

Published by NY Research Press
118-35 Queens Blvd., Suite 400,
Forest Hills, NY 11375, USA
www.nyresearchpress.com

The Science and Engineering of Materials
Edited by Heather Dale

International Standard Book Number: 978-1-63238-641-0 (Hardback)

Cataloging-in-Publication Data

The science and engineering of materials / edited by Heather Dale.
    p. cm.
Includes bibliographical references and index.
ISBN 978-1-63238-641-0
1. Materials science. 2. Materials. I. Dale, Heather.
TA403 .S35 2019
620.11--dc23

Contents

Preface

It is often said that books are a boon to mankind. They document every progress and pass on the knowledge from one generation to the other. They play a crucial role in our lives. Thus I was both excited and nervous while editing this book. I was pleased by the thought of being able to make a mark but I was also nervous to do it right because the future of students depends upon it. Hence, I took a few months to research further into the discipline, revise my knowledge and also explore some more aspects. Post this process, I begun with the editing of this book.

Materials science is an interdisciplinary study of the engineering of materials. It integrates principles of chemistry, metallurgy, ceramics and solid state physics. This discipline is concerned with the understanding of the properties, structure and manufacturing as well as the design of materials. Studies in these dimensions are useful for advancing the techniques of nanotechnology, biomaterials and metallurgy. It is also involved in developing analyses and investigations in forensics and failure detection in engineering processes. This book elucidates the concepts and innovative models around prospective developments with respect to the science and engineering of materials. It presents studies and researches performed by experts across the globe and foregrounds the practical applications and ramifications of the theories relevant to this discipline. Those who want to develop a thorough understanding of this field will be greatly benefited by this book.

I thank my publisher with all my heart for considering me worthy of this unparalleled opportunity and for showing unwavering faith in my skills. I would also like to thank the editorial team who worked closely with me at every step and contributed immensely towards the successful completion of this book. Last but not the least, I wish to thank my friends and colleagues for their support.

Editor

Analysis of Mechanical Behavior and Microstructural Characteristics Change of ASTM A-36 Steel Applying Various Heat Treatment

Hasan MF*

Department of Industrial Engineering and Management, Khulna University of Engineering and Technology, Khulna-9203, Bangladesh

Abstract

In this study ASTM A-36 (mild steel) is selected as specimen for testing various mechanical properties and microstructure change. The effects of heat treatment on the mechanical properties and microstructure characteristics change of selected specimen are analyzed. Annealing, hardening and tempering are the most important heat treatment processes often used to change mechanical properties of engineering materials. The purpose of heat treating is to analyze the mechanical properties of the steel, usually ductility, hardness, Yield strength, tensile strength and impact resistance. The heat treatment develops hardness, softness, and improves the mechanical properties such as tensile strength, yield strength, ductility, corrosion resistance and creep rupture. These processes also help to improve machining effect, and make them versatile. The mechanical properties can easily be modified by heat treating to suit a particular design purpose. In the present study, selected samples are heat-treated at certain temperature above the austenitic region and quenched in order to investigate the effect on the mechanical properties microstructure of the mild steel. The changes in mechanical behavior and microstructure as compared with unquenched samples are explained in terms of changes in tensile strength. Results showed that the mechanical properties of mild steel can be changed and improved by various heat treatments for a particular application.

Keywords: Heat treatment; Mild steel; Mechanical properties; Microstructure; Universal testing machine; Rockwell hardness tester

Introduction

The subject of mechanical testing of materials is an important aspect of engineering practice. Today, more concern is being given to the interpretation of test results in terms of service performance, as well as giving reliable indications of the ability of the material to perform certain types of duty. Mechanical tests are also employed in investigational work in order to obtain data for use in design to ascertain whether the material meets the specifications for its intended use. Heat treatment is defined as an operation or combination of operations involving heating and cooling of a metal or alloy for this case involving the mild steel in the solid state in such ways as to produce certain microstructure and desired mechanical properties (hardness, toughness, yield strength, ultimate tensile strength, Young's modulus, percentage elongation and percentage reduction). Annealing, normalizing, hardening and tempering are the most important heat treatments often used to modify the microstructure and mechanical properties of engineering materials particularly steels. Annealing is defined as a heat treatment that consists of heating to and holding at a suitable temperature followed by cooling at an appropriate rate, most frequently applied in order to soften iron or steel materials and refines its grains due to ferrite-pearlite microstructure; it is used where elongations and appreciable level of tensile strength are required in engineering materials. Hardening is the heat treatment processes in which increases the hardness of a steel piece by heating it to a certain temperature and then cooling it rapidly to room temperature. Tempering is the process of imparting toughness at the cost of its hardness to an already hardened piece of steel by reheating it to a certain temperature and then cooling it rapidly. The temperature of heating depends on the toughness to be imparted and hardness to be reduced. In normalizing, the material is heated to the austenitic temperature range and this is followed by air cooling. This treatment is usually carried out to obtain a mainly pearlite matrix, which results into strength and hardness higher than in as received condition. It is also used to remove undesirable free carbide present in the as-received sample [1].

Steel is an alloy of iron with definite percentage of carbon ranges from 0.15-1.5% [2], plain carbon steels are those containing 0.1-0.25% [3]. Steel is mainly an alloy of iron and carbon, where other elements are present in quantities too small to affect the properties. The other alloying elements allowed in plain-carbon steel are manganese and silicon. Steel with low carbon content has the same properties as iron, soft but easily formed. As carbon content rises, the metal becomes harder and stronger but less ductile and more difficult to weld. There are two main reasons for the popular use of steel: (1) It is abundant in the earth's crust in form of Fe_2O_3 and little energy is required to convert it to Fe. (2) It can be made to exhibit great variety of microstructures and thus a wide range of mechanical properties. Although the number of steel specifications runs into thousands, plain carbon steel accounts for more than 90% of the total steel output. The reason for its importance is that it is a tough, ductile and cheap material with reasonable casting, working and machining properties, which is also amenable to simple heat treatments to produce a wide range of properties [1]. The purpose of heat treating carbon steel is to change the mechanical properties of steel, usually ductility, hardness, Yield strength, tensile strength and impact resistance. The standard strengths of steels used in the structural design are prescribed from their yield strength. Most engineering calculations for structure are based on yield strength. The heat treatment develops hardness, softness, and improves the mechanical properties (such as tensile strength, yield strength, ductility, corrosion resistance and creep rupture. These processes

*Corresponding author: Hasan MF, Department of Industrial Engineering and Management, Khulna University of Engineering and Technology, Khulna-9203, Bangladesh, E-mail: ferdaus2k11@gmail.com

also help to improve machining effect, and make them versatile. They are found in applications such as train railroads, beams for building support structures, reinforcing rods in concrete, ship construction, tubes for boilers in power generating plants, oil and gas pipelines, car radiators, cutting tools etc. [3]. The mild steel or called low carbon steel as the main component to through the process of the heat treatment where it containing several characteristic. The general range of mild steel is 0.05% to 0.35%. Mild steel is a very versatile and useful material. It can be machined and worked into complex shapes has low cost and good mechanical properties. It is forms the vast bulk of the steels employed for general structural fabrication, sheet metal and so on. Bolts and studs are supposed to be made from mild steel (up to 0.25% carbon) with characteristic toughness and ductility.

Various Microstructures

Prediction of microstructure transformations is prerequisite for successful prediction of mechanical properties after a heat treatment and of generation of stresses and strains during a heat treatment. Phase transformation modeling is one of the main challenges in modeling of heat treatment [4]. During annealing, softening processes are under way in the microstructure and, in some cases, recovery and recrystallization take place as well. Naturally, the morphology of carbides changes as well [5].

Ferrite

It is α-iron (B.C.C.) having not more than 0.025% carbon in solid solution. It is major constituent in low carbon steels and wrought iron. Its hardness varies from 50 to 100 B.H.N. Its upper tensile strength is about 330 MN/m^2 and percentage elongation about 40. It can be easily cold worked [6].

Cementite

It is iron carbide, with 6.67% carbon. Its upper tensile strength is about 45 MN/m^2 and hardness about 650 B.H.N. It is white in color and is brittle. It occurs in steels which have been cooled slowly. It is magnetic below 250°C. In steels containing carbon less than 0.8% it is present as a component of another constituent, "pearlite". In steels containing more than 0.8% carbon it exists as a grain boundary film [6].

Pearlite

In its microstructure it consists of alternate laminations of ferrite and cementite. It contains about 0.8% carbon in iron. It is the strongest constituent of steel. Its hardness is about 180 B.H.N., ultimate tensile strength about 920 MN/m^2 and percentage elongation about 5% [6].

Austenite

It is a solid solution of carbon in ý-iron (F.C.C.) containing a maximum of 2% carbon at 1130°C. It is tough and non-magnetic. It exists in plain carbon steels above upper critical temperature. Elements like chromium and manganese in steel preserve all or some of austenite down to 0°C. Austenite consists of polyhedral grains showing twins [6].

Martensite

In plain carbon steel it is obtained by quenching from above upper critical temperature. It is the hardest constituent obtained in given steel. It shows a fine needle-like microstructure. Its hardness is about 700 B.H.N. It is unstable and disappears on reheating the steel. It is magnetic and less tough than austenite. It is considered to be highly stressed α-iron supersaturated with carbon [6].

Materials and Methods

Sample of ASTM A-36 mild steel was purchased from a local market located in Khulna, Bangladesh. All specimens of mild steel of dimensions 8 × 8 × 8 mm was cut using power hacksaw. The chemical composition of the mild steel sample was determined as given in Table 1. Standard tensile and impact specimens were made from ASTM A-36 mild steel sample using lathe machine. Samples were subjected to different heat treatment: annealing, normalizing, hardening, and tempering in accordance to ASM International Standards [7]. Heat treated specimens were tested for mechanical properties. The heat treatment conditions are listed in Table 2. Four specimens were prepared for each heat treatment type.

Material composition

The chemical composition of ASTM A-36 steel shown in Table 1.

Working steps

The following steps were carried out in our experimental investigation:

I. Samples of ASTM A36 steel were prepared for mechanical properties test.

II. After that the following specimens were heat treated in the furnace for reaching the austenization temperature (850-900°C) of the following specimens.

III. Then the specific heat treatment operation like hardening, annealing and normalizing had been done.

IV. For specific heat treated specimen, the change of mechanical properties was determined in the following (2.1) using appropriate methods.

V. Metallographic tests were carried out to observe the changes in microstructures after heat treatment.

For annealing: In this case the specimen was put in the furnace for 910°C and we kept it in this situation for approximately 70 minutes. After that it was cooled in a heap of ashes so that it was cooled down at a very slow rate.

For hardening: In this case the specimen was put in the furnace for 910°C and we kept it in this situation for approximately 30 minutes. After that it was cooled in water so that it was cooled down very quickly.

For normalizing: In this case the specimen was put in the furnace for 910°C and we kept it in this situation for approximately 70 minutes. After that it was cooled in room temperature (Air).

For tempering: In this case the specimen was put in the furnace for 450°C and we kept it in this situation for approximately 70 minutes. After that it was cooled in room temperature (Air).

Iron family	C%	Si %	Mn %	S %	P%	Fe %	Cu%
Mild steel	0.29	0.28	0.10	0.10	0.04	98.14	0.2

Table 1: Chemical composition of mild steel.

Condition	Annealed	Normalized	Hardened	Tempered
Temperature, °C	910	910	910	450
Holding time, min	70	70	30	70
Cooling medium	Ash	Air	Water	Air

Table 2: Summary of heat treatment process.

Determination of mechanical properties

Mechanical properties (hardness, tensile strength, toughness, yield strength, elongation and percentage of elongation) of the treated and untreated samples are determined using standard methods. For hardness testing, oxide layers formed during heat treatment were removed by stage-wise grinding and then polished. Average Rockwell Hardness Number (BHN) readings were determined by taking two hardness readings at different positions on the samples, using a Standard Rockwell hardness tester and tensile test using universal testing machine. Impact energy was recorded using the Izod impact tester. For tensile properties, tensile specimens were loaded into a 2000 kg Mosanto Tensiometer hooked up to a data logger. Load-elongation data were recorded and converted into stress-strain graphs. Yield strength, ultimate (tensile) strength, Young's modulus and ductility (% elongation and reduction) are determined based on these graphs, in accordance with ASTM standard test procedures (ASTM A-36) [8-10].

Microstructure examination

Microstructure examination of the treated and untreated samples was carried out. Each sample was carefully grounded progressively on emery paper in decreasing coarseness. The grinding surface of the samples were polished using Al_2O_3 carried on a micro clothe. The crystalline structure of the specimens were made visible by etching using solution containing 5% Cupric cloride, 8% HCl acid and 87% methylated spirit on the polished surfaces. Microscopic examination of the etched surface of various specimens was undertaken using a metallurgical microscope with an inbuilt camera through which the resulting microstructure of the samples were all photographically recorded with magnification of 400 [11,12].

Results and Discussions

Effect of heat treatment on mechanical properties

The effect of heat treatment (annealing, normalising, hardening, and tempering) on the mechanical properties (ultimate tensile strength, hardness, toughness, percentage elongation, and percentage reduction) of the treated and untreated samples is shown in Table 3. The function of annealing is to restore ductility and also removes internal stresses but its Brinell Hardness Number is less than hardening because here carbon get more time to react with oxygen in the atmosphere for slow cooling rate. The function of hardening is to increase the hardness of the specimen and so its Brinell hardness number is larger than annealing and normalizing because here carbon cannot get more time to react with oxygen (for quick cooling rate), so carbon is trapped with the specimen and formed martensite.

Normalizing does not soften the steel to the extent it is done by annealing and also it does not restore ductility as much as is done by annealing. Its Brinell Hardness Number is less than hardening but greater than annealing. Tempering imparts toughness at the cost of its hardness to an already hardened piece of steel by reheating it to a certain temperature and the cooling it rapidly.

The untreated samples value of mechanical behavior was noted as follows: tensile strength 402.45 MPa, yield strength 220.03 MPa, hardness 69.80 HRC, toughness J, percentage of elongation 23.16%, percentage of reduction 56.24%, young modulus 207.88 Gpa, yield strength 217.31 N/mm². Comparing the mechanical properties of annealed sample with the untreated sample, annealed sample showed that lower tensile strength (389.34 MPa), yield strength 212.54 MPa and hardness (62.15 HRC) and increase in reduction in area (25.22%),

elongation (64.12%), modulus of elasticity (302.32 GPa). The decrease in tensile strength and hardness can be associated with the formation of soft ferrite matrix in the microstructure of the annealed sample by cooling. The mechanical properties of the normalized specimen are found to be 452.13 MPa, 242.26 MPa, 120.36 HRC, 63.23% and 22.70% for tensile strength, yield strength, hardness, percentage reduction and percentage elongation, respectively. The increase in tensile strength and hardness as compared to annealed and untreated sample was due to proper austenising temperature at 910°C and higher cooling rate, which resulted in decrease in elongation, which was lower than those obtained for untreated and annealed samples due to pearlite matrix structure obtained during normalization of ASTM A-36 steel.

The mechanical properties of the hardened sample revealed that it had the highest value of tensile strength 734.32 MPa, yield strength 278.11 MPa and highest hardness (293.4 HRC) were obtained. The specimen was austenite at 910°C for 30 minutes and then water quenched. This treatment increased the tensile strength and hardness but there was massive reduction in elongation and reduction in area 6.90%, and 37.39%, respectively.

The mechanical properties of tempered sample showed that the tensile strength, yield strength, hardness, percentage reduction and percentage elongation were 421.76 MPa, 232.78 MPa, 100.01 HRC, 69.01% and 23.20%, respectively. Comparing the mechanical properties of tempered sample with hardened sample, it was found that there was decrease in tensile strength and hardness at tempering temperature 450°C while the percentage elongation and percentage reduction increased which can be associated to the graphitization of the precipitated carbides that resulted in the formation of ferrite at tempering temperature of 450°C. This showed that tempering temperature improved the degree of tempering of the martensite, softening the matrix and decreased its resistance of plastic deformation. However, the test results showed that annealing treatment gave an elongation superior to any other heat treatment studied. The variability in ultimate tensile strength, percentage elongation, percentage reduction hardness and toughness of treated and untreated ASTM A-36 steel are shown in Figures 1-5, respectively.

The value of tensile strength were observed to be in the order; hardened > normalized > tempered > untreated < annealed, possibly as a result of the refinement of the primary phase after the subsequent cooling processes. The value of hardness was observed to be higher for the hardened steel specimen. The hardness of the steel increases both with cooling rate and pearlite percentage. The reason being that martensite is one of the strengthening phases in steel. The increase in the hardness was due to the delay in the formation of pearlite and martensite at a higher cooling rate. The yield strength value for the hardened specimen was also observed to be greater than that of normalized and annealed specimens, while the normalized specimen also has a greater value than that of tempered and annealed specimen. It was also observed from the graphs that for all the heat treated

Mechanical properties						
Heat Treatment	Tensile Strength (Mpa)	Hardness (BHN)	Percentage Elongation (%)	Percentage Reduction (%)	Yield Strength (Mpa)	Young Modulus (GPa)
Untreated	402.45	69.8	23.16/15	56.24	220.03	207.88
Annealed	389.34	62.15	25.22	64.12	212.54	302.32
Normalised	452.13	120.36	22.7	63.23	242.26	288.12
Hardened	734.32	293.4	6.9	37.39	278.11	632.47
Tempered	421.76	100.01	23.2	69.01	232.78	293.63

Table 3: Mechanical properties of heat treated and untreated ASTM A -36 steel.

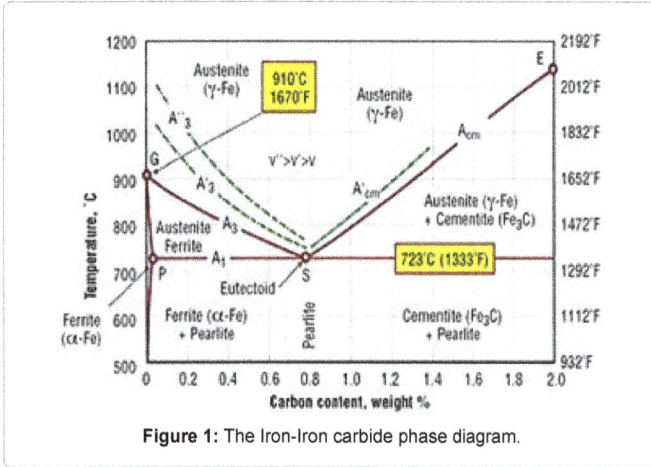

Figure 1: The Iron-Iron carbide phase diagram.

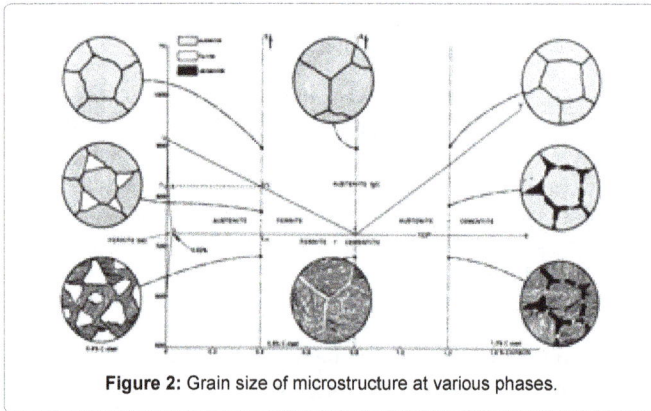

Figure 2: Grain size of microstructure at various phases.

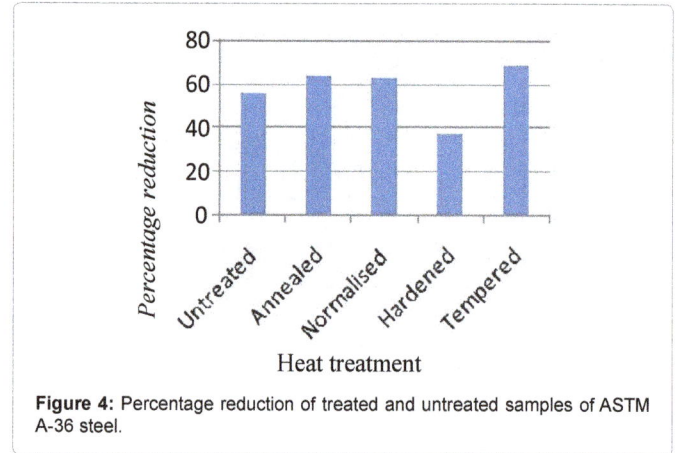

Figure 3: Tensile strength of treated and untreated samples of ASTM A-36 steel.

constituents, which are ferrite (white) and pearlite (black). The light coloured region of the microstructure is the ferrite and the dark region is the pearlite. The microstructure of the annealed sample is shown in Figure 7. As it can be seen in Figure 7, the ferrite grains had undergone complete recrystallization and these constituted the major portion of the microstructure the annealed low carbon steel with stress free matrix. At 910°C the deformed structure was fully homogenized and during the slow cooling from austenizing range to room temperature the final microstructure consisted of fine ferrite grains in which the pearlite was more uniformly distributed.

Figure 8 shows the microstructure of the normalized ASTM-A36 mild steel. The normalized sample showed that the shape and size of the original austenite grains were influenced to a remarkable extent. The sample revealed a pearlite matrix in which shorter graphite flakes

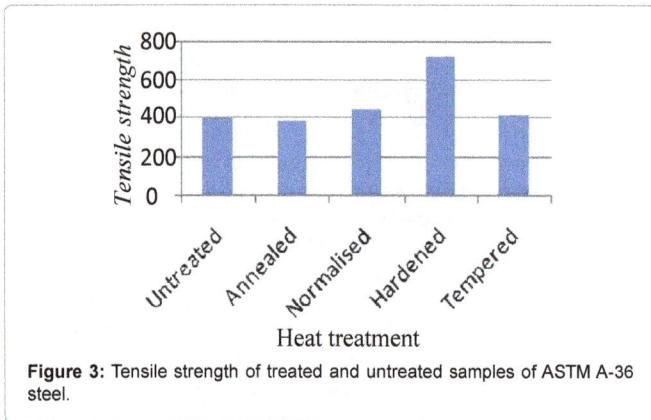

Figure 4: Percentage reduction of treated and untreated samples of ASTM A-36 steel.

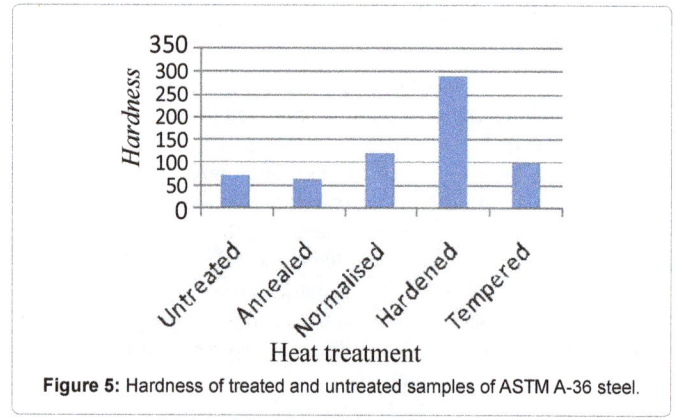

Figure 5: Hardness of treated and untreated samples of ASTM A-36 steel.

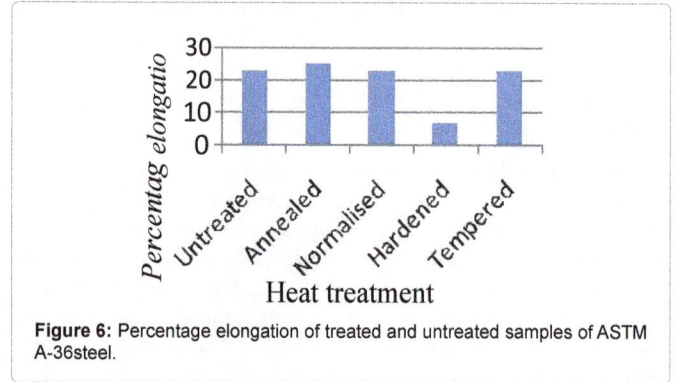

Figure 6: Percentage elongation of treated and untreated samples of ASTM A-36steel.

specimens, except for the hardened specimen, there were tremendous increase in the toughness of the material which indicates that hardened material, though have a very high tensile stress, but at the expense of its toughness, hence where toughness is a major concern. However if strength is also desired along with hardness, this should not be done. It is seen that annealing causes a tremendous increase in % elongation (ductility). It can be clearly seen comparing all the heat treatment processes, optimum Combination of Ultimate Tensile Strength, Yield Strength, % elongation as well as hardness can be obtained through austempering only.

Effect of heat treatment on microstructure

The microstructure of untreated specimen (Figure 6) has two major

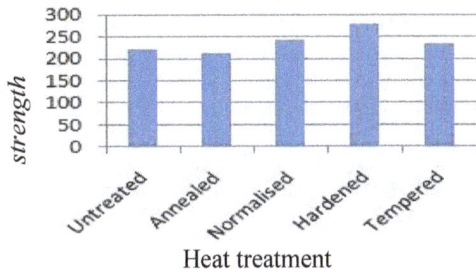

Figure 7: Yield Strength of treated and untreated samples of ASTM A-36 steel.

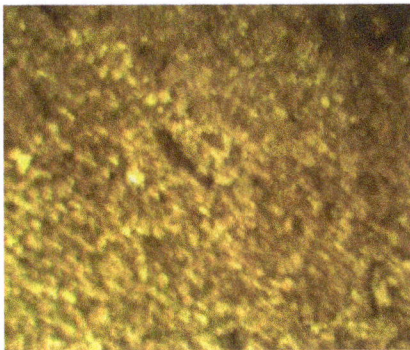

Figure 8: Microstructure of untreated ASTM A-36 steel (X400).

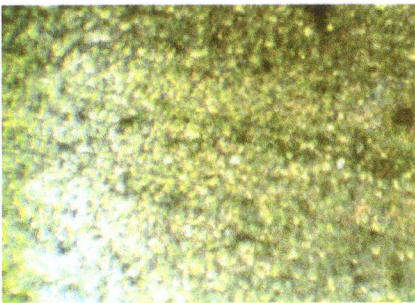

Figure 9: Microstructure of annealed ASTM A-36 steel (X400).

than in annealed sample existed. It was observed that there was many short graphite flakes surrounded with patches of uniformly distributed pearlite grains as seen in Figure 8.

Figure 9 shows the massive martensite structure of hardened sample, when medium carbon steels are rapidly quenched from its austenite temperature to room temperature, the austenite will decompose into a mixture of some medium carbon martensite and fewer pearlite as a result of this microstructure which is hard, hence, there was increase in tensile strength, hardness and reduction in ductility of the material (Figure 10) [11].

The microstructure of hardened and tempered at 450°C is shown in Figures 11 and 12. A highly recrystallized ferrite grains (white dotted areas) with some secondary graphite site was observed. This micrograph revealed that the microstructure of tempered specimen consisted of a number of appreciable carbide particles precipitated out from the matrix, which indicated that the precipitate carbide

particles decomposed by a process of solution in ferrite matrix [12]. The summary of the observed microstructure of treated and untreated ASTM-A36 steel is given in Table 4.

In the microstructures of these specimens we have indicated the carbon-saturated region with arrow marks (Figures 7 and 8) and these help to find out the differences among the microstructures of untreated, annealed, normalized hardened and tempered specimens comparatively. As slow cooling is done in annealing so it transforms austenite to soft pearlite and also mixed with ferrite or cementite and this cementite increases the brittleness of the steel. Normalizing converts soft steel to moderate hard steel. In this case cooling rate is faster than annealing and for this reason, when the specimen is cooled in room temperature then ferrite and cementite are formed but their quantity is less. So the specimen is enhanced with considerable ductility by reducing its brittleness. In hardening process austenite structure is directly formed into martensite structure for fast cooling. Actually the rapid cooling converts most of the austenite into martensite which

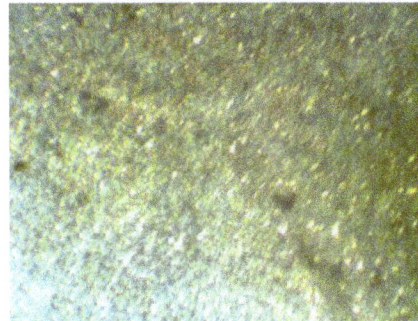

Figure 10: Microstructure of normalised ASTM A-36 steel (X400).

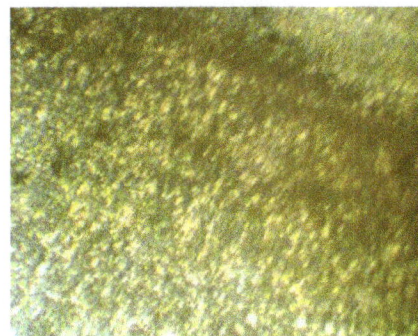

Figure 11: Microstructure of hardened ASTM A-36 steel (X400).

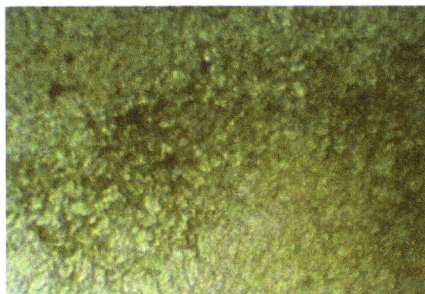

Figure 12: Microstructure of tempered ASTM A-36 steel (X400).

Heat treatments	Microstructure Developed
Untreated	Graphite Flakes in ferrite and pearlite matrix
Annealed	Graphite flakes in ferrite matrix
Normalizing	Graphite flakes in pearlite matrix
Hardened	Graphite flakes in martensite matrix
Tempered	Graphite flakes in martensite matrix with recrystallized ferrite grains

Table 4: Summary of microstructure of treated and untreated ASTM A-36 steel.

is a hard constituent and more stable than austenite at ordinary temperatures. In tempering process austenite structure is directly formed into martensite structure matrix with recrystallized ferrite grains.

Conclusions

From the results obtained, it can be said that mechanical properties depends largely upon the various form of heat treatment operations and cooling rate. Hence depending upon the properties and the applications that may be required for any design purpose, a suitable form of heat treatment should be adopted. For high ductile and minimum toughness, annealed mild steel will give satisfactory results. According to the results of investigation on the effect of heat treatment on mechanical properties and microstructure of ASTM-A36 mild steel, the following conclusions were made: Tensile strength, yield strength and hardness of low carbon ASTM A-36 steel increased with plastic deformation while ductility and impact strength decreased due to strain hardening effect. Normalization treatment had also resulted in higher tensile strength and hardness than annealed samples. This treatment is recommended as final treatment after manufacturing. The tempered samples gave an increase in tensile strength and hardness than untreated sample as a result of formation of tempered martensite and resultant ferrite structure that were obtained. Hardened sample had the highest tensile strength and hardness with lowest ductility and impact strength when compared to other heat treated samples. Hardening is strongly recommended when the strength and hardness

are the prime desired properties in design. The mechanical properties of ASTM A-36 steel can be altered through various heat treatments. The results obtained confirmed that improvement in mechanical properties that can be obtained by subjecting ASTM A-36 steel to different heat treatments investigated in this study.

References

1. Dell KA (1989) Metallurgy Theory and Practical Textbook. American Technical Society, Chicago.

2. John VB (1980) Introduction to Engineering Materials. (2nd Ed), Macmillan Publishing Company Ltd.

3. Alawode AJ (2002) Effects of Cold Work and Stress Relief Annealing Cycle on the Mechanical Properties and Residual Stresses of Cold-Drawn Mild Steel Rod. Mechanical Engineering Department, University of Ilorin, Nigeria.

4. Funatani K, Totten GE (1997) Present Accomplishments and Future Challenges of Quenching Technology. Proceedings of the 6th International Federation of Heat treatment and Surface Engineering Congress, IFHTSE, Kyongju, Korea.

5. Nam WJ, Bae CM (1999) Coarsening Behavior of Cementite Particles at a Subcritical temperature in a medium Carbon Steel. Scripta Materialia 41: 313-318.

6. Manchanda VK, Narang GBS (2005) Materials and Metallurgy. (6th Ed), Khanna Publishers.

7. ASTM E18 (2008) Standard Test Method for Rockwell Hardness of Metallic Materials. American Society of Testing and Materials.

8. ASTM E23 (2008) Standard Test Method for Notched Bar Impact Testing of Metallic Materials. American Society of Testing and Materials.

9. ASTM E8 (2008) Standard Test Method for Tension Testing of Metallic Materials. American Society of Testing and Materials.

10. Jokhio MH (1991) Effect of Retained Austenite on Abrasive Wear Resistance of Carburised SAE 8822H Steel. Mehran University of Engineering and Technology, Jamshoro.

11. Hu HJ, Zhai ZY, Li YY, Gong XB, Wang H, et al. (2015) The simulation researches on hot extrusion of super-fined tube made of magnesium alloys. Russian Journal of Non-Ferrous Metals 56: 196-205.

12. Gong X, Li H, Kang SB, Cho JH, Li S (2010) Microstructure and Mechanical Properties of Twin-roll Cast Mg-4.5Al-1.0Zn Alloy Sheets Processed by Differential Speed Rolling. Materials and Design 31: 1581-1587.

Electrochemical Studies of Aluminium 7075 Alloy in Different Concentration of Acid Chloride Medium

Pruthviraj RD[1]* and Rashmi M[2]

[1]R&D Centre, Department of Chemistry, Raja Rajeshwari College of Engineering, Bangalore, India
[2]Department of Chemistry, Sri Krishna Institute of Technology, Bangalore, India

Abstract

The corrosion behavior of Al 7075 was investigated in Hydrochloric acid over a range of acid concentration electrochemical techniques like Tafel extrapolation and electrochemical impedance spectroscopy were started and the studies have revealed that the corrosion rate of Al 7075 samples increase with increase in concentration of Hydrochloric acid in the medium.

Keywords: Al 7075; Acid solutions; EIS; Polarization

Introduction

Corrosion of structural elements is a major issue for any industry because of the chemical environment of chemical processing. Al 7075 is special class of ultra-high strength metals that differ from conventional Aluminium in that they are hardened by a metallurgical reaction. Recently, the needs of high reliable substances of high strength and high ductility are gradually increased with the development of aerospace industry [1]. The characteristics of this grey and white steel are high ductility, formability; high corrosion resistance, high temperature strength, ease of fabrication; weldability and maintenance of an invariable size even after heat treatment. The corrosion rate of Al 7075 in acid solutions such as sulphuric acid, hydrochloric acid, formic acid, and stearic acid are substantial. Heat treatment affects the corrosion rate. Critical and passive current densities increase as the structure is varied from fully annealed to fully age [2]. Several technical papers covering alloy design, material processing, thermo-mechanical treatments, welding, strengthening mechanisms, etc., have been published. This aluminium have emerged as alternative materials to conventional quenched and tempered steels for advanced technologies such as aerospace, nuclear and gas turbine applications. They frequently come in contact with acids during cleaning, pickling, descaling, acidizing etc. Most of the reported studies were conducted on corrosion of various metals and alloys in HCl and H_2SO_4 medium. Phosphoric acid is also used in pickling delicate, costly components and precision items where rerusting after pickling has to be avoided [3-9]. But no literature seems to be available which reveal corrosion behavior of Al 7075 in acid medium. So it is intended to study the corrosion behavior of Al 7075 in Hydrochloric acid medium [10-15].

Experimental Part

Composition of Al 7075(weight%)

7075 aluminum alloy's composition roughly includes 5.6-6.1% zinc, 2.1-2.5% magnesium, 1.2-1.6% copper, and less than a half percent of silicon, iron, manganese, titanium, chromium, and other metals.

Material

Rectangular shape of metals were cut from the plate and covered with Teflon tape in such a way that, the area exposed to the medium is 1 cm². These metals were polished as per standard metallographic practice, followed by polishing with emery papers, finally on polishing wheel using legated alumina to obtain mirror finish, degreased with acetone, washed with double distilled water and dried before immersing in the corrosion medium [16,17].

Medium

Standard solutions of Hydrochloric acid having concentration 0.1 N, 0.05 N, 0.025 N, were prepared by diluting analytical grade (Nice) Hydrochloric acid with double distilled water. Experiments were carried out at Laboratory temperature.

Electrochemical measurements

Tafel polarization studies: Electrochemical measurements were carried out by using an electrochemical work station, CH Instrument (USA). Tafel plot measurements were carried out using conventional three electrode Pyrex glass cell with platinum counter electrode and Ag/AgCl electrode as reference electrode. All the values of potential are therefore referred to the SRE [18,19]. Finely polished Al 7075 specimens of 1 cm² surface area were exposed to corrosion medium of different concentrations of Hydrochloric acid (0.1 N, 0.05 N, 0.025 N) at Laboratory temperature and allowed to establish a steady state open circuit potential [20-23]. The potentiodynamic current potential curves were recorded by polarizing the specimen to -250 mV catholically and +250 mV anodically with respect to open circuit potential (OCP) at scan rate of 5 mVs⁻¹.

Electrochemical impedance spectroscopy studies (EIS)

Electrochemical impedance spectroscopy (EIS), which gives early information about the electrochemical processes, at the metal solution interface, has been used in many reports on the corrosion studies. The corrosion behavior of the Al 7075 was also obtained from EIS technique using electrochemical work station, CH Instrument (USA). In EIS technique a small amplitude ac signal of 10 mV and frequency spectrum from 100 kHz to 0.01 Hz was impressed at the OCP and

---

*Corresponding author: Pruthviraj RD, R&D Centre, Department of Chemistry, Raja Rajeshwari College of Engineering, Bangalore, India
E-mail: pruthvirajrd@gmail.com

impedance data were analyzed using Nyquist plots [24]. The charge transfer resistance, Rt was extracted from the diameter of the semicircle in Nyquist plot. In all the above measurements, at least three similar results were considered and their average values are reported. The scanning electron microscope images were recorded to establish the interaction of acid medium with the metal surface using JEOL JSM-6380LA analytical scanning electron microscope (CPRI) (Figure 1).

The potentiodynamic polarization parameters like corrosion potential (Ecorr), corrosion current (icorr), polarization resistance (Rp), anodic and cathodic slopes (ba and bc), and corrosion rate are calculated from Tafel plots. These results indicate that with increase of both concentration of HCl and solution the Rp value decreases, polarization curves are shifted to high current density region indicating increase in corrosion rate. The nature of polarization curves predicts active corrosion behavior at each temperature and concentration of HCl. It is observed from these results that the corrosion potential is shifted to noble values as the concentration of HCl is increased [25,26]. This is in accordance with Murralidharan who proposed dependence of Ecorr and Icorr on solution parameters. The positive shift in the

corrosion potential, Ecorr, indicates that the anodic process is much more affected than the cathodic [27]. With increase of solution temperature ba and bc are almost unchanged indicating no change in the mechanism of corrosion with temperature. The increasing corrosion rate with increasing temperature is in agreement with the observation reported by Jones that, in open system, the corrosion rate of iron increases with temperature up to 80°C. This can also be explained by the characteristics of the cathodic process of hydrogen evolution in acidic solutions. The hydrogen evolution overpotential decreases with increase in temperature that leads to increase in cathodic reaction rate [28,29].

At the interface of Al and acid electrolyte, the dissolution of Al can be written as follows:

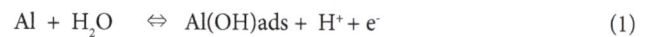

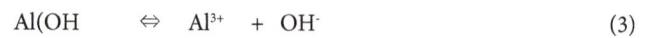

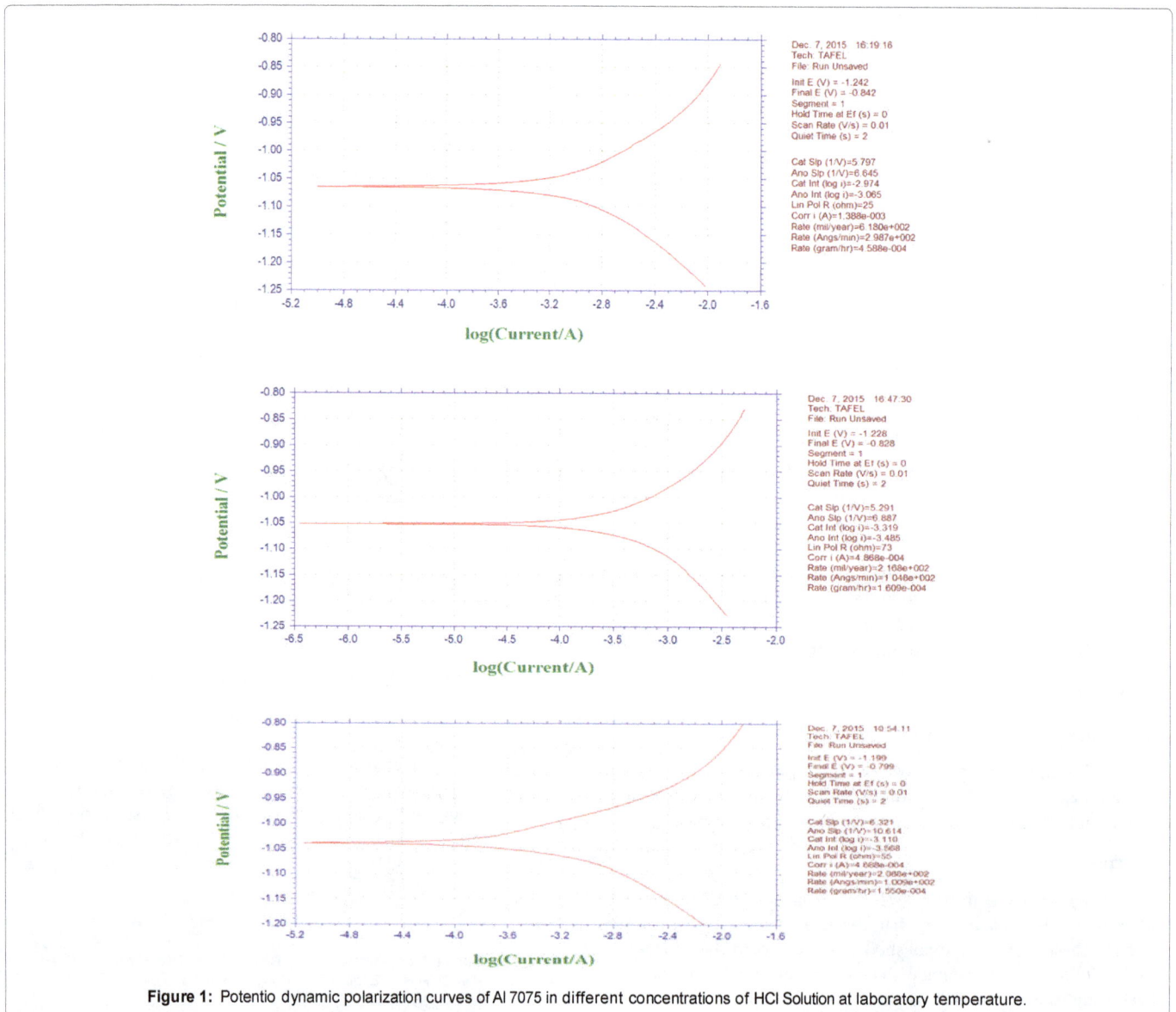

$$Al + H_2O \Leftrightarrow Al(OH)ads + H^+ + e^- \tag{1}$$

$$Al(OH)ads \rightarrow Al(OH)^+ + e^- \tag{2}$$

$$Al(OH \Leftrightarrow Al^{3+} + OH^- \tag{3}$$

Figure 1: Potentio dynamic polarization curves of Al 7075 in different concentrations of HCl Solution at laboratory temperature.

At medium and high concentrations of Hydrochloric acid, precipitation of Aluminium chloride occurs at the interface as follows:

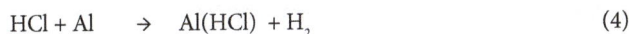

$$HCl + Al \rightarrow Al(HCl) + H_2 \tag{4}$$

$$Al(HCl) \rightarrow Al_3(Cl)_2 + HCl \tag{5}$$

Electrochemical impedance spectroscopy

The corrosion behavior of Aluminium 7075 specimens was also investigated by EIS in various concentrations of Hydrochloric acid at Laboratory temperature. The impedance spectra recorded are displayed as Nyquists plots for Al 7075 specimen as a function of concentration of acid as shown in Figure 2.

The depressed semicircles of the Nyquist plots suggest the distribution of capacitance due to in homogeneities associated with the electrode surface. In order to analyse the impedance spectra containing one capacitive loop, the equivalent circuit given in Figure 3 is used, which has been used previously to model Aluminium/acid interface. The capacitive loops are not perfect semicircles, because the Nyquist plots obtained in the real system represent a general behavior where the double layer at the metal solution interface does not behave as an ideal capacitor. The fact that impedance diagrams have a semicircular appearance shows that the corrosion of Al 7075 is control by charge transfer process. The intersection of capacitive loop with the real axis on the high frequency region represents the ohmic resistances of corrosion product films and the solution enclosed between the working electrode and the reference electrode, Rs. Rt represents the charge transfer resistance whose value is a measure of electron transfer across the surface and is inversely proportional to corrosion rate. In evaluation of Nyquists plot, the difference in real impedance at lower and higher frequencies is considered as charge transfer resistance R_t. The diameter of the semicircle decreases as acid concentration increases indicating increase in corrosion rate.

Where Rt is charge transfer resistance obtained from Nyquists plots. A is area of cross section of material under observation ba and bc are Tafel constants. The results obtained by EIS method at various concentrations of are similar to that of Tafel polarization results.

Scanning Electron Microscope Studies (SEM)

The SEM images of freshly polished surface of Al 7075 samples of are given in Figure 4 and which show uncorroded surface with few scratches due to polishing. The surface morphology of the Al 7075 samples was examined by SEM immediately after corrosion tests in Hydrochloric acid medium. The SEM image of corroded Al 7075 sample in Figure 4 shows degradation of alloy, with more or less uniform attack. In the case of corroded samples this degradation is highly pronounced as shown above seems to be concentrated more on and around the grain boundaries. The intermetallic precipitation at grain boundary may be responsible for the higher rate of corrosion.

Conclusions

Based on results of investigation, the following conclusions may be drawn that the corrosion rate of Al 7075 specimen in Hydrochloric acid medium is substantial. The corrosion rates of the specimens are influenced by concentration of Hydrochloric acid medium. The corrosion rate of the specimens under investigation increases with

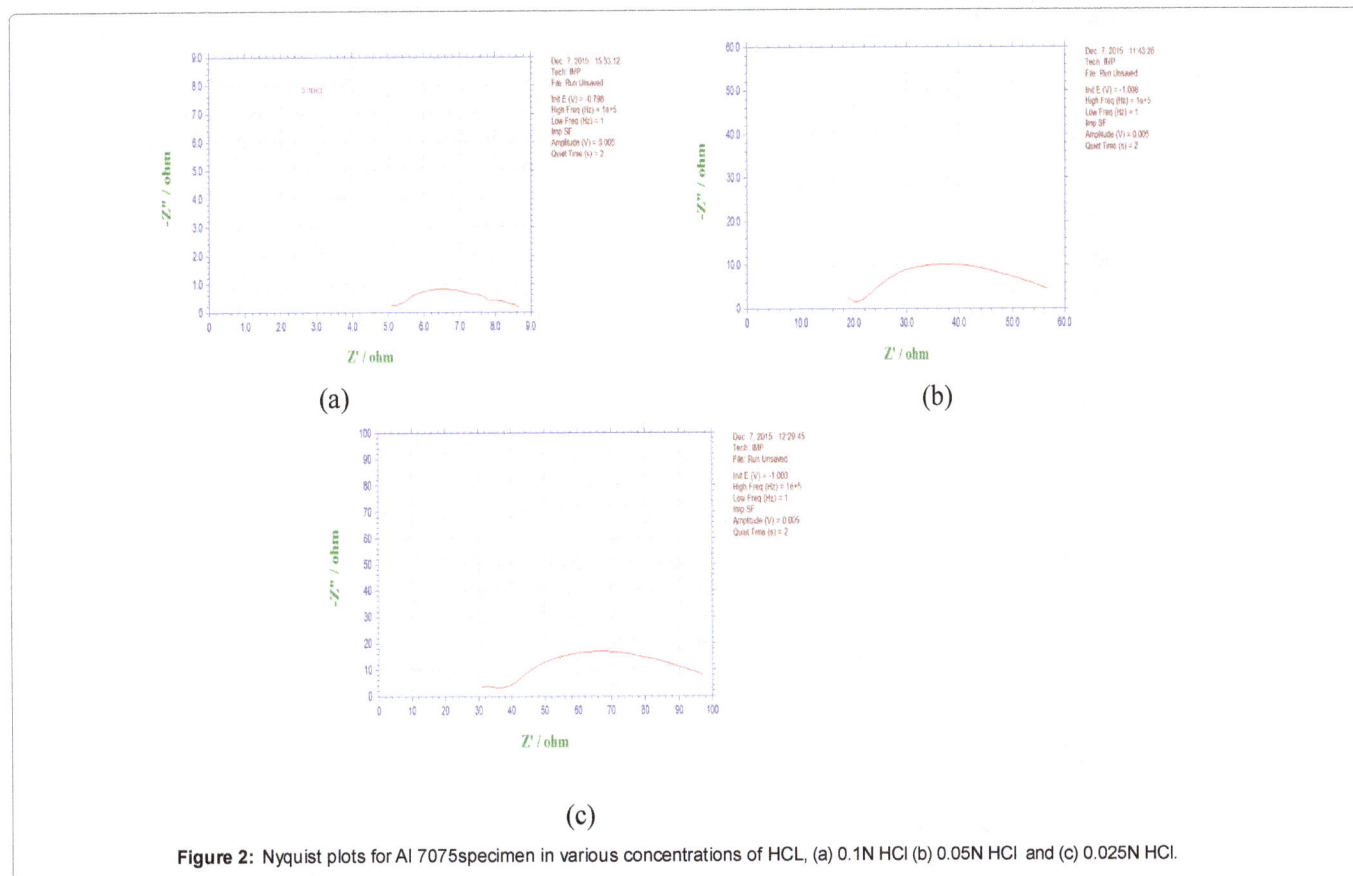

Figure 2: Nyquist plots for Al 7075specimen in various concentrations of HCL, (a) 0.1N HCl (b) 0.05N HCl and (c) 0.025N HCl.

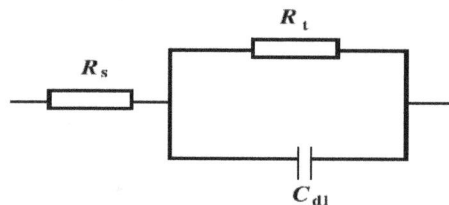

Figure 3: Equivalent circuit forth metal/acid interface (R_S solution resistance, R_t charge transfer resistance, C_{dl} double layer capacitance).

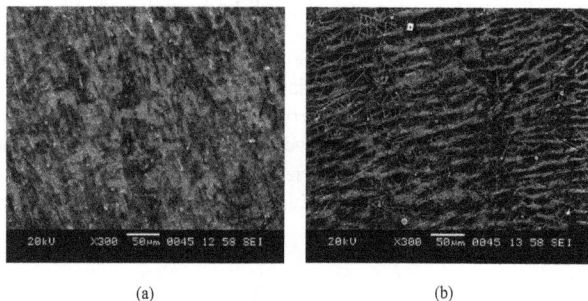

(a)   (b)

Figure 4: SEM images of Al 7075 specimen (a) Fresh lypolished surface (b) Corroded surface after polarization experiment in HCL.

increase in concentration of Hydrochloric acid.

Acknowledgment

The authors gratefully acknowledge VGST, Govt. of Karnataka for financial support under the project CISEE /2014-15/ GRD No. 325. Also express their gratitude to Management and Principal, Raja Rajeswari College of Engineering, Bangalore for providing infrastructural facilities.

References

1. Sastry KY, Narayanan R, Shamantha CR, Sunderason S, Seshadri SK, et al.(2003)Stress corrosion cracking of maraging steel weldments. Mat Sci and Techn 19: 375-381.

2. (1990) ASM Hand Book. (10th Ed)ASM International.

3. Lee DG, Jang KC, Kuk JM, Kim IS (2005) The influence of niobium and aging treatment in the 18% Ni maraging steel. J of Mat Proc Techn 162: 342-349.

4. Krick WW, Covert RA, May TP (1968) Mechanical behavior of crystalline solids at elevated temperature. Met Eng Quart 8: 31

5. Dean SW, Copson HR (1965) Selective electrogenerative oxidation of benzyl alcohol with platinum–graphite packed-bed anodes. Corrosion 21: 95.

6. Bellanger G, Rameau JJ (1996) Effect of slightly acid pH with or without chloride in radioactive water on the corrosion of maraging steel. J Nuclear Mat 228:24-37.

7. Bellanger G (1994) Effect of carbonate in slightly alkaline medium on the corrosion of maraging steel. J Nuclear Mat 217:187-193.

8. Poornima T, Nayak J, Shetty AN (1964) Data bulletin on 18% Ni maraging steel. The International Nickel Company, INC.

9. Rezek J, Klein IE, Yhalom J (1997) Electrochemical properties of protective coatings on maraging steel. Corros Sci 39: 385-397.

10. Sinha PP (1999) IIM Metal News.

11. Muralidharan VS, Rajagopalan KS (1979) Kinetics and mechanism of corrosion of iron in phosphoric acid. Corros Sci 19: 205-207.

12. Quartarone G, Battilana M, Bonaldo L, Tortato T (2008) Investigation of the inhibition effect of indole-3-carboxylic acid on the copper corrosion in 0.5 M H_2SO_4 .Corros Sci 50: 3467-3474.

13. El-Sayed A (1997) Towards the realistic silicon/carbon composite for Li-ion secondary battery anode. J Appl Electrochem 27: 94.

14. Speller FN (1951) Corrosion: Causes and Prevention. (3rd Ed)McGraw Hill, USA.

15. Larabi L, Harek Y, Benali O, Ghalem S (2005) Hydrazide derivatives as corrosion inhibitors for mild steel in 1 M HCl. Prog Org Coat 54: 256-262.

16. Gunasekaran G, Chauhan LR (2004) Eco friendly inhibitor for corrosion inhibition of mild steel in phosphoric acid medium. Electrochim Acta 49: 4387-4395.

17. Almeida E, Pereira D, Figueiredo MO, Lobo VMM, Morcillo M (1997) The influence of the interfacial conditions on rust conversion by phosphoric acid. Corros Sci 39: 1561-1570.

18. El-Neami KKH, Mohamed AK, Kenawy IM, Fouda AS (1995) Inhibition of the corrosion of iron by oxygen and nitrogen containing compounds. Monatsh Chem J 126: 369-376.

19. Morad MS (1999) Influence of propargyl alcohol on the corrosion behaviour of mild steel in H3PO4 solutions. Mat Chem Phy 60:188-195.

20. Abd Ei-Rehim SS, Ibrahim MAM, Khaled KF (1999) 4-Aminoantipyrine as an inhibitor of mild steel corrosion in HCl solution. J Appl Electrochem 29: 593-599.

21. Shoesmith DW (1999) Metals Handbook. (9th Ed) ASM International.

22. Oguzie EE, Li Y, Wang FH (2007) Effect of surface nanocrystallization on corrosion and corrosion inhibition of low carbon steel: Synergistic effect of methionine and iodide ion. Electrochim Acta 52: 6988-6996.

23. Azhar ME, Mernari B, Traisnel M, Bentiss F, Lagrenee M (2001) Corrosion inhibition of mild steel by the new class of inhibitors [2,5-bis(n-pyridyl)-1,3,4-thiadiazoles] in acidic media. Corros Sci 43: 2229-2238.

24. Cheng S, Chen S, Liu T, Chang X, Yin Y (2007) Carboxymethylchitosan + Cu^{2+} mixture as an inhibitor used for mild steel in 1 M HCL. Electrochim Acta 52: 5932-5938.

25. Ozcan M, Dehri I, Erbil M (2004) Organic sulphur-containing compounds as corrosion inhibitors for mild steel in acidic media: correlation between inhibition efficiency and chemical structure. Appl Surf Sci 236: 155-164.

26. Lrabi L, Harek Y, Traisnel M, Mansri (2004) Synergistic Influence of Poly(4-Vinylpyridine) and Potassium Iodide on Inhibition of Corrosion of Mild Steel in 1M HCl. J Appl Electrochem 34: 833-839.

27. Qu Q, Jiang S, Bai W, Li L (2007) Effect of ethylenediamine tetraacetic acid disodium on the corrosion of cold rolled steel in the presence of benzotriazole in hydrochloric acid. Electrochim Acta 52: 6811-6820.

28. Hamdy HH, Essam A, Mohammed AA (2007) Inhibition of mild steel corrosion in hydrochloric acid solution by triazole derivatives: Part I. Polarization and EIS studies. Electrochim Acta 52: 6359-6366.

29. Ozcan M, Dehri I (2004) Electrochemical and quantum chemical studies of some sulphur-containing organic compounds as inhibitors for the acid corrosion of mild steel. Prog Org Coat 51: 181-187.

3

Determining the Amount of Hydrogen in Thin Films Well $Si_{1-X}g_{ex}$: H (X = 0 ÷ 1) for Electronic Devices

Najafov BA* and Abasov FP

Institute of Radiation Problems of Azerbaijan Nationale Academy of Science, Baku, Azerbaijan Republic, Russia

Abstract

Possibilities of plasma chemical deposition of a-$Si_{1-x}Ge_x$:H (x=0 ÷ 1) films undoped and doped with PH_3 or B_2H_6 have been analyzed from the viewpoint of their application in *p-i-n* structures of solar cell. The optical properties are considered, and the amount of hydrogen contained in those films is determined. The film properties are found to strongly depend on the film composition and the hydrogenation level. The number of hydrogen atoms in the films is varied by changing the gas mixture composition, and IR absorption in *a*-Si:H and *a*-Ge:H films is measured. The *a*-Si:H and a-$Si_{0.88}Ge_{1.2}$:H films were used to fabricate three-layer solar with an element area of 13 sm² and an efficiency (ξ) of 9.5%.

Keywords: Thin film; Amorphous silicon; Solar cells; Efficiency; Optical properties

Introduction

Si alloy film and have different structural phases. The most interesting, are in an amorphous matrix of them are crystalline grains. Such alloys a0072e manufactured by various methods and under various process conditions. For films of amorphous hydrogenated silicon a-Si: H, formed by cyclic deposition annealing in hydrogen plasma, the effect Staeblera-Wronski is weak [1]. Golikova [2] also noted the virtual absence Staeblera-Wronski effect in nanostructured films as-Si: H. Crystallization films and silicon-Si: H is carried out by various methods: prolonged annealing in vacuum at 600°C, rapid thermal processing [3], the laser annealing [4] and ion implantation [5].

This single junction p-i-n a-Si: H solar cell deposited on a glass substrate coated with transparent conductive oxide (TCO) and aluminium back contact exhibited 2.4% conversion efficiency. In order to increase the output voltage of a-Si:H solar cells the concept of a stacked (also called multi-junction) solar cell structure was introduced.

Due to a high absorption coefficient of a-Si:H in the visible range of the solar spectrum, 1 micrometer (μm) thick a-Si:H layer is sufficient to absorb 90% of usable solar light energy. Low processing temperature allows using a wide range of low-cost substrates such as glass sheet, metal or polymer foil. These features has made a-Si:H a promising candidate for low-cost thin-film solar cells.

The a-Si:H has the same short-range order as the single crystal silicon but it lacks the long range order. The small deviations in bonding angles and bonding lengths between the neighbouring atoms in a-Si:H lead to a complete loss of the locally ordered structure on a scale exceeding a few atomic distances. The resulting atomic structure of aSi:H is called the continuous random network. Due to the short-range order in the continuous random network of a-Si:H, the common semiconductor concept of the energy states bands, represented by conduction and valence bands, can still be used.

Mobility of charge carriers and the efficiency of doping in these films is higher than in the a-Si: H, and the optical absorption coefficient higher than that of crystalline silicon. Films and $Si_{1-x}Ge_x$: H, a-$Si_{1-x}S_x$: H are effective and cheap material for the manufacture of solar cells and other electronic devices [6,7]. Therefore these films of receiving and changing their conductivity type are urgent tasks. In Colder [8,9] with the change that the substrate temperature is increased nanocrystal growth. It was found that the average grain size (d) and the proportion of crystal grain volume (Vc) decreases with increasing concentration of PH_3. When doped with boron, increasing B_2H_6 concentration value (d) does not change, and Vc decreases. Passivating properties and hydrogen - Ge: H worse than a-Si: H, so in general the films and fotoeffektivnity- $Si_{1-x}Ge_x$: H, somewhat lower than in a-Si: H [10,11].

Hydrogen atoms, and play an important major role in the structure of the film. The purpose of this work - is to determine the amount of hydrogen in the film and measuring its electro-physical properties, as well as the creation of electronic devices based on the films and $Si_{1-x}Ge_x$: H (x=0 ÷ 1).

Experimental Part

Thin films of a-$Si_{1-x}Ge_x$: H (x=0 ÷ 1) obtained by plasma deposition using gaseous mixtures of H_2 + SiH_4, He + GeH_4 in various proportions. The details for the preparation of films are shown in the above mentioned theory [12,13]. The plasma is created by RF field mainly inductive coupling. The film thickness was 0,1 ÷ 1,0 mm. Measured absorption coefficient (α), refraction (n), reflection (R), transmission (T), the band gap (E) for each sample, using the Tauc model [14]. The optical absorption was studied at room temperature as described in [15,16] 21-IR spectrometer.

Results and Discussion

The concentration of hydrogen in the films as-Si1-xGex: H, (x=0 ÷ 1) is determined by the method of Brodsky et al. [15,16]:

$$N = \frac{AN_A}{(\Gamma/\xi)}\int \frac{\alpha(\omega)}{\omega}d\omega \qquad (1)$$

*Corresponding author: Najafov BA, Institute of Radiation Problems of Azerbaijan Nationale Academy of Science, AZ1143, 9, B. Vahabzade str., Baku, Azerbaijan Republic, Russia, E-mail: bnajafov@rambler.ru

Where, Avogadro number of N- and (D/£) integral force hydride with the unit cm^2/mol (H/ξ)=3,5. If the absorption width is denoted by Aw and the center frequency ω_0, then when $\Delta\omega/\omega_0 \leq 0,1$, after approximation with an accuracy of ± 2%, the equation (2) can be written as follows:

$$N = \frac{AN_A}{(\Gamma/\xi)\omega_0}\int\frac{\alpha(\omega)}{\omega}d\omega \quad (2)$$

Where, ε- dielectric constant. For Si ε=12; Ge ε=16. In equation (3) before the integral expression denoted by AS, and - the total absorption of stretching modes for each film, then in determining the concentration of hydrogen (NH) obtain a general expression in abbreviated form:

$$N_H = A_S J_S \quad (3)$$

Ratio AS - films for a-Si: H, is in the tension mode 1.4 *10^{20} cm$^-$ ^{2}. The absorption coefficient (α) for said frequency (2100 cm^{-1}) is 8*10^{-1} ÷ 3*10^2 cm^{-1} with NH=10^{21} ÷ 72.1*10^{22} cm^{-3}. For films as-Ge: H=1.7*10^{20} cm^{-2}. Clearly, equation (3), also describes an oscillating fashion in connection stretching films as-Si: H, a-Ge: H and a-Si$_{1-x}$Ge$_x$: H. Estimates of relative hydrogen bonding to the hydrogenated amorphous as-Si$_{1-x}$Gex: H: NSi-H.

$$P = \left\{\frac{N_{Si-H}}{N_{Ge-H}}\right\}\frac{x}{1-x} \quad (4)$$

Where, NGe-H - the hydrogen concentration in the a-Si: H and a-Ge: H (in cm^3). Equation (3) can be rewritten for (wagging mode) films and rocking fashion-Si: H and a-Ge: H [17,18]. Thus the value of NSi-H and NGe-H determined from the equation (3) for the fashion swing in the following form:

$$N_H = A_W J_W \quad (5)$$

Where, the total absorption rocking fashion for films and-Si: H and a-Ge: H. For these films A$_w$=1.6* 10^{19} cm^{-2} and A$_w$=1.1*10^{19} cm^{-3}, respectively. Knowing NGe-H (where, for a film-Ge: H, A$_w$ = 1.6*10^{19} cm^{-2} and α=5 *10^1 cm^{-1}) to calculate the concentration of hydrogen in the NH and the film-Si$_{1-x}$Ge$_x$: H for expression:

$$N_H = N_{Ge-H}^{wag}\left\{\frac{\int_{streets}\left(\frac{\alpha_1(\omega)}{\omega}\right)d\omega}{\int_{streets}\left(\frac{\alpha_2(\omega)}{\omega}\right)d\omega}\right\} \quad (6)$$

Where, number of connections defined by modes of oscillation in a clean well-Ge: H, whose value is calculated according to the equation (5). The second factor in the expression for NH (integral ratio of infrared absorption maxima) is stretching vibrational mode of the sample in a clean and well-Ge: H. To calculate the ratio using integrated peak corresponding Ge-H (2000 cm^{-1}) and a film-Si$_{1-x}$Ge$_x$ stretching vibrational mode: H.

From these data we can estimate the strength of the oscillator in the film as well-Si$_{1-x}$Ge$_x$: H ratio:

$$\Gamma = J_S/J_W ,$$

Where, $J_S \approx J_S^{Ge} + J_S^{Si}$, $J_W \approx J_W^{Ge} + J_W^{Si}$. The values - $J_S^{Ge}, J_S^{Si}, J_W^{Ge}, J_W^{Si}$ are integral acquisitions fashion stretching and rocking fashion, respectively oscillator strength f=0.51 (x=0) and r=0.13 (x=1). The maximum value of P=4.16 to x=0.40. Table 1 shows the characteristic parameters of amorphous films a-Si$_{0,60}$Ge$_{0,40}$:H. Figure 1 shows the distribution of hydrogen across the film thickness d: 1-specific method

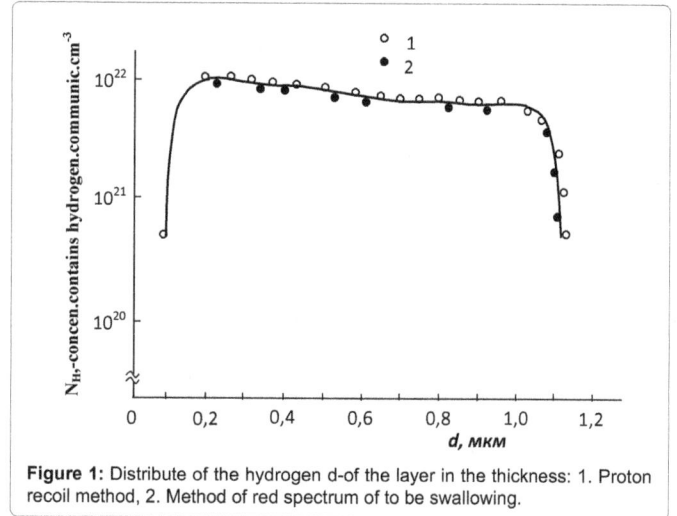

Figure 1: Distribute of the hydrogen d-of the layer in the thickness: 1. Proton recoil method, 2. Method of red spectrum of to be swallowing.

of recoil protons 2 by infrared absorption spectrum. It can be seen that sufficiently uniform distribution of hydrogen. Note that the value of NH, defined by recoil protons (MOS) and IR spectroscopy agree within 2-3 atm. %.

The optical properties of the films

Dependence on hv makes it possible to determine the width of the band gap [14,16] for each film.

In all the studied films of the optical absorption coefficient of the edge is described by the relation:

$$\alpha hv = B(hv - E_0)^2 \quad (7)$$

Where, α=5·10^4 ÷ 10^5 cm^{-1}, E$_0$ is- optical band gap for each film, B- proportionality factor. The value is determined depending $(\alpha hv)^{1/2}$ on hv extrapolation for each sample. The quadratic dependence [7] obtained theoretically for Tauc model [14], which describes the density of states of the mobility gap. The value for x=0 ÷ 1 is from 527 to 343 eV 1 sm$^{1/2}$, respectively, and for films-Si$_{1-x}$Ge$_x$: H (x=0 ÷ 1), E$_0$=1.14 ÷ 1.86 eV. We use the well-known relation, the absorption coefficient - is determined by the following equation [2,4]:

$$T = \frac{(1-R_1)(1-R_2)(1-R_3)\exp(-\alpha d)}{(1-R_2 R_3)\left\{1-\left[R_1 R_2 + R_1 R_2(1-R_2)^2\right]\right\}\exp(-2\alpha d)} \quad (8)$$

Here we assume that

$$R_1 = \left|(n-1)^2 + k_0^2\right|/\left|(n+1)^2 + k_0^2\right|$$
$$R_2 = \left|(n-n_1)^2 + k_0^2\right|/\left|(n+n_1)^2 + k_0^2\right| \quad (9)$$
$$R_3 = \left|(n_1-1)\right|/\left|(n_1+1)\right|^2$$

For weakly absorbing regions of the world, k$_0$ shows light attenuation in the substrate. Note that the film thickness d, is determined in this case from the respective extrema transmission or reflection of the interference fringes.

From equation (8), the absorption coefficients are defined as follows:

$$T = \frac{kx}{a(1-bx)^2} \quad (10)$$

$$aT(1-bx)^2 = kx,$$

$$k(1-R_1)(1-R_2)(1-R_3),$$

$$a = 1 - R_2R_3,$$

$$b = R_1R_2 + R_1R_3(1-R_2)^3,$$

$$m = aTb, \; n = aT,$$

$$mx^2 + kx - n = 0,$$

$$e^{-ad} = \alpha = \frac{-k \pm \sqrt{k^2 + 4mn}}{2m}; \frac{-k \pm \sqrt{k^2 + 4mn}}{2m} > 0.$$

Then,

$$e^{-ad} = \alpha = \frac{2m}{k \pm \sqrt{k^2 + 4mn}},$$

$$\alpha d = \ln \frac{2m}{k \pm \sqrt{k^2 + 4mn}}, \tag{11}$$

$$\alpha = \frac{1}{d} \cdot \frac{2m}{k \pm \sqrt{k^2 + 4mn}}.$$

Equation (11) is a working formula to determine the optical absorption coefficients for the films in weakly absorbing regions of the spectrum [19].

In a strongly-absorbing regions of the spectrum, and. Then the equation (8) can be rewritten as follows:

$$T = \frac{(1-R)^2 e^{-\alpha d}}{1 - R^2 e^{-2\alpha d}}; \tag{12}$$

$$x = e^{-\alpha d}; \; x^2 = e^{-2\alpha d}; \; T = \frac{(1-R)^2 x}{1 - R^2 x^2};$$

Then,

$$T - R^2 T x^2 = (1-R^2)x, \tag{13}$$

$$R^2 T x^2 + (1-R^2)x - T = 0, \tag{14}$$

$$x = -\frac{(1-R^2) \pm \sqrt{(1-R)^4 + 4R^2 T^2}}{2R^2 T},$$

$$x = e^{-\alpha d} > 0.$$

$$x = \frac{\sqrt{(1-R)^4 + 4R^2 T^2} - (1-R)^2}{2R^2 T},$$

$$e^{-\alpha d} = \frac{\sqrt{(1-R)^4 + 4R^2 T^2} - (1-R)^2}{2R^2 T},$$

$$e^{\alpha d} = \frac{2R^2 T}{\sqrt{(1-R)^4 + 4R^2 T^2} - (1-R)^2}, \tag{15}$$

$$\alpha = \frac{1}{d} \ln \left[\frac{1}{2T} \sqrt{(1-R)^4 + 4R^2 T^2} + (1-R)^2 \right],$$

$$\alpha = \frac{1}{d} \ln \frac{1}{2} \left(\sqrt{\left[\frac{(1-R)^2}{T} \right]^2 + 4R^2} + \left[(1-R)^2 \right] \right),$$

$$p = \frac{(1-R)^2}{T}; \; \alpha = \frac{1}{d} \ln \left[\frac{1}{2} \left(\sqrt{p^2 + 4R^2} + p \right) \right] \tag{16}$$

This formula is working to determine the optical absorption coefficient in strongly absorbing regions of the spectrum. Accordingly, the refractive indices are defined using or in the form of the following:

$$nd = \frac{\lambda_m \lambda_{m-1}}{2(\lambda_{m-1} - \lambda_m)}$$

$$\Delta n = \frac{c}{2\pi^2} \int \left[\frac{\alpha(v)}{v^2} \right] dv \tag{17}$$

Creating a solar cell

The research results show that the film and-Si1-xGex: H (x≥0,20) can be used as high-quality material in semiconductor electronics. For this purpose, we have developed a three-layer element based on cascade-type dual-layer elements. The three-layer element is made of two-layer element consisting of two elements based on a-Si: H with p-i-n junction and the p-i-n element i- layer of a film-Si$_{0,88}$Ge$_{0,12}$: H (Figure 2). The thicknesses of the i- layers to the upper two transitions were selected so as to comply with the conditions of equality of short-circuit current of the lower element. Short-circuit current was about half the value for the element with one p-i-n junction. Open-circuit voltage is increased, and short circuit current is reduced with increasing number of superimposed layers. This method can increase the number of layers (create an n-layered element). Note that for each item produced i- layer 0.5 microns thick. The area of each element was 1.3 cm². When a three-layer solar cells must be observed uniform thickness and area of each element. The material of the substrate steel was chosen and used as a coating ZrO$_2$ with 80% transmissivity. Covering ZrO$_2$ simultaneously plays the role of the upper (front) contact. The thickness of the layers a-Si: H p- and n- type was about 300 and 400 Å, respectively [20].

For films doping amount of B$_2$H$_6$ and PH$_3$ gas mixtures was varied between 0.1 and 0.5%, respectively. After deposition of the amorphous semiconductor layers deposited by evaporation ZrO$_2$ film thickness of about 500 Å. Upper contacts used for Ni/Ag, to lower - stainless steel substrate. Items covered sunlight source provided AM-1 (100 mW/

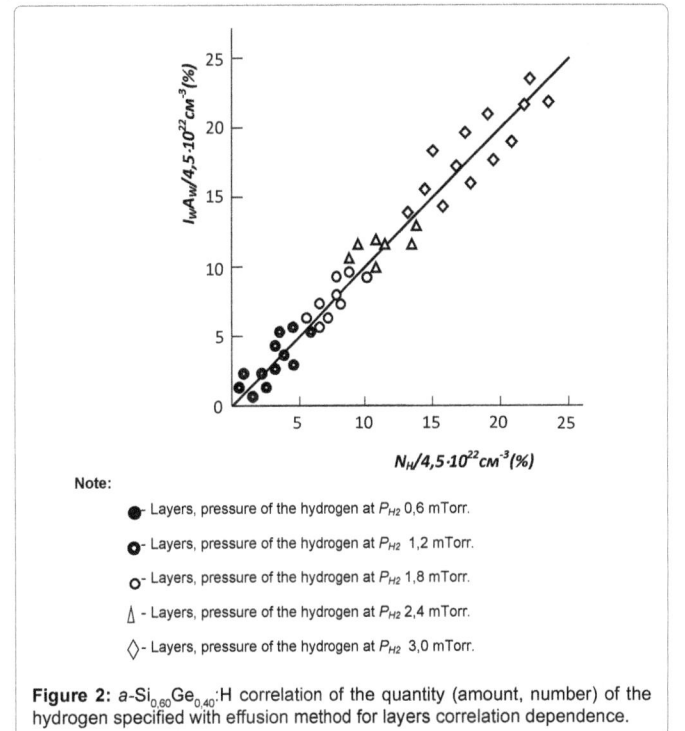

Note:

● - Layers, pressure of the hydrogen at P_{H2} 0,6 mTorr.

◉ - Layers, pressure of the hydrogen at P_{H2} 1,2 mTorr.

○ - Layers, pressure of the hydrogen at P_{H2} 1,8 mTorr.

△ - Layers, pressure of the hydrogen at P_{H2} 2,4 mTorr.

◇ - Layers, pressure of the hydrogen at P_{H2} 3,0 mTorr.

Figure 2: a-Si$_{0,60}$Ge$_{0,40}$:H correlation of the quantity (amount, number) of the hydrogen specified with effusion method for layers correlation dependence.

cm^2). Short circuit current for sandwich elements was 8.5 mA/cm^2, the open circuit voltage of ~2.25 B, the filling factor of ~ 0.50 and n (efficiency) ~ 9.5% (Figure 3). Efficiency for single-layer and two-layer element is 7% and 8.9%, respectively. Collection efficiency of carriers at different wavelengths is defined as:

$$Y(\lambda) = \frac{J_f(\lambda)}{eN(\lambda)} \qquad (18)$$

J$_f$ where (λ) is the density of the photocurrent (10 mA/cm^2), N (λ) - the number of photons incident on a unit area per second, e- free charge carriers. For elements with the above structures is calculated short-circuit current in the assumption of complete exhaustion of all-shells, in the absence of forward bias. Thus, the short-circuit current for the first, second and third elements are given by the following expressions:

$$I_{Sc1} = q \int_0^{1,24/E_{01}} (1-R)N_{ph}\exp(-\alpha_n W_n)\left[1 - \exp(-\alpha_1 W_1)\right]d\lambda \qquad (19)$$

$$I_{Sc2} = q \int_0^{1,24/E_{02}} (1-R)N_{ph}\exp(-2\alpha_n W_n - \alpha_1 W_1 - \alpha_p W_p)\left[1 - \exp(-\alpha_2 W_2)\right]d\lambda \quad (20)$$

$$I_{Sc3} = q \int_0^{1,24/E_{03}} (1-R)N_{ph}\exp(-3\alpha_n W_n - 2\alpha_n W_n - 2\alpha_p W_p - \alpha_1 W_1 - \alpha_2 W_2)\left[1-\exp(-\alpha_3 W_3)\right]d\lambda \quad (21)$$

Here, the field distribution within the layer, respectively, - the number of photons incident on the surface of the elements, - the reflectance of the film, - the absorption coefficient for each layer elements. The open circuit voltage for the cascade elements with two and three transitions represented in the form:

$$V_{oc}(//) = 0.5(E_{01} + E_{02}) \qquad (22)$$

$$V_{oc}(III) = 0.5(E_{01} + E_{02} + E_{03}) \qquad (23)$$

The duty cycle for all elements specified by the value of 0.5. Short-circuit current of the cascade element with two transitions is given by the smaller of the values or Short-circuit current of the cascade element with three transitions is determined by the smallest value of, or Efficiency multijunction cascade elements is given by:

$$\eta(i) = 0.5 \times 0.5\left(\sum_i E_{0i}\right)\frac{I_{sc}(i)}{P_{in}} \qquad (24)$$

Where, i=2 or 3 indicates the number of layers, P_{in} the power of the incident light on the surface of the elements, its value is 100 mV/cm^2 [10] - respectively, the band gap for each I th layer. To improve the solar cell to η is required to increase the number of layers to reduce the

area of the elements, the choice of metal wires, metal contacts to reduce resistance and others.

Measurements of spectral sensitivity are usually produced with a constant white light illumination, the intensity of which corresponds to normal operating conditions (AM-100 ~ 1 MW/cm^2), simultaneously modulating element falls calibrated monochromatic radiation. The photo-current, and its dependence on the wavelength of monochromatic radiation are measured in the shorted circuit with the help of a synchronized amp.

To determine the effectiveness of gathering important knowledge of the electric field that is transmitted to the element. It has been observed that the device configuration dependent collection efficiency is shifted from the red light in the blue region of the spectrum.

The wavelength dependence of the number of photons is calculated using equation (27). It is known that the photon energy and momentum of the corresponding electromagnetic waves with a frequency and wavelength in a vaccum are:

$$W = h\nu = \frac{hc}{\lambda}\;;\; P^* = \frac{h\nu}{c} = \frac{h}{\lambda},$$

Where, h is equal to Planck's constant. Malia frequencies - the predominant role played by the wave properties at high - corpuscular properties of light. Photoelectric effect (photoelectric effect) is the process of interaction of electromagnetic radiation with matter, in which the energy of the photon is transferred to an electron of an atom of matter. In addition, these properties of substances exist mechanical action produced by electromagnetic waves in the fall for what is called a surface pressure of light. If - energy electromagnetic radiation normally feed on a surface of unit area per 1 sec, c is the velocity of propagation of light waves in a vacuum, R- coefficient of surface reflection of light, the pressure P - light on the surface is equal to:

$$P = (1+R)N\frac{h\nu}{c} = P^*/c\;(1+R) \qquad (25)$$

Light pressure (P) is given by equation (19) and is denoted in the following form:

$$P^* = \frac{W}{S} = \frac{hc}{\lambda S}\frac{N}{t}, \qquad (26)$$

N- number of incident photons at the time. W- energy photon incident on all the wavelengths of the body surface. P * - a pulse of light falling on the surface of 1 cm^2 for 1 sec. Then the force of the incident light is determined by the pressure in the following form:

$$F = P \cdot S$$

$$F = \frac{hcNS(1+R)}{\lambda Stc} = \frac{hN(1+R)}{\lambda t}$$

$$F\lambda t = hN(1+R)$$

$$N = \frac{F\lambda t}{h(1+R)}, \qquad (27)$$

F - Light pressure force (F=10^{-8}H) on the surface (S=1 cm^2), λ- the length of the incident wave, h - Planck's constant - the fall of light for 1 sec, with energy and its value is N$h\nu$ - photons, with each impulse photon is hn/c. On reflection radiation - and wavelength, the number of incident photons is 10^{17} ÷ 10^{18} m^{-2}s^{-1}, R=0.2 ÷ 0.8; λ=300 ÷ 900 nm. R- reflects the ability of the elements at λ=300 ÷ 900 nm. It is also possible with other means to determine the number of photons:

$$P = \frac{A}{t} = \frac{Nh\gamma}{t} \qquad (28)$$

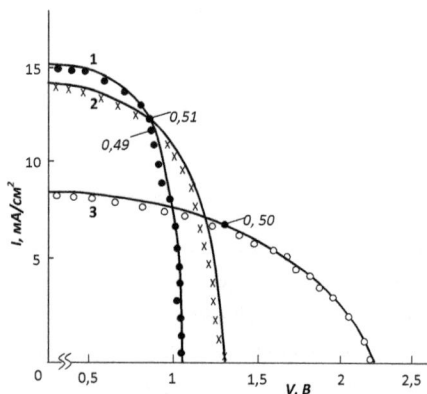

Figure 3: The current-voltage characteristics of solar cells: 1. cascade type УH/a-Si:H a-Si$_{0,60}$Ge$_{0,40}$:H; 2. a-Si$_{1-x}$C$_x$:H/a-Si:H; 3. нк-a-Si:H.

Here, the work function of the A-rays of light in the fall and is, the P power light radiation and is 100 mW/cm^2. From (22) we can determine the number of photons in the following form:

$$N = \frac{P\lambda t}{hc} \qquad (29)$$

Example:

$$N = \frac{F\lambda t}{h(1+0.2)} = \frac{10^{-8}H \times 9 \times 10^{2} \times 10^{-9}M}{6.62 \times 10^{-34}\,\text{dj.cek} \times 1.2} = 1.3 \times 10^{17}\,\text{fot}\,/\,M^{2}c$$

$$N = \frac{F\lambda t}{h(1+0.8)} = \frac{10^{-8}H \times 3 \times 10^{2} \times 10^{-9}M}{6.62 \times 10^{-34}\,\text{dj.cek} \times 1.8} = 10^{18}\,\text{fot}\,/\,M^{2}c$$

Conclusion

By plasma deposition using gas mixtures H$_2$ + SiH$_4$; H$_2$ + GeH$_4$ obtained in different proportions and thin film-Si$_{1-x}$Ge$_x$: H (x=0 ÷ 1). It is shown that the absorption coefficient for visible light and the band gap increases with increasing silicon content. The carrier mobility in the film photoconductivity and a-Si: H is greatly reduced when a germanium content of 40 atm. %. Films based on a-Si: H and a-Si$_{0,88}$Ge$_{0,12}$: H solar cells are made and established monolayer, two-layer and three-layer structure; measure their performance. It was found that for single, double and triple layer structures with an area of 1.3 cm^2 element η is 7; 8.9; 9.5%, respectively. For the three-layer cell collection efficiency peaks shifted to longer wavelengths. The structures obtained when illuminated by light in the wavelength range 0.3 ÷ 1.1 micron for 120 hour, no degradation was observed. The depletion region, where an internal electric field is created, represents only a tiny part of the wafer. Most electron-hole pairs are generated in the bulk of the electrically neutral p-type region. Electrons, which are the minority carriers in the p-type region, diffuse towards the p-n junction and in the depletion region of the junction the electrons drift to the n-type layer under the influence of the internal electric field. Hence, it is shown that the structure of the solar cells on the basis of well-Si$_{0,88}$Ge$_{0,12}$: H and a-Si: H are effective and further efforts to improve the quality of films and higher-efficiency of the efficiency they are urgent tasks. So, we can conclude that the above study is very useful in various applications like hydrogenated amorphous silicon, hydrogenated microcrystalline silicon, deposition of thin-film silicon.

References

1. Afanasyev VP, Gudowski AS, Nevedomsky VN (2002) The effect of heat treatment on structure and properties of the films as well-Si: H, obtained by cyclic deposition.

2. OA Golikova, MM Kazanin, VH Kudoyarova (1998) Effect Staeblera-Wronski, depending on the position of the Fermi level and the structure of undoped amorphous hydrogenated silicon.

3. MM Mezdragina, AV Abramov, GM Macina (1998) Effects of the atomic hydrogen in the annealing atmosphere on the properties of amorphous hydrogenated silicon films and the parameters p-i-n structures on their basis.

4. Golikova OA, Bogdanova EV, Babahodzhaev US (2002) Crystallization of amorphous hydrogenated silicon films deposited at different conditions. Semiconductors 36: 1180-1183.

5. OA Golikova, AN Kuznetsov, VH Kudoyarova, MM Kazanin (1997) Features of the structure of amorphous hydrogenated silicon films deposited by the decomposition of silane at a constant current in the magnetic field.

6. Joannopoulos JD, Lucovsky G (1988) The physics of hydrogenated amorphous silicon. Springer, USA.

7. Najafov BA (2006) Amorphous and microcrystalline semiconductors.

8. Colder H, Rizk R, Morales M, Marik P, Vicens J, et al. (2005) Influence of substrate temperature on the growth of nanocrystalline silicon carbide in reactive magnetron sputtering. J Appl Phys 98: 44-49.

9. King RR, Zaw DC, Edmontson KM, Fetzer CM, Kinzey GS, et al. (2007) 40% efficient metamorphic GaInP/GaInAs/GeGaInP/GaInAs/Ge multijunction solar cells. Appl Phys Lett 90: 231-239.

10. Hamakav Y (1986) Amorphous semiconductors and devices based on them. Metallurgy.

11. Mei JJ, Chen H, Shen W, Dekkers HFW (2006) Optical properties and local bonding configurations of hydrogenated amorphous silicon nitride thin films. Appl Phys 100: 107-114.

12. Najafov BA (2007) Producing films of amorphous hydrogenated silicon carbide-and-of Si$_{1-x}$G$_x$: H (x=0 ÷ 1) for photovoltaic inverters. Solar Energy 11: 177-179.

13. Najafov BA, Isakov GI (2005) Preparation and doping amorphous Si-based films and Ge. ISIAEE Solar Energy 24: 79-82.

14. Tauc J, Grigorovici R, Vancu A (1966) Optical properties and electronic structure of amorphous germanium. J Non-Cryst Sol 15: 627-637.

15. Brodsky MN, Cardona M, Cuomo JJ (1977) Infrared and raman spectra of the silicon hydrogen bonds in amorphous silicon prepared by glow discharge and sputtering. Phys Rev B 16: 3556-3581.

16. Najafov BA, Isakov GI (2005) Optical properties of amorphous films of solid solution and-Si$_{1-x}$Ge$_x$: H with different hydrogen concentrations. J Appl Spectrosc 72: 396-402.

17. Mott SNF, Davis EA (1982) Electronic processes in non-crystalline materials. Oxford University Press, USA.

18. Najafov BA (2000) Electrical properties of hydrogenated amorphous Ge0.90Si0.1:Hx films. Semiconductors 34: 1330-1333.

19. Rudder RA, Cook JW, Fucovsky G (1984) High photoconductivity in dual magnetron sputterd and germanium alloy films. Appl Phys Lett 45: 887-889.

20. Najafov BA, Fiqarov VR (2010) Hydrogen content evaluation in hydrogenated nanocrystalline silicon and its amorphous alloys with germanium and carbon. Int J Hydrogen Energy 35: 4361-4367.

4

Effect of HF Welding Process Parameters and Post Heat Treatment in the Development of Micro Alloyed HSLA Steel Tubes for Torsional Applications

Udhayakumar T* and Mani E

Corporate Technology Centre, Tube Investments of India Ltd, Chennai-600 054, India

Abstract

The aim of the present study is to investigate the effect of High frequency (HF) Electric resistance welding process parameters and post heat treatment on the torsional fatigue life of micro alloyed HSLA steel tubes. Micro alloyed grades exhibit higher strength and formability owing to the presence of fine recrystallized ferritic grains due to thermo mechanical treatment and presence of alloying elements like Vanadium, Niobium and Titanium. Welded tubular components made of micro alloyed HSLA steel grades are highly emerging and manufacturing them remains quite challenging. Weld bond width, HAZ width and bond angle are the significant factors that directly influences the Weld quality and strength. The effect of key welding parameters like Heat input, welding temperature, squeeze roll pressure, Vee-angle, Vee-length and Impeder diameter on the above mentioned significant factors was analysed. Narrow bond, Minimum HAZ width with pronounced Hour glass pattern and optimum bond angle resulted in superior bond strength and formability of HSLA tubes. Microstructural characterization of the samples was carried out using Light optical microscopy and Scanning electron microscopy. Residual stress was determined using X-ray diffract meter and tube slitting method. Higher tensile residual stress of magnitude 200 MPa was observed in the weld region. Since such high magnitude of tensile residual stress is detrimental to torsional fatigue life, stress relieving of the tubes was carried out at different subcritical temperatures 650°C and 700°C with soaking time of 45 minutes. Without significant drop in the tensile properties, compressive residual stress of magnitude 129 MPa was observed at a particular stress relieving cycle. As an effect of stress relieving heat treatment below Ac1 temperature, there is a significant improvement in the Torsional fatigue life of the HSLA steel. Thus, High Frequency welded micro alloyed HSLA steel tubes with enhanced torsional fatigue performance were successfully developed.

Keywords: HSLA steel; High frequency (HF) electric resistance welding; Stress relieving; Residual stress; Fatigue test

Introduction

In automobile industry, High strength low alloy steels were primarily developed to replace the conventional carbon steels in order to improve the strength to weight ratio without affecting its performance and efficiency. HSLA steels have unique properties, such as high strength, formability, excellent ductility, good weld ability and good low temperature impact toughness. It has fine ferritic microstructure with small amount of pearlite less than 10% volume fraction [1]. Among the different categories of HSLA steel, micro alloyed ferrite-pearlitic steel has small amount of alloying elements like Vanadium, Niobium and Titanium and it forms strong carbides and carbo nitrides for the precipitation strengthening, grain refinement and transformation temperature control [2]. Generally welding of HSLA steel tube is very difficult and it requires rigid forming stands, accurate forming rolls and efficient welding parameters. High Frequency tube welding is one of the most forgiving industrial processes and it is possible to produce high bond strength welded tubes for many applications. High frequency welding is electric resistance welding which works on the electromagnetic induction principle mainly and it is associated with proximity effect and skin effect. Because of skin effect and proximity effect, increasing the frequency helps to concentrate intense current closer to the strip edges. This combined effect results in less metal being heated, using less current, which translates into higher efficiency. In general, frequencies used for tube welding range from 100 kHz to 800 kHz for the ferrous material. Apart from Frequency, Induction current coil, impeder design plays a big role in increasing the weld efficiency. Vee length, Vee angle and impeder design and position helps to heat the strip corners first and it leads to the Hour glass shape in the Heat affected zone [3,4]. In HF induction welding of HSLA steel, the better weldment can be obtained by applying sufficient cold forming, controlling the strip width, strips edge quality, applying sufficient and uniform squeeze out pressure, optimizing and monitoring the effective weld parameters/heat input, avoiding negative inside bead and optimizing post heat treatments [5]. Weldment microstructure of the HSLA steel depends on the cooling rate, heat input and the material cleanliness. As a consequence of lower heat input and faster cooling rate, finer grain microstructure consisting of acicular ferrite, upper bainite and small clusters of low carbon marten site formed in the weldment and it possess high weld strength and ductility [6,7].

Heat affected zone width and the hardness transverse along the weldment depends on the heat input and the cooling rate in the welding process [8]. Low temperature impact toughness was found high in the HAZ region of the lower heat input weldment, in the Niobium added HSLA steel. Lower heat input reduces the prior austenite grain coarsening by controlling Ac3 temperature, which resulted in very high toughness [9-11]. Also, by the effect of alloying elements, recrystallization temperature got increased in the HSLA steel [12]. Reheating the heat treated HSLA steels in the intercritical temperatures

*Corresponding author: Udhayakumar T, Corporate Technology Centre, Tube Investments of India Ltd, Chennai-600 054, India
E-mail: Thendralarasu@tii.murugappa.com

with high cooling rates results in the microstructure of marten site with retained austenite on grain boundaries, leads to poor impact toughness. Lowering the rate of cooling leads to high impact toughness because of its upper bainite and pearlitic microstructure [13].

In HSLA steel, various research works have been done to improve the weldment characteristics, low temperature impact toughness and increasing the strength level etc. and it becomes new and challenging to study the torsional fatigue characteristics of HSLA steels in automobile application. The present investigation in this material is to study the effect of high frequency electric resistance welding parameters and the post heat treatment influence on the torsional applications.

Experimental and Simulation Methods

Material

Hot rolled high strength low alloy steel with the strength level of 800 MPa and the chemical composition of which is listed in Table 1 was used for the study. Micro alloying of the steel with addition of Niobium and Titanium and thermo mechanical treatment by the steel supplier has resulted in precipitation strengthening and grain refinement in the raw material. Microstructure of the micro alloyed HSLA steel (Figure 1) shows very fine ferrite grains of size less than 10 μm.

Energy Dispersive spectroscopic analysis was done in the raw material, at the precipitates in the grain boundary (Figure 2). Analysis confirms the presence of Ti and Nb precipitates in the HSLA steel.

High frequency welding of tubes

HF welded steel tube is normally made from rolled steel coils. Rolled steel coils were slit into different widths according to tube dimensions. At the first stage, the strip is gradually cold formed into a circular tube shape through series of rolls. Subsequently the edges of the strip are then heated to a welding temperature by the high frequency induction coil. The heated edges are then squeezed and pressure is applied to form a forged weld (Figure 3).

Weld bond width, HAZ width and bond angle are the significant factors that directly influences the Weld quality and strength. The effect of key welding parameters like Heat input, welding temperature, squeeze roll pressure, Vee-angle, Vee-length and Impeder diameter on the above mentioned significant factors was analysed and parameter optimization was carried out.

Stress relieving heat treatment

Stress relieving annealing at different temperatures like 650°C and 700°C in the sub critical range below Ac1 temperature, to study the effect of post heat treatment on the fatigue life of the component. Stress relieving Heat treatment was carried out in electric heating muffle furnace without controlled atmosphere. The primary objective of stress relief annealing is to reduce the residual stress produced due to welding and forming.

Residual stress measurement

HF Electric resistance welding of the HSLA strips is intended to induce residual stresses in the final tube. The residual stress plays a significant role in the torsional fatigue life of a component. Two methods were used to experimentally determine the circumferential (hoop) residual stresses in the tubes:

Slitting method for tubes

X-ray Diffract meter

Slitting Method is one of the destructive techniques that rely on the introduction of an increasing cut to a part containing residual stresses. In slitting method, a narrow cut of progressive depth is introduced into a part containing residual stresses and the relieved strain is measured in terms of tube dimension. In this technique, the cutting may alter the original residual stresses through temperature rise and plastic deformation near the cut. Hence EDM cutting is preferred for slitting the tube

Residual stress is calculated using the formula:

$$S = \left(\frac{Et}{1-\mu^2} \right) * \left(\frac{Df - Do}{Df * Do} \right)$$

Element	Wt.%
Carbon	0.074
Silicon	0.231
Manganese	1.62
Phosphorus	0.015
Sulphur	0.0025
Chromium	0.017
Aluminium	0.051
Titanium	0.127
Niobium	0.059
Molybdenum	0.103
Nickel	0.006

Table 1: Chemical composition of the steel.

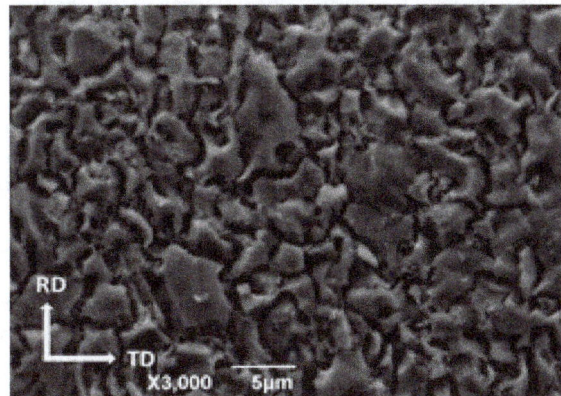

Figure 1: SEM micrograph of the raw material.

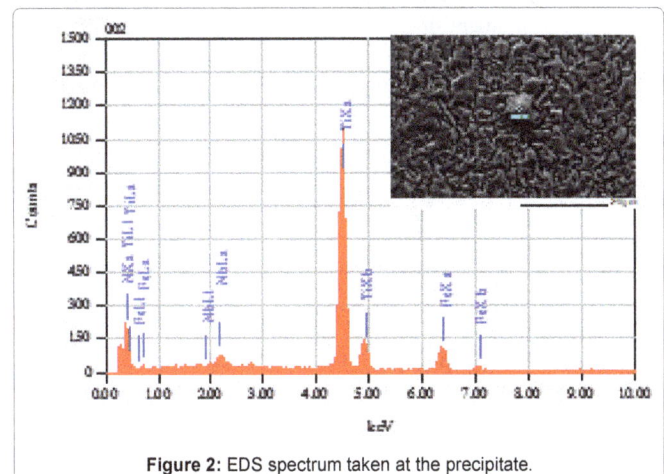

Figure 2: EDS spectrum taken at the precipitate.

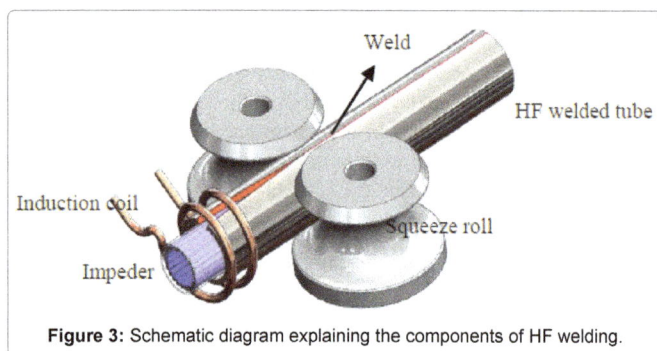

Figure 3: Schematic diagram explaining the components of HF welding.

S: Residual stress in the circumferential direction, MPa

E: Modulus of Elasticity, MPa

μ: Poisson's ratio

T: Thickness of the tube, mm

Do: Mean Outer diameter of tube before slitting, mm

Df: Mean Outer diameter of tube after slitting, mm

XRD is a quantitative and precise technique used for residual stress measurement. Portable residual stress analyser iXRD Proto was used to determine the circumferential residual stress at the surface of tube. In x-ray diffraction residual stress measurement, the strain in the crystal lattice is measured, and the residual stress producing the strain is calculated, assuming a linear elastic distortion of the crystal lattice. Testing was done based on SAE HS 784.

Torsional fatigue testing

Instron Servo hydraulic rotary actuator (Figure 4) was used to carry out fatigue testing of the HSLA tubes in the non-heat treated and post heat treated condition. Prior to torsional fatigue testing, welded tubes were formed into a particular profile as per design requirement and the ends of the tube were welded with suitable fixtures for actuator. Actuator head is connected with one end of the component and the other end remains fixed. Testing was done based on the design specification, which states maximum stress condition as half the yield strength of the tube.

Simulation using ANSYS: Simulation was done at different Torsional moments 300 Nm, 400 Nm and 500 Nm and the maximum stress resulted are shown in Figures 5-7. The result shows that 400 Nm torsional moment develops maximum stress of value 358 MPa, which is roughly half of yield strength of this tube. Hence 400 Nm was selected as the input torsional moment for the experimentation of fatigue testing.

The aim of the study is to investigate the effect of HF welding process parameters and post heat treatment on the torsional fatigue life of micro alloyed HSLA steel tubes.

Results and Discussion

HF welding parameter optimization

Micro alloyed HSLA steel tubes were manufactured by the conventional HF welding route. To obtain very high weld bond strength and weld quality, several iterations were carried out. Major parameters like Strip width, power, speed, Vee-length and Impeder diameter were considered. Vee-angle maintained was 2 degrees.

Table 2 explains the different trials conducted by varying the key parameters to produce HF welded HSLA steel tube of outer diameter of 88.9 mm and thickness of 2.9 mm, with very high weld bond strength.

Critical V forming test was developed to assess the weld quality. For this test, specific tool was manufactured with respect to the tube dimension (Figure 8). The tube was formed to the profile shown in Figure 9 such that the weld region falls in maximum strain region. Maximum strained region was initially identified through simulation using Finite element analysis (LS Dyna software). Simulation of the forming proves that maximum strain falls in the corners of the V profile (Figure 9). Tube samples of 100 mm length were taken and formed with the weld in the maximum strain position.

Samples from each trial were subjected to V-forming as a qualifying test and results are monitored. Trial H samples results in line with our expectation and it shows the weld bond strength and formability. Trial H samples withstood maximum load of 156 kN during max strain condition. Initially trial resulted in very wide HAZ width (Figure 10), but increasing the impeder diameter from 45 mm to 60 mm improves the concentration of current in strip edges and avoids heat loss. Optimised Vee-length and the increased mill speed resulted in the pronounced hour glass pattern; narrow bead width and required bond angle of 60-65 deg. Bead width of the resultant tubes measured around 994 μm.

Material characterization

Following optimisation, welded HSLA steel tube with high bond strength was identified and taken for micro structural characterisation and mechanical properties assessment. Zeiss Optical microscope and JEOL Scanning electron microscope were used for characterization. Picral and 2% Nital reagents etchants were used to view the weld flow and microstructure respectively. Base region has very fine ferrite grains lesser than 10 μm (Figure 11a). Weld region has the mix of microstructure – Low carbon martensite, bainitic ferrite and polygonal ferrite (Figure 11b). Presence of Bainitic ferrite improves the toughness of weld region. Polygonal ferrite and bainitic ferrite was observed in HAZ region (Figure 11c)

Stress relieving of the HF welded tube samples was done below the A_1 critical temperature at 650°C, and 700°C with the soaking time of 45 min. The very high recrystallization temperature of the material rules out possibility of grain coarsening during SRA Microstructural analysis of the stress relived samples also confirms the same.

Mechanical properties of the HSLA steel tube in the HF welded condition and after stress relieving heat treatment was evaluated as per the ASTM E8 and the results are shown in Table 3

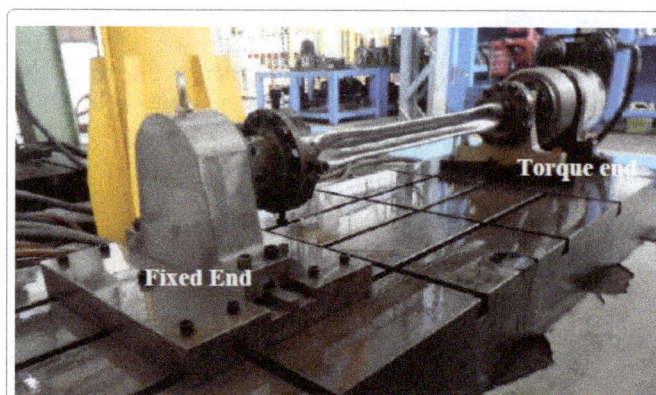

Figure 4: Image of torsional fatigue test rig.

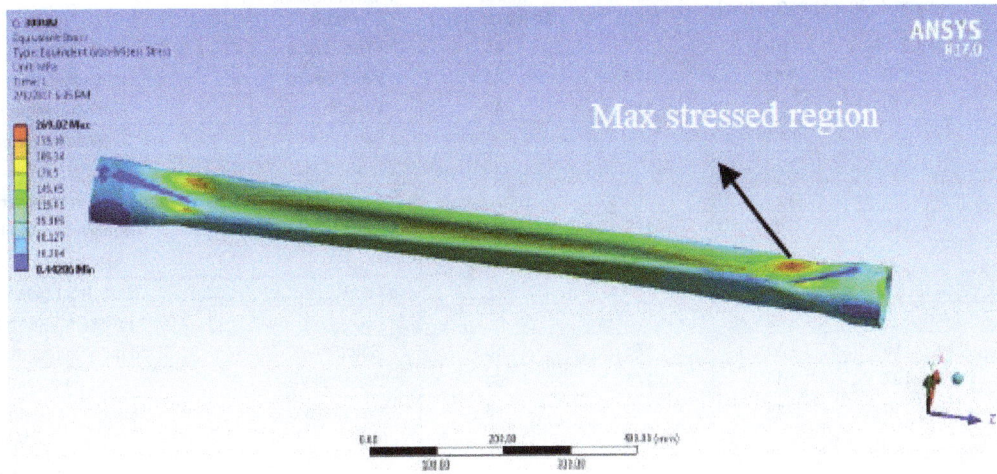

Figure 5: Stress analysis of the twisted model – 300 Nm.

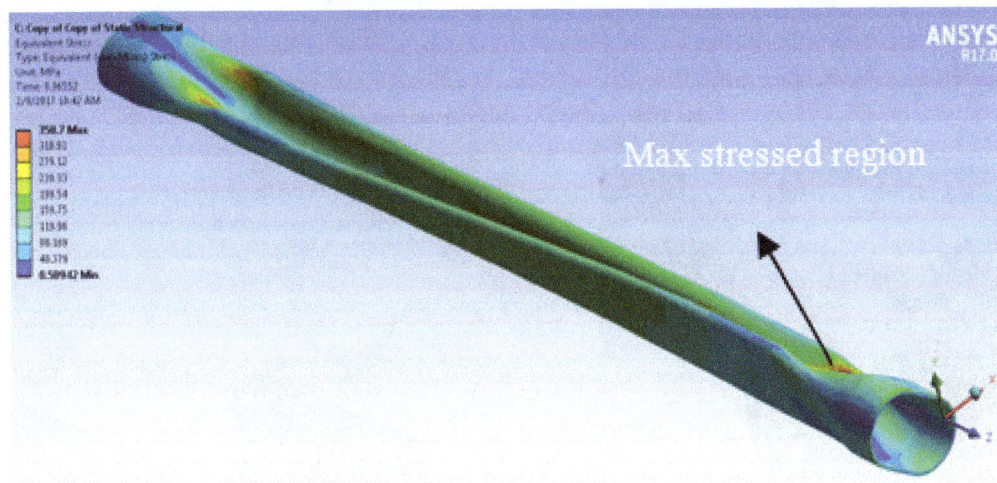

Figure 6: Stress analysis of the twisted model – 400 Nm.

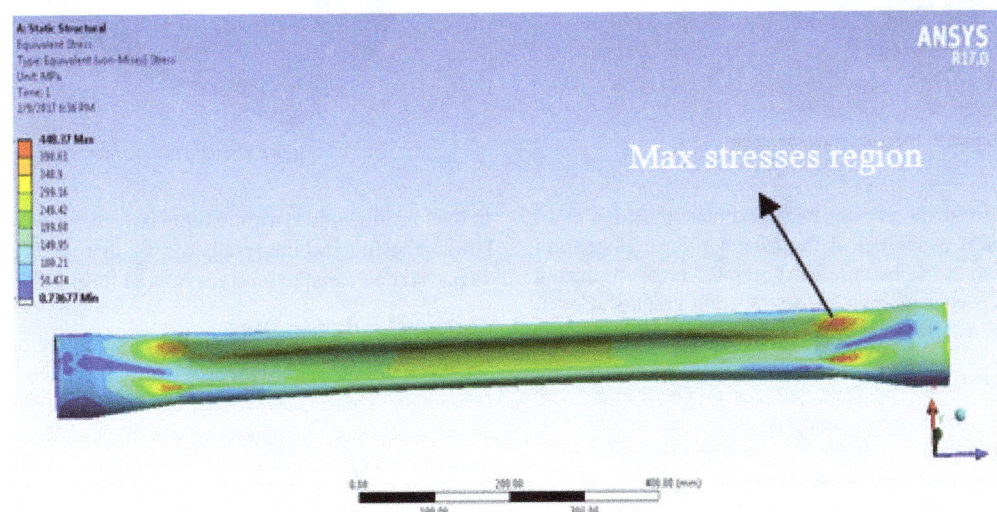

Figure 7: Stress analysis of the twisted model – 500 Nm.

Trial	Strip width (mm)	Power (kW)	Speed (mpm)	Vee length* (mm)	Impeder diameter (mm)	V forming test result	Remarks
A	276.5	102-104	10.2	175	45	Weld failure	Insufficient material flow
B	278	110	11	175	45	Weld failure	Insufficient material flow
C	279	110	11	175	45	Weld failure	Sufficient material flow
							Wider weld bead
							Poor hour glass pattern
							tube failed in V-forming operation
D	279	110	11	175	60	Weld failure	Sufficient material flow
							Narrow weld bead
							Poor hour glass pattern
							Tube Failed in V-forming operation
E	279	118	11	175	60	Stitch mark	Stitch mark defect after V-forming
F	279	102	11	175	60	Weld failure	Paste weld
G	279	110	11	145	45	Weld failure	Material flow is sufficient
							Narrow weld bead
							Poor hour glass pattern
							Tube failed in critical V-forming
H	279	110	11	145	60	OK	Material flow is sufficient
							Narrow weld bead
							Pronounced hour glass pattern
							Tubes passed in critical V-forming

*Vee length is the distance between centre of coil and apex.

Table 2: HF welding process parameter optimization.

Figure 8: Image of the forming tool.

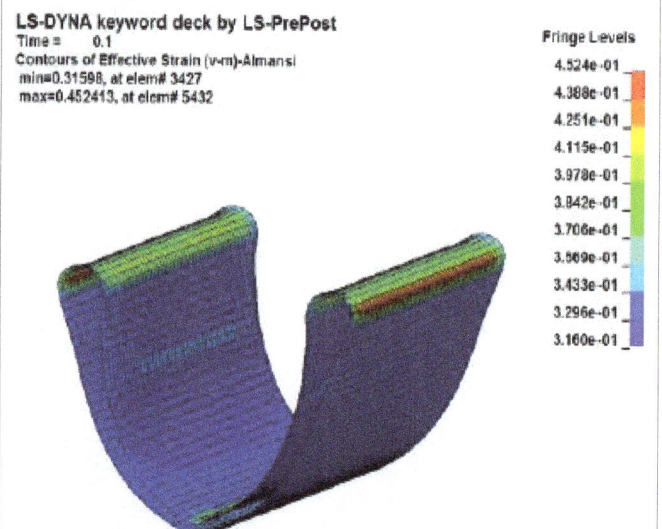

Figure 9: Strain plot of V-formed tube.

As a consequence of stress relieving heat treatment, there is a slight drop in the mechanical properties of the tube. Uniform elongation increases to 10% from 5.5%, because of the grains recovery during stress relieving. This is evident from the stress strain graph shown in Figure 12. Weld ductility also improves on stress relieving.

Residual stress measurement

The results of residual stress analysis are given in Table 4. Measurements were carried out at the base and weld region in the circumferential (hoop) direction using XRD and slitting method.

Higher magnitude of tensile residual stress was observed in the base material of the micro alloyed HSLA tube. This might be attributed to the forming of the strips to tubular form. Stress relaxation occurs on stress relieving and the results convey that the magnitude of residual stress has decreased to greater extent in the post heat treated tubes.

Torsional fatigue testing

As a design requirement, HF welded tubes were mechanically formed to a particular profile before the torsional fatigue testing. Torsional fatigue testing was done at the maximum stress level of half the yield strength as the design criteria. Maximum stress of 358.7 MPa was developed in the tubes for the torsional moment of 400 Nm was determined through ANSYS simulation. Testing was carried out in the formed condition i.e. component form.

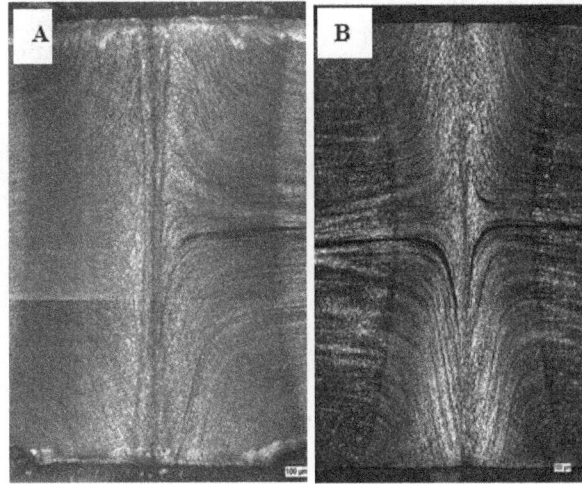

Figure 10: Optical micrographs showing the weld region: (A) Trial C sample; (B) Trial H sample.

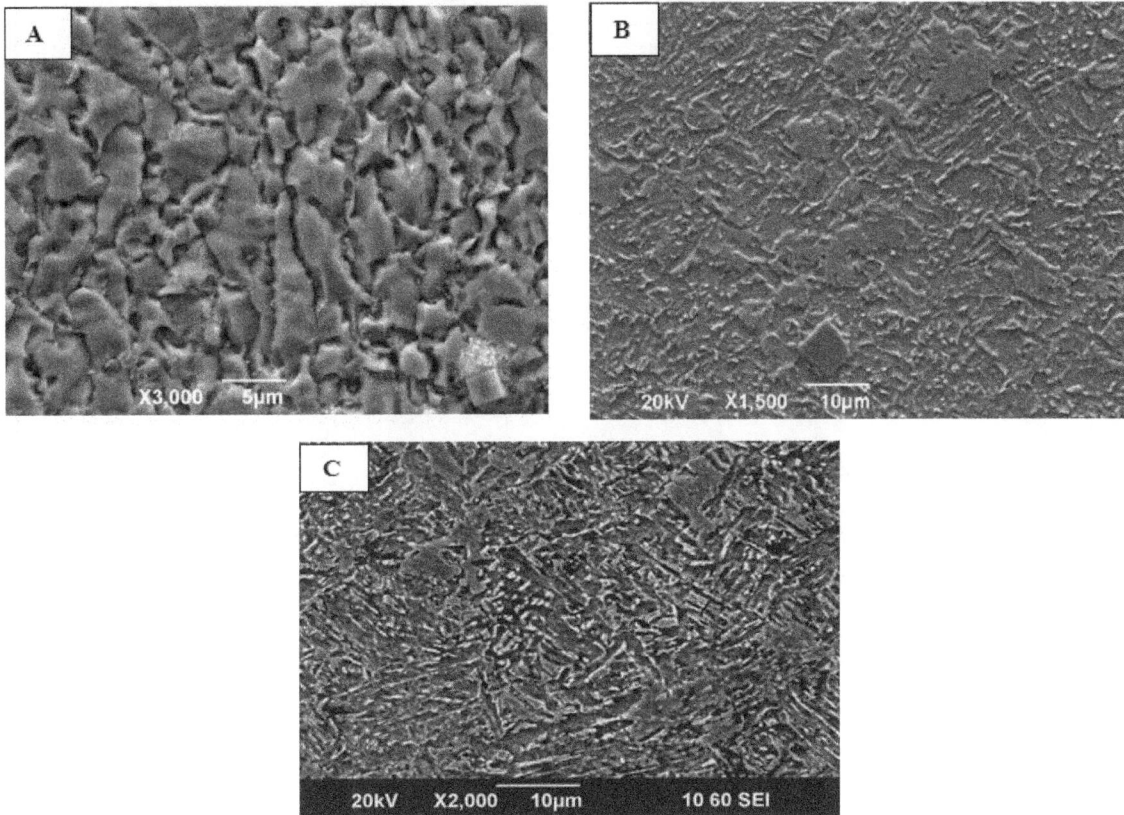

Figure 11: SEM micrographs showing the microstructure of different regions in the HF welded micro alloyed HSLA tube: (A) Base; (B) HAZ; (C) Weld.

Sample condition	Stress relieving temperature[a]	YS[b] (MPa)	UTS[b](MPa)	% of Uniform elongation[b]	% of elongation[b]
As HF welded	-	790	845	5.5	6
As Stress relieved	650°C	775	815	9.5	11
	700°C	750	790	10	17

[a]Stress relieving was performed for soaking time 45 minutes followed by air cooling.
[b]The reported values are average of three tensile samples (ASTM E8).

Table 3: Mechanical properties of the micro alloyed HSLA tubes.

Test Parameters: Torsional Moment-400 Nm; Frequency-2 Hz

Failure occurred in the base material region of the HF welded HSLA steel tube component and the crack has initiated in the maximum stressed region, as indicated in the simulation results (Figure 6). Table 5 explains the fatigue life of the HF welded tubes.

Stress relieved samples were tested in the same test condition and the samples crossed 1000000 cycles without failure. This explains the effect of stress relieving heat treatment of the HSLA steel in improving the torsional fatigue life.

Conclusions

The objective of the study is to investigate the effect of HF welding process parameters and post heat treatment on the torsional fatigue life of micro alloyed HSLA steel tubes. The following conclusions can be made from the study:

1. The role of HF welding parameters on the weld quality of micro alloyed HSLA tubes is clearly evident from the work. Optimum combination of factors like Strip width, Vee-length, Impeder diameter, Heat input and Vee-angle leads to superior weld characteristics

- Pronounced Hour Glass pattern
- Narrow Weld Bead width
- Optimum Weld Bond angle
- High Bond Strength

2. Critical V forming test proves to be a best qualitative test to assess the weld quality of tubes. It reveals the importance of development of application related testing methods in manufacturing industries.

3. Mechanical properties like YS, UTS did not vary significantly

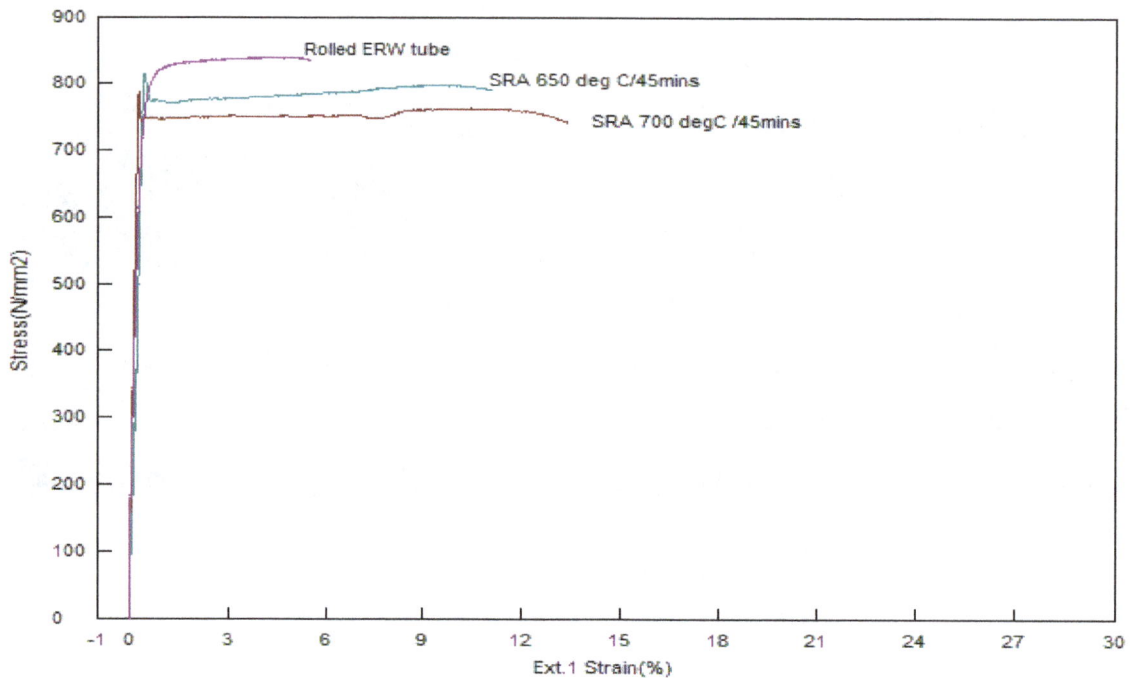

*SRA: Stress Relieving Annealing.

Figure 12: Stress-strain graphs of As HF welded and SRA* samples.

Sample condition	Circumferential residual stress(MPa)		Nature of stress
	XRD	Slitting method	
As HF welded	195 ± 04*	195	Tensile
650°C SRA[1]	-70 ± 5*	-2	Compressive
700°C SRA[1]	-129 ± 7*	-5	Compressive

*Residual stress measurement was carried out in the base material region.

[1]SRA stands for Stress relieving annealing.

Table 4: Residual stress values – XRD and Slitting method.

Sample condition	Number of samples tested	Number of cycles to failure
HF welded	3	398791
		238808
		219882

Table 5: Results of torsional fatigue test.

between HF welded and post heat treated tubes. Improvement in Uniform elongation is observed in the post heat treated tubes. Hence stress relieving heat treatment enhances the formability of the HSLA tubes. It is advantageous, because yield strength of the material directly influences the fatigue life.

4. Magnitude of tensile residual stress at the base region of HF welded tubes is 195 MPa whereas the magnitude of residual stress drops to 129 MPa in 700°C stress relieved sample and it is compressive in nature. Thus stress relieving of HSLA tubes aids in elimination of residual stress.

5. The Significant improvement in fatigue life of HSLA steel tubes heat treated sub critically (below A1) is mainly attributed to the absence of tensile residual stresses, which was developed in the HF welded tube due to forming and welding operation. Thus, residual stress plays an important role in the torsional endurance life of micro alloyed HSLA steel tubes and stress relieving heat treatment of the micro alloyed HSLA tubes enhances its fatigue life.

Acknowledgements

The Authors are thankful to Mr. Natarajan, Technical Advisor, Corporate Technology Centre and Mr. Krishna Srinivas, Senior Vice President, Corporate Technology Centre, Tube investments of India Ltd for their valuable suggestions and guidance throughout the project. The authors are grateful to the organisation Tube investments of India ltd for providing the complete support for this project.

References

1. Lee CH, Bhadeshia HKDH, Lee HC (2003) Effect of plastic deformation on the formation of accicular ferrite. Material Science Engineering A 360: 249-57.

2. ASM International, Materials selection, Alloying-understanding the basics #06117G, ISBN: 978-0-87170-744-4.

3. Wright J, Equipment EH (1999) Optimizing Efficiency in HF Tube Welding Processes. Electronic Heating Equipment, Inc. Bonney Lake. USA. s/f.

4. Haga H, Aoki K, Sato T (1980) Welding phenomena and welding mechanisms in highfrequency electric resistance welding-1st report. Welding Journal 59: 208s-212s.

5. Babakri KA (2010) Improvements in flattening test performance in high frequency induction welded steel pipe mill. Journal of Materials Processing Technology 210: 2171-2177.

6. Nathan SR, Balasubramanian V, Malarvizhi S, Rao AG (2015) Effect of welding processes on mechanical and microstructural characteristics of high strength low alloy naval grade steel joints. Defence Technology 11: 308-317.

7. Kang J, Wang C, Wang GD (2012) Microstructural characteristics and impact fracture behaviour of a high-strength low-alloy steel treated by intercritical heat treatment. Materials Science and Engineering: A 553: 96-104.

8. Reddy GM, Mohandas T, Papukutty KK (1998) Effect of welding process on the ballistic performance of high-strength low-alloy steel weldments. Journal of materials processing technology 74: 27-35.

9. Shi Y, Han Z (2008) Effect of weld thermal cycle on microstructure and fracture toughness of simulated heat-affected zone for a 800MPa grade high strength low alloy steel. Journal of materials processing technology 207: 30-39.

10. Quesnel DJ, Meshii M, Cohen JB (1978) Residual stresses in high strength low alloy steel during low cycle fatigue. Materials Science and Engineering 36: 207-215.

11. Show BK, Veerababu R, Balamuralikrishnan R, Malakondaiah G (2010) Effect of vanadium and titanium modification on the microstructure and mechanical properties of a micro alloyed HSLA steel. Materials Science and Engineering: A 527: 1595-1604.

12. Zhang YQ, Zhang HQ, Li JF, Liu WM (2009) Effect of heat input on microstructure and toughness of coarse grain heat affected zone in Nb micro alloyed HSLA steels. Journal of Iron and Steel Research, International 16: 73-80.

13. Davis CL, King JE (1993) Effect of cooling rate on intercritically reheated microstructure and toughness in high strength low alloy steel. Materials Science and technology 9: 8-15.

5

Electrocatalytical Oxidation of Methanol on (Zn or Cu/Natural Phosphate/Iron) Electrode

Kouider N[1], Bengourram J[1], Mabrouki M[1], Gamouh A[2], Forsal I[3] and Chtaini A[2]*

[1]*Laboratoire Génie Industriel, Faculté des Sciences et Techniques de Béni Mellal, Morocco*
[2]*Equipe d.Electrochimie Moléculaire et Matériaux Inorganiques, Faculté des Sciences et Techniques de Béni Mellal, Morocco*
[3]*Laboratoire de Chimie Organique et Analytique, Faculté des Sciences et Techniques de Béni Mellal, Morocco Cruz do Sul, Brazil*

Abstract

Two new electrodes were prepared and characterized electrochemically; Zn/NP/Iron and Cu/NP/Iron, for a direct methanol fuel cell (DMFC). Themorphology and structure of the catalyst layer were analyzed by optical microscopy. The catalyst coating layer shows an alloy character. The results show that, the oxidation of methanol is catalysed by the formation of copper oxides on the surface of the Cu / NP / Iron.

Keywords: Electrodeposit; Methanol oxidation; Catalyst; SQWV

Introduction

Fuel cells are efficient and environmentally acceptable conversion devices. Electric current is generated in the fuel cell by the direct electrochemical oxidation of either hydrogen (proton exchange membrane fuel cell, PEM) or methanol (Direct Methanol Fuel Cell, DMFC). The electrochemical processes that yield energy are essentially pollution free. Water formed during the operation of the device is beneficial in space travel and submarines. Applications of fuel cells are diverse ranging from stationary (individual homes or district schemes) or mobile (transportation as cars, buses, etc.), mobile phones and lap top computers [1,2]. Hydrogen is currently the only practical fuel for use in the present generation of fuel cells. The main reason for this is this is its high electrochemical reactivity compared with that of the more common fuels from which it is derived, such as hydrocarbons, alcohols, or coal. Also, its reaction mechanisms are now rather well understood [3,4] and are characterized by the relative simplicity of its reaction steps, which lead to no side products. Pure hydrogen is attractive as a fuel, because of its high theoretical energy density, its innocuous combustion product (water), and its unlimited availability so long as a suitable source of energy is available to decompose water. One of the disadvantages of pure hydrogen is that it is a low density gas under normal conditions, so that storage is difficult and requires considerable excess weight compared with liquid fuels.

Methanol has been considered for fuel cell power generation for a number of years because it can be processed into a hydrogen-rich fuel gas fairly easily and efficiently by steam or auto thermal reforming. Methanol, as a liquid fuel is easily transported and stored in comparison to hydrogen gas.

The methanol fuel has a superior specific energy density (6000 (Wh/Kg) in comparison with the best rechargeable battery, lithium polymer and lithium ion polymer (600 (Wh/Kg) systems. This means longer conservation times using mobile phones, longer times for use of laptop computers and more power available on these devices to support consumer demand. Another significant advantage of the direct methanol fuel cells over the rechargeable battery is its potential for instantaneous refuelling [5-7].

In this work we propose to coat stainless steel surfaces with another material as a natural phosphate phase in order to form satisfactory modified electrodes for eventually anode for methanol fuel cell. To fulfill the purpose of strong adhesion of natural phosphate on stainless steel, coating by electrochemical technique is interesting as a simple technique and can be achieved at room temperature. Based upon the solubility product constant of natural phosphate at 25°C, it can be estimated that the matrices can be easily deposited on stainless steel substrate by electrochemical methods.

Experimental

Reagent

A Natural Phosphate (NP) used in this work was obtained in the Khouribga region (Morocco) [8]. Prior to use, this material was treated by techniques involving attrition, sifting, calcinations (900°C), washing, and recalcination.

Sample preparation

Stainless steel (type 316 L) plate was cut into rectangular strips with typical dimensions of 1 cm². The chemical composition of stainless steel investigated in this work is: (wt%) C ≤ 0.02, Mo: 3.5-4.5, Cr: 24-26, Ni: 6-8, N ≥ 0.25, Fe remainder. Strips were abraded with SiC paper in successive grades from 400, 600 up to 1200 grit and then cleaned in distilled water and dried.

The current was maintained by a galvanostat with a function generator. The anode electrode was a platinum wire, and a stainless steel electrode was used as cathode. Then, the electrodes were immersed in a glass chamber containing electrolyte of natural phosphate and/or ZnSO$_4$, and subjected to anodic oxidation by applying 100 mA for 6 hours NP and 30 min for Zn electrodeposition.

Modified electrodes characterization

Optical microscope in reflection was used to observe the morphology of samples. Electrochemical experiments were carried out with a voltalab potentiostat (model PGSTAT 100, Eco Chemie

*Corresponding author: Chtaini A, Equipe d.Electrochimie Moléculaire et Matériaux Inorganiques, Faculté des Sciences et Techniques de Béni Mellal, Morocco
E-mail: a.chtaini@usms.ma

B. V., Utrecht, The Netherlands) driven by the general purpose electrochemical systems data processing software (voltalab master 4 software). The electrochemical cell was configured to work with three- electrodes, using NP and/or Zn-NP modified stainless steel as the working, platinum plate for counter, and saturated calomel (SCE) as reference electrodes. Impedance measurements were made in the frequency range between 10 mHz and 100 KHz with five points per decade at the corrosion potential. A sine wave with 10 mV amplitude was used to perturb the system. The Nyquist diagrams obtained were automatically controlled by computer programs. The electrolytical solution used (0.1M H_2SO_4) as prepared from distilled and deionised water. All tests were performed at 25°C.

Results and Discussion

Zn/NP/Iron electrode

The surface structure of prepared electrodes (NP/Iron and Zn/NP/Iron) was observed using optical reflexion microscopy (Figure 1). The phosphate film deposited on the surface of the iron appears continuously and forms a non homogeneous matrix, consisting of crystals of the order of 100 μm and 400 μm. In the case of Zn/NP/Iron, the film exhibited a porous microstructure with micro pores, which were relatively well separated and homogeneously distributed over

the surface. The film formed is continued and not disintegrated from substrate surface.

Methanol oxidation at modified electrodes

The potentiostatic polarization curves and table of measurement results for the test specimen, registered in 0.1M H_2SO_4 solution containing a small amount of methanol (0.5 μmol), are shown in Figure 2 and Table 1 respectively. We can see in Table 1, the polarization resistance value drops slightly in the presence of methanol, which can be explained by the activity of the new electrode with respect to

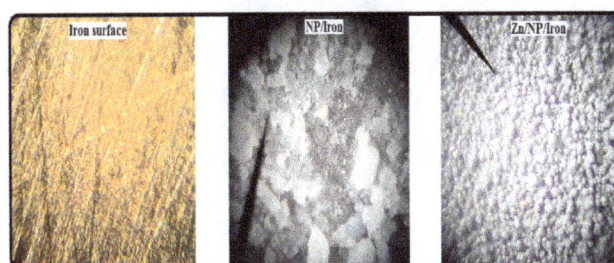

Figure 1: Typical AFM images of electrodeposited films into stainless steel plates.

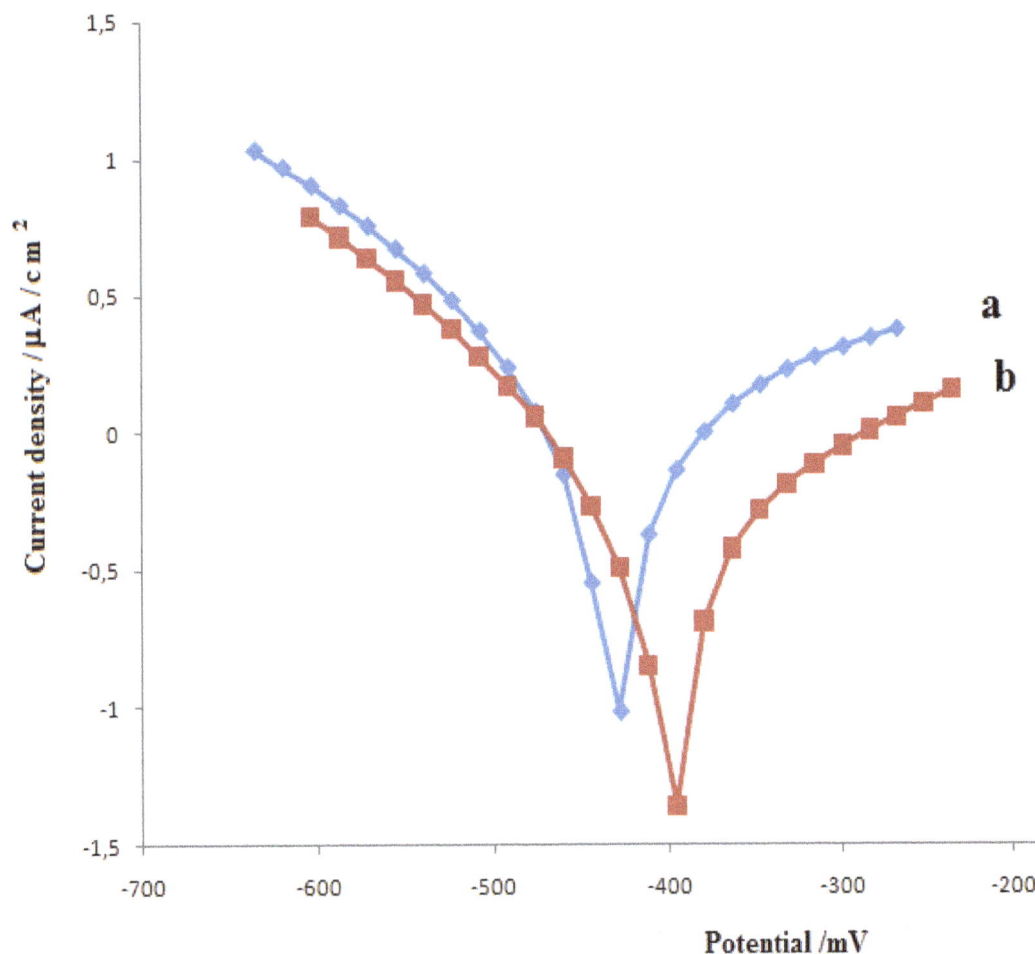

Figure 2: Polarization curves of Zn/NP/Iron electrode in 0.1M H_2SO_4 solution, (a) without methanol, (b) with methanol.

Samples	E(i=0) Mv	Rp ohm.cm²	βa mV	βc mV
Zn/NP/Iron without methanol	-356,6	131,69	389,1	-215,5
Zn/NP/Iron in presence of methanol	-391,3	92,83	436,5	-221,1

Table 1: Summary of electrode polarization results for the methanol oxidation.

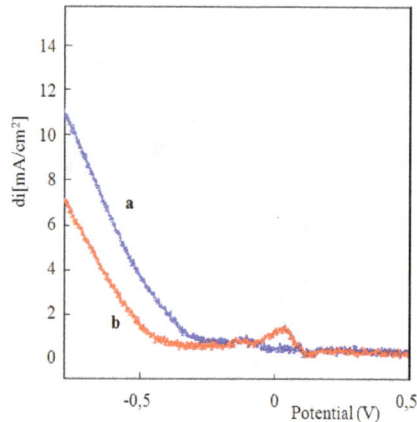

Figure 3: Square wave voltammetry in N_2 saturated 0.1M H_2SO_4, recorded for Zn/NP/Iron electrode in presence of methanol (0.5 μmol, curve b), and in absence of methanol (curve a).

the methanol oxidation. Polarization resistance (Rp) is the transition resistance between the electrodes and the electrolyte. In other words, an increased resistance to the flow of current in a voltaic cell is caused by chemical reactions at the electrodes. Polarization results in a reduction of the electric potential across the voltaic cell. An electrode is polarized when its potential is forced away from its value at open circuit or corrosion potential. Polarization of an electrode causes current to flow due to electrochemical reactions that it induces at the electrode surface.

Figure 3 shows the square wave voltammograms (SWV) recorded for Zn/NP/Iron electrode in 0.1M H_2SO_4 solution containing 0.5 μmol of methanol. The Figure shows the existence of a well-defined peak at potential value about 0.1 V. It was suggested that this peak corresponds to methanol oxidation.

The Nyquist plots for the prepared electrode, recorded in H_2SO_4 solution containing or not methanol, are presented in Figure 4. In both cases, the locus of Nyquist plots is regarded as one part of semi-circle, who is struggling to close, due to the very high resistance.

Cu/NP/Iron electrode

The surface structure of Cu/NP/Iron electrode was observed using reflexion optical microscopy (Figure 5). As we can see, the copper is deposited on the active sites, which are scattered along the phosphate surface. Clusters of copper, deposited electrochemically on the phosphate matrix have a porous and roughness structure.

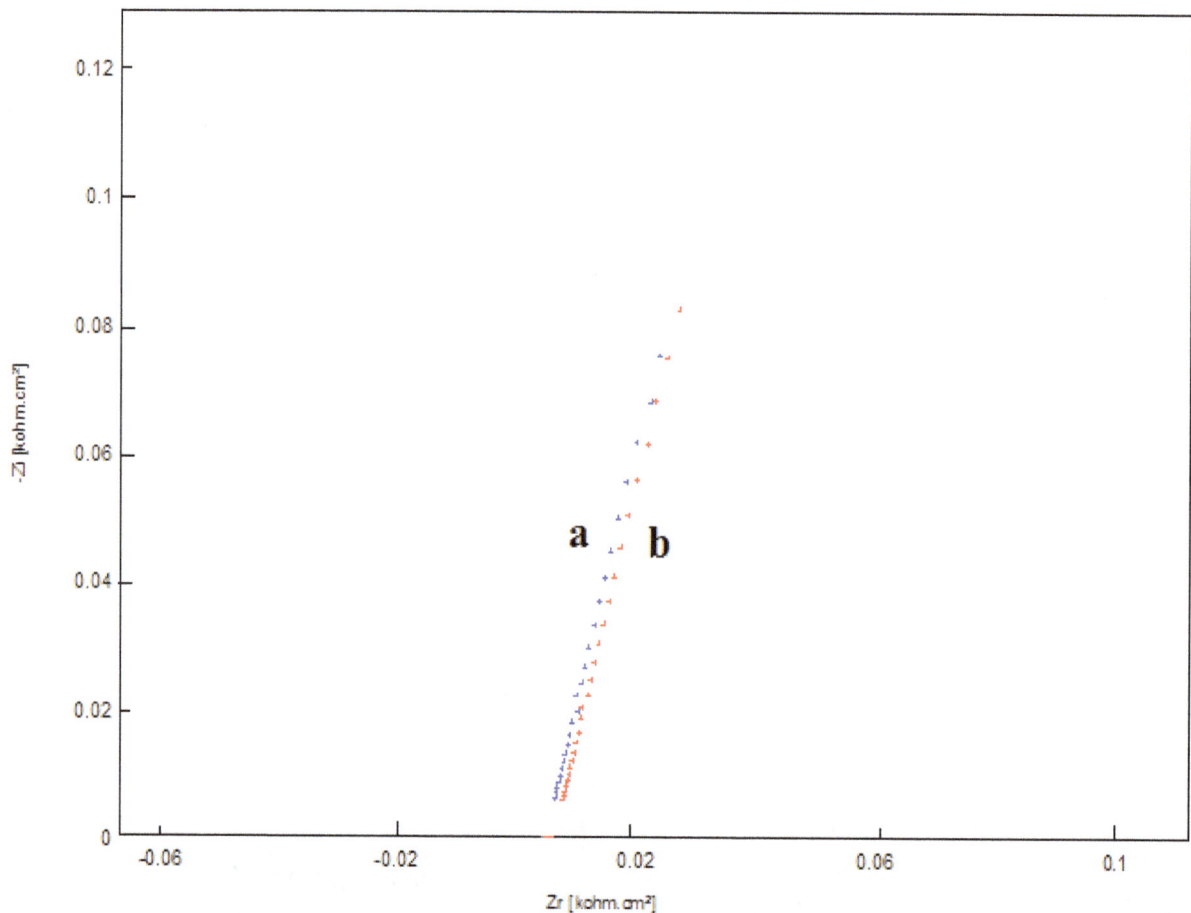

Figure 4: Electrochemical impedance spectroscopy for Zn/NP/Iron electrode, in 0.1M H_2SO_4, containing methanol 0.5 μmol (curve b), and without methanol (curve a).

In contrast to zinc, the copper deposited on the surface of the steel electrode, exhibits a slight increase of the polarization resistance at the electro oxidation of methanol (Table 2). To explain this phenomenon, we appealed to the SWV, shown in Figure 5. We can find that the presence of methanol in the electrolytic solution causes the growth of the copper oxidation, which appears to 0.1 V. This leaves suggest that the oxidation of phenol is stimulated by the copper oxides. The locus of Nyquist plots (Figures 6 and 7) has the shape of semi-circle, in the absence and presence of methanol, the diameter of the semi-circle correspond to the polarization resistance, which means that the methanol oxidation reaction is catalyzed by the copper oxide.

Conclusion

The current study introduces novel anodes for the electro-catalytic oxidation of methanol in H_2SO_4 medium. The Zn/NP/Iron and Cu/NP/Iron anodes prepared by the potentiostatic deposition respectively, of

Echantillon S	E(i=0) mV	Rp ohm. cm²	I Corr µA/ cm²	βa mV	βc mV
Acier316L/ NP/Cuivre sans Méthanol	-439,5	109,30	511,7	340,1	-195,5
Acier 316L/ NP/Cuivre Avec Méthanol	-424,1	113,64	337,3	244,9	-171,1

Table 2: Summary of electrode polarization results for the methanol oxidation.

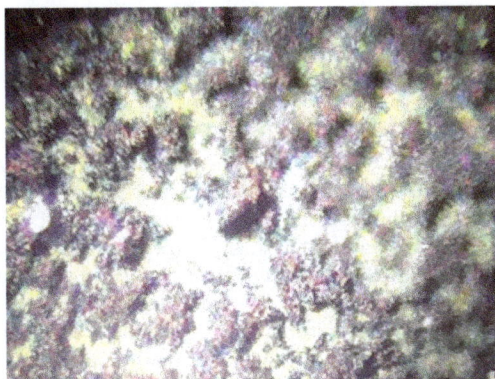

Figure 5: Typical AFM images of electrodeposited films into stainless steel plates.

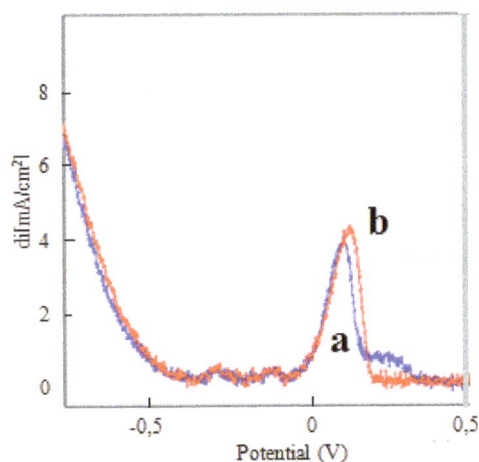

Figure 6: Square wave voltammetry in N_2 saturated 0.1M H_2SO_4, recorded for Zn/NP/Iron electrode in presence of methanol (0.5 µmol, curve b), and in absence of methanol (curve a).

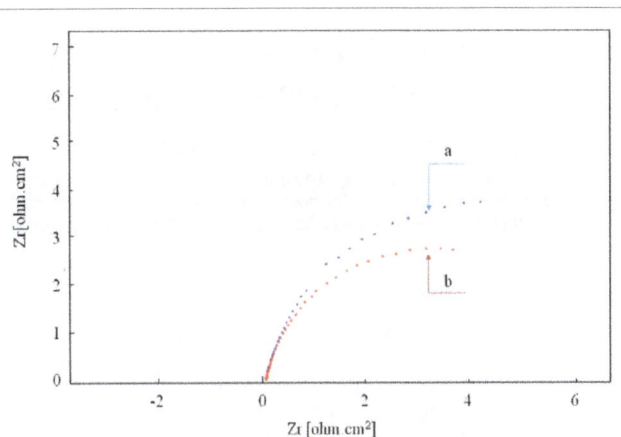

Figure 7: Electrochemical impedance spectroscopy for Cu/NP/Iron electrode, in 0.1M H_2SO_4, containing methanol 0.5 µmol (curve b), and without methanol (curve a).

natural phosphate and zinc or copper on iron surface are good catalysts for methanol electro-oxidation. In the case of copper this reaction is promoted by the formation of copper oxides on the surface.

References

1. Neel PJ (2008) MVC Sastry Hall, NCCR.

2. Carabineiro SAC, Thompson DT (2007) Catalytic Applications of Gold Nanotechnology in Nanoscience and Technology. In: Heiz U, Landman U (eds.) Nanocatalysis. Springer-Verlag, Berlin Heidelberg.

3. Appleby AJ, Foulkes FR (1989) Fuel Cell Handbook. Van Nostrand Reinhold, New York.

4. Chtaini A (1993) Thesis, Poitiers (French).

5. Baglio V, Di Blasi A, Modica E, Creti P, Antonucci V, et al. (2006) Electrochemical Analysis of Direct Methanol Fuel Cells for Low Temperature Operation. Int J Electrochem Sci 1: 71-79.

6. Shukla AK, Christensen PA, Hamnett A, Hogarth MP (1995) A vapour-feed direct-methanol fuel cell with proton-exchange membrane electrolyte. J Power Sources 55: 87-91.

7. Antolini E, Salgado JRC, Santos LGRA, Garcia G, Ticianelli EA , et al. (2006) Carbon supported Pt-Cr alloys as oxygen-reduction catalysts for direct methanol fuel cells. J Appl Electrochem 36: 355-362.

8. Natural phosphate (NP) Comes Khouribga Region (Morocco). It is readily available from CERPHOS, Casablanca, Morocco.

6

Attenuation Performance of Polymer Composites Incorporating NZF Filler for Electromagnetic Interference Shielding at Microwave Frequencies

Ahmad AF[1]*, Abbas Z[2], Obaiys SJ[3] and Abdalhadi DM[2]

[1]Materials Processing and Technology Laboratory, Institute for Advance Material, Universiti Putra Malaysia, Serdang, Selangor Darul Ehsan, Malaysia
[2]Department of Physics, Faculty of Science, Universiti Putra Malaysia, Serdang, Selangor Darul Ehsan, Malaysia
[3]School of Mathematical and Computer Sciences, Heriot-Watt University Malaysia, Putrajaya, Malaysia

Abstract

Polymer composites have been thoroughly explored for future electromagnetic interference (EMI) applications owing to their unique combination of electrical, mechanical, and optical properties. The composition, morphology, and surface characteristics of the filler material play critical roles in regulating the composite activity. We studied the formation, synthesis and EM attenuation properties of nickel zinc ferrite (NZF) + Polycaprolactone (PCL) micro-composites that were prepared via the conventional mixed oxide (CMO) technique. Compared with other preparation routes, CMO may provide the advantages of a simple process and the ability for mass production and controlled product formation. A rectangular waveguide connected to a vector network analyser coaxial cable was employed to measure the scattering parameters [S] for use in determining the attenuation values of NZF+PCL substrates for a variety of NZF% values. Measurement tests showed a simultaneous increment in the attenuation value with the filler percentage. NZF+PCL samples of 1-mm thickness were able to attenuate microwave frequencies by up to ~3.33 dB, where the highest attenuation magnitude of 8.599 dB over a large area was attributed to the 12.5% NZF filler content at 12 GHz. Thus, a low transmission of waves resulted from the high shielding effectiveness (SE) values that showed a maximum 6.86 dB EM interference. Scanning electron microscopy (SEM) was utilized to analyse the average particle size (1.45 μm) of the filler powder.

Keywords: Microwave; Rectangular waveguide; Composites; FTIR; SEM and T/R coefficients

Introduction

Different engineering applications require different minimum values of shielding effectiveness (SE), where the selection of the shielding materials is an important factor in their design. Knowledge of the behaviour of a material placed in an electromagnetic interference (EMI) field is of immense importance, particularly for military hardware, electronics, communication, and industrial applications and shielding [1,2]. It is vital to understand that the transmission coefficient measurements enable an attenuation analysis for materials of good microwave absorption. Attenuation is a function that can be affected by a group of factors. The SE of a material depends on both the conductivity and permittivity values, but for high-frequency shielding, the conductivity dominates. Materials with high conductivity values provide excellent SE results [3]. Commonly, a composite consists of filler and matrix, where the filler is usually surrounded by matrix materials to keep the resulting composite in position. Matrix materials must be flexible, lightweight, corrosion resistant, and of lower cost than metals. Polycaprolactone (PCL) is an excellent microwave-absorbing material and well known as a material for EMI SE in both near and far fields [4]. The enumerated characteristics such as the simplicity of fabrication, light weight, low cost, and excellent insulation properties play a major role in the design of advanced materials that are suitable for a variety of applications, such as electrodes, sensors and electrical devices of high frequency [5-7]. The properties of the polymer composites are affected by factors such as the inherent characteristics of each component, contact, dimension and shape of the fillers, and the nature of their interfaces [8]. Ferrites are very good dielectric materials whose main constituents are oxygen, iron and one or more metallic elements such as Ni and Zn that have many applications at microwave frequencies [9]. The properties of nickel zinc ferrite (NZF) composites can be tailored by controlling the preparation conditions and the amount of metal ion substitution. Their useful characteristic properties such as electrical resistivity, low dielectric loss, and chemical stability

enable them to contribute in both the domestic and industrial sectors [10]. A composite of a ferrite material with a good polymer matrix will not only reduce the cost and enhance the structural properties of the outcomes but also increase the ability to control the EMI properties, particularly when operating at the high-frequency range [9,11]. Ferrite-polymer composites are widely studied due to their novel properties of simple preparation, light weight, low cost, better electronic properties, good optical properties, environmentally friendliness and high stability in air that make them highly suitable for many applications, such as EMI SE and microwave absorption [12].

Herein, the synthesis of (2.5%, 5%, 7.5%, 10%, and 12.5%) NZF+ PCL substrate of 1mm thickness that show great EM propagation and affect the absorption properties at 8-12 GHz is achieved. The microwave properties of the samples were investigated using a rectangular waveguide connected with a vector network analyser coaxial cable. The total volume fraction of the composites was set to 15 g and divided into 1 g, 2 g, 3 g, 4 g and 5 g, with respect to the NZF%.

Experimental

Preparation of nickel zinc-ferrite composites

In this work, the formation of Ni-Zn-ferrite material was carried

---

*Corresponding author: Ahmad AF, Materials Processing and Technology Laboratory, Institute for Advance Material, Universiti Putra Malaysia, Serdang, Selangor Darul Ehsan, Malaysia, E-mail:ahmad_al67@yahoo.com

out using the conventional mixed oxide (CMO) method, which is a highly favoured route in commercial ferrite production due to its relative simplicity, scalability, high mass production and economic efficiency. The oxide raw materials of NiO (99.7% purity), ZnO (99.9% purity) and Fe_2O_3 (99.7% purity) were mingled well together and then weighed according to the zinc ferrite ($Ni_x Zn_{1-x} Fe_2O_4$) stoichiometric equation [13].

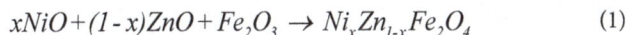

$$xNiO + (1-x)ZnO + Fe_2O_3 \rightarrow Ni_xZn_{1-x}Fe_2O_4 \qquad (1)$$

The mixture was then carefully ground in a ball mill and exposed to heat to ensure the particle homogeneity of the powder. After that, the milled powder was moulded and pressed into the required shape for the final sintering process of 900-1000°C for 10 hours to obtain the crystalline structure of the materials under test.

Preparation of NZF+PCL composites

The NZF+PCL composites were prepared by the melt blending technique using a Thermo Haake Poly Drive extruder with a three-phase motor with a drive of 1.5 kW, 3×230 V, 40 A and speed range of 0-120 rpm. The resulting crude blends have to undergo hot-pressing to prepare a thin film for each blend. Compounding samples of 15 g each were prepared with different percentages of NZF and PCL with a rotation speed of 50 rotations per minute (rpm). The NZF+PCL composites were preheated for 10 minutes with upper and lower platen temperatures of 80°C, which is close to the PCL melting point (60-70°C). A venting time of 10 seconds was allowed to release the bubbles and reduce the voids, and the samples were then pressed at the same temperature for another 10 minutes. Finally, a cooling pressing of 110 kg/cm² was carried out for another 10 minutes at 25°C for the best substrate fabrication result. Figure 1 shows the prepared NZF+PCL substrates, and Table 1 presents the different compositions that were utilized in the experimental step.

Preparation of substrates

The samples were carefully fabricated to fit the rectangular waveguide dimensions best and to prevent any scratch or crack that may alter the measurement results. The substrates were prepared by placing 10 g of the blends into a mould of 8-10 cm² dimensions and 1 mm thickness. The samples were then restricted to suit the internal waveguide dimensions perfectly and remove any possibility of an air gap around the sample walls. A vector network analyser (VNA) (Agilent 8750B) based waveguide measurement technique was utilized to measure the S-parameters of the two-port network formed by placing the substrate inside the rectangular waveguide. The material properties were studied in the X-band (8-12GHz) regions of the microwave frequency spectra. Figure 2 shows the experimental technique of this work.

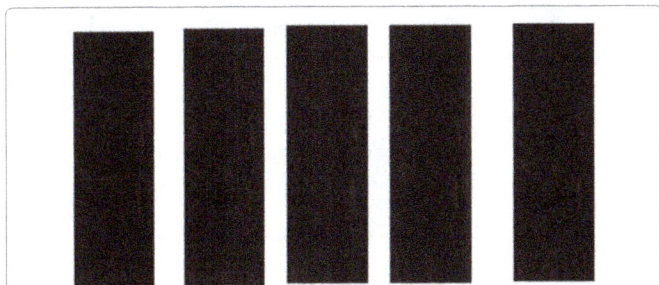

Figure 1: Prepared substrate of NZF+PCL micro-composites.

NZF%	PCL %
2.5	97.5
5	95
7.5	92.5
10	90
12.5	87.5

Table 1: Percentages of NZF and PCL in the prepared composites.

Figure 2: A rectangular waveguide connected to a VNA for the measurement set up.

Figure 3: SEM micrograph of $N_{0.5}Z_{0.5} Fe_2O_4$ micro-particles prepared via CMO technique.

Results and Discussion

Morphological properties of NZF micro-particles

The prepared NZF fillers were analysed using scanning electron microscopy (SEM) (S-3400N) with a field emission gun and an accelerating voltage of 10 kV. A careful observation of the SEM image presented in Figure 3, focusing on the variety of signals, reveals a good arrangement of particles. The spherical structure of the synthesized NZF particles shows an average size of approximately 69 to 286 mm, which confirms the successful preparation.

Measurement of the scattering parameters [S]

It is known that the EM properties can be calculated from the scattering parameters [S]. The boundaries of the materials under test (MUT) are defined, and the reflection coefficients (S_{11}) and transmission coefficients (S_{21}) can then be measured accurately [14]. In the transmission/reflection (T/R) method of a waveguide, the samples were inserted in a piece of transmission line, and the properties of the material were deduced based on the rejection of the material and the transmission through the material [15]. This method is commonly performed in the measurement of the EMI properties of materials due to the field-focusing ability of the microwave guides that enables accurate measurements at microwave frequencies, where the EM waves propagate through the microwave guides at 8-12 GHz [16].

The measurement was performed at the connectors of the waveguides. However, the specimen was considered as a two-port network whose S_{11} and S_{21} values used a "thru reflect line" (TRL) calibration that needs to be performed to set the planes of the incident and reflected waves at the ends of the waveguides rather than that at the connectors [17]. After the calibration step, the accuracy of the calibration technique was evaluated by measuring the [S] of air (without sample) and Teflon (PTFE) samples. Figure 4 presents the [S] curves of the standard materials, where the values of $|S_{21}|$ were higher than those of $|S_{11}|$. Theoretically, the value of $|S_{21}|$ should be close to unity.

The responses of a network to external circuits can also be described by the input and output waves. The input [a] and output [b] waves at port 1 and port 2 are denoted as a_1, a_2, b_1 and b_2, respectively. These parameters (a_1, a_2, b_1, and b_2) may be voltage or current, and in most cases, we do not distinguish between whether they are voltage or current [11]. The relationships between [a] and [b] are often described by [S] values, Where,

$$[b] = \begin{bmatrix} S_{11} & S_{12} \\ S_{21} & S_{22} \end{bmatrix} \begin{bmatrix} a_{11} & a_{12} \\ a_{21} & a_{22} \end{bmatrix}; \quad or \quad [b] = [S][a] \quad (2)$$

It is known that the S_{11} and S_{21} parameters can be measured directly at microwave frequencies. Generally, the complex form of [S] can be defined as

$$\left. \begin{array}{l} S_{qq} = \dfrac{b_q}{a_q} \ (q = 1, 2) \\[2ex] S_{pq} = \dfrac{b_p}{a_p} \ (p \neq q; \ p = 1, 2; \ q = 1, 2) \end{array} \right\} \quad (3)$$

Then, the S_{11} values are defined as

$$\Gamma_q = S_{qq} = \dfrac{b_q}{a_q} \quad (4)$$

The S_{21} values are defined by

$$T_{q \to p} = S_{qq} = \dfrac{b_p}{a_p} \quad (5)$$

[S] can then be obtained by the combinations of these four waves according to Eq. (3), where the parameters a_1, a_2, b_1, and b_2 are normally used to measure the corresponding four waves. The signal separation

devices ensure that the four waves are measured independently [11]. The obtained VNA results of S_{21} were then used to compute the attenuation results of the MUT by the following equation:

$$Attenuation\,(dB) \ = -20 \log(S_{21}) \quad (6)$$

It is known that several actions such as transmission, reflection and absorption are performed as EM radiation falls on a shielding material [18]. The total EMI SE (SE_T) is defined as the summation of all of the contributors to the SE (absorption loss (SE_A), reflection loss (SE_R), and multiple reflection loss (SE_M)):

$$SE_T = SE_A(dB) + SE_R(dB) + SE_M(dB) \quad (7)$$

For a single layer of shielding material, when SE_A is ≥10 dB, then $SE_M = 0$, so it can be ignored [18]. Thus, the SE_T in Eq. (7) can be written as

$$SE_T = SE_A(dB) + SE_R(dB) \quad (8)$$

The incident EM wave power inside the shielding material can be estimated as

$$SE_R = -10 \log(1 - R) \quad (9)$$

$$SE_R = -10 \log\left(\dfrac{T}{1 - R}\right) \quad (10)$$

where the transmittance (T) value is equal to $(S_{21})^2$, and the reflectance (R) is $(S_{11})^2$.

While the basic mean-value ($\overline{x}$) analysis of attenuation, SE_R, SE_A and SE_T values for different percentages of NZF+PCL composites based excel program is calculated by

$$\overline{x} = \dfrac{\sum\limits_{i=1}^{n} x_i}{n}; \qquad n = 200 \quad (11)$$

Figures 5 and 6 show proportional graphs of the measured $|S_{11}|$

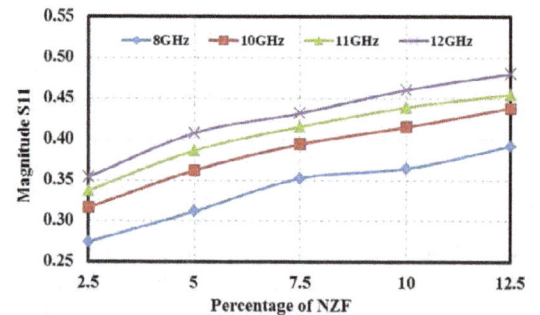

Figure 5: Percentage of NZF filler content vs transmission coefficient values in the microwave frequency range.

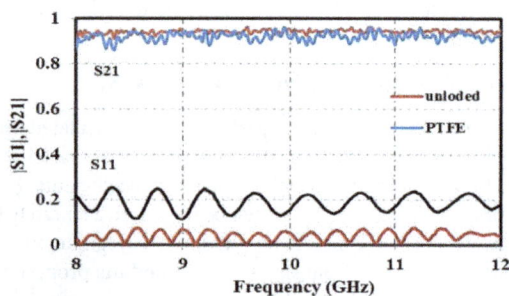

Figure 4: Measured $|S_{11}|$ and $|S_{21}|$ of air and PTFE at microwave frequencies (8-12 GHz).

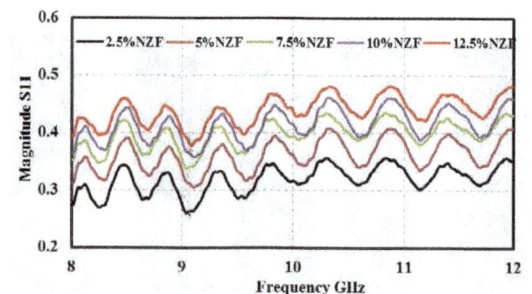

Figure 6: Measured $|S_{11}|$ in the microwave frequency range for all samples.

values with the NZF micro-filler percentage and the frequency, respectively. The highest $|S_{11}|$ value of ~0.48 was recorded for the highest NZF% of 12.5 and the maximum frequency of 12 GHz for all samples under test.

Oppositely, Figures 7 and 8 shows an inversely proportional relationship between the filler composition and frequency range and the $|S_{21}|$ values, where increases in the filler content and frequency range both reduce the $|S_{21}|$ values. These results demonstrate the impedance mismatch theory, where materials with a higher permittivity exhibit lower transmission coefficient values [19]. This might be attributed to the perfect dispersion of the filler in the matrix, as the homogeneity of the particle dispersion in the matrix will tend to increase or decrease the transmission of the EM radiation depending on the dispersion of the NZF particles in the matrix.

To calculate the attenuation of the various NZF+PCL micro-composites, all the obtained $|S_{21}|$ results were input into Eq. 6. The calculated results, as shown in Figure 9, confirm the proportional relationship between the attenuation outcomes of the NZF+PCL micro-composites and the filler composition. From Figure 9, it can be clearly observed that the lowest attenuation value was calculated for the

2.5% NZF micro-particle filler, whereas the highest attenuation result was observed for the 12.5% NZF micro-particle filler.

Further analysis is provided in a graph of the filler content against the attenuation values that is presented in Figure 10. The highest attenuation value (~8.6 dB) was obtained from the highest filler content (12.5%). Thus, a reduction in the $|S_{21}|$ values was clearly observed for the higher valves of NZF-filler content, as depicted in Figure 7. Based on the above, Table 2 provides a tabulated summary of the mean attenuation values, highlighting the proportional relationship between the attenuation magnitude and NZF filler composition.

The exact statistical algorithm based $\overline{x}$ analysis has provided efficient results that allow us to properly analyze the different percentages of NZF+PCL composites. Figure 11 shows a graph of the mean attenuation values of the NZF+PCL micro-composites against the filler content. The attenuation values constantly increased with the NZF filler content.

EMI SE application

Figure 12 shows the variation in the SE_T values due to the alteration of the NZF loading, SE_R and SE_A of the NZF+PCL at 8-12 GHz. The

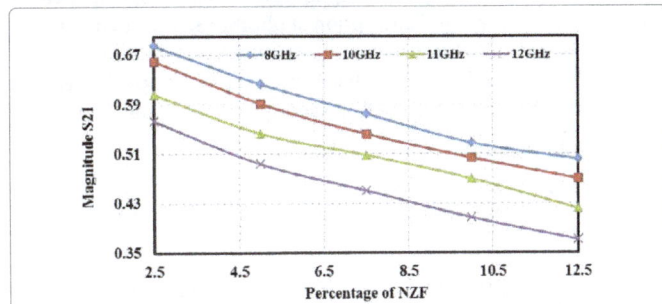

Figure 7: Percentage of NZF filler content vs measured $|S_{21}|$ values at 8-12 GHz.

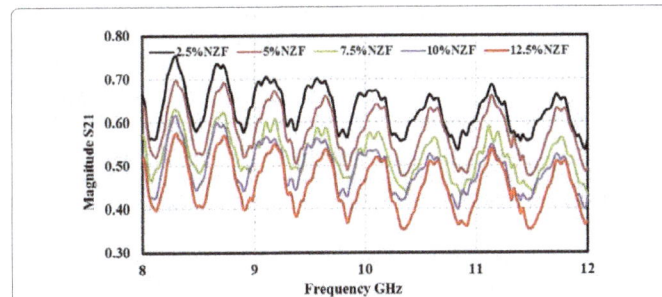

Figure 8: Measured $|S_{11}|$ values for all samples at the X-band frequency.

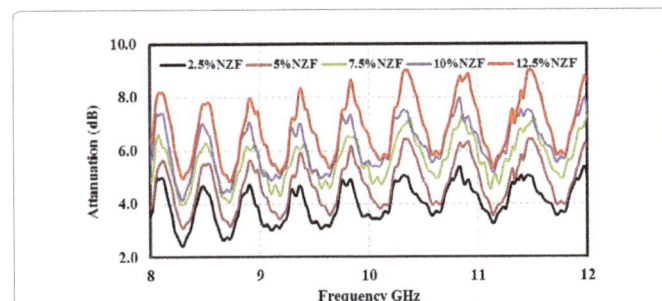

Figure 9: Calculated attenuation results of different NZF+PCL compositions at 8-12 GHz.

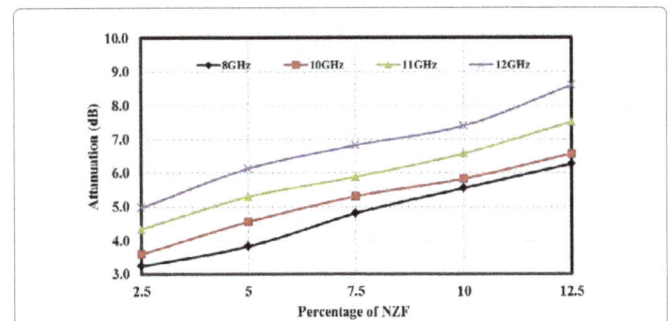

Figure 10: Percentage NZF filler contents vs attenuation in the selected frequency range.

NZF%	8 GHz	10 GHz	11 GHz	12 GHz
2.5	3.245	3.607	4.346	4.978
5	3.835	4.562	5.301	6.133
7.5	4.811	5.316	5.889	6.822
10	5.549	5.827	6.57	7.396
12.5	6.263	6.569	7.511	8.599

Table 2: Mean values of attenuation for different percentages of NZF+PCL micro-composites at microwave frequencies.

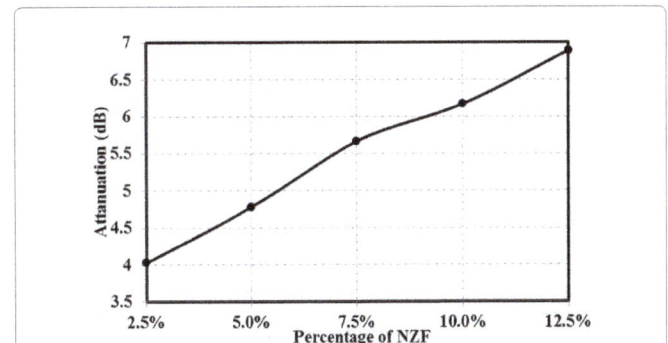

Figure 11: Mean attenuation of NZF+PCL micro-composites vs NZF-filler content.

difference between SE_A and SE_R increases with the NZF content, suggesting that the absorption contribution to the EM SE increases with the NZF loading increment. The primary mechanism of the EMI shielding is usually a reflection of the EM radiation incident on the shield, which is a consequence of the interaction of the EMI radiation with the free electrons on the surface of the shield. Absorption is usually a secondary mechanism of EMI SE, whereby electric dipoles in the shield interact with the EM waves in the radiation [11,18]. Another reason for such a variation between the SE_R and SE_A values may be the interfacial polarization of the PCL by the NZF, which increases the absorption component. The values of SE_A increase from ~ 2.78 dB at 2.5% to ~6.86 dB at 12.5% loading. Based on the above, the EMI SE results are mostly independent of the frequency in the X-band.

From Figure 13, the EMI SE value increased dramatically with a slight NZF vol. % increment. In fact, the highest EMI SE of approximately 6.86 dB was achieved from 12.5% NZF in the composite at a particular frequency in the X-band region. Moreover, it was found that SE_A increases much faster than SE_R as the NZF content increments.

NZF+PCL composites can be used for many shielding applications by adjusting the filler content [9]. For example, an addition of only 2.5% NZF filler in the NZF+PCL composite already satisfies the minimum 3.339 dB SE requirement for the aircraft structural shielding of an antenna based on Wireless Avionics Intra-Communications specifications [20]. However, the scope of this paper is to determine the minimum percentage of filler for a sample of 1 mm thickness to obtain a maximum 6.86 dB SE. The mean values of SE_R, SE_A and SE_T for the composites which calculated by $\overline{x}$ in Eq. 11 for different NZF% are listed in Table 3 below, where the widely used $\overline{x}$ statistical analysis are reasonably accurate.

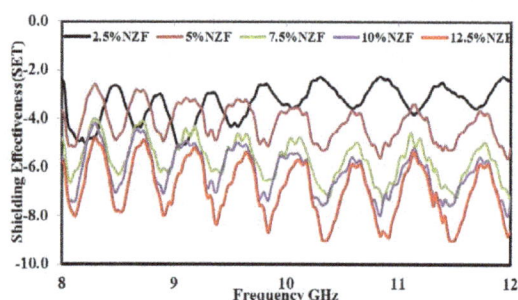

Figure 12: SE_T as a function of frequency measured in the 8-12 GHz range of different NZF+PCL composites.

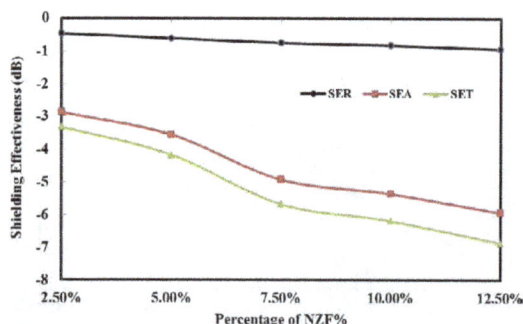

Figure 13: Comparison of SE_T, SE_R, and SE_A in the 8–12 GHz range for NZF+PCL composites.

NZF%	SE_R(dB)	SE_A(dB)	SE_T(dB)
2.5	-0.469	-2.87	-3.339
5	-0.613	-3.556	-4.169
7.5	-0.749	-4.917	-5.665
10	-0.824	-5.344	-6.168
12.5	-0.937	-5.923	-6.86

Table 3: Mean SE_R, SE_A and SE_T of NZF + PCL composites.

Conclusions

The shielding effectiveness is defined as the process of using specialized materials to reduce the EMI fields or waves that enter a specific enclosure. The shielding performance highly depends on type, size and thickness of the utilized materials along with the frequency range. In this work, NZF+PCL micro-composite structures have been successfully synthesized for potential SE and absorption applications. The EM propagation and attenuation properties of these composites in waveguides were theoretically and experimentally investigated in the 8-12 GHz range of frequency. For microwave characterization, the effects of various NZF compositions on the attenuation of the NZF+PCL composites were calculated using the measured $|S_{21}|$ result based rectangular waveguide technique and were implemented in shielding effectiveness and absorption applications. The attenuation of the different NZF+PCL micro-composites showed that the attenuation magnitude increased with the filler content, which confirmed the correlation between the attenuation value and the NZF-filler content.

Acknowledgements

The authors sincerely extend their gratitude to Universiti Putra Malaysia for providing financial support and facilities for the completion of this work.

References

1. Canan D, Papila M (2010) Dielectric behavior characterization of a fibrous-ZnO/PVDF nanocomposite. Polym Composite 31: 1003-1010.

2. Xiaogu H, Zhang J, Lai M, Sang T (2015) Preparation and microwave absorption mechanisms of the Ni-Zn ferrite nanofibers. J Alloy Compd 627: 367-373.

3. Ahmad AF, Abbas Z, Obaiys SJ, Ibrahim NA, Zainuddin MF (2015) Microwave characterization of bio-composites materials based finite element and nicholson–ross–weir methods. Malaysian Journal of Science 34: 180-184.

4. Geetha S, Satheesh Kumar KK, Rao CRK, Vijayan M, Trivedi DC (2009) EMI shielding: methods and materials—A review. J Appl Polym Sci 112: 2073-2086.

5. Ahmad FA, Abbas Z, Obaiys SJ, Ibrahim N, Hashim M, et al. (2015) Theoretical and numerical approaches for determining the reflection and transmission coefficients of OPEFB-PCL composites at X-band frequencies. PloS one 10: e0140505.

6. Elammaran J, Hamdan S, Rahman MR, Bin Bakri MK (2014) Comparative study of dielectric properties of hybrid natural fiber composites. Procedia Engineering 97: 536-544.

7. Harun B (2011) Complex permittivity, complex permeability and microwave absorption properties of ferrite–paraffin polymer composites. J Magn Magn Mater 323: 1882-1885.

8. Roman K (2012) Thermal and electrical properties of nanocomposites, including material properties. TU Delft, Delft University of Technology.

9. Ahmad AF, Zulkifly A, Obaiys SJ, Ibrahim NA, Hashim M, et al. (2016) Permittivity properties of nickel zinc ferrite-oil palm empty fruit bunch-polycaprolactone composite. Procedia Chemistry 19: 603-610.

10. Khoon TF, Hassan J, Mokhtar N, Hashim M, Ibrahim NA (2011) Dielectric behavior of $Ni_{0.1}Zn_{0.9}Fe_2O_4$-polypropylene composites at low microwave frequencies. Solid State Science and Technology 19: 207-213.

11. Fahad A, Abbas Z, Obaiys SJ, Ibrahim N, Yakubu A (2016) Dielectric behavior of OPEFB reinforced polycaprolactone composites at X-band frequency. Int Polym Proc 31: 18-25.

12. Liangchao L, Liu H, Wang Y, Jiang J, Xu F (2008) Preparation and magnetic

Attenuation Performance of Polymer Composites Incorporating NZF Filler for Electromagnetic Interference...

33

properties of Zn Cu Cr La ferrite and its nanocomposites with polyaniline. J Colloid Interf Sci 321: 265-271.

13. Cedomir J, Nikolic AS, Gruden-Pavlovic M, Pavlovic MB (2012) Mechano–chemical synthesis of stoichiometric nickel and nickel–zinc ferrite powders with Nicolson–Ross analysis of the absorption coefficients. J Serb Chem Soc 77: 497-505.

14. Paula D, Luiz A, Rezende MC, Barroso JJ (2011) Experimental measurements and numerical simulation of permittivity and permeability of teflon in X band. Journal of Aerospace Technology and Management 3: 59-64.

15. Harun B (2012) Electromagnetic propagation and absorbing property of ferrite-polymer nanocomposite structure. Progress in Electromagnetics Research M 25: 269-281.

16. Bayrakdar H (2011) Magnetically modified biocells in constant magnetic field. J Magn Mater 323: 1882-1885.

17. Doug R (2008) ARFTG 50 year network analyzer history.

18. Xingcun Colin T (2016) Advanced materials and design for electromagnetic interference shielding. CRC Press.

19. Pozar DM (2009) Microwave engineering. John Wiley and Sons, USA.

20. WAIC (2012) Agenda Item 1.17 update and status on implementing of a regulatory framework for WAIC. Presentation for ICAO Regional Meeting, Lima, Peru.

7

Evaluation of Mechanical Properties of Medium Carbon Low Alloy Forged Steels Quenched in Water, Oil and Polymer

Chandan BR* and Ramesha CM

Department of Mechanical Engineering, MS Ramaiah Institute of Technology, Bangalore, Karnataka, India

Abstract

Medium carbon low alloy forged steels (EN 18, EN19, EN 24, and EN25) have been investigated with respect to their mechanical properties after heat treatment. For heat treatment solutionizing temperature of 855°C with a soaking period of 60 min was used. Thereafter quenching was carried out in three media, viz., Step water, oil and polymer (polyethylene glycol) separately. The quenched samples were step tempered at 575°C and at 220°C sequentially for 60 min each. Hardness, tensile strength, Charpy impact strength and metallographic were carried out on the untreated and heat treated specimens. The heat treated specimens showed higher hardness (10-30%), higher strength (20-100%) and higher impact energy (20-160%). The specimens quenched in poly ethylene glycol exhibited the best mechanical properties. The heat treated specimens had a structure of fine tempered martensite with small amount of bainite.

Keywords: EN steels; Heat treatment; Polymer quenching; Forged steels; Impact energy

Introduction

Medium carbon steels are similar to low-carbon steels except that the carbon ranges from 0.30 to 0.50% and manganese from 0.60 to 1.65%. The uses of medium carbon steels include shafts, axles, gears, crankshafts, couplings and forgings. Steels of 0.40 to 0.60% C ranges are used for rails, railway wheels and rail axles. Therefore medium carbon low Alloy steels (MCLA) are commonly used for engineering applications in automobiles, aircraft and transportation industries. They are readily hot forged usually at temperatures ranging from 1065°C to 1230°C [1]. In industry, quenching in oil or polymers is resorted to especially for heavy sections to avoid quench crack commonly observed during water quenching. Choice of quenching medium for the MCLA forged steels depends on the composition of the steel in addition to the size and shape of the parts. Of the conventional quenching media, (water, oil and polymer), water quenching and in some cases even oil quenching may result in quench cracks in thick sections. Further though mineral oils exhibit good cooling capacity for the majority of alloy steels, they are relatively expensive and toxic. The synthetic polymer quenchant, polyethylene glycol (PEG) [2] has the advantages of lesser risk of cracking and less distortion, resulting in better mechanical properties compared to water, oil and brine solutions.

Microalloyed steels, are designed to provide better mechanical properties and/or greater resistance to atmospheric corrosion than conventional carbon steels. They are not considered to be alloy steels in the normal sense because they are designed to meet specific mechanical properties rather than a chemical composition (steels have yield strengths greater than 275 MPa, or 40 ksi). The chemical composition of a specific steel may vary for different product thicknesses to meet mechanical property requirements. The steels in sheet or plate form have low carbon content (0.05 to-0.25% C) in order to produce adequate formability and weldability, and they have manganese content up to 2.0%. Small quantities of chromium, nickel, molybdenum, copper, nitrogen, vanadium, niobium, titanium, and zirconium are used in various combinations [3]. The mechanical properties of microalloyed steel result, however, from more than just the mere presence of microalloying elements. Austenite conditioning, which depends on the complex effects of alloy design and rolling techniques, is also an important factor in the grain refinement of hot-rolled steels [4]. Grain refinement by austenite conditioning with controlled rolling methods has resulted in improved toughness and high yield strengths in the range of 345 to 620 MPa (50 to 90 ksi). The present study is aimed at determining a comparative evaluation of mechanical properties achieved in MCLA steels (EN steels) with different quenching media with standard step tempering procedure [5-8].

Materials and Methods

Four steels, viz., EN 18, EN19, EN 24, and EN25 in the normalized condition were procured from Mumbai market. The compositions were analysed for confirmation and Table 1 gives the results of the composition check.

Mineral oil (SAE 320 gear oil) and polymer {Polyethylene glycol, $H-(O-CH_2-CH_2)n-OH$ [where n represent the average number of oxyethylene groups]} from Bangalore market to serve as quenchant.

Test specimen preparation

A set of specimens was prepared for Tensile, Impact, Hardness and Microstructural analyses. The standards used for samples to carry out the various tests are listed out in Table 2.

Heat treatment/quenching and step tempering

An electric furnace with maximum temperature of 1200°C was used for both solutionizing and step tempering. The heat treat temperatures were kept same for all four steels studied, as shown in Table 3. After solutionizing, the samples were directly quenched in water, oil [9] or

*Corresponding author: Chandan BR, Department of Mechanical Engineering, MS Ramaiah Institute of Technology, Bangalore, Karnataka, India
E-mail: chandanbrmech@gmail.com

Element	EN18	EN19	EN24	EN25
C	0.380	0.393	0.431	0.350
Mn	0.700	0.660	0.605	0.700
P	0.012	0.014	0.023	0.04
S	0.012	0.019	0.030	0.04
Si	0.192	0.253	0.283	0.40
Ni	--	--	1.395	2.80
Cr	0.70	1.043	0.978	0.80
Mo	--	0.202	0.207	0.65
Fe	bal	bal	bal	bal

Table 1: Chemical Composition of steels studied.

Test	Standard used
Hardness Test	ASTM 92
Tensile Test	ASTM E-8
Charpy Test	IS: 1499

Table 2: Standards used [3].

Process	Temp °C	Soaking time
Hardening	855	60 min
Tempering I	575	60 min
Tempering II	220	60 min

Table 3: Temperature and soaking ftime [4].

polymer solution. The quenched samples were thereafter tested for their mechanical properties.

Mechanical tests

A standard Brinell Hardness Tester was used for measurement of indentation hardness. The tests were conducted using a 10 mm diameter steel ball and 3000-kg load. The tensile tests were carried out using an electrically powered Hounsfield tensometer with a capacity of 20 KN. Impact energy to failure was found using a Charpy impact tester.

Results and Discussions

Table 4 shows the mechanical properties of the as–received and heat treated steel samples with the three quenchants. Figures 1-4 are plots of variations of mechanical properties when different quenchants are used. Figure 1 shows the variation of hardness for the four EN steels when the quenchant is changed from water to oil and then polymer. It is observed that the maximum hardness value of 362 BHN is obtained in EN 25 with water quenching in general, water quenching has the maximum impact on hardness followed by oil quenching and polymer quenching as the least defect [10-12]. However, in case of EN 19 polymer quenching resulted in highest hardness of 286 BHN. Figure 2 depicts the defect of varying the quenchant on the UTS of EN steels. The variation of the UTS generally follows the same behaviour as the variation of hardness. It is observed that once again the highest UTS value is obtained in water quenched EN 25 steel (1280 MPa) which is more than twice the value (580 MPa) for the as received steel. Figure 3 depicts the change of impact energy of the four EN steels considered. Once again it is evident that step tempering after quenching for hardening has improved impact energy of steels except in case of water quenching of EN18 and EN19. This may appear as in congruent with Figure 4 where the percentage of elongation of the forged steels is highest in the as received condition in the lower in the heat treated conditions (except for oil quenching).

The explanation for higher impact energy (and therefore higher toughness) of the heat treated steels lies in 20-50% increase strength values while the elongation is reduced by maximum 20% only.

Effect of quenching in different media on the microstructure of EN steels

The microstructural investigation was performed using a Carl Zeiss optical microscope. In sequence, the steps include sectioning, mounting, coarse grinding, fine grinding, polishing, etching and

EN Series	Sample uenching medium	Tempering Temperature (°C)	BHN	Tensile strength MPa)	Impact Energy, J	% El
EN 18	As-received	-	188	580	42	32
	Water	575,220	270	1050.2	35	28
	Oil	575,220	232	1020.7	68	34
	Polymer	575,220	238	952.8	78	27
EN 19	As-received	-	252	900	56	31
	Water	575,220	280	1201.8	37	27
	Oil	575,220	252	1150.3	55	33
	Polymer	575,220	286	1135.9	88	26
EN 24	As-received	-	270	920	55	29
	Water	575,220	362	1220.4	60	26
	Oil	575,220	340	1060.2	75	31
	Polymer	575,220	260	1198.3	100	25
EN 25	As-received	-	290	1020	45	26
	Water	575,220	375	1280.3	60	25
	Oil	575,220	352	1240.2	80	29
	Polymer	575,220	315	1239.9	112	28

Table 4: Shows the Mechanical properties of as-received and quenched steel samples.

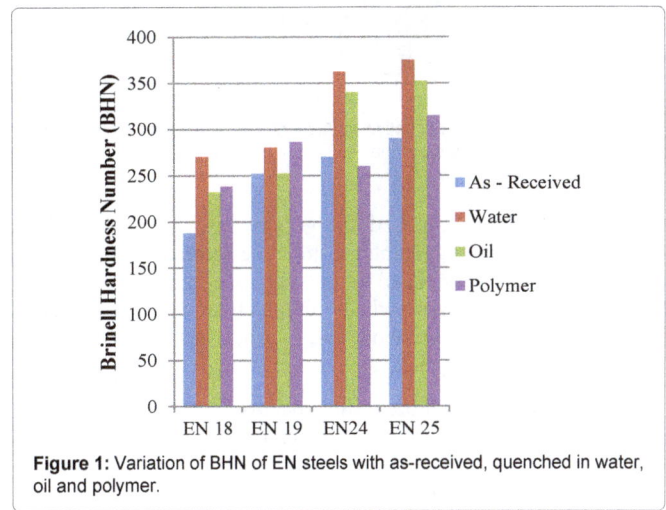

Figure 1: Variation of BHN of EN steels with as-received, quenched in water, oil and polymer.

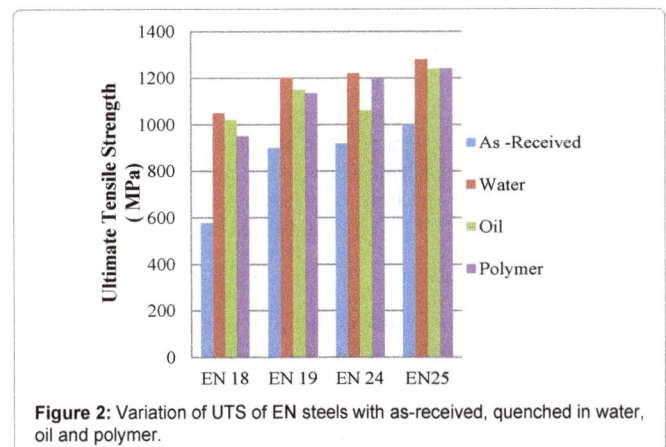

Figure 2: Variation of UTS of EN steels with as-received, quenched in water, oil and polymer.

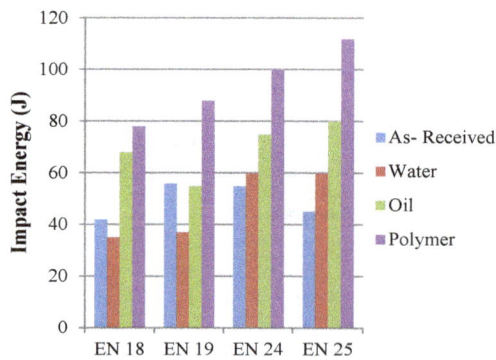

Figure 3: Variation of Impact Energy of EN steels with as-received, quenched in water, oil and polymer.

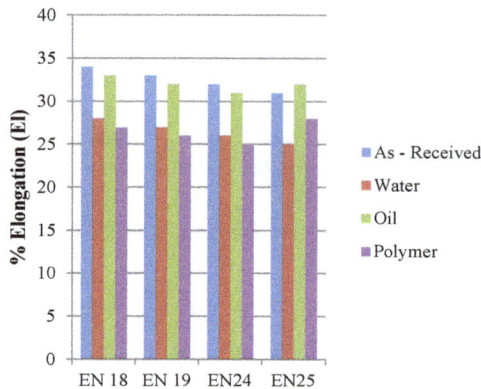

Figure 4: Variation of Elongation in Area of EN steels with as-received, quenched in water, oil and polymer.

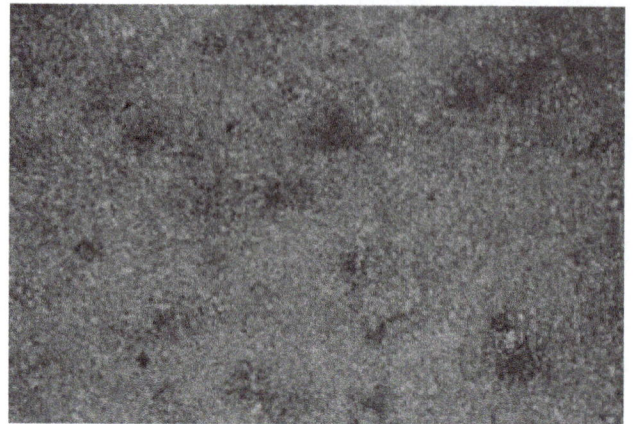

Figure 5a: EN 18 water quenched.

Figure 5b: EN 18 Oil quenched.

Figure 5c: EN 18 Polymer quenched.

microscopic examination, and the general procedure followed by earlier investigators was employed [6]. The samples were polished using a series of emery papers of grit size varying from 1000 µm to 1500 µm. The samples were etched with nital solution, 100 mL ethanol and 1-10 mL nitric acid for about 10-20 seconds before observation in the optical microscope [13-17].

The mechanical properties of microalloyed steels result, however, from more than just the mere presence of microalloying elements. Austenite conditioning, which depends on the complex effects of alloy design and rolling techniques, is also an important factor in the grain refinement of hot-rolled steels. Grain refinement by austenite conditioning with controlled rolling methods has resulted in improved toughness and high yield strengths in the range of 345 to 620 MPa (50 to 90 ksi).

Figures 5-8 are the photomicrographs of EN18 and EN19, EN 24 and EN 25, respectively. Figure 5a-5c shows the microstructures of EN 18 in the water quenched, oil quenched and polymer quenched conditions respectively. Similarly, Figures 6a-6c, 7a-7c and 8a-8c depict the microstructures for EN 19, EN24 and EN25 in different conditions.

Observations

The microstructure of water-quenched specimens appears to be a mixture of martensite and lower bainite. For oil-quenched specimens, a mixture of upper bainite, lower bainite and some martensite is obtained. In case of polymer quenched steels, the microstructure consists of fine tempered martensite with a small amount of ferrite.

Figure 6a: EN 19 water quenched.

Figure 7a: EN 24 water quenched.

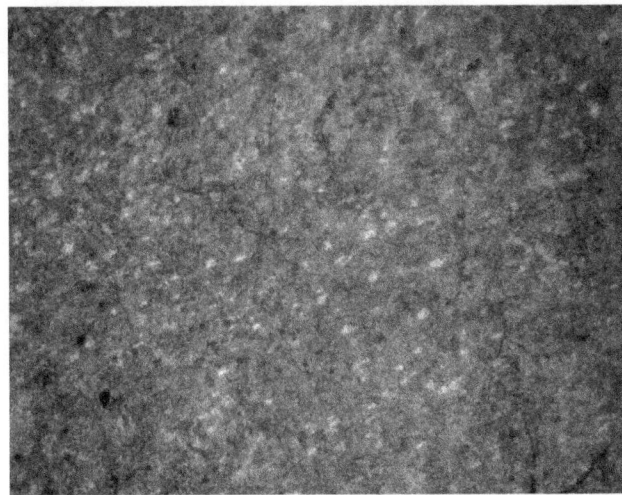

Figure 6b: EN 19 Oil quenched.

Figure 7b: EN 24 oil quenched.

Figure 6c: EN 19 polymer quenched.

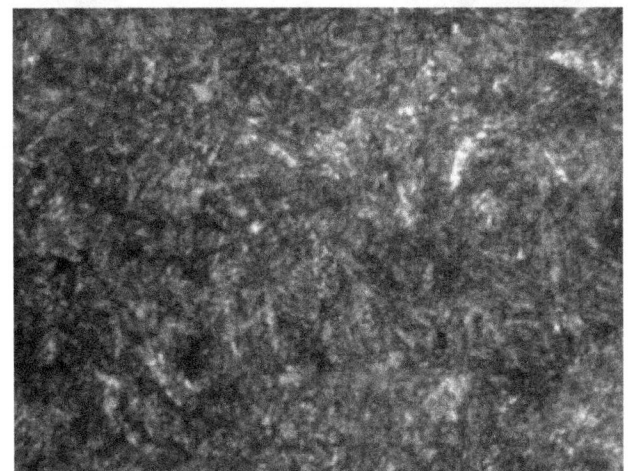

Figure 7c: EN 24 polymer quenched.

Figure 8a: EN 25 water quenched.

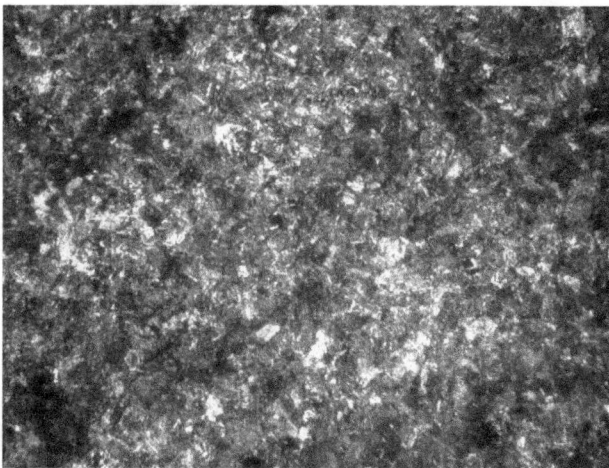

Figure 8b: EN 25 oil quenched.

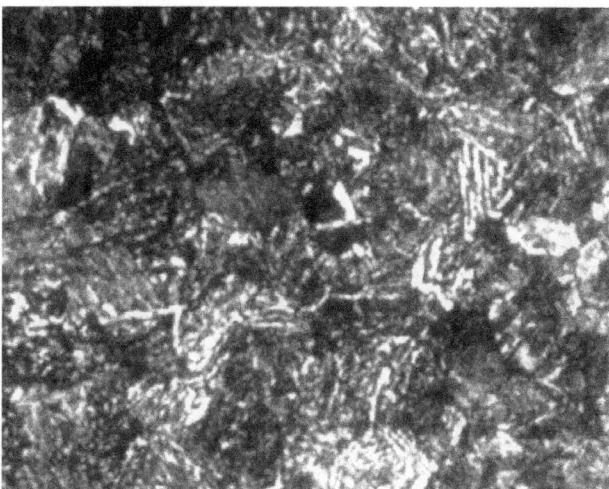

Figure 8c: EN 25 Polymer quenched.

Thus polymer quenching would improve ductility, toughness and impact strength values.

Conclusions

1. It has been established that polymer can also be used as a quenching medium for MCLA forged steels.

2. The study has shown that using of water, oil and polymer as quenchants improves the mechanical properties when compared to the untreated steels.

3. Polymer quenching improves the ductility of the steel because of its lower and uniform cooling rate compared with water and oil; also there is lesser risk of cracking and distortion in the parts. The uniform low cooling rates also result in better mechanical properties for the polymer quenched steels.

4. Microstructural analysis corroborates the changes in mechanical properties observed.

Acknowledgements

The authors are thankful to the management of M.S. Ramaiah Institute of Technology, Bengaluru for facilitating this research work.

References

1. Philip TV, Thomas J, Cafffery M (1961) Properties and selection - Iron, Steels and high Performance Alloys. ASM Hand Book Vol-1, ASM International, Ohio.

2. Momoh M, Bamike BJ, Saliu AM, Adeyemi OA (2015) Effects of Polyethylene Glycol on the Mechanical Properties of Medium Carbon Low Alloy Steel. Nig J Tech Devlop 12.

3. Designation: E8/E8M – 09. Standard Test Methods for Tension Testing of Metallic Materials.

4. Becherer BA, Witheford TJ (1961) Heat Treating of Ultra-high-strength Steels. ASM Hand Book Vol-4, ASM International, Ohio.

5. Odusote JK, Ajiboye TK, Rabiu AB (2012) Evaluation of Mechanical Properties of Medium Carbon Steel Quenched in Water and Oil. AUJT 15: 218-224.

6. Zipperian DC (2016) Pace Technologies, Metallographic Specimen Preparationbasic.

7. Ahmed OJ (2011) Study the effect of polymer solution and oil quenchants on hardening automotive camshaft. J Thi-Qar University 6: 134-146.

8. Ericsson T (1991) Principle of Heat treating of Steels. ASM Handbook, Ohio.

9. Classification and Designation of Carbon and Low Alloy Steel (1990). ASM Handbook.

10. Carbon Steel Handbook (2007) Electric Power Research Institute, Palo Alto, California.

11. Philip TV, Mccaffrey TJ (1990) Ultrahigh strength steels. ASM Handbook, Ohio.

12. Ramesha CM (2003) A study on suitability criteria of steels with lower alloy contents for semi critical application maintaining reliability and structural integrity by process modifications 2003-2010. ASM Handbook "Heat Treating".

13. Eshraghi-Kakhki M, Golozar MA, Kermanpur A (2011) Application of polymeric quenchants in heat treatment of crack-sensitive steel mechanical parts: Modeling and experiments. Materials and Design 32: 2870-2877.

14. Higgins AR (2004) Engineering Metallurgy - Part 1 - Applied Physical Metallurgy. (7th Ed) Edward Arnold, England.

15. Khanna OP (2009) Material Science and Metallurgy. Dhanpat Rai Pub (P) Ltd.

16. Martin JW, Doherty RD, Cantor B (1997) Stability of Microstructure in Metallic Systems (2nd edition). Cambridge: Cambridge University Press, UK.

17. Ndaliman MB (2006) An Assessment of Mechanical properties of Medium Carbon Steel under Different Quenching Medium. AUJT 10:100-104.

8

Effects of Heat Treatment on the Mechanical Properties of the Vanadis 4 Extra and Vanadis 10 Tool Steels

Baykara T* and Bedir HF

Mechanical Engineering Department, Doğuş University, Istanbul, Turkey

Abstract

Vanadis tool steels which are a trademark of the Uddeholm AB Company are high vanadium content (along with chromium and molybdenum) steels with unique mechanical properties such as very high wear resistance along with a good machinability, dimensional stability and grind ability. They are widely used in blanking operations, stamping, and deep drawing, cutting and slitting blades. Microstructural features of Vanadis steels are directly depended upon the distribution of the carbide grains. Based upon the carbon and vanadium contents of the Vanadis tool steels, wear test results and micro hardness values are correlated with the resulting microstructural features.

Keywords: Tool steels; Vanadis; Powder metallurgy; Wear resistance

Introduction

Vanadis tool steels which are a trademark of Uddeholm Company are high vanadium content (along with chromium and molybdenum) steels with unique mechanical properties such as very high wear resistance along with a good machinability, dimensional stability and grind ability. Vanadis steels are manufactured using powder metallurgical routes and they offer a combination of high wear resistance, high hardness and good toughness [1]. Vanadis tool steels have a very homogeneous microstructure and highly refined grain distribution compared to other conventional steels. Based on such characteristics, they are widely used in blanking operations, stamping, and deep drawing, cutting and slitting blades [1,2].

Microstructural features of Vanadis steels are directly depended upon the type of carbides and the matrix phase. The distribution and the size of carbides as eutectic and secondary carbides effect the tool performance and tool life. In this regard, the conditions of heat treatment have a determining impact on such demanding properties. Undissolved carbide grains have also unique effect on the wear resistance due to inhibiting role on the coarsening of the austenite grains.

It is reported that there are two major types of carbides in Vanadis steel following austenitizing, i.e., M_7C_3 secondary carbides (dissolved in the austenite) and eutectic MC carbides (stable up to 1150°C along with small spherical carbides as alloyed cementite [2,3]. Laser hardening method is another technique to improve the properties of such tool steels [4,5].

In this study, the effect of heat treatment on the Vanadis 4 extra and Vanadis 10 grade tool steels are investigated. Following the annealing and quenching cycles on the steel samples, abrasive wear behavior, microstructural changes and micro hardness were determined. Based upon the carbon and vanadium contents of the Vanadis tool steels, wear test results and micro hardness values are correlated with the resulting microstructural features [6].

Material and Methods

Vanadis steel parts were provided by Uddeholm Company. Steel samples were machined properly to shape them into cylindrical forms for wear tests. Each specimen machined properly and reduced diameters till fit the pin on disc sample attached. Lengths have reduced enough approximately 15-25 mm and diameters reduced 4, 5-7, 5 mm

after machining. Thereafter, the Vanadis 4 extra samples were annealed at 1000°C for 9 minutes and Vanadis 10 at 800°C for 5 minutes and quenched into water at room temperature (heat treatment procedure was selected according to Uddeholm AB Materials Safety Data Sheet). All the measurements were determined on both as-received (witness samples) and heat treated samples. Chemical compositions of the samples of Vanadis 4 extra and Vanadis 10 manufactured with powder metallurgical routes are shown in Table 1.

Wear tests

Wear tests were conducted on the as-received and heat treated samples using the Pin-On-Disk (POD) wear test rigs. AISI H13 steel disc is used as the sliding platform for wearing under dry conditions. The steel disc is also heat treated and has the hardness value of 36.3 Rockwell HRC. 20 N loads are applied for Vanadis 4 extra samples and 10 N loads are applied for Vanadis 10 samples.

In all the tests, sliding diameter, the rate and the duration of tests were about 140 mm, 3000 rev/min and 10, 20, 30, 40, 50 and 60 min period of time respectively. Weights of the specimens before and after the wear tests were measured by an electronic balance which has (±0.1 mg) sensitivity to determine the weight loss and consequently the wear rates.

Micro hardness tests

Vickers micro hardness values for both as-received and heat treated samples were determined using a Vickers micro hardness testing unit HV 0.2. Average values of five measurements were recorded.

Steel	C	Si	Mn	Cr	Mo	V	Fe
Vanadis 4	1.4	0.4	0.4	4.7	1.5	3.7	Balance
Vanadis 10	2.9	0.5	0.5	8	1.5	9.8	Balance

Table 1: Chemical composition of the samples in mass fractions (wt%).

*Corresponding author: Tarik Baykara, Mechanical Engineering Department, Doğuş University, Acıbadem, Kadıköy, Istanbul, Turkey
E-mail: tbaykara@dogus.edu.tr

Microstructural investigation

Microstructural analysis of the samples of the Vanadis 4 extra and Vanadis 10 both untreated and heat treated were conducted using a scanning electron microscope, SEM Philips XL 30 SFEG with resolution 1.5 nm at 10 kV. Standard metallographic sample preparation route (grinding, polishing, fine polishing and etching) was used to prepare the samples for SEM analysis.

Results

Weight loss (in mg) vs. time (in min.) graph is shown in Figures 1 and 2 demonstrating the effect of heat treatment on the wear resistance which shows approximately average 54% and 76% decrease in weight loss in 40 min for the heat treated Vanadis 4 extra and Vanadis 10 samples respectively. It should be noted that the wear loss data demonstrates significant decrease in weight loss after 40 and 50 min due to strain hardening effect in between the surfaces of Vanadis samples and AISI H13 steel disc.

Table 2 shows the results of these measurements.

Micro hardness values tabulated in Table 2. Indicate more than 160% increase in micro hardness in Vanadis 4 samples while Vanadis 10 as-received and heat-treated samples show only a slight and insignificant increase in micro hardness values.

Discussion

SEM micrographs of the Vanadis 10 samples both in untreated and heat treated conditions are shown in Figures 2 and 3. Figures 3

and 4 illustrate the microstructures of the as-received and heat treated Vanadis 4 extra and Vanadis 10. As can be seen in Figures 3b and 4b, the as-received structures of Vanadis 4 extra and Vanadis 10 consist of carbide grains (eutectic carbides, MC and secondary carbides, M_7C_3 and other finer grain size carbides) embedded in a ferrite matrix. MC type carbides appear darker since they contain higher amount of carbon while other carbides reveal brighter appearance due to less amount of carbon and more iron, chromium (6). Figures 4a and 4b show no hardening effects in the microstructures as in the case of micro hardness data of the Vanadis 10 samples given in Table 2. Micrographs of these samples reveal no quenching structures meaning that no carbide grains are dissolved. However, in the Vanadis 10 microstructure, larger size grains are noticeable and following the heat treatment some grain growth can be seen. Such a grain growth mechanism could lead much higher wear resistance after the heat treatment.

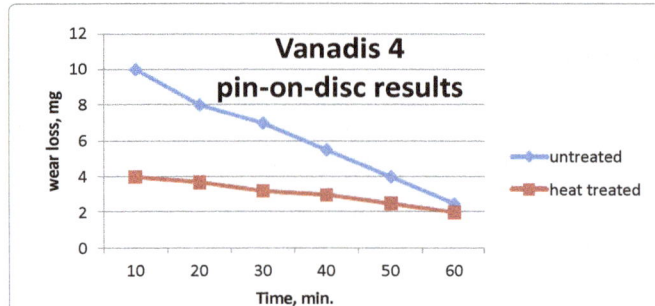

Figure 3a: SEM Micrographs of the Vanadis 4 extra samples (×5000 mag.); (left) heat treated; (right) untreated.

Figure 3b: SEM Micrographs of the Vanadis 4 extra samples (×20000 mag.); (left) heat treated; (right) untreated.

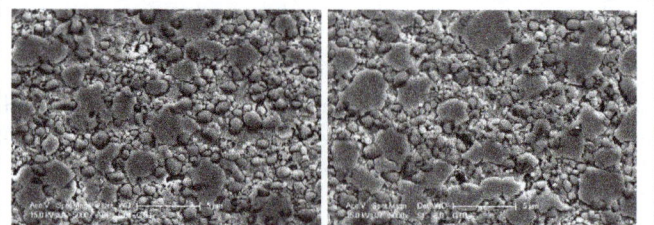

Figure 4a: SEM Micrographs of the Vanadis 10 samples (×5000 mag.); (left) heat treated; (right) untreated.

Figure 1: Pin-on-Disc wear test results for untreated and heat treated Vanadis 4 extra samples.

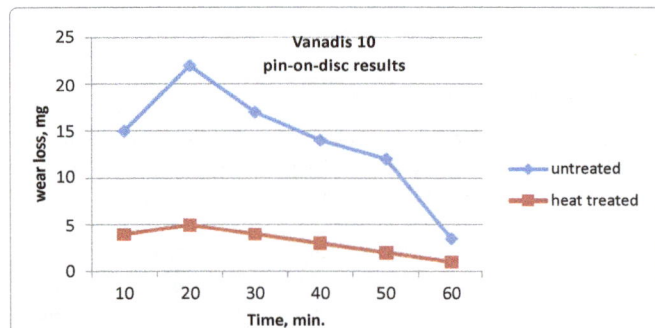

Figure 2: Pin-on-Disc wear test results for untreated and heat treated Vanadis 10 samples.

Vanadis 4 extra		Vanadis 10	
As received	Heat Treated	As received	Heat treated
278.7 ± 15.5	728.4 ± 7.5	313.6 ± 6.5	315 ± 6.2

Table 2: The results of microhardness measurements, HV 0.2.

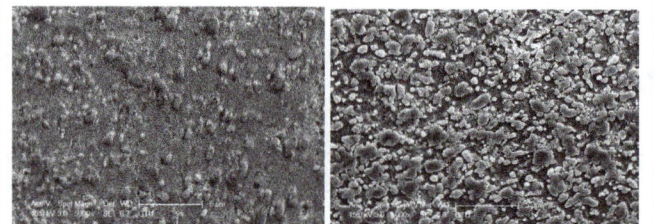

Figure 4b: SEM Micrographs of the Vanadis 10 samples (×20 000 mag.); (left) heat treated; (right) untreated.

On the other hand, there can be seen considerable changes in the microstructures of the Vanadis 4 extra samples with heat treatment (Figures 3a and 3b). In this microstructure, secondary carbides, M_7C_3 are dissolved in austenite while MC type carbide grains are still stably distributed in the matrix. Characteristic martensitic needle-like structure and retained austenite can be observed in the Figure 3b micrograph along with dissolved carbide grains in ferrite matrix. A distinct increase in the micro hardness values (app. 161% increase following the heat treatment) is given in Table 2 for the Vanadis 4 extra while only a very slight increase in the Vanadis 10 is recorded. Weight loss data in Figure 1 also support such a drastic micro hardness increase.

Conclusion

- The inhomogeneous microstructure of the Vanadis 4 extra which consists of needle-like marten site, carbide grains and most probably retained austenite forms following the heat treatment. A distinct increase in micro hardness values and considerable increase in wear resistance would be the results of such microstructural formation.

- Considerable increase in wear resistance of the heat treated Vanadis 10 would be due to grain growth of carbide grains embedded in the ferrite matrix. Larger size carbide grains may have an inhibiting role against the wear mechanism in between the surfaces of Vanadis 10 samples and AISI H13 steel disc.

Acknowledgement

Authors would like to thank to Suleyman Demirel University, Mechanical Engineering Department for their help in using pin on disc facilities. Authors would like to thank to Gebze Technical University, Scanning Electron Microscopy Laboratories for their help in SEM analysis.

References

1. Uddeholm AB (2013) Uddeholm Vanadis 10 Superclean.

2. Sobotova J, Jurci P, Adamek J, Salabova P, Prikner O, et al. (2013) Diagnostics of the Microstructural Changes in Sub-zero Processed Vanadis 6 P/M Ledeburitic Tool Steel. Materials and Technology 47: 93-98.

3. Ocak M (2015) An Investigation On Mechanical And Microstructural Property Changes, Due To Different Heat Treatment, of Vanadis 4 Cold Work Tool Steel Produced By Powder Metallurgy. (Doctoral dissertation, Fen Bilimleri Enstitüsü).

4. Surzhenkov A, Kulu P, Viljus M, Vallikivi A, Latokartano J (2010) Microstructure and wear resistance of the laser hardened pm tool steel vanadis 6. In Proceedings of the International Conference of DAAAM Baltic Industrial Engineering.

5. Köse G, Mutlu Z (2015) Laser hardening of tools steels. Doğuş University.

6. Jurči P, Dománková M, Ptačinová J, Hudáková M (2015) Phase constitution of sub-zero treated Vanadis 6 tool steel. Metal 2015, Jun 3rd-5th, Brno, Czech Republic, EU.

9

Evaluation of Physico-Chemical, Thermal and Mechanical Properties of Sintered Graphite and Mesophase Formulations

Deepak Kumar G[1] and Singh Arora SP[2]*

[1]Department of Chemistry, Indian Institute of Space Science and Technology (IIST), Trivandrum, Kerala, India
[2]Scientist/Engineer-SD, Indian Space Research Organisation (ISRO), Govt. of India, Tamil Nadu, India

Abstract

Carbon/graphite materials are widely applied to the areas of seal and friction materials. These applications exploit the exceptional properties of carbon/graphite, such as its excellent mechanical behavior at high temperature and low reactivity. However, Owing to high open porosity, and low strength conventional carbon/graphite materials prepared using pulverized coke as filler and pitch as binder fall across enormous challenge. Therefore, it is necessary to reduce the open porosity and increase in the strength along with thermal conductivity. Using natural graphite as filler instead of coke to prepare carbon blocks can avoid high-temperature graphitization and repeated dipping owing to the high degree of graphitization and high thermal stability of natural graphite; thus it can simplify the working process, save energy and reduce cost. Carbonaceous mesophase has inherent properties, such as a high softening point, high fluidity, high carbon yield, and graphitizability, which make it suitable for the fabrication of polygranular carbon materials. Furthermore, it can be used as a binder. The sintering was performed in a static furnace with argon atmosphere and compared with the same compound sintered in passage furnace with hydrogen and nitrogen atmosphere. The analysis of the properties of the tested material was performed with the aid of metallography using a scanning electron microscope, which verified the particle size distribution, chemical elements and pores present. However, for the graphite powder and zinc stearate, present in smaller percentages were disregarded its influence on the physical properties of the compound generated. Compressibility and compaction are parameters that indicate and describe the behavior of metal powders as they are compressed. The ability of a powder densification is related to compressibility. Already compaction is defined as the stability of the structure of the pressed compacted to a certain working pressure. Therefore, the natural graphite as filler and mesophase pitch as binder can be able to produce high density graphite blocks. The main aim of this work is to produce isotropic graphite material and evaluation of its properties such as high density, high thermal conductivity, high electrical resistivity, low coefficient of thermal expansion, and high compressive strength from natural graphite powder mesophase pitch.

Keywords: Natural graphite; Mesophase pitch; Density; Thermal conductivity; Electrical resistivity; Coefficient of thermal expansion; Compressive strength

Introduction

High-density artefacts of carbon and graphite have been widely applied in various fields because of their excellent properties. They are now recognized to be essential for the most advanced technologies. So far, many studies of the production of high density carbon artefacts have been reported. The processes for preparing carbon artefacts are typically classified into two categories. Procedures in the first category use filler coke and binder pitch as starting materials. The process consists of moulding the mixture by CIP (Cold Isostatic Press), carbonization (baking) followed by impregnations, and graphitization into the final product. The carbonization and impregnation are often repeated for densification. The second process consists of preparation of self-adhesive carbonaceous grains, moulding them without any additional binding substances, carbonization, and graphitization. No necessity of the binder and the impregnation steps is a characteristic of this process. Carbon in small quantities is added to iron, 'Steel' is obtained. Since the influence of carbon on mechanical properties of iron is much larger than other alloying elements. The atomic diameter of carbon is less than the interstices between iron atoms and the carbon goes into solid solution of iron. As carbon dissolves in the interstices, it distorts the original crystal lattice of iron. This mechanical distortion of crystal lattice interferes with the external applied strain to the crystal lattice, by mechanically blocking the dislocation of the crystal lattices.

Conventionally, carbon blocks are fabricated from the filler of coke and the binder of pitch via a series of complex processing steps involving mixing of raw materials, molding, many times of dipping and pyrolysis, and graphitization at elevated temperatures. The mesophase pitches exhibited high coke yield, low softening point, and high fluidity.

Mesophase pitches are considered to be good starting material for many industrial and advanced carbon products, such as carbon fibres, needle coke, graphite electrodes, C-C composites, fine-grained sintered carbons, Li-ion battery anodes, mesocarbon microbeads, carbon foam and plasma-facing components for fusion devices etc. Carbonaceous mesophase is a self-sintering precursor, which is able to produce polygranular carbons and graphites. The sinterability of mesophase derives from its intrinsic thermoplastic properties that allow mesophase to be moulded and compacted, giving rise to high-density and high-strength artefacts. Several researchers studied the self-sinterability of mesocarbon microbeads (MCMB) for preparation of high-density isotropic carbon [1,2]. Yongzhong Song et al. [3] prepared graphite seal materials from mesocarbon microbeads. Silicon

*Corresponding author: Singh Arora SP, Scientist/Engineer-SD, Indian Space Research Organisation (ISRO), IPRC (M), Govt. of India, Tamil Nadu, India
E-mail: mikkiiarora@rediffmail.com

Evaluation of Physico-Chemical, Thermal and Mechanical Properties of Sintered Graphite and Mesophase...

43

carbide (SiC) is one of the most promising structural materials due to its superior thermomechanical properties, such as high chemical and thermal stability, good chemical inertness, high thermal conductivity, high hardness, low density, and low coefficient of thermal expansion. Silicon carbide is a compound of silicon and carbon with chemical formula SiC. It occurs in nature as mineral moissanite. Because of the rarity of natural moissanite, most silicon carbides are synthetic. SiC is part of a family of materials that exhibit a one-dimensional polymorphism called polytypism. Thanks to its structure, an almost infinite number of SiC polytypes are possible, and more than 200 have already been discovered. With the introduction of nanotechnology and manipulation at nanoscale, new opportunities have been opened. One-dimensional (1D) nanostructures such as nanowires or nanotubes have gained much interest in fundamental research as well as tremendous potential applications. Among many materials, SiC-based 1D nanostructure has very interesting physical, chemical, and electronic properties. SiC is a wide band gap semiconductor, from 2.2 eV to 3.4 eV depending on the polytype with a high electronic mobility and thermal conductivity. One of the other merits of SiC is that it can be used at high temperatures because its noncongruent melting point reaches 3100°K and its maximal operating temperature is considered to be 1200°K. Moreover, SiC is resistant to corrosive environments because of its chemical inertness. These facts justify the high application of SiC in a wide range of areas, such as electronics, heating elements, and structural materials. More particularly, SiC nanoparticles have been introduced in CMCs to enhance the mechanical and thermal properties of ceramic materials such as Al_2O_3 and mullite. However, the thermo plasticity of the mesophase is usually too high, and consequently, after being moulded into rigid pieces, the mesophase deforms and distorts in subsequent carbonizations. It is necessary, therefore, to modify thermoplastic properties of the mesophase before sintering. A process widely used to reduce plasticity in carbon precursors is oxidative stabilization at moderate temperatures. Isao Mochida et al. [4] prepared the self-adhesive carbonaceous grains through the oxidation of the mesophase pitches synthesized from naphthalene. Fanjul et al. [5] prepared high-density polygranular carbons and graphites from carbonaceous mesophase, to study the changes in the chemical composition of the mesophase obtained from a coal-tar pitch and a naphthalene-based pitch at different stages of oxidative stabilization with air (200-300°C), and the carbonization behaviour of samples stabilized at different temperatures below 1000°C.

Related Work

In view of the above the objective of this analysis is focused on the development of high density graphite material from a mixture of graphite powder and mesophase pitch with the target properties as depicted in Table 1 for possible application as pump seals at cryogenic temperatures. Attempts will also be made to design and fabricate a

die for shaping of the formulation into a seal to establish the shaping capability. Our system aims at the automatic detection of text.

Selection and characterisation of starting raw materials, milling and mixing followed by the physico-chemical, morphological and thermal properties evaluation

Commercially available graphite powder and mesophase pitch (the precursor for graphite) binder were ball milled. The particle morphology before and after milling was observed under Scanning Electron Microscope (Hitachi, Tokyo, Japan). The particle size was measured using Laser Diffraction Technique (Malvern, UK). The mixtures have shown the flaky and irregular morphology (Figure 1). The graphite and mesophase pitch shows crystalline and amorphous nature, respectively. Graphite-mesophase pitch mixtures with graphite to mesophase pitch ratio of 90:10 (designated as GM-I), 80:20 (designated as GM-II) and 70:30 (designated as GM-III) were prepared. The mixtures contain particle with sizes in the range of 4-15 μm [6,7].

Compaction of formulations by die, compaction experiments, characterization and heat treatments of the compacts

Compaction behavior was elucidated by feeding ~0.5 g of the mix individually into the compaction die with 10 mm diameter (Figure 2) and compacted using Universal Testing Machine (Instron, 4483) (Figure 3) [8]. Load-displacement values were recorded using a computer interface. In all the three mixes while carrying out the compaction process initially the powder was compacted to an initial pressure of 5 MPa the displacement values of the top punch were recorded. All mixes were compacted at a maximum pressure of 550 MPa with definite increment of pressure and for every pre-set value of these pressures the displacements of the top punch from the initial position are recorded [9,10].

Experimental Results

Plot of variation of the thickness of the compacts for Mix-I, Mix-II and Mix-III plotted by the method of periodic loading and unloading under compaction process are shown in Figure 4. The curve depicted in Figure 4 demonstrates the compaction process and associated elastic stresses on the compact. When the sample is under compaction pressure the strain energy is stored in the pellets, on release of compaction stresses the pellet expands in axial direction while the pellet is still constrained within the die walls.

Density of the ejected compacts for each mix was also determined by measurement of weight and dimensions. Mix-I with the maximum graphite content (90%) exhibited a spring back of 12%, however, attained a maximum density of 89% of the theoretical density (2.205 g/cc). In case of Mix-II and Mix-III exhibited densities of 86 and 85% of theoretical density, where the theoretical densities are 2.150 and 2.095 g/cc respectively. As the mesophase pitch content in the mix increases from 10% to 30% it facilitates compaction by better bonding and hence reduces the spring back. It is evident from Table 2 that after normalizing

S.no	Properties	Required values
1	Density (g/cc)	1.75 to 2.10
2	Surface conditions	The surface should be free from cracks, delaminations, blisters etc.
3	Presence of anisotropic pyrographite	Minimum
4	Presence of particulates	Nil
5	Thermal conductivity (W/m/k)	30 to 50
6	Linear coefficient of thermal expansion (K^{-1})	$(6.5 \text{ to } 8.5)*10^{-6}$
7	Electrical resistivity (Ohm m)	$(3.08 \text{ to } 4.36)*10^{-5}$
8	compressive strength (MPa)	190 to 360

Table 1: Target properties of seal material.

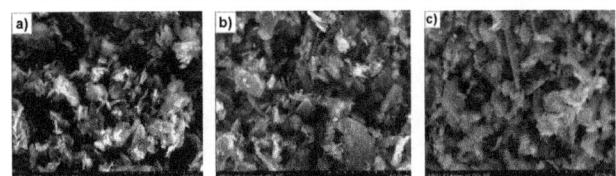

Figure 1: Morphology of (a-c) Mix-I, Mix-II and Mix-III.

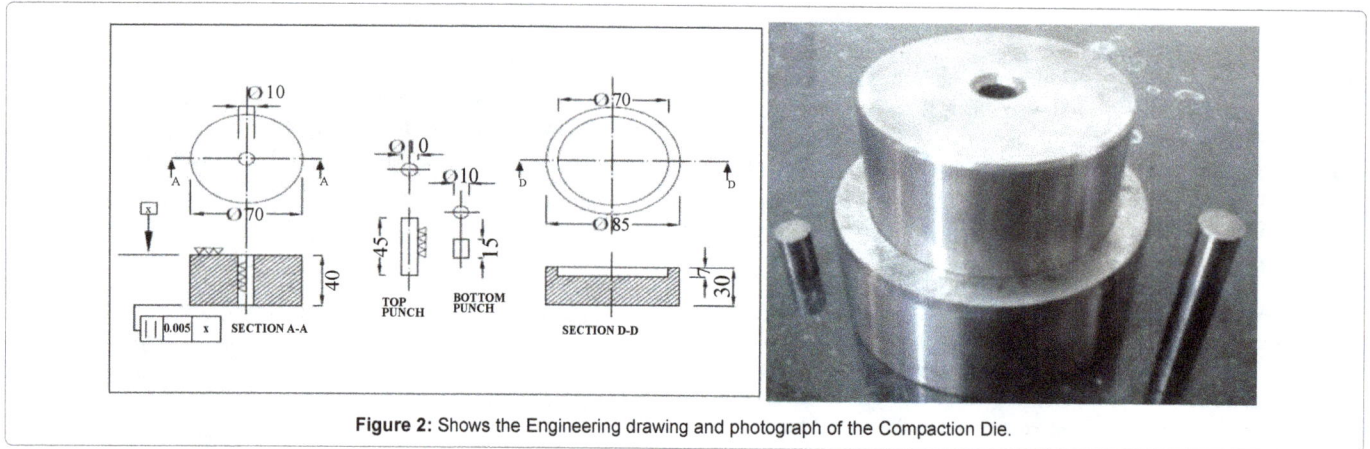

Figure 2: Shows the Engineering drawing and photograph of the Compaction Die.

Figure 3: Shows the hydraulic press and compaction die along with bottom and top punches.

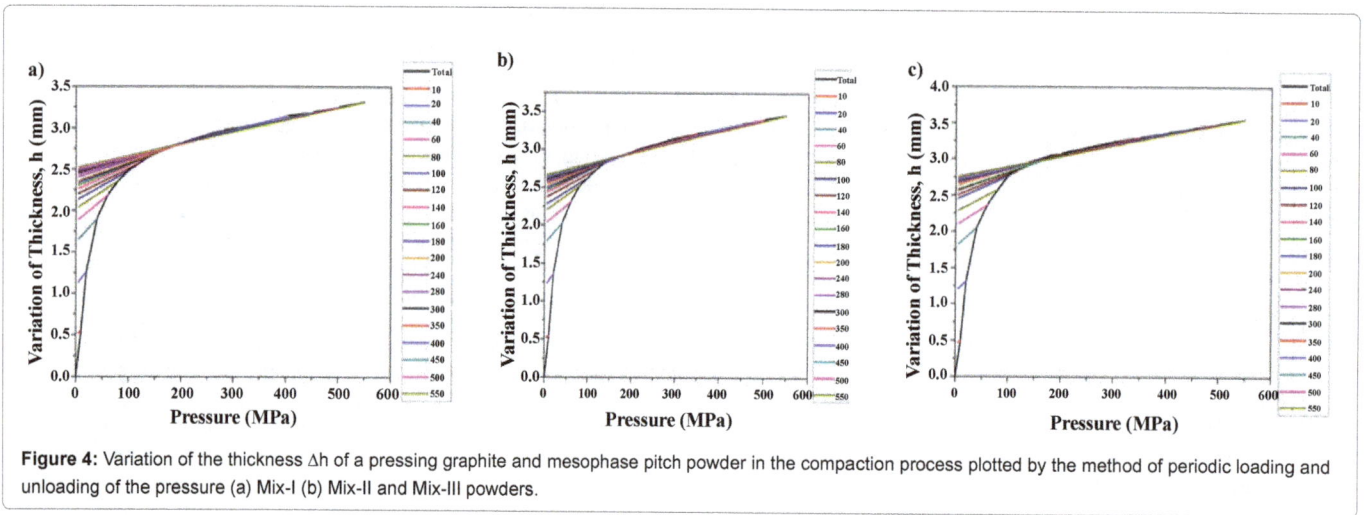

Figure 4: Variation of the thickness Δh of a pressing graphite and mesophase pitch powder in the compaction process plotted by the method of periodic loading and unloading of the pressure (a) Mix-I (b) Mix-II and Mix-III powders.

by the rule of mixtures, though spring back is less, densities achieved for Mix-II and Mix-III are less which is not desirable for compaction processing. Though spring back is significant by 12% in case of Mix-I there is no introduction of any lamination defects.

It is also interesting to note that though spring back is significant the samples are defect free as is evidenced by the high density value. The study reveals that a mix of graphite and mesophase pitch in the ratio of 90: 10 is the best choice for the compaction processing.

All the cold compacted samples were subjected to carbonization at 1200°C in argon atmosphere. We have observed development of cracks

throughout all the samples (Figure 5) due to the escape of organics even with very low heating rates and also due to the too high thermoplasticity of mesophase pitch. Therefore, it is necessary to modify the thermo-plastic properties of mesophase pitch before sintering to eliminate the cracks formation. A process widely used to reduce plasticity in carbon precursors is oxidative stabilization at moderate temperatures. The compacted samples of all the three mixtures were heat treated at a temperature of 250°C for oxidative stabilization purpose. The compact obtained from GM-I mixture produced the sample without any crack after the carbonization (Figure 6). However, the sample densities observed is in the range of 1.65 to 1.7 g/cc. Alternatively, the mixtures were subjected to oxidative stabilization before compaction. Subsequent carbonization results in crack free samples only for GM-I.

Exploration of advanced compaction techniques such as Hot Pressing as shaping options followed by graphitization experiments at high temperatures >2000°C

In view of this GM-I mixture was subjected to hot pressing at temperature of 1200°C and at pressures of 12, 15 and 18 MPa. The samples exhibited a maximum density of 1.7 g/cc under hot compaction under 15 MPa, however, resulted in lamination cracks beyond 15 MPa. The low density of 1.7 g/cc compared to cold compaction even after simultaneous application of temperature and pressure can be attributed to the low pressure involved in hot pressing due to the strength limitation of the graphite dies used. The hot pressed samples were graphitized at 2500°C in a resistive heating graphite furnace. Unlike carbonized samples, the graphitized samples have shown highly crystalline graphite phase with an improved thermal conductivity of 48.26 W/m/K complementing the density and XRD results. A significant decrease in the electrical resistivity has also been observed which can be attributed to the above cited reasons. A thermal expansion value has exhibited only marginal increase. However, there is a significant increase in compressive strength achieving a maximum of 60 MPa and a low coefficient of friction 0.08 which are in good agreement with the improved density values. The properties of hot-pressed and graphitized samples are listed in Table 3.

Fabrication of the die followed by compaction experiments to establish the shaping suitability of the raw mix into seals

To study the adaptability for shaping a die has been fabricated and GM-I was compacted at a pressure of 500 MPa into the circular ring for seal application by die (Figure 7). The photograph of the fabricated rings is shown in Figure 7.

Conclusion

Comparative evaluation of compaction behavior of graphite-mesophase pitch mixtures of graphite to mesophase pitch ratio of 90:10 (GM-I), 80:20 (GM-II) and 70:30 (GM-III) prepared by planetary ball

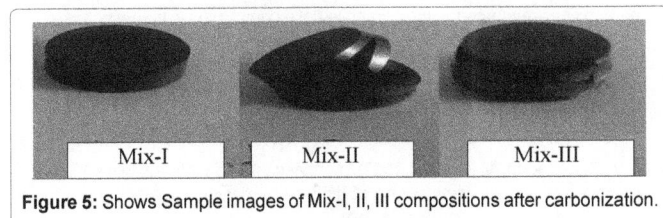

Figure 5: Shows Sample images of Mix-I, II, III compositions after carbonization.

Figure 6: Shows the sample image of Mix-I with no Crack after carbonization.

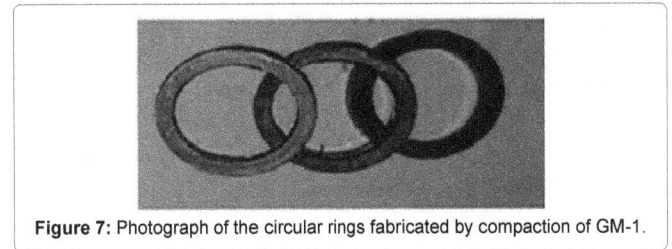

Figure 7: Photograph of the circular rings fabricated by compaction of GM-1.

Sample	Compaction pressure (Mpa)	Green density (g/cc)	Theoretical density *(g/cc)	Theoretical density%
Mix-I	550	1.984	2.205	89
Mix-II	550	1.8525	2.105	86
Mix-III	550	1.784	2.095	85

*Estimated based on the rule of mixtures (Graphite -2.26(g/cc) and Mesophase pitch-1.71 (g/cc))

Table 2: Green densities of sample.

S. no	Properties	Hot pressed (carbonised sample)	Graphitized sample
1	Density (g/cc)	1.7	1.78
2	Surface conditions	Free from cracks, delaminations, and blisters	Free from cracks, delaminations, and blisters
3	Presence of particulates	Nil	Nil
4	Thermal conductivity(W/m/k)	12.23	48.26
5	Linear cofficient of thermal expansion (k^{-1})	12*10^{-6}	13.4*10^{-6}
6	Electrical resistivity (Ohm m)	5.475*10^{-5}	3.21*10^{-5}
7	Compressive strength	49	60

Table 3: Properties of hot pressed and graphitized samples.

S. no	Properties	Values targeted	Values achieved
1	Density (g/cc)	1.75 to 2.10	1.78
2	Surface conditions	Surface should be free from cracks, delaminations, blisters etc.	Free from cracks, delaminations and blisters
3	Presence of anisotropic pyrographite	Minimum	-
4	Presence of particulates	Nil	Nil
5	Thermal conductivity(W/m/k)	30 to 50	48.26
6	Linear coefficient of thermal expansion (k^{-1})	(6.5 to 8.5) *10^{-6}	13.4*10^{-6}
7	Electrical resistivity (Ohm m)	(3.08 to 4.36)*10^{-5}	3.21*10^{-5}
8	Compressive strength	190 to 360	60

Table 4: Properties of Achieved by the graphitized sample.

milling were studied by uni-axial loading and unloading technique using a universal testing machine. Spring back on release of compaction stresses are found to be higher in case of GM-I as evident from the plots exhibiting elastic interaction in comparison to GM-II and GM-III formulations. Higher normalized density values are achieved in case of GM-I in spite of significant spring back can be attributed to the optimum concentration of mesophase pitch in the formulation.

Though cold compaction at a high pressure of 500 MPa has resulted in higher density values, exhibited cracks while heat treatment. Simultaneous application of temperature and pressure by hot pressing has yield carbonized samples with relatively low densities without cracks. Graphitization of hot pressed compact at 2500°C has shown a maximum density of 1.78 g/cc and enhancement of thermal, mechanical, wear and electrical properties as revealed by Table 3.

Properties of high density graphite proposed to develop from graphite and mesophase pitch in the study and the properties achieved are summarised as below. Table 4 clearly indicates that the properties such as density, electrical resistivity, thermal conductivity could be achieved or closes to the required values.

Future Directions

Minimisation of presence of anisotropic pyrographite could not be attempted in the study and the variation in the properties can be attributed to the random orientation of graphite crystals. Future work can be focused on addressing this issue by carrying out hot isostatic pressing.

References

1. Gao Y, Song H (2003) Self-sinterability of mesocarbon microbeads (MCMB) for preparation of high-density isotropic carbon. Journal of Materials Science 38: 2209-2213.

2. Yuan G, Li X (2012) Graphite blocks with preferred orientation and high thermal conductivity. Carbon 50: 175-182.

3. Song Y, Zhai G (2004) Carbon/graphite seal materials prepared from mesocarbon microbeads. Carbon 42: 1427-1433.

4. Mochida I (1994) Self-adhesive carbon grains oxidatively prepared from naphthalene derived mesophase pitch for mould of high density. Carbon 32: 961-969.

5. Fanjul F, Granda M, Santamar R, Menéndez R (2003) Pyrolysis behaviour of stabilized self-sintering mesophase. Carbon 41: 413-422.

6. Zhao Y, Liu Z (2013) Microstructure and thermal/mechanical properties of short carbon fiber-reinforced natural graphite flake composites with mesophase pitch as the binder. Carbon 53: 313-320.

7. Zhong B, Zhao GL (2014) Binding natural graphite with mesophase pitch: A promising route to future carbon blocks. Materials Science & Engineering A 610: 250-257.

8. Fang MD, Tseng WL, Jow JJ (2012) Improving the self-sintering of mesocarbon-microbeads for the manufacture of high performance graphite-parts. Carbon 50: 906-913.

9. Mochida I, Korai Y (2000) Chemistry of synthesis, structure, preparation and application of aromatic-derived mesophase pitch. Carbon 38: 305.

10. Liub Z, Guo Q, Shi J, Zhai G, Liu L (2008) Graphite blocks with high thermal conductivity derived from natural graphite flake. Carbon 46: 414-421.

10

Effect of Temperature of Electron Beam Evaporated CdSe Thin Films

Sahuban Bathusha MS[1], Chandramohan R[1]*, Vijayan TA[1], Saravana Kumar S[1], Sri Kumar SR[2], Ayeshamariam A[3] and Jayachandran M[1]

[1]*Research Department of Physics, Sree Sevugan Annamali College, Devakottai, India*
[2]*Department of Physics, Kalasalingam University, Krishnankoil, India*
[3]*Department of Physics, Khadir Mohideen College, Adiramapattinam, India*

Abstract

CdSe thin films were deposited on a glass substrate by using electron beam evaporation technique. The as deposited films were annealed from 100°C to 300°C with an increment of 100°C. Morphological, structural and optical characterization of the films was carried out by using scanning electron microscope (SEM), X-ray diffraction (XRD), ultraviolet-visible (UV-Vis) spectroscopy; and Fourier transform infrared spectroscopy. The X-ray diffraction pattern that the film has a cubic phase with preferred orientation (100), the grain size was found to be in the range of 29-46 nm. SEM results reveal that film grains are polycrystalline in nature covered the whole surface of the substrate.

Keywords: CdSe thinfilm; X-ray diffraction (XRD); Scanning electron microscope; FTIR

Introduction

Semiconductor nano crystallites prepared as powder or thin film form is of recent interest for many novel applications [1,2]. It is used as an n-type window layer material in thin film solar cells, and is a suitable candidate for photovoltaic applications [3]. The deposition of CdSe thin films was made using different deposition methods, EBE is one of the most promising methods for making high quality thin films for photovoltaic applications because it is an efficient and reasonably cost effective method [4]. Electron beam evaporation technique is very useful owing to low consumption of material, high deposition rate and low cost of operation. The reaction was carried out in open atmosphere with one-pot by using selenium dioxide to replace selenium or its other hazardous, expensive and unstable precursors. This new precursor of selenium has been successfully implemented to develop a new aqueous method to obtain fluorescent thin films. The diffraction patterns and other morphological characterization indicate that as deposited thin films have a pure cubic cadmium selenide structure with spherical shape. Small particle size in the range 29-46 nm was achieved. The main advantage of the present thin films are that it is low cost, greener, water soluble precursors; the capping molecule itself acts as reducing agent and is easily up-scalable for larger quantity synthesis. The optical properties and morphological analysis of the CdSe were evaluated. The crystalline semiconductors are being used in electronic, optoelectronic and solar energy conversion devices very often. Notably in the field of bio photonics, long excitation wavelength and high order nonlinear absorptions are strongly preferred. So far, lot of researches has been carried out on controlling the size, shape and crystal structure as these affect their optical properties. The optical properties of the semiconductor nanomaterials can be tuned by the band gap engineering, which could be achieved by varying their shapes and size. The optical properties and morphological analysis of the CdSe were evaluated. The crystalline semiconductors are being used in electronic, optoelectronic and solar energy conversion devices very often. Notably in the field of bio photonics, long excitation wavelength and high order nonlinear absorptions are strongly preferred. So far, lot of researches has been carried out on controlling the size, shape and crystal structure as these affect their optical properties. The optical properties of the semiconductor nanomaterials can be tuned by the band gap engineering, which could be achieved by varying their shapes and size.

Low dimensional semiconductor structures, usually called nanocrystals or quantum dots possess unique features having importance in the field of science and technology [5]. During the last two decades, optical properties of semiconductor nanoparticles have been extensively studied due to their unique size-dependent properties which originate mainly from quantum confinement effect [6-11]. The electronic properties of solids are determined by occupation of the bands and by the absolute values of the forbidden gap between the completely occupied and the partly unoccupied or the empty bands. If all the bands at T=0 are either occupied or completely free, material will show dielectric properties. The highest occupied band is called valence band and the lowest unoccupied band is called conduction band.

Optical nonlinearities in bulk semiconductors have been extensively investigated for potential applications in photonics. In recent years, there has been intense research on the nonlinear optical properties of nanometer-sized semiconductors and fabrication techniques for these small particles. A large enhancement of the nonlinear coefficient in the semiconductor crystallites is redicted by theoretical considerations that are based on the quantum size effects of the carriers in the crystallites. In the nano-regime, quantum confinement produces exciton resonances those are sharper than the corresponding ones in the bulk semiconductors, resulting in large optical nonlinearities.

Many studies were devoted to CdSe QDs due to their excellent optical properties, narrow band gap and a variety of optoelectronic conversion properties compared to bulk CdSe [12,13]. In light of this and UV-visible absorption has been one of the most important measurements to investigate the optical properties of CdSe QDs [14]. In addition, considerable progress has been made in the synthesis of CdSe QDs to produce CdSe QDs with excellent optical properties [5,15]. The process needs to be operated in nitrogen atmosphere. However, little

*Corresponding author: Chandramohan R, Research Department of Physics, Sree Sevugan Annamali College, Devakottai-630303, India
E-mail: rathinam.chandramohan@gmail.com

knowledge has been obtained for the formation mechanism of CdSe QDs using this technique [6]. Furthermore, the cost of large-scale synthesis of CdSe QDs is very high for such expensive TOP solvents. Furthermore, TOP is hazardous, unstable and not environmentally friendly. Recently, a new method has been developed for the synthesis of CdSe QDs without TOP solvents [9]. Furthermore, the process can be operated in open atmosphere [10]. In this work, the CdSe films have been prepared by EB evaporation technique at room temperature and annealed at different temperatures. The structural, optical, electrical and surface morphological properties of the films were characterized by XRD, SEM, UV-visible and FTIR measurements. The results were tabulated and discussed in detail. The Stoichiometry, surface morphology and optical property changes with temperature were discussed elaborately.

Experimental

Thin films of CdSe were prepared by EBE technique using a HINDHI-VAC vacuum unit (model: 12A4D) fitted with electron beam power supply (model: EBG-PS-3K) on glass substrates at different substrate temperatures in the range of RT to 300°C in steps of 100°C. Well degreased microscopic glass plates have been employed as the substrates in the present work. 500 mg of spectroscopically pure Cadmium selenide (99.99%) was mixed well using a pestle and mortar. The mixture was pressed into pellets by hydraulic method to get pellet with a pressure of 500 kg/cm^2, which was used as the source material for evaporation. The pellet was taken in a graphite crucible and kept in water cooled copper hearth of the electron gun. The pelletized CdSe targets were heated by means of an electron beam collimated from the dc heated tungsten filament cathode. The surface of the CdSe pellet was bombarded by 180° deflected electron beam with an accelerating voltage of 5 kV and a power density of~1.5 kW/cm^2. The evaporated species from CdSe pellet were deposited as thin films on the substrates in a pressure of about 1×10^{-5} mbar (or the films were deposited with 5 kV and 10 mA under a vacuum of 10^{-6} Torr). Each substrate was placed normal to the line of sight from the evaporation source at a polar angle to avoid shadow effects and also to obtain uniform deposition. The different preparation parameters such as source to substrate distance (15 cm) and partial pressure (0.5×10^{-5} millibar), have been varied and optimized for depositing uniform, well adherent and transparent films. The rate of evaporation (0.5 nm/s) was used to deposit all CdSe films. CdSe films were prepared at room temperature (RT), and annealed at different temperatures to study the effect of Oxygen Pressure on the structural, optical, and morphological properties. The Film thickness was measured by the stylus Profilometer (Mitutoyo). The films were characterized by X-ray diffraction (XRD) studies using CuKα radiation form an X'pert Pro PANanalytical XRD unit. Optical studies were made at room temperature using a Hitatchi-330 UV-Vis-NIR spectrophotometer. The elemental composition was found using an energy-dispersive X-ray (EDX) spectrometer attached with the HITACHI Model S-3000H SEM instrument.

Results and Discussion

XRD analysis

The XRD analysis was carried out to study the temperature effect on the material properties of CdSe films. The thickness of the films was found to be in the range of 250-270 nm. The Grain sizes, dislocation density, strain and lattice constant of CdSe films have been calculated and their variations could be viewed from their graph. The Figure 1 showed the XRD patterns of CdSe thin films deposited at RT and that films were annealed at different temperatures. A strong peak was

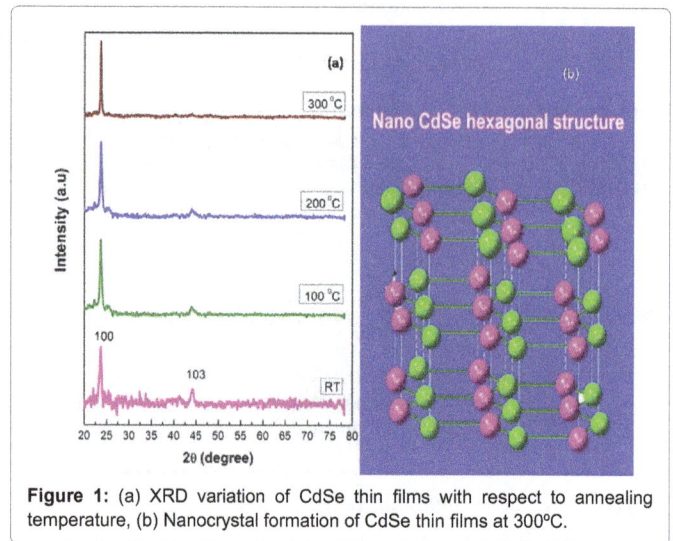

Figure 1: (a) XRD variation of CdSe thin films with respect to annealing temperature, (b) Nanocrystal formation of CdSe thin films at 300°C.

observed at 25.3° (2θ) corresponding to (100). Both the signals confirm the CdSe structure belongs to hexagonal lattice and also validated as per the JCPDS (08-0459) data. When the annealing temperature was increased from 100° to 300°C, the peak intensity of (100) plane was increased further and attained maximum. Such results exposed the fact that, the hexagonal structure was stabilized at 300°C. The same result has been observed in the CdSe films deposited by molecular beam epitaxy technique in the earlier work [11]. From the analysis, it was also observed that, in the room temperature the CdSe film was polycrystalline nature.

The Figure 2 showed the possible nanocrystals formation of the CdSe compound which was analysed by using the Gaussian software. As per the XRD spectral data at all noted temperatures, the title compound belongs to hexagonal lattice. In this hexagonal frame, it was observed that, the Cd and Se atoms were situated alternatively and formed the hexagonal ring. Thus the entire structure was constructed during the thin film coating. The possible formation of the hexagonal form was shown in Figure 2. Normally, the CdSe possess hexagonal primitive structure; in this case the S was replaced by Se and making CdSe. So, the hexagonal structure was not changed by the replacement of the Se. Usually, the condition of the hexagonal structure is a=b≠c (a=0.434 Å=b and c=0.699 Å), it can be concluded that the present compound possess hexagonal primitive structure. Further, the crystal structure of CdSe does not get altered after the annealing of thin films. Hence it is concluded that CdSe atoms are deposited on the surface of glass substrate loading stress causes the observed changes in the lattice parameters 'a' and 'c'. The average crystallite size from (100) hexagonal peak is calculated using Scherer's relation, which is found to be in the range of 29-46 nm.

Lattice parameter analysis

Figures 3a-3d showed the variation of crystal size, dislocation density and strain of CdSe film deposited at three different annealing temperatures. The crystal size was calculated using Scherrer's equation.

$$D = \frac{0.94\lambda}{\beta\cos\theta} \tag{1}$$

Where, D is the grain size, λ is the wavelength of X-rays (Cu Kα radiation – 1.5406 Å), β is the full width at half maximum and θ is the Bragg angle. The crystal size of the film at RT, was found to be 29 nm

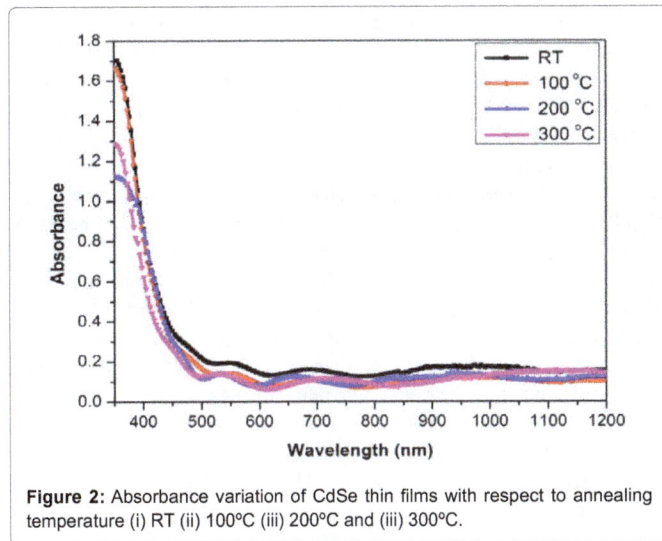

Figure 2: Absorbance variation of CdSe thin films with respect to annealing temperature (i) RT (ii) 100°C (iii) 200°C and (iii) 300°C.

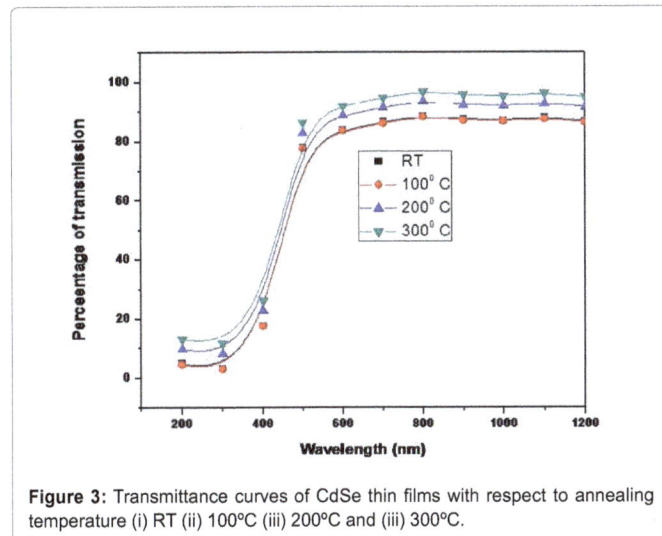

Figure 3: Transmittance curves of CdSe thin films with respect to annealing temperature (i) RT (ii) 100°C (iii) 200°C and (iii) 300°C.

whereas at 300°C, it was identified to be 46 nm. This view indicated that, the crystal size of CdSe was increased with respect to temperature. Thus, the crystal size of CdSe can be tuned by varying the substrate temperature. The values of dislocation density and lattice strain were calculated using the following relations,

Dislocation density $(\delta) = 15\beta\cos\theta/4aD$ (2)

lattice strain $(\varepsilon) = \beta\cos\theta/4$ (3)

The dislocation density and strain values are decreased when the temperature is increased which was tabulated. Generally, when the dislocation density of the film was decreased the crystal size increases with respect to annealing temperature [16].

The Strain is inherent and natural mechanism of annealed materials of thin films was obtained. Due to the large number of grain boundaries and the concomitant short distance between the grains, the intrinsic strain associated with interface of the lattice are always present in thin films. Moreover, the increasing surface energy contributes the variation of magnitude of strain. Similar results have been observed in this case and it was decreased with the increasing of annealing temperature. It was evident that, when the annealing temperature

increased to 300ºC, the strain in the lattice utilized for the completion of hexagonal structure. The lattice parameters values of the hexagonal structure (planes) were calculated from the following equation;

$$\frac{1}{d_{hkl}^2} = \frac{4}{3}\left[\frac{h^2 + hk + k^2}{a^2}\right] + \frac{l^2}{c^2} \qquad (4)$$

Normally, the strain creates local deviation of lattice constants from its bulk value which is size dependent [17]. From the tabulated result of the lattice parameters, it was observed that the lattice parameter values were very close to the standard values and these values were fluctuated slightly due to the stress movement along the lattice. Generally the CdSe crystal has Zinc blende (Cubic) structure of space group F4̄3m or Wurtzite (Hexagonal) structure of space group P6₃mc. The XRD patterns of CdSe thin films on glass substrate at RT, 100, 200 and 300°C deposition temperature by electron beam evaporation technique with thickness in the range of 300 nm shown in Figure 1 clearly indicates the increase of crystallinity of the film with the increasing of deposition temperature. A sharp peak is observed mostly in all films close to 2θ at 25.3° corresponding to (100) plane of the hexagonal phase (JCPDS 08–0459). The CdSe films deposited by Electron Beam Evaporation (EBE) technique at various substrate temperatures ranging from room temperature to 300°C showed single peak corresponding to highly oriented hexagonal structure of polycrystalline nature. The effect of substrate temperature on the microstructure of CdSe films is summarized in Table 1. It is observed that the grain size of CdSe can be tuned between 29 nm to 46 nm by varying the substrate temperature. Grain size was calculated using Debye-Scherrer's equation (1).

Transmittance studies

The transmission spectra of CdSe thin films were recorded in the wavelength range 450 nm to 2500 nm at different substrate temperatures and were shown in Figure 4. Usually, the Transmittance spectra have given the information regarding the optical features, band structure and the result of deposition temperatures of semiconductor films. The absorption co-efficient (α) and direct or in-direct band gap were estimated from the transmission spectra. In order to find out the nature of the band gap (direct or in-direct band gap) of the CdSe films, α and hν values were used to fit-in with the following equation for finding direct band gap,

$$(\alpha h\nu) = A (h\nu - E_g)^{1/2} \qquad (5)$$

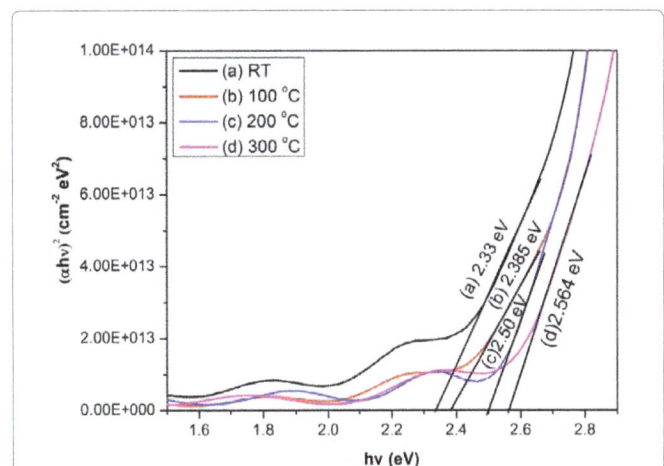

Figure 4: Bandgap variation of CdSe thin films with respect to annealing temperature (i) RT (ii) 100°C (iii) 200°C and (iii) 300°C.

Where, E_g is the band gap, α is the absorption coefficient, A is the constant and hv is the photon energy. The plots of $(\alpha hv)^2$ versus (hv) were shown in Figures 5a-5d. The band gap values can be found on straight line portions which cut the hv axis (x-axis) on extrapolation. From the graph, it was observed that, the presence of straight line portions in the high energy region confirms the direct band gap nature of the CdSe films. The calculated E_g values were 2.33, 2.385, 2.50 and 2.564 eV of the CdSe films deposited on glass substrates at RT, 100, 200 and 300°C respectively. From these values order it was observed that, the E_g value decreases with increasing of substrate temperature.

For direct band gap, where E_g is the bandgap of the CdSe films, α is the absorption coefficient, A is the constant and hv is the photon energy. The $(\alpha hv)^2$ versus (hv) plots Figure 3c, for all the films deposited on glass substrates, show straight line portions which cut the hv axis (x-axis) on extrapolation giving the band gap values. It is observed that all graphs for the films deposited at different substrate temperatures have straight line portions in the high energy region which confirms the direct band gap nature for all the CdSe films prepared by EB evaporation technique here. The values of E_g are listed in Table 1. It is observed that the E_g value increased with increasing substrate temperature. Which, in turn, depend on the increase of grain/particle size of the CdSe films with increasing substrate temperature, such observations have been reported for vacuum evaporated films. When the substrate temperature increases the band gap value of the CdSe films decreased. Such an observation has been reported for vacuum evaporated films [18]. When the substrate temperature was increased from RT to 200°C, a band gap variation in the range of 2.33-2.564 eV was observed for the vacuum evaporated CdSe films [19]. A decrement in band gap from 2.33 to 2.564 eV was observed when the substrate temperature was raised for the chemically deposited CdSe thin films [20,21]. After the deep screening of the literature, it was found that, the variation of the band gap is inversely proportional to the substrate temperature. The decrement of the band gap is also favour for the electrical conductivity of the CdSe film. Whenever the electrical conductivity of the material increases, the resistivity of the same is decreased. So it can be concluded that, the conductivity and resistivity of the material can be controlled by the substrate temperature [22-24].

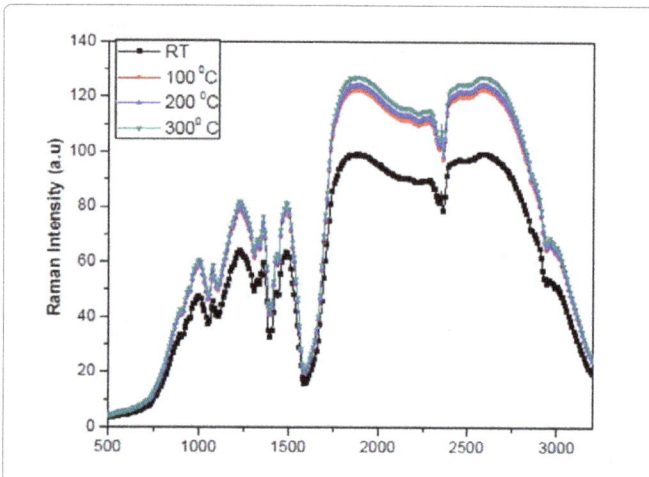

Figure 5: FTIR analysis of CdSe thin films with respect to annealing temperature (i) RT (ii) 100°C (iii) 200°C and (iii) 300°C.

The electrons are excited from the CdSe valence band to conduction band by absorbing the energy equal to or greater than their bandgap

Samples	d-spacing	2 theta	JCPDS 08-0459 A°		JCPDS Lattice parameter r A°		Plane	Crystal size (nm)	Band Gap (eV)
RT	3.72	23.901	4.299	7.01	4.34	6.99	100	29	2.17
100°C	3.72	23.901			4.37	6.99	100	32	2.221
200°C	3.72	23.901			4.34	6.98	100	38	2.227
300°C	3.72	23.901			4.35	6.97	100	46	2.292

Table 1: Structural and optical band gap values of CdSe thin films at (i) RT (ii) 100°C (iii) 200°C and (iv) 300°C.

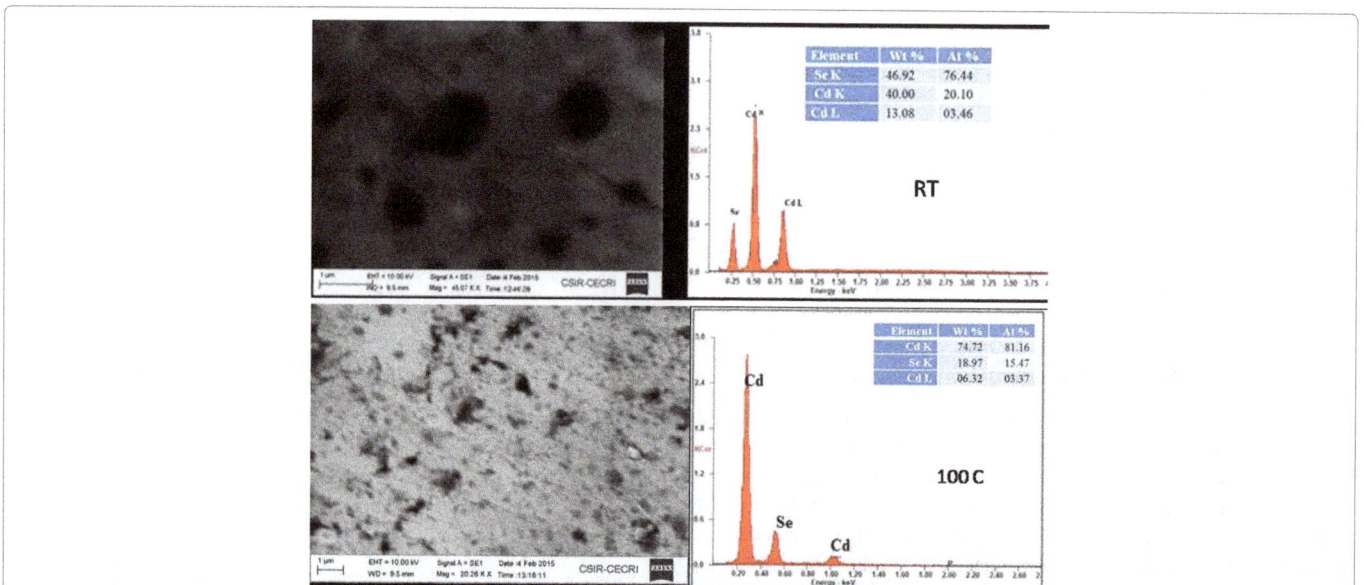

Figure 6: Scanning Electron Microscope and EDAX studies of CdSe thin films with respect to annealing temperature (a) RT (b) 100°C.

energy. Subsequent relaxation of these photo-excited electrons to some surface states or levels is followed by radiative decay enabling the luminescence invisible region [25,26].

FT-IR Analysis

The present compound was made up two atoms such as Cadmium and Selenide and its FTIR spectra are shown in Figures 6a-6d. Both the atoms were bonded and form CdSe molecule. In a crystal, there were number of molecules. When the energy was applied to the molecule whether in bulk or nano, the Cd-Se molecule executes the vibrational motion two three forms. Such as stretching and bending vibrations. The bending vibration was further can be divided in to two vibrations such as in plane and out of plane bending vibration. The entire molecule was vibrated according to their degrees of freedom. In this present case, the Cd-Se can execute only one Cd-Se stretching mode. Normally, the metal atoms stretching vibrations are observed in the region of 500-3000 cm^{-1} [27,28]. The films thickness was in the range of 400-500 nm with a growing time of 1 hr. The material obtained was characterized by optical absorption, SEM (Scanning Electron microscope), Energy Dispersive X-ray analysis (EDAX) and X-ray diffraction (XRD). XRD analysis reveals that, as deposited films are polycrystalline in nature with the orientation in the (100) direction. The films revealed the homogeneous and uniform film growth in the RT & 300°C regions without pinholes. No perceptible cracks are there and the films are well covered all over the substrate [29]. At 300°C, the films show few discontinuities. The grains are of nano size and distributed all over the surface in all the films. The presence of Cd and Se are determined by EDAX spectrum [30].

Acknowledgement

Partial Financial support by granting Major Research Project by UGC, New Delhi is gratefully acknowledged by authors RCM and TAV.

References

1. Lokhande CD, Ubale AU, Patil PS (1997) Thickness dependent properties of chemically deposited Bi2S3 thin films. Thin Solid Films 302: 1-4.

2. Hodes G (2003) Chemical solution deposition of semiconductor films. CRC Press, USA.

3. Hankare PP, Chate PA, Sathe DJ, Jadhav BV (2010) Investigation of ordering and disordering in B-doped Ni76Al24 by residual resistometry. J Alloys Compd.

4. Sulthan Kissinger M, Jayachandran K, Sanjeeviraja P (2008) Structural and optical properties of electron beam evaporated CdSe thin films. Bulletin of material science 30: 547- 551.

5. Bullen CR, Mulvaney P (2004) Metal–organic polyhedra-coated Si nanowires for the sensitive detection of trace explosives. Nano Lett 44: 2303.

6. Zhengtao D, Li C, Fangqiong T, Bingsuo Z (2005) A new route to zinc-blende cdse nanocrystals: mechanism and synthesis. J Phys Chem B 109: 16671-16675.

7. Yordanov GG, Yoshimura H, Dushkin CD (2008) Fine control of the growth and optical properties of CdSe quantum dots by varying the amount of stearic acid in a liquid paraffin matrix. Coll Surf A: Physicochem Eng Aspects 322: 177-182.

8. Mao H, Chen J, Wang J, Li Z, Dai N, et al. (2005) Photoluminescence investigation of CdSe quantum dots and the surface state effect. Physica E 27: 124-128.

9. Aichele T, Robin IC, Bougerol C, André R, Tatarenko S, et al. (2007) Structural and optical properties of CdSe quantum dots induced by amorphous Se. J Cryst Growth 301: 281.

10. Rosenthal SJ, Mcbride J, Pennycook SJ, Feldman LC (2007) Synthesis, surface studies, composition and structural characterization of CdSe, core/shell and biologically active nanocrystals. Surf Sci Rep 62: 111-157.

11. Hyugaji M, Miura T (1985) Dynamic mechanism of fast switching in light valve using liquid crystal. Jpn J Appl Phys 24: 950.

12. Peng XG, Wickham J, Alivisatos AP (1998) Optical activity and optical anisotropy in photomechanical crystals of chiral salicylidenephenylethylamines. J Am Chem Soc 120: 5343.

13. Zhou XP, Shao Z, Kobayashi Y, Wang X, Ohuchi N, et al. (2000) ZnO growth on Si by radical source MBE. J Crystal Growth 214: 52.

14. Zhu JJ, Palchik O, Chen S, Gedanken A (2000) Magnesium K-Edge NEXAFS spectroscopy of chlorophyll a in solution. J Phys Chem B 104: 7344.

15. Hambrock J, Birkner A, Fischer RA (2001) Synthesis of CdSe nanoparticles using various organometallic cadmium precursors. J Mater Chem 11: 3197-3201.

16. Edelestein AS, Camarata RC (1998) Nanomaterials synthesis properties and application. CRC Press, USA.

17. Nirmal M, Norris DJ, Kuno M, Bawendi MG, Efros AL, et al. (1995) Observation of the "Dark Exciton" in CdSe quantum dots. Phys Rev Lett 75: 3728.

18. Shallan MS, Muller R (1990) Coupling between planes and chains in YBa2Cu3O7: a possible solution for the order parameter controversy. Solar Cells 28: 185.

19. Pathinetan Padiyan D, Marikani A, Murali KR (2002) Influence of thickness and substrate temperature on electrical and photoelectrical properties of vacuum-deposited CdSe thin films. Mat Chem Phys 78: 51-58.

20. Kale RB, Lokhande CD (2000) Thickness-dependent properties of chemically deposited CdSe thin films. Mat Chem Phys 62: 103-108.

21. Sarangi SN, Sahu SN (2004) CdSe nanocrystalline thin films: composition, structure and optical properties. Physica E 23: 159-167.

22. Datta J, Bhattacharya C, Bandyopathyay S (2006) Synthesis and characterization of electro-crystallized Cd–Sn–Se semiconductor films for application in non-aqueous photoelectrochemical solar cells. Appl Surf Science 252: 7493-7502.

23. Anonymous (1999) Nanoscope Command Reference Manual. Digital Instruments. Santa Barabra, CA, USA.

24. Luo M, Jiang Y, Xu C, Yang X, Burger A, et al. (2006) Optical and electrical characterization of cadmium selenide doped with cobalt. J Phys Chem Solids 67: 2596-2602.

25. Chaure S, Chaure NB, Pandey RK (2005) Self-assembled nanocrystalline CdSe thin films. Physica E 28: 439-446.

26. Kaviyarasu K, Ayeshamariam A, Manikandan E, Kennedy J, Ladchumananandasivam R, et al. (2016) Characterization of GaAs1−ySby grown by molecular beam epitaxy. Materials Science and Engineering B 9: 210.

27. Liu Y, Claus RO (1997) Blue light emitting nanosized TiO2 colloids. J Am Chem Soc 119: 5273-5274.

28. Thompson TL, Yates JT (2006) Surface science studies of the photoactivation of TiO2 new photochemical processes. Chem Rev 106: 4428-4453.

29. Kalbacova M, Macak J, Schmidt-Stein F, Mierke C, Schmuki P (2008) TiO2 nanotubes: photocatalyst for cancer cell killing. Physica Status Solidi (RRL)-Rapid Research Letters 2: 194-196.

30. Irimpan L, Krishnan B, Nampoori V, Radhakrishnan P (2008) Luminescence tuning and enhanced nonlinear optical properties of nanocomposites of ZnO–TiO2. J Colloid Interface Sci 324: 99-104.

11

Analytical and Numerical Analysis of Functionally Graded Heat Conduction Based on Dirichlet Boundary Conditions

Essa S*

Department of Civil Engineering, Erbil Technical Engineering College, Erbil, Iraq

Abstract

An analytical and numerical solution for the one dimensional of heat conduction in a slab exposed to different temperature at both ends is presented. The distribution of heat throughout the transient direction obeys to functionally graded (FG) temperature based on Dirichlet boundary conditions. The variation of functionally graded temperature can be described by any form of continuous function. In this case, where the external heat fluxes are not directly definite based on the Dirichlet or mixed boundary conditions, the fluxes that concluded over the slab faces are free to vary until the equilibrium condition is reached. By numerically solving the resulting heat-conduction equation, the distribution of temperature which vary with time through the slab is obtained. The obtained analytical results are presented graphically and the influence of the gradient variation of the temperature on shape formed with changed time is investigated.

Keywords: Functionally graded temperature; Heat conduction; Heat flux; Crank Nicolson

Introduction

Functionally graded materials (FGMs) are defined as the perfect materials in mechanical, thermal and corrosive resistant properties. Different from fiber-matrix laminated composites, FGMs do not have the problems of de-bonding resulting from large inter-laminar and thermal stresses. The impression of FGMs was firstly introduced by Japanese researchers in the mid-1980s, as ultra-high temperature resistant materials for various engineering fields for instance aircraft, space vehicles, and nuclear reactors. FGMs consist of two or more materials which microscopically inhomogeneous and spatial composite materials such as a pair of ceramic–metal. The mechanical properties of the material structure changes gradually throughout the thickness with varying continuously and smoothly from top to the bottom surface. Noda [1] showed many topics range from thermoelastic to thermoinelastic problems. He suggested that temperature dependent properties of the material should be taken into account in order to achieve more accurate analysis.

For a historical era, FGMs studying focused on the analyses of thermal stress in the ceramic coatings, static deformation, and forced vibration. Noda and Jin [2,3] presented a steady thermal stress for a crack elastic solid based on nonhomogeneous infinite, and concluded that effects of the thermal stress intensity for cracks in FGMs. Cho and Oden [4] used a Crank–Nicolson and Galerkin scheme to investigate a parametric study of thermal stress characteristics. Reddy [5] proposed a theoretical formulation and finite element models for the analysis of functionally graded plates (FGPs). Praveen and Reddy [6] studied responses of FGPs based on static and dynamic thermoelastic responses and concluded that, the differences of pure ceramic or metal plates depend on responses of FGPs. Yang and Shen [7] presented a free and forced vibration analyses of functionally graded plates subjected to impulsive lateral loads under thermal environments. Dirichlet boundary conditions (DBCs) generally hold two forms: homogeneous Dirichlet boundary conditions (HDBCs) and inhomogeneous Dirichlet boundary conditions (IDBCs). The former can be considered as a special case of the latter with zero imposed value. Zhang and Zhao [8] developed a weighted finite cell method (FCM) with high computing accuracy which extended to define boundary value function so that the inhomogeneous Dirichlet boundary conditions (IDBCs) are imposed exactly.

Theoretical Formulation

There are many models for expressing the variation of material properties in FGMs. The most commonly used of these models is the power law distribution. In this study, the new expression of heat transfer through in slab is assumed based on two parameters as:

$$\Phi(x) = \Phi_0 \left[\frac{3}{2}(x+0.5)^2 - \left(mx^2 - 0.5e^{kx} \right) \right] \tag{1}$$

Where, m and k are the parameters which are used to define the variation. ϕ_0 is a constant and related to the value of the variation function $\phi(x)$ at the left surface of the slab $(x=a)$ by:

$$\Phi_0 = \varphi_0 \frac{1}{0.375 + 0.5e^{ka} + a\left(1.5 + (1.5-m)a\right)} \tag{2}$$

Here, $\varphi_0 = \phi(x)$ is initial value of the variation at $x=a$ Eq. (1) is a nonlinear function and its variation (shape) is controlled by using two parameters. One may observe that, the adjustment of the parameters m and k is not easy for describing the desired variations. Figure 1 shows the variation of temperature of metal in the transverse direction of slab.

Exact Solution of the Heat Equation

Consider a one-dimensional diffusion equation which is a partial differential equation for the temperature $T(x,t)$:

$$c(x)\frac{\partial T}{\partial t} = \frac{\partial}{\partial x}\left(\kappa(x)\frac{\partial T}{\partial x} \right) + S(x,t) \tag{3}$$

Where, $c(x)$ is the specific heat of the material, κ is the constant of proportionality (thermal conductivity) of the material and $S(x,t)$ represents a given source of heat energy per unit volume. The equation simplifies when κ and c are independent of position:

*Corresponding author: Essa S, Department of Civil Engineering, Erbil Technical Engineering College, Erbil, Iraq, E-mail: saad_khalis@epu.edu.krd

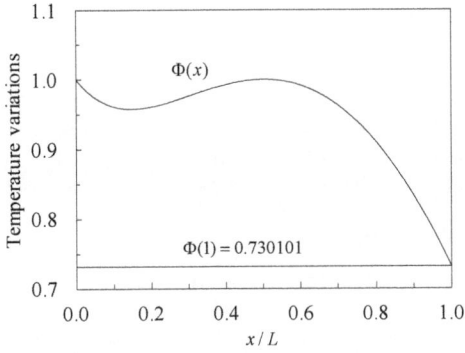

Figure 1: Temperature variation along the transverse position.

$$\frac{\partial T}{\partial t} = \chi \frac{\partial^2 T}{\partial x^2} + \frac{S(x,t)}{c} \qquad (4)$$

Where, χ is the material thermal diffusivity,

$$\chi = \frac{\kappa}{c} \qquad (5)$$

The initial temperature in the slab heat equation is a first-order PDE in time:

$$T(x,0) = T_0(x) \qquad (6)$$

The second order equation in space requires two boundary conditions to satisfy the solution. By putting the faces in a suitable thermal contact at a specified temperature $T_1(t)$ and $T_2(t)$ so the Dirichlet boundary conditions are achieved as:

$$T(0,t) = T_1(t)$$
$$T(L,t) = T_2(t) \qquad (7)$$

The equilibrium solution $\phi(x)$ that satisfies these boundary conditions, as well as the time-independent heat equation:

$$0 = \frac{\partial}{\partial x}\left(\kappa(x)\frac{\partial \Phi(x)}{\partial x}\right) + S(x,t) \qquad (8)$$

Using the direct integration (analytically), equation goes to general solution:

$$\Phi(x) = C_1 + \int_0^x \left(\frac{C_2 - \int_0^{x'} S(x'')dx''}{k(x')}\right)dx' \qquad (9)$$

The constants C_1 and C_2 are chosen to match the boundary conditions. For a temperature balance to exist, there can be no net heat energy infused into the section by the source: $\int_0^L S(x'')dx'' = 0$), else the temperature should increase or decrease as the overall energy content in the slab varies with time. The relation between the equilibrium temperature and the deviation from equilibrium $\Delta T(x,t)$ is:

$$T(x,t) = \Phi(x) + \Delta T(x,t) \qquad (10)$$

By Substitution Eq. (10) into the heat equation Eq. (3) and application of Eq. (8) indicates that $\Delta T(x,t)$ satisfies the homogeneous heat equation:

$$c(x)\frac{\partial \Delta T}{\partial t} = \frac{\partial}{\partial x}\left(\kappa(x)\frac{\partial \Delta T}{\partial x}\right) \qquad (11)$$

With homogeneous boundary conditions, and initial condition:

$$\Delta T(x,0) = T_0(x) - \Phi(x) \qquad (12)$$

We will only consider the case where κ and C are constants, so that Eq. (11) becomes the diffusion equation with no heat source,

$$\frac{\partial \Delta T}{\partial t} = \chi \frac{\partial^2 \Delta T}{\partial x^2} \qquad (13)$$

We look for a solution of the form $\Delta T(x,t) = f(t)\psi(x)$. Substituting this expression into Eq. (13) and dividing by $f(t)\psi(x)$ yields,

$$\frac{1}{f(t)}\frac{\partial f}{\partial t} = \frac{\chi}{\psi(x)}\frac{\partial^2 \psi}{\partial x^2} \qquad (14)$$

Which can be separated into two ODEs with respect to independent x and t. By putting of each separated equations of Eq. (14) are equal to a constant λ,

$$\frac{1}{f(t)}\frac{\partial f}{\partial t} = \lambda \qquad (15)$$

$$\frac{\chi}{\psi(x)}\frac{\partial^2 \psi}{\partial x^2} = \lambda \qquad (16)$$

After applying Dirichlet boundary conditions, solution of Eq. (16) for the eigenmodes is,

$$\psi(x) = D\sin\frac{n\pi x}{L} \qquad (17)$$

And;

$$\lambda = \lambda_n = -\chi(n\pi/L)^2, \quad n = 1,2,3,\cdots \qquad (18)$$

The solution of Eq. (15) provides the time-dependent amplitude for each eigenmode as,

$$f(t) = Ae^{\lambda t} \qquad (19)$$

Therefore, the general solution of the heat equation away from equilibrium is,

$$\Delta T(x,t) = \sum_{n=1}^{\infty} A_n e^{\lambda_n t}\sin\frac{n\pi x}{L} \qquad (20)$$

The constant D is absorbed into the Fourier coefficient A_n which can be found by matching the initial condition, Eq. (12),

$$\Delta T(x,0) = \sum_{n=1}^{\infty} A_n \sin\frac{n\pi x}{L} = T_0(x) - T_{eq}(x) \qquad (21)$$

Equation (21) is a Fourier sine series and the coefficient A_n is determined as,

$$A_n = \frac{2}{L}\int_0^L [T_0(x) - \psi(x)]\sin\frac{n\pi x}{L}dx \qquad (22)$$

Dirichlet boundary conditions in a uniform slab are applied to get equations (20) and (22) which is the solution for the deviation from equilibrium. In addition, Eq. (22) is solved numerically by using trapezoidal method which can be seen in the numerical example.

The Crank-Nicolson Method

An implicit scheme of Crank-Nicolson which is based on the central approximation of Eq. (13) at the point $\left(x_i, t_j + \frac{1}{2}\Delta t\right)$ as shown in Figure 2,

$$\frac{T_i^{j+1} - T_i^j}{2\frac{\Delta t}{2}} = \chi \frac{(T_{i+1}^{j+\frac{1}{2}} - 2T_i^{j+\frac{1}{2}} + T_{i-1}^{j+\frac{1}{2}})}{\Delta x^2} \qquad (23)$$

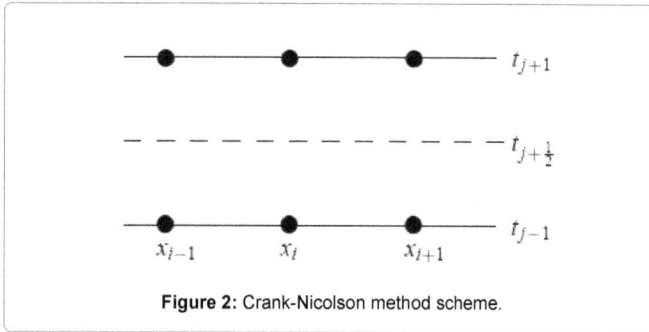

Figure 2: Crank-Nicolson method scheme.

Space derivative approximation which is used for is just an average of approximations in points (x_i, t_j) and (x_i, t_{j+1}),

$$\frac{T_i^{n+1} - T_i^n}{\Delta t} = \chi \frac{(T_{i+1}^{n+1} - 2T_i^{n+1} + T_{i-1}^{n+1}) + (T_{i+1}^n - 2T_i^n + T_{i-1}^n)}{2\Delta x^2} \quad (24)$$

Introducing $\alpha = \chi \Delta t / \Delta x^2$ one can rewrite Eq. (24) as,

$$-\alpha T_{i+1}^{j+1} + 2(1+\alpha)T_i^{j+1} - \alpha T_{i-1}^{j+1} = \alpha T_{i+1}^j + 2(1-\alpha)T_i^j + \alpha T_{i-1}^j \quad (25)$$

The terms which appear in the right-hand side of Eq. (25) are known. Hence, Eq. (25) form a tridiagonal linear system $(AT=b)$ which can be solve simultaneously to find the temperature at every node at any point in time.

The complex stability analysis will be procedure in some simple steps. The equation (24) can be written as,

$$\frac{T_i^{n+1} - T_i^n}{\Delta t} = \frac{\chi}{2\Delta x^2}\left(T_{i+1}^{n+1} - 2T_i^{n+1} + T_{i-1}^{n+1} + T_{i+1}^n - 2T_i^n + T_{i-1}^n\right) \quad (26)$$

The equation (26) has a consistency,

$$\frac{\chi}{2}\left(T_{i+1}^{n+1} - 2T_i^{n+1} + T_{i-1}^{n+1} + T_{i+1}^n - 2T_i^n + T_{i-1}^n\right) = \frac{\chi}{2}\left(\left.\frac{\partial^2 T}{\partial x^2}\right|_i^{n+1} + O\left(\Delta x^2\right) + \left.\frac{\partial^2 T}{\partial x^2}\right|_i^n + O\left(\Delta x^2\right)\right)$$
$$= \chi \left.\frac{\partial^2 T}{\partial x^2}\right|_i^{n+\frac{1}{2}} + O\left(\Delta t^2 + \Delta x^2\right) \quad (27)$$

$$\frac{T_i^{n+1} - T_i^n}{\Delta t} = \left.\frac{\partial T}{\partial t}\right|_i^{n+\frac{1}{2}} + O\left(\Delta t^2\right) \quad (28)$$

To make a stability analysis, the scheme will be written into two form stages:

$$\frac{T_i^{n+\frac{1}{2}} - T_i^n}{\frac{\Delta t}{2}} = \chi \frac{T_{j+1}^n - 2T_j^n + T_{j-1}^n}{\Delta x^2} \rightarrow Z_1(\theta) = 1 - 4\chi \frac{\Gamma}{2}\sin^2\frac{\theta}{2} \quad (29)$$

$$\frac{T_i^{n+1} - T_i^{n+\frac{1}{2}}}{\frac{\Delta t}{2}} = \chi \frac{T_{j+1}^{n+1} - 2T_j^{n+1} + T_{j-1}^{n+1}}{\Delta x^2} \rightarrow Z_2(\theta) = \frac{1}{1 + 4\chi \frac{\Gamma}{2}\sin^2\frac{\theta}{2}}$$

Where, $\Gamma = \dfrac{\Delta t}{\Delta x^2}$ and is called of grid ratio.

Therefore;

$$w^{n+\frac{1}{2}} = Z_1(\theta)w^n, \qquad w^{n+1} = Z_2(\theta)w^{n+\frac{1}{2}} = \underbrace{Z_1 Z_2}_{Z(\theta)}w^n \quad (30)$$

Eq. (30) leads to the consequence that $Z(\theta) \leq 1$, so this scheme is an unconditionally stable and $\|\varepsilon\| = O\left(\Delta x^2 + \Delta t^2\right)$ then this system has a second order of convergence. The consistency error is obtained by substituting the exact solution T in the discrete system. This means the numeric solution $T_{\Delta x}$ converges to exact solution in a given norm if $\|\varepsilon_{\Delta x}\| = \|T - T_{\Delta x}\|$ satisfies $\lim_{\Delta x \to 0}\|\varepsilon_{\Delta x}\| = 0$.

The consistency between the solutions of the continuous and discrete problems does not guarantee also have a tendency to zero. On the other hand, the difference between the differential and discrete operators on a smooth sufficient function tends to zero.

Numerical Example

In this section, the distribution of temperature throughout the transverse direction of slab is tested based on assumed temperature gradient. The properties of slab are selected as the parameters in Eq (1): $k = -4.210933$, $m = 2.743578$ and $\Phi_0 = 1.142857$ to ensure the start point at the left side equal 1.0 and the end point goes to 0.730101. Other properties, L=1.0 m, n=40, χ=1.0 m²/sec, Δt=0.0001 sec and Δx=L/n. The difference between an analytical and trapezoidal method used for integral part in Eq. (22) and the values applied to find the temperature in Eq. (10) based on time step t=0.02sec as shown in Table 1.

Figures 3-6 represent the shape of temperature distribution which taking with increasing time (t=0.02, 0.1, 0.2 and 0.4). By fixing the temperature variation based Crank-Nicolson (CN) method, the comparison with the exact is done with showing two-time step t=0.3 sec and t=0.4 sec as in Figures 7 and 8. For example, the time elapsed to get (CN) temperature variation for 0.5, 0.55, 0.575 and 0.6 are respectively: 0.535268, 0.519372, 0.566319 and 0.505062.

Conclusions

In this paper, simple method to solve the transient response problem of a functionally graded temperature variation in the 1-D

x/L	Analytical	Trapezoidal
0.0	1.000000	1.000000
0.1	0.588196	0.588655
0.2	0.294891	0.295522
0.3	0.129512	0.130024
0.4	0.0610078	0.061314
0.5	0.0489602	0.049157
0.6	0.0727702	0.073010
0.7	0.139666	0.140043
0.8	0.272146	0.272607
0.9	0.479501	0.479836
1.0	0.730101	0.730101

Table 1: Comparison between analytical and trapezoidal method for heat equation based on time step t=0.02 sec that integral appears in Eq.22.

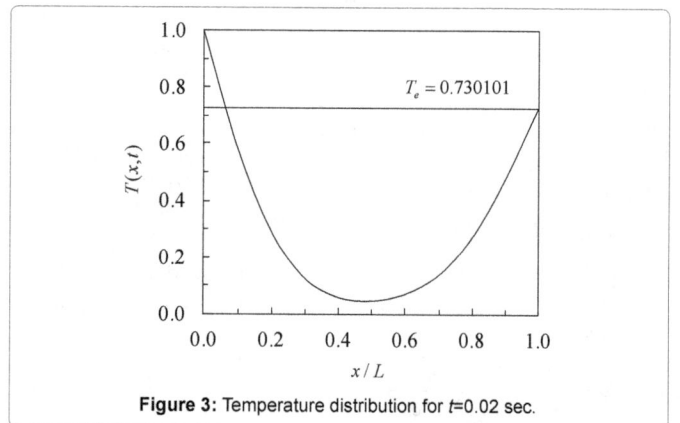

Figure 3: Temperature distribution for t=0.02 sec.

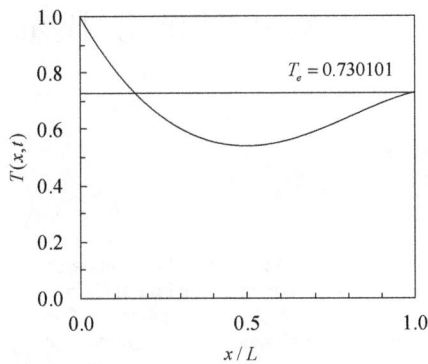

Figure 4: Temperature distribution for *t*=0.1 sec.

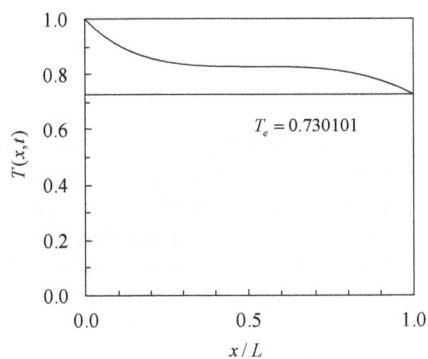

Figure 5: Temperature distribution for *t*=0.2 sec.

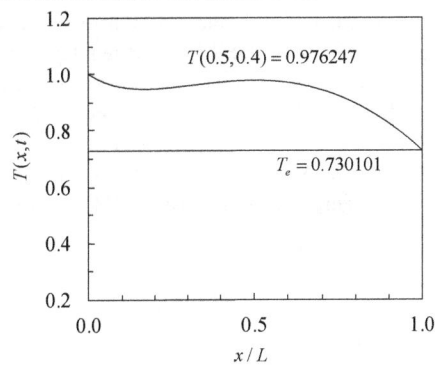

Figure 6: Temperature distribution for *t*=0.4 sec.

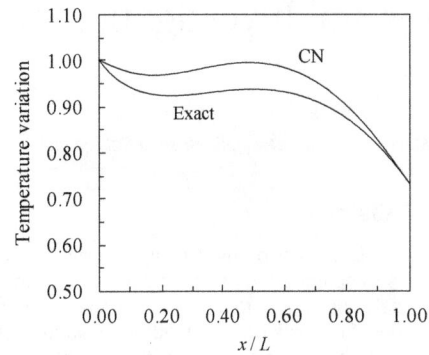

Figure 7: Temperature variation for *t*=0.3 sec based on exact and Crank Nicolson method.

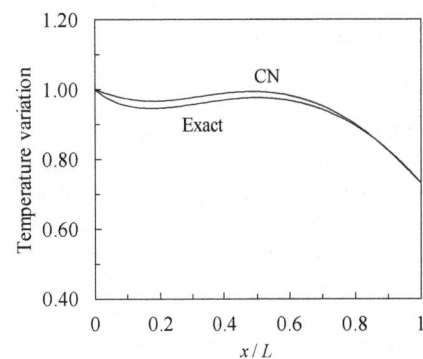

Figure 8: Temperature variation for *t*=0.4 sec based on exact and Crank Nicolson method.

the assumed model can be calculated at any position with the given Dirichlet boundary condition.

References

1. Noda N (1991) Thermal stress in materials with temperature-dependent properties. Appl Mech Rev 44: 383-397.

2. Noda N, Jin ZH (1993) Steady thermal stress in an infinite non-homogeneous elastic solid containing a crack. J Thermal Stresses 16: 181-196.

3. Noda N, Jin ZH (1993) Thermal stress intensity factors for a crack in a strip of functionally graded material. Int J Solids Struct 30: 1039-1056.

4. Cho JR, Oden JT (2000) Functionally graded material: a parametric study on thermal-stress characteristics using the Crank–Nicolson–Galerkin scheme. Comput Methods in Appl Mech Eng 188: 17-38.

5. Reddy JN (2000) Analysis of functionally graded plates. Int J Numer Meth Eng 47: 663-684.

6. Praveen GN, Reddy JN (1998) Nonlinear transient thermoelastic analysis of functionally graded ceramic–metal plates. Int J Solids Struct 35: 4457-4476.

7. Yang J, Shen HS (2002) Vibration characteristics and transient response of shear-deformable functionally graded plates in thermal environments. J Sound and Vibration 255: 579-602.

8. Zhao X, Lee YY, Liew KM (2009) Thermoelastic and vibration analysis of functionally graded cylindrical shells. Int J Mech Sci 51: 694-707.

slab has been proposed. The assumed temperature is embedded into the diffusion equation with no source of heat then exact solution is obtained. In addition, the integral part of A_n coefficient in Eq. (22) is solved numerically using trapezoidal method and the difference is exhibited in Table 1. In this table, the differences of temperature between the two methods are too small. Crank-Nicolson (CN) method is applied to solve the diffusion equation based on the given variation $\phi(x)$. The assumption of variation is selected for two variables k and m. The temperature at any point goes to take the assumed variation with time. By fixing the numerical method at the shape of variation function, the times that the exact solution is needed to get the numerical solution are computed. Using these methods, the time required to reach

12

Experimental Investigation on Behavior of Friction Materials in Brake Liners

Asvath R* and Suresh M

Department of Mechanical Engineering, Sri Krishna College of Engineering and Technology Coimbatore, India

Abstract

Brake linings have complex microstructure and consists of different components. The purpose of this research is to optimize the use of brake liner frictional materials with new formulation incorporating micaceous iron oxide and calcium carbonate as a filler material and comparison of improvement in frictional value with other filler materials. This new formulation was tested under various operating conditions with a chase machine friction testing rig. The friction test rig was programmed with different test loops to perform test as per SAE J661. The average coefficient of friction, coefficient of friction with changing speed, temperature and load, thickness loss and weight loss were accounted. Formulation exhibits better thermal stability and wear resistance to the latter.

Keywords: Brake liners; Chase machine; Filler materials; Mix formulation; Wear characteristic

Introduction

The brake friction materials play an important role in the braking system. During breaking it convert the kinetic energy to thermal energy by friction due to the braking process. In this instinct the brake liners should poses the constant coefficient of friction at any operating conditions. Apart from this, it should also possess various resistance to heat and water, low wear rate and high thermal stability, but this is impossible to achieve all this in practical conditions. For this reason, the research investigated the filler materials like calcium carbonate, micaceous iron oxide with the abrasive and binders to give a better structural reinforcement. This filler material is low in cost for the manufacturing process and only less quantity is required compared to filler quantity used earlier. Followed by, this new formulation is tested for coefficient of friction. In which various stages including baseline test, fade, and wear and recovery test are performed with friction testing machine. From the measurements, the analysis was carried out to check the behavior of friction in brake lining comparing with the regular formulation.

Related Works

Tsang studied testing of various friction materials with the aid of chase testing machine and inertia brake dynamometers testing. The results from both test method was compared which resulted inconsistent [1]. Peter surveyed the brake materials and additives used for the formulation of the commercial brake and to indicate their typical properties prepared for U.S Department of energy [2]. Sterle and Klob studied the brake friction materials and analyzed the wear the characteristics of the materials especially with organic materials in brake [3]. Herring studied and developed a mechanism for the fade in the organic brake linings, which resulted in high frictional stability of the brake liner were improved at very high speed [4]. Rukiye and Nurettin described the tribological parameter such as wear resistance and frictional stability depending upon the test temperature and number of braking involved [5,6].

Experimental Setup

Sensors, load cell, pressure controller and blower

The main drive to the test drum is provided by 10 HP motor through drive pulley and belt arrangement to give a rotational speed of 417 rpm ± 5%. The 1" square sample piece is mounted inside the drum with easily retractable arms. The 11" (279 mm) inside diameter brake drum is made of cast iron and have pearlitic grain structure. The drum will have average Brinell hardness between 179 and 229. Heating of the drum is by electrical heaters and temperature control is achieved by connecting a centrifugal blower to the control circuit. These are the setup that can be seen in the Figure 1.

An air cylinder, through a pressure regulator the vertical 150 lb load is applied. The resulting friction force and the applied vertical load are measured by load cells. The Drum temperature is measured by IR thermometer. The speed of rotation of drum is measured using rpm speed sensor. Along with a SSR (solid state relay) is provided for the entire heater coils (A1, A2 and A3), blower exhaust fan, solenoid-1 and solenoid -2 shown in the Figure 2 respectively. RS232 port inputs the values into the picKit processor with loop controlled operation, the results are manipulated. All the major parameters like applied load, friction force, Drum temperature and drum RPM etc are recorded.

Testing Program

Operating conditions

For this test desired measuring points are needed and followed by that various operating conditions are chosen according to the requirement,

1. The compressor is turned on and the pressure is set by adjusting the pressure regulator screw at around 10-12 psi, so that a load of 150 lb (68 kg +/- 10%) is applied on the sample.

*Corresponding author: Asvath R, Student, Department of Mechanical Engineering, Sri Krishna College of Engineering and Technology, Coimbatore, India, E-mail: asvathravichandran@yahoo.in

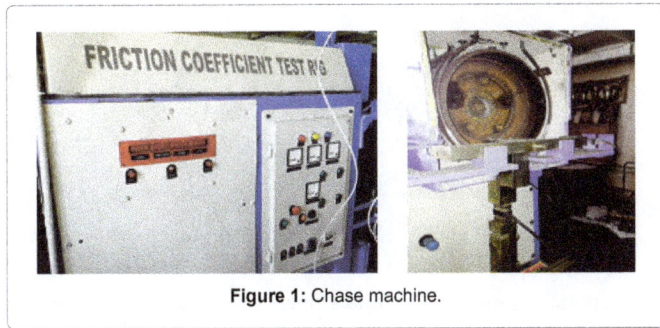
Figure 1: Chase machine.

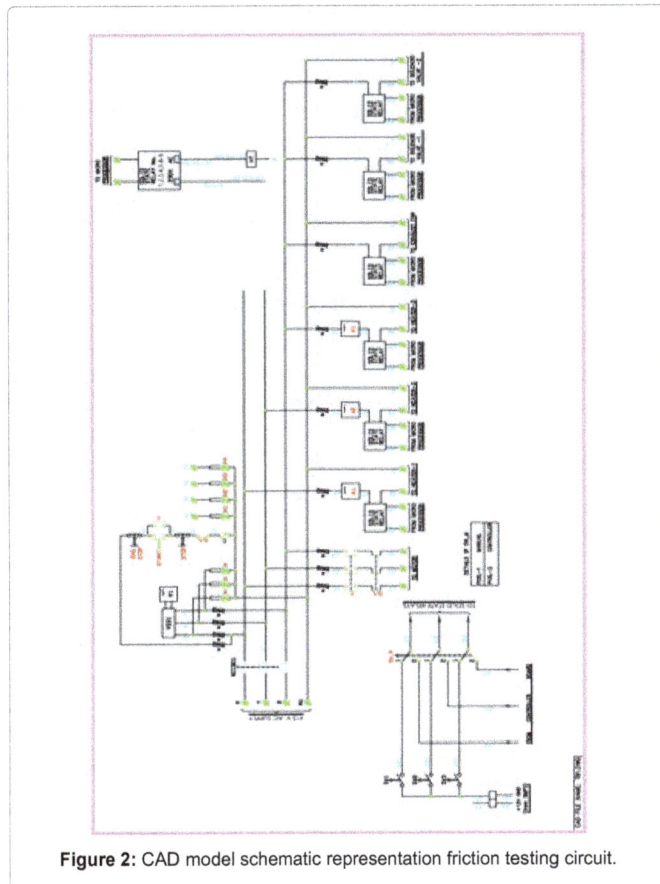
Figure 2: CAD model schematic representation friction testing circuit.

2. The motor is switched on from the control panel shown in the Figure 1 and the frequency setting on the variable frequency distributor made around 18.25 Hz. This will give 420 rev/min on the brake drum end.

Friction testing stages with description

Base line run: This stage runs continuously for 10 seconds brake ON and followed by 20 seconds brake OFF at 150 lb/sq.inch load, 417 rev/min of the test drum for 20 allocations. The test starts at 93°C (200°F) and maintains temperature of 200°F throughout the test (+/- 10%).

First fade run: After baseline run, this stage runs with continuous drag at 150 lb load with heater ON and cooler (blower) is turned OFF. The test starts the operation at 93°C and runs for either 10 minutes or until the temperature reaches 550°F. Whichever occurs first, the coefficient of friction is recorded on drum temperature at

50°F temperature interval. The time required to reach 550°F are also recorded.

First recovery run: Immediately on the completion of the first fade run, the heater turns OFF automatically and then cooler turns ON. In this instinct it records 10 seconds (load) allocations at each 37.8°C (100°F) interval during from 500°F to 200°F.

Wear measurement: Once the recovery stage gets completed, the wear stage runs for 20 sec brake ON and 10 sec brake OFF condition. A constant load of 150 lb pressure and 417 rev/min is maintained for 20 allocations. The test starts at drum temperature of 400°F (+/-5 %) and maintain the same temperature throughout.

Second fade run: On the completion of the second wear test. Cool drum to 200°F and then run continuously brake drag at 150 lb pressure 417 rev/min. With heater ON and cooling OFF, start at 200°F and run for either 10 minutes or until temperature raises to 650°F. which ever occurs first take simultaneous reading of friction and temperature at 50°F interval record time to reach 650°F.

Second recovery run: Immediately on completion of the second fade turn OFF heater and turn ON cooling, then make 10 seconds brake allocations at each 100°F interval during cooling from 600°F to 200°F repeat second base line run as first run.

Formulation Involved

To analyze the role of filler materials in coefficient of friction value improvement, two samples were prepared from formulation involved with the change in filler materials (the formulations are shown in Tables 1 and 2. The purpose of adding filler materials is to improve the coefficient of friction. The function of filler with resin will enhance

SI.No	Material	% in the formula
1.	Friction dust	8.0
2.	Barytes	7.0
3.	Resin	22.0
4.	Graphite	24.0
5.	Steel fiber	18.0
6.	Iron powder	4.0
7.	Alumina	1.0
8.	Micaceous iron oxide	8.0
9.	Marble dust	4.0
10.	Tire dust	3.0
11.	Friction dust black	1.0
	Total	100.0

Table 1: Sample A formulation.

SI.No	Material	% in the formula
1.	Friction dust	8.0
2.	Barytes	7.0
3.	Resin	22.0
4.	Graphite	24.0
5.	Steel fiber	18.0
6.	Iron powder	4.0
7.	Alumina	1.0
8.	Calcium Carbonate	8.0
9.	Marble dust	4.0
10.	Tire dust	3.0
11.	Friction dust black	1.0
	Total	100.0

Table 2: Sample B formulation.

the structure reinforcement of the brake lining composition. Barium Sulphate (barytes) is used as additional filler materials in both the formulation. As the micaceous iron oxide are basically fine particles with high surface area. An additional quantity of resin was added to produce perfect bond. In the first sample micaceous iron oxide is incorporated and in the latter with calcium carbonate respectively. These samples were made in size of 1" square piece and tested shown in Figure 3.

Results and Analysis

Composite materials are tested under various operating cycles. From the results it is inferred that in sample B the coefficient of friction is not stable at high temperature compared to the sample A. From the fade and recovery stage it clearly shows that micaceous iron oxide properties reveal better wear resistance. The coefficient of friction is 7% increase in this case compared to the calcium carbonate added sample. However there are some problems when adding conventional micaceous iron oxide caused mainly by variations in quality, which lead to unreliable results. Noise levels were reduced during test cycles performed in sample A. The commercial brake liner friction coefficient is about 0.55 and less. With the addition of appropriate filler additive friction materials, the new formulation improvement of coefficient of friction is around 9% to the commercial brake liners. The weight loss and thickness loss calculated for both the samples, the wear rate is high for the sample B and in the case of sample A it is less due to density is higher in number. The availability of calcium carbonate is better alternative to the barytes, which is less expensive. In order to improve the formulation still higher, the organic materials added should be reduced. Overall this formulation has significantly improved

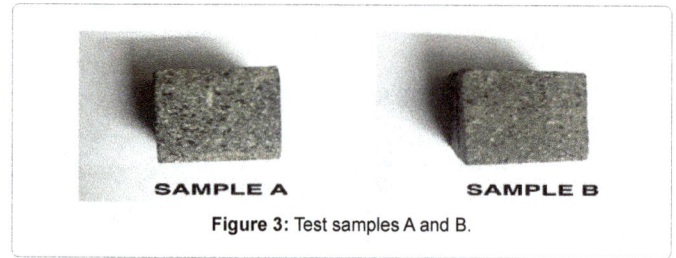

Figure 3: Test samples A and B.

the important properties of brake liner friction materials like heat dissipation, better reinforcement of the composition and also improved the coefficient of friction with reduced noise level.

In Figure 4, At the ambient temperature the baseline stage however shows low coefficient of friction but after 6 applications. Both the samples coefficient of friction increases consistently. Then at 17th application sample B falls down in coefficient of friction whereas sample A the coefficient of friction value is quite constant

In the Table 3 the baseline 1 and 2 throughout the 20 allocations of constant loading and maintain a temperature of 200°F. The sample A in both the cases show higher friction coefficient than the sample B. Also the value drops at last 5 applications for sample B but the former one is gradually increasing.

From the Figure 5, both the samples curve steeply increases. It can be inferred that material behaves good consistently up to 16 applications. But sample A there is no drop in consistency, it peaks upto the scale.

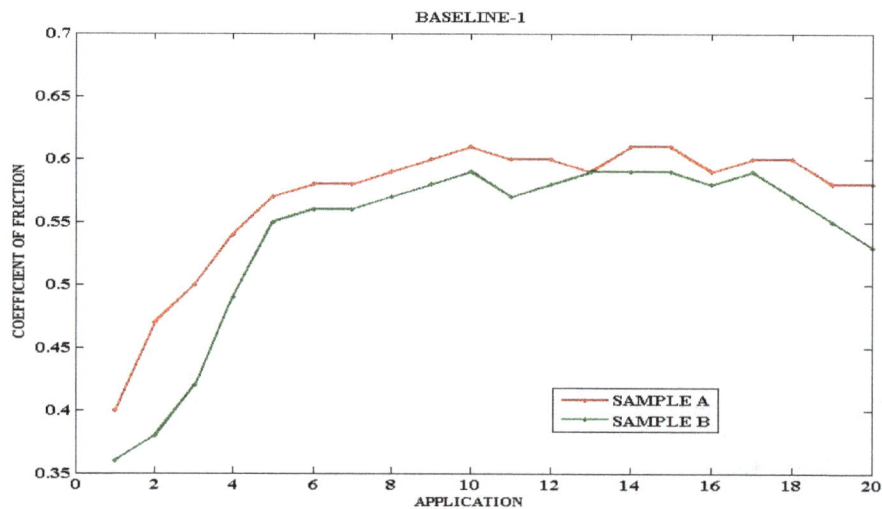

Figure 4: Plot comparison between the samples in baseline-1.

	Baseline-1						Baseline-2					
	Temp (°F)		Force-F		μ		Temp (°F)		Force-F		μ	
	Samples								Samples			
	A	B	A	B	A	B	A	B	A	B	A	B
1	213.2	215.8	2.0	0.9	0.40	0.36	210.2	209.5	16.0	14.8	0.45	0.43
5	205.4	203.5	18.0	17.2	0.54	0.54	208.3	195.6	17.5	17.6	0.56	0.53
10	218.4	216.3	17.5	16.7	0.61	0.59	215.7	204.8	17.5	16.5	0.55	0.51
15	210.5	207.5	18.5	18.7	0.61	0.59	209.4	223.5	17.5	17.4	0.59	0.58
20	208.1	205.1	19.0	18.3	0.58	0.53	212.6	219.6	17.7	17.6	0.61	0.56

Table 3: Baseline stage 1 and 2.

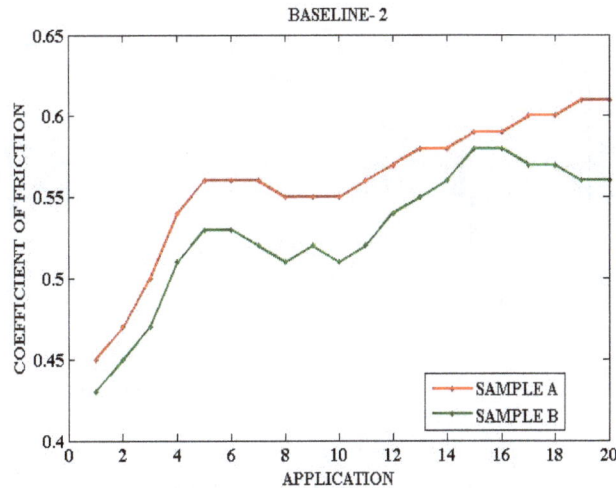

Figure 5: Plot comparison between the samples in baseline-2.

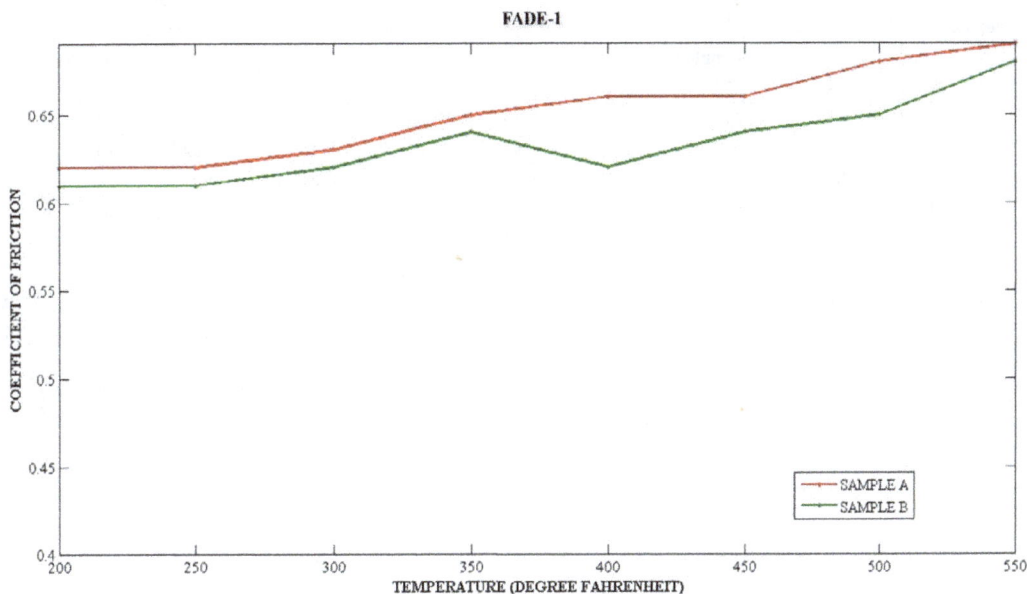

Figure 6: Plot comparison between the samples in fade-1.

It is evident from Figure 6, the coefficient of friction is consistent and steps up considerably. But it is inferred that sample B drops in coefficient of friction after 350°F whereas sample A shows consistent progress, actually the curve gets increasing in coefficient of friction after 400°F.

The transient temperature behavior of both sample were significantly same at 390°F. But just in this case, it falls down and somehow the physical property of material is good at higher range of temperature for sample B. The former performs consistent than latter shown in Figure 7.

The temperature is consistently increased 50°F for every loading condition and time is recorded for each interval temperature reaching time. It is inferred that sample A that frictional value increases at a gradual scale, however sample B drops at intermediate temperature.

At higher temperature both samples result in better frictional values shown in the Table 4.

From the Figure 8, it clearly depicts that sample A friction characteristic behavior changes at 400°F. In this recovery stage the sample B deviates a lot in coefficient of friction with changing temperature. In Figure 9, both the samples correspond to same coefficient of friction at the intermediate temperature. But the sample A is strictly linear and sample B drops at 300°F, then it regains its coefficient of friction after 400°F. This sample B shows higher wear rate but it shows quite better wear rate at higher range of temperature. From the Table 5 it clearly depicts that the frictional values gradually drops down for both the samples at a considerable rate but the sample A recovers earlier corresponding to the sample B.

Wear characteristics of both the samples show similar result shown

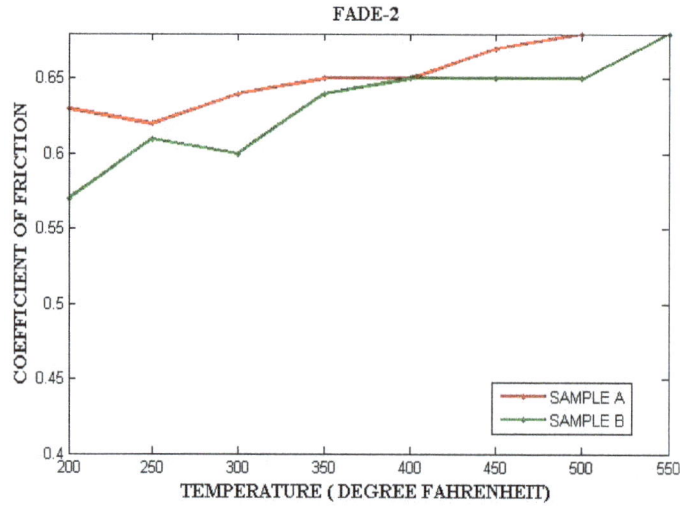

Figure 7: Plot comparison between the samples in fade-2.

Temp (° F)	Fade-1						Fade-2					
	Time(sec)		Force-F		μ		Time(sec)		Force-F		μ	
	Samples						Samples					
	A	B	A	B	A	B	A	B	A	B	A	B
200	0	0	19.4	19.2	0.62	0.61	0	0	19.5	19.4	0.63	0.57
250	15	15	19.2	19.2	0.62	0.61	17	16	19.3	19.0	0.62	0.61
300	30	30	19.2	19.0	0.63	0.62	32	31	19.3	19.4	0.64	0.60
350	52	49	19.3	18.8	0.65	0.64	49	50	19.6	19.6	0.65	0.64
400	70	73	19.4	19.0	0.66	0.62	74	75	19.2	19.3	0.65	0.65
450	107	104	19.6	19.2	0.66	0.64	103	107	19.6	19.2	0.67	0.65
500	160	158	19.5	19.3	0.68	0.65	158	159	19.2	19.6	0.68	0.65
550	281	279	19.8	20.0	0.69	0.68	280	268	20.0	20.0	0.68	0.68

Table 4: Fade stages 1 and 2.

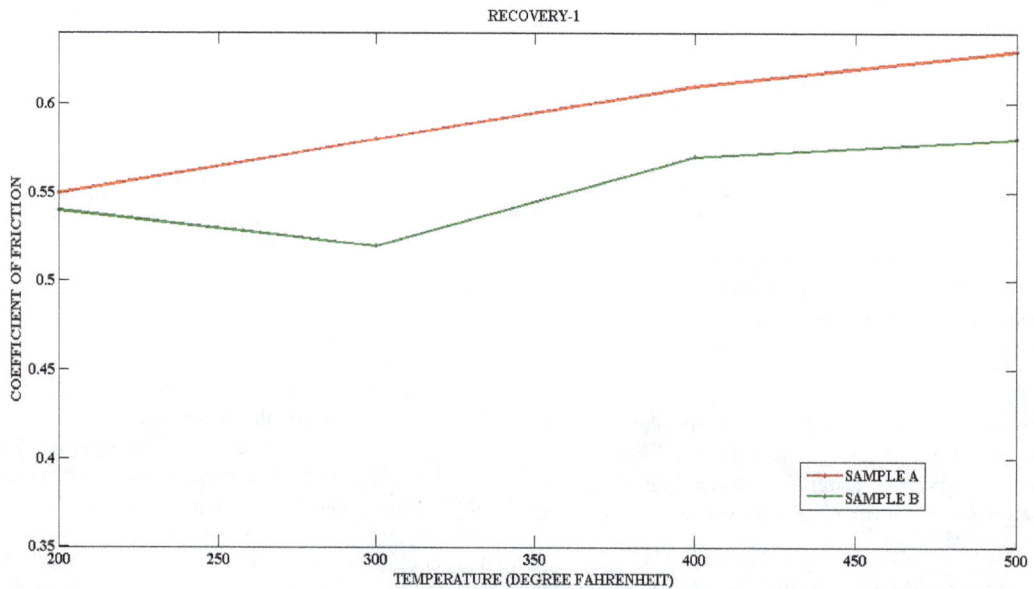

Figure 8: Plot comparison between the samples in recovery-1.

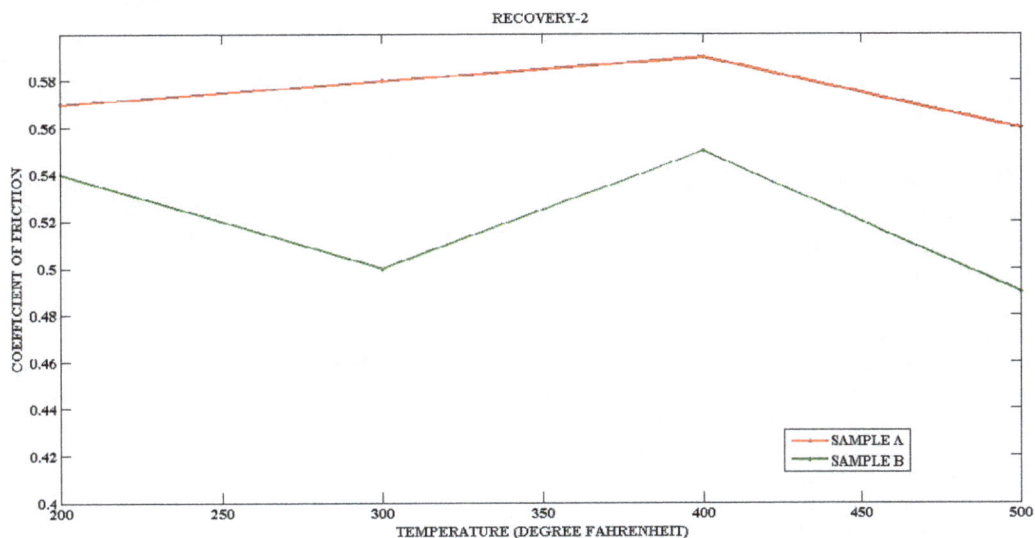

Figure 9: Plot comparison between the samples in recovery-2.

Temp (° F)	Recovery-1				Recovery-2			
	Force-F		μ		Force-F		μ	
	Samples				Samples			
	A	B	A	B	A	B	A	B
500	19.00	18.70	0.63	0.58	17.60	15.70	0.56	0.49
400	18.90	18.40	0.61	0.57	18.00	17.40	0.59	0.55
300	18.30	17.7	0.58	0.52	17.80	16.90	0.58	0.50
200	18.00	17.9	0.55	0.54	17.50	17.80	0.57	0.54

Table 5: Recovery stages 1 and 2.

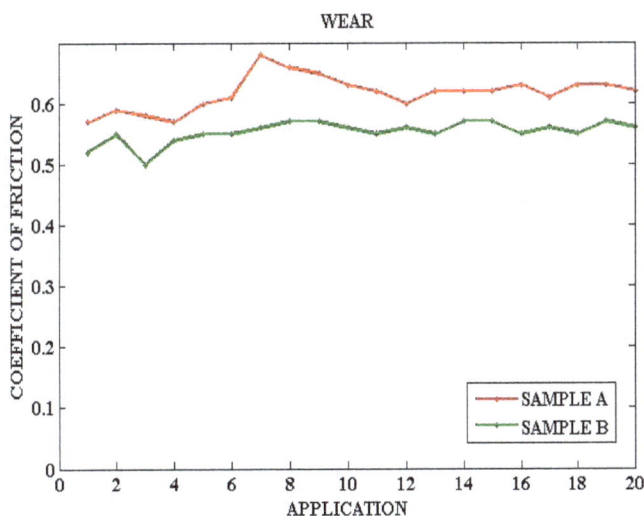

Figure 10: Plot comparison between the samples in wear.

in Figure 10. The coefficient of friction is consistent for both the samples. After ten applications, it depicts that wear is higher for sample B and picks up at later cycle. As in the case of sample A frictional stability is better to the latter. In Table 6 the wear characteristics shown for the 20 allocations. From the overall allocations the frictional value starts from nominal value and increases in a gradual scale for both the samples. In the sample A frictional value resulted is higher compared to the sample B. It is evident that sample A shows the better wear characteristics.

Conclusion

This test results has shown the friction materials behavior of brake lining with comparison to different fillers materials in combination to

WEAR					
	Force-F		Load-N		μ
	Samples				
	A	B	A	B	A	B
1	18.9	17.6	34.3	33.7	0.64	0.52
10	18.5	17.1	31	32.8	0.63	0.52
20	19.2	18.9	33.5	33.3	0.59	0.57

Table 6: Wear characteristics of both samples.

various operating test cycles in chase machine. From the test it is evident that sample A shows better frictional stability and wear characteristics, Also the temperature dependent behavior of the friction materials. With the addition of this filler materials the early presence of asbestos materials are now totally eradicated. Development of the friction material require significant amount of laboratory testing and on board vehicle testing. In future testing has to be performed with wide range of temperature and real time testing has to be done to know the actual life of the brake liner. The coefficient of friction what obtained in the chase machine is useful to judge the quality of the liner and to check the best formulation used. This case reveals only the critical value of friction; various operating conditions must be implemented in the computer controller to test the brake liners in the future.

Acknowledgement

I gratefully acknowledge Grips N' Brakes Automation, Coimbatore for technical support and providing the research facilities. I would also like to thank to Mr. S. Sanjeevraj, Managing Director (Grips N' Brakes Automation, Coimbatore) for his help and dedication toward my research work and related research, also for their support and excellent co-operation.

References

1. Tsang PHS, Jacko MG, Rhee SK (1985) Comparison of Chase and Inertia Brake Dynamometer Testing of Automotive Friction Materials. ASME Wear of Materials 129-137.

2. Blau PJ (2001) Composition, Functions and Testing of friction brake materials and their additives.

3. Sterle WO, Klob H (2007) Towards the better understanding of brake friction materials. Wear 263: 1189-1201.

4. Herring JM (1967) Mechanism of Brake Fade in Organic Brake Linings. Society of Automotive Engineers Transaction.

5. Ertan R, Yavuz N (2010) An experimental study on the effects of manufacturing parameters on the tribological properties of brake lining material. Wear 268: 1524-1532.

6. Anderson AE (1980) Wear of brake materials in wear control handbook. ASME.

13

Effect of Al_2O_3 Nanoparticles Doping on the Microwave Dielectric Properties of CTLA Ceramics

Dou Z, Wang G, Zhang E, Ning F, Zhu Q, Jiang J and Zhang T*

Hubei Collaborative Innovation Center for Advanced Organic Chemical Materials, Ministry of Education Key Laboratory for the Green Preparation and Application of Functional Materials and School of Material Science and Engineering, Hubei University, Wuhan, China

Abstract

$CaTiO_3$-$LaAlO_3$ (CTLA) ceramics with Al_2O_3 nanoparticles were prepared by conventional two-step solid-state reaction, and the sintering behavior and the microwave dielectric proprieties were investigated. The results show that the doping of Al_2O_3 nanoparticles can promote the growth of uniform grain, increase the density and $Q \times f$ value, while it has no obvious influence on the dielectric constant (~45). CTLA ceramics with 0.05 wt% Al_2O_3 nanoparticles doping, sintered at 1420°C for 4h exhibited the optimal microwave dielectric properties, and CTLA ceramics with 0.2 wt% Al_2O_3 nanoparticles doping sintered at 1380°C for 4h exhibited the optimal microwave dielectric properties, respectively. A proper increase of Al_2O_3 nanoparticles content can reduce the sintering temperature.

Keywords: CTLA ceramics; Al_2O_3 nanoparticles doping; Sintering behavior; Microwave dielectric proprieties

Introduction

With the development of telecommunication technology, microwave dielectric ceramics with light mass, miniaturization, integration and thermal stability have been intensively studied in the field of telecommunication materials during the last several years [1,2]. Medium dielectric constant microwave ceramics, which have the properties including moderate dielectric constant, low dielectric loss and near-zero temperature coefficients of the resonant frequencies, have been applied in wireless microwave devices in a wide range such as resonator, filter, dielectric antenna and dielectric guided wave circuit. $CaTiO_3$ ceramics with orthorhombic perovskite structure have the microwave dielectric properties of $\varepsilon_r = 170$, $Q \times f = 3500$ GHz and $\tau_f = 800$ ppm/°C, while $LaAlO_3$ ceramics with rhombohedral perovskite structure have the microwave dielectric properties of $\varepsilon_r = 23.4$, $Q \times f = 68,000$ GHz and $\tau_f = -44$ ppm/°C. $CaTiO_3$-$LaAlO_3$ ceramics (CTLA) prepared from calcined $CaTiO_3$ and $LaAlO_3$ were considered as the typical microwave dielectric ceramics with the middle dielectric constant.

Previous studies have demonstrated that the crystal structure of CTLA ceramics transformed from orthorhombic to rhombohedral as the increase of $LaAlO_3$ content [3]. Hou et al. [4] have reported the sintering behavior of CTLA microwave dielectric ceramics heavily depends on the preparation temperature. Jiang et al. [2] investigated the solid-state reaction mechanisms of CTLA ceramics by X-ray diffraction and thermogravimetric/differential scanning calorimetric analysis techniques. Modified perovskite ceramics by replacing the A/B-site of CTLA were reported, such as $0.7CaTiO_3$-$0.3[La_xNd_{(1-x)}]AlO_3$ [5] and $xCaTiO_3$-$(1-x)La[Ga_{(1-\delta)}Al_\delta]O_3$ [6]. However, both the dielectric loss and sintering temperature of CTLA ceramics cannot meet the requirement of application.

Al_2O_3 showed most considerable microwave dielectric properties of $\varepsilon_r = 10$, $Q \times f = 500,000$ GHz, $\tau_f = -60$ ppm/°C, thus the microwave dielectric properties of CTLA ceramics may be enhanced by the Al_2O_3 doping. Ravi et al. [7] studied the densification, structure, and microwave dielectric properties of $0.7CaTiO_3$-$0.3LaAlO_3$ ceramics doped with 0.25 wt% Al_2O_3 and sintered at 1500°C, and the ceramics exhibited better dielectric properties of $\varepsilon_r = 46$, $Q \times f = 38,289$ GHz, and $\tau_f = 12$ ppm/°C compared with those of the non-doped sample ($\varepsilon_r = 41$ and $Q \times f = 26,618$ GHz). In the work of Yao et al. [8], the dielectric property test revealed that the $Q \times f$ value of the BNT ceramics with Al_2O_3 additive has been improved by 47%, compared with that of the undoped BNT ceramics. Meanwhile, Al_2O_3 were added to suppress the reduction of Ti^{4+} ions so as to improve the microwave dielectric properties of $Ba_{4.2}Sm_{9.2}Ti_{18}O_{54}$ ceramics by Yao et al. [9] reported. So, Al_2O_3 is considered as a candidate to improve the microwave dielectric properties of CTLA ceramics.

The nanoparticles ratio of surface area is larger, compared with the bulk ratio of surface area. With the reduction of dimensions, nanoparticles provide a tremendous driving force during the sintering, particularly at elevated temperatures and hence the surface energy of the nanoparticles substantially affects the interior bulk properties of the host materials [10].

With doping of these smaller particles, the ceramics can be sintered at lower temperatures over shorter time than ceramics doping with larger particles [11]. Thus, nanoparticles doped can increase the density, and reduce the sintering temperature. For example, it is found that the addition of CeO_2 nanoparticles to the $MgTiO_3$ ceramics leads to improvement of the relative density, reduction in sintering temperature [10].

In a word, Al_2O_3 nanoparticles are an excellent additive with excellent dielectric properties. In this work, the addition of Al_2O_3 nanoparticles was used to improve the sintering behavior and the microwave dielectric properties of CTLA ceramics, and the effects of Al_2O_3 nanoparticles addition on the sintering behavior and microwave dielectric properties of CTLA ceramics were investigated and discussed in detail.

***Corresponding author:** Zhang T, Hubei Collaborative Innovation Center for Advanced Organic Chemical Materials, Ministry of Education Key Laboratory for the Green Preparation and Application of Functional Materials and School of Material Science and Engineering, Hubei University, Wuhan 430062, China E-mail: zhangtj@hubu.edu.cn

Experimental Procedures

Specimen preparation

Al_2O_3 nanoparticles doped $0.67CaTiO_3$-$0.33LaAlO_3$ ceramics, CTLA-xAl_2O_3 (x = 0 wt%, 0.05 wt%, 0.1 wt%, 0.2 wt%, 0.4 wt%, 0.8 wt%), were prepared using a conventional two-step solid-state reaction from high-purity $CaCO_3$ (99.3%), TiO_2 (99.9%), La_2O_3 (99.99%), micron Al_2O_3 (94%) and nano-size Al_2O_3 (99.999%). $CaTiO_3$ was calcined at 1090°C for 6h and $LaAlO_3$ was calcined at 1200°C for 5h. The fully calcined powders ($CaTiO_3$, $LaAlO_3$) and nano-size Al_2O_3 were weighed stoichiometrically and ground using ball milling for 8h, dried at 120°C for 12h, and then mixed with a 5 wt% polyvinyl alcohol solution as a binder. The final particles were uniaxial pressed 12 mm diameter and 6 mm in high pellets. The pellets were debindered at 650°C and then sintered at the temperature of 1360°C ~1450°C for 4h in air. Pellets were cooled from the sintering temperature to 1000°C at a rate of 2°C/min, and then the ceramics were allowed to cool inside the furnace naturally.

Property characterization and theoretical analysis

The crystal structure of the sintered samples was characterized using X-ray powder diffraction (XRD) using a Bruker advanced X-ray powder diffractometer and Cu Kα radiation. The apparent densities of the ceramics were measured using the Archimedes method. The microstructures of the ceramics were examined with a JEOL JSM-7100F scanning electron microscope (SEM). The microwave dielectric properties of the ceramic samples were measured with an Agilent E5071C network analyzer using the cavity reflection method, and τ_f was measured between -20°C and 65°C and was defined as

$$\tau_f = \frac{f_{(T_2)} - f_{(T_1)}}{f_{(T_1)}(T_2 - T_1)} \qquad (1)$$

Where,

$f_{(T_1)}$ is the resonant frequency at -20°C.

$f_{(T_2)}$ is the resonant frequency at 65°C.

Results and Discussion

Figure 1 presents the XRD patterns of CTLA-xAl_2O_3 ceramics sintered at 1420°C for 4h. A single phase of perovskite structure was confirmed over the entire compositional range. It can be seen that pure orthorhombic perovskite phase with space group $Pnma$ (62) (PDF#52-1773) were identified. At the range of 0~0.4 wt%, the diffraction peaks didn't shift almost. But at the doping level of x = 0.8 wt% the diffraction peaks slight shifted towards higher angles, because the Al^{3+} ions with smaller ionic radius (R_{Al}^{3+} = 0.535 Å) substituted Ti^{4+} ionic with larger ionic radius ($RTi4+$ = 0.605 Å) in the CTLA ceramics.

Figure 2 shown the densities of CTLA-xAl_2O_3 ceramics at different sintering temperatures. As is shown, the density of CTLA-xAl_2O_3 ceramics increased firstly. The density of ceramics with low Al_2O_3 nanoparticles doping amount (x < 0.1 wt%) reached a maximum value at the temperature of 1420°C, while the density of ceramics with high content ($x \geq$ 0.2 wt%) reached the maximum value at a lower temperature (1380°C). After that the densities decreased as the sintering temperature continues to rise. At x = 0.05 wt%, the CTLA ceramics sintered at 1420°C obtained the maximum density of 4.669 g/cm^3. And at x = 0.2 wt%, the CTLA ceramics sintered at 1380°C also obtained a higher density of 4.663 g/cm^3. It is well known that the nanoparticles have large ratio of surface area, high surface energy, low

melting point, and the smaller the particle size, the faster the sintering rate [10,12,13]. When the Al_2O_3 nanoparticles dispersed in the grain boundaries, it can promote mass transfer in the sintering process, reduce the surface energy of the system, accelerate the reaction, and then lead to the increase of the density. Similar, to previous reports [10,11], when the sintering temperature is higher than the certain value, the grains appeared abnormal growth and resulted in the lower density. Therefore, adding appropriate Al_2O_3 nanoparticles can increase CTLA ceramic density effectively, and reduce the sintering temperature.

The SEM of Al_2O_3 nanoparticles modified CTLA ceramics sintered at 1420°C was given in Figure 3. As shown in the Figure 3, the surface of CTLA ceramics modified by Al_2O_3 nanoparticles was smooth and densification with the pores in the grain and there was no obvious secondary phase for the composition. The SEM images of the CTLA ceramics with smooth grain and uniform size was shown in Figure 3a. By adding a small amount of nanoparticles (Figure 3b), the surface energy of the grain boundary can be improved, and the material transfer can be accelerated, which promote the uniformity growth of grain size. With the increase of the content of Al_2O_3 nanoparticles, the decline of onset sintering temperature and higher sintering temperature makes the grain abnormal growth due to two times recrystallization in the Figures 3c-3f. At the same time, with the increase of Al_2O_3 nanoparticles content, the grain boundary becomes vague, this is due to the excess introduction of Al_2O_3 nanoparticles particles, for nanoparticles cannot completely diffuse, forming aggregate in the grain boundary surface. Therefore, Al_2O_3 nanoparticles can accelerate the mass transfer rate of CTLA system and promote the grain size more uniform. But the excess addition cause grains abnormal growth and resultantly affects the density of CTLA ceramics.

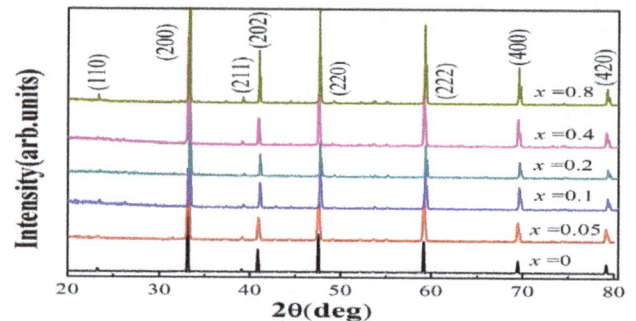

Figure 1: XRD patterns of CTLA-xAl_2O_3 ceramics sintered at 1420°C for 4 h.

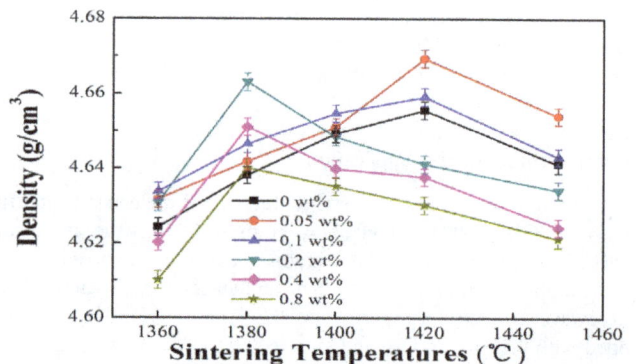

Figure 2: The densities of CTLA-xAl_2O_3 ceramics at different sintering temperatures.

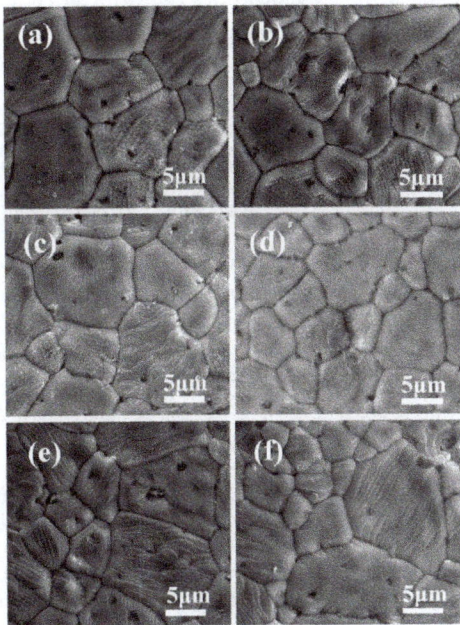

Figure 3: SEM images of CTLA-xAl$_2$O$_3$ ceramics sintered at 1420°C for 4 h (a) x = 0wt%, (b) x = 0.05 wt%, (c) x = 0.1 wt%, (d) x = 0.2 wt%, (e) x = 0.4 wt%, (f) x = 0.8 wt%.

Figure 4 shows the dielectric constant (ε_r) of CTLA-xAl$_2$O$_3$ ceramics sintered at different temperatures ranged from 1360 to 1450°C. For a given composition, the dielectric constant of CTLA ceramics increased with the increase of the sintering temperatures. When the sintering temperature is higher, the ion polarization ability of ionic crystal is stronger than at the lower temperature. At the same sintering temperatures, the dielectric constant increases first and then decreases as the increase amount of Al$_2$O$_3$ nanoparticles. At x = 0.05 wt%, the dielectric constant reaches the maximum. It was known that the dielectric constant depended significantly on the relative density at microwave frequencies [14], therefore, the dielectric constant increases first and then decreases, following the change of the density. With the increase of Al$_2$O$_3$ nanoparticles amount, the dielectric constant fluctuated slightly around 45.

The $Q{\times}f$ values of the CTLA-xAl$_2$O$_3$ ceramics sintered at different temperatures are shown in Figure 5. With the increase of the sintering temperature, $Q{\times}f$ values of CTLA-xAl$_2$O$_3$ ceramics showed the same trend with the density. The $Q{\times}f$ values first increased and then decreased, but $Q{\times}f$ values of different Al$_2$O$_3$ nanoparticles content ceramics reached maximums at different sintering temperature. The $Q{\times}f$ of CTLA ceramics with 0.05 wt% Al$_2$O$_3$ content reached the maximum value (38,215 GHz) at the sintering temperature of 1420°C, while the $Q{\times}f$ of CTLA ceramics with 0.2 wt% Al$_2$O$_3$ content reached the maximum value (37,346 GHz) at the sintering temperature of 1380°C. This indicates that doping Al$_2$O$_3$ nanoparticles can reduce the sintering temperature, while the influence on the dielectric loss is subtle. The trend of the $Q{\times}f$ values as a function of Al$_2$O$_3$ nanoparticles content (Figure 6) was similar to the trend of the density, which illustrates that the $Q{\times}f$ values of CTLA-xAl$_2$O$_3$ ceramics were affected significantly by the relative density.

The relationship between the temperature coefficient of resonant frequency (τ_f) and ε_r of the CTLA-xAl$_2$O$_3$ ceramics sintered at 1420°C are shown in Figure 6. As is shown in Figure 6, both τ_f and

ε_r increase and then decrease with the increase of the content of the Al$_2$O$_3$ nanoparticles. Because the τ_f of Al$_2$O$_3$ is -60 ppm/°C, the τ_f of the CTLA-xAl$_2$O$_3$ ceramics should be increased to the negative direction. However, at x = 0.05 wt%, τ_f reached a maximum value of 3.65 ppm/°C. So there are other reasons for this situation. Harrop [15] differentiated the Clausius-Mossotti equation to justify physically the following empirical relationship

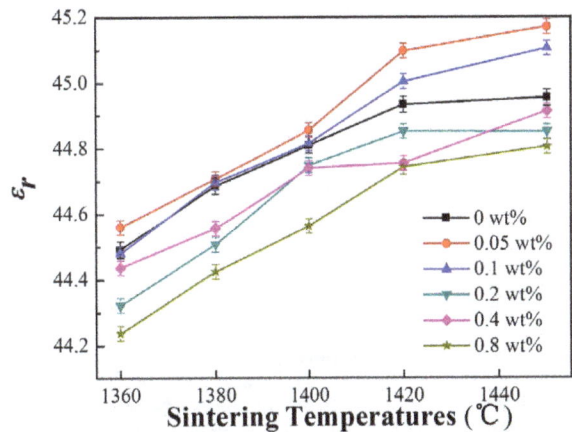

Figure 4: ε_r of CTLA-xAl$_2$O$_3$ ceramics sintered at different temperatures.

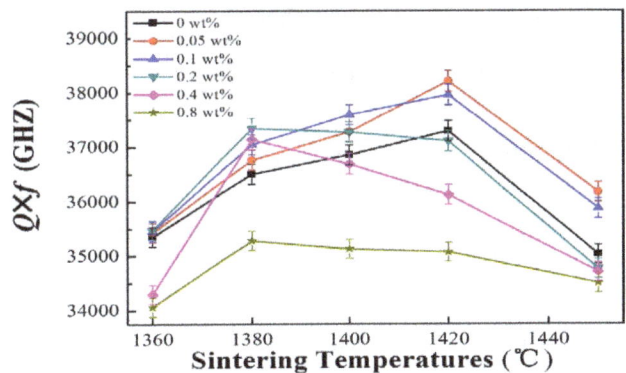

Figure 5: $Q \times f$ values of CTLA-xAl$_2$O$_3$ ceramics sintered at different temperatures.

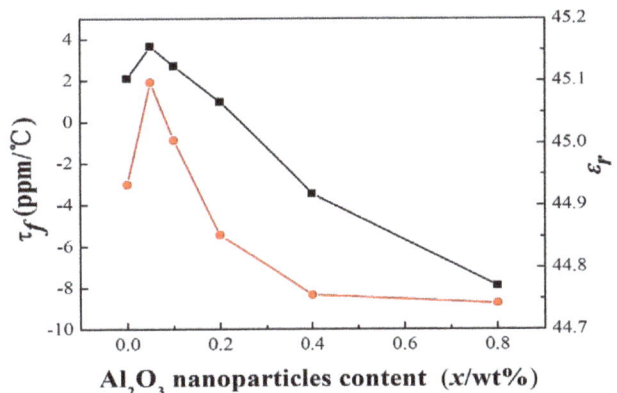

Figure 6: τ_f of CTLA ceramics with different Al$_2$O$_3$ nanoparticles content sintered at 1420°C.

$$\tau_c = -\alpha_L \varepsilon_r \qquad (2)$$

Where,

τ_c is the temperature coefficient of capacitance, and

α_L is the linear expansion coefficient of the dielectric materials.

It is related to the temperature coefficient of permittivity (τ_ε) by [16]

$$\tau_\varepsilon = \tau_c - \alpha_L \qquad (3)$$

τ_f is identified as

$$\tau_f = -\left(\frac{1}{2}\tau_\varepsilon + \alpha_L\right) \qquad (4)$$

Furthermore, α_L is the linear expansion coefficient of the CTLA ceramics, and it is known to be approximately constant. It follows that:

$$\tau_f \propto -\alpha_L \varepsilon_r \qquad (5)$$

It illustrated that τ_f mainly depends on ε_r in the case of CTLA-xAl$_2$O$_3$ ceramic system, so the τ_f increases firstly and then decreases with the change of dielectric constant (ε_r).

The microwave dielectric properties of the 0.67CaTiO$_3$-0.33LaAlO$_3$ ceramics doped with x wt% Al$_2$O$_3$ nanoparticles are shown in Table 1. It can be seen that CTLA ceramics with 0.05 wt% Al$_2$O$_3$ nanoparticles doping sintered at 1420°C for 4h exhibited the optimal microwave dielectric properties of εr = 45.10, $Q{\times}f$ = 38,215 GHz, and τf = 3.65 ppm/°C, and at x= 0.2 wt%, the sintering temperature of CTLA ceramics was decreased from 1420°C to 1380°C, and the ceramics exhibited the optimal microwave dielectric properties of εr = 44.51, $Q{\times}f$ = 37,346 GHz, and τ_f = 4.87 ppm/°C.

Conclusions

Al$_2$O$_3$ nanoparticles were adopted to modify the CTLA ceramics. The sintering behavior and microwave dielectric proprieties of the CTLA ceramic samples were studied. The results show that the doping of Al$_2$O$_3$ nanoparticles has no obvious influence on the dielectric constant (~45). The Al$_2$O$_3$ nanoparticles, as a mass transfer media, promoted the growth of uniform grain, and reduced dielectric loss availably. The temperature coefficient of resonant frequency of the ceramics showed the same dependence on Al$_2$O$_3$ nanoparticles content as the dielectric constant. CTLA ceramics with 0.05 wt% Al$_2$O$_3$ nanoparticles doping sintered at 1420°C for 4h exhibited the optimal microwave dielectric properties of ε_r = 45.10, $Q{\times}f$ = 38,215 GHz, and τ_f = 3.65 ppm/°C. Compared with the undoped CTLA ceramics, the $Q{\times}f$ value has been improved by 2.5 %. At x= 0.2 wt%, the sintering temperature of CTLA ceramics was decreased from 1420°C to 1380°C, and the ceramics exhibited the optimal microwave dielectric properties of ε_r = 44.51, $Q{\times}f$ = 37,346 GHz, and τ_f = 4.87 ppm/°C. The results showed that the proper amount of Al$_2$O$_3$ nanoparticles content reduced the sintering temperature and slightly improved the $Q{\times}f$ value.

References

1. Jiang J, Fang D, Lu C, Dou Z, Wang G, et al. (2015) Solid-state reaction mechanism and microwave dielectric properties of CaTiO$_3$–LaAlO$_3$ ceramics. J Alloys Compd 638: 443-447.

2. Dou Z, Jiang J, Wang G, Zhang F, Zhang T (2016) Effect of Ga^{3+} substitution on the microwave dielectric properties of 0.67 CaTiO$_3$–0.33LaAlO$_3$ ceramics. Ceram Int 42: 6743-6748.

3. Khalyavin DD, Salak AN, Senos AMR, Mantas PQ, Ferreira VM (2006) Structure sequence in the CaTiO$_3$-LaAlO$_3$ microwave ceramics-revised. J Am Ceram Soc 89: 1721-1723.

4. Hou G, Wang Z, Zhang F (2011) Sintering behavior and microwave dielectric properties of (1−x)CaTiO$_3$-xLaAlO$_3$ ceramics. J Rare Earths 29: 160-163.

5. Liang F, Ni M, Lu W, Fan G (2013) Microwave dielectric properties and crystal structures of 0.7CaTiO$_3$–0.3[La$_x$Nd$_{(1-x)}$]AlO$_3$ ceramics. J Alloys Compd 568: 11-15.

6. Liang F, Ni M, Lu W, Feng S (2014) Crystal structure and microwave dielectric properties of CaTiO$_3$–La[Ga$_{(1-\delta)}$Al$_\delta$]O$_3$ ceramics system. Mater Res Bull 57: 140-145.

7. Ravi GA, Azough F, Freer R (2007) Effect of Al$_2$O$_3$ on the structure and microwave dielectric properties of Ca$_{0.7}$Ti$_{0.7}$La$_{0.3}$Al$_{0.3}$O$_3$. J Eur Ceram Soc 27: 2855-2859.

8. Yao X, Lin H, Zhao X, Chen W, Luo L (2012) Effects of Al$_2$O$_3$ addition on the microstructure and microwave dielectric properties of Ba$_4$Nd$_{9.33}$Ti$_{18}$O$_{54}$ ceramics. Ceram Int 38: 6723-6728.

9. Yao X, Lin H, Chen W, Luo L (2012) Anti-reduction of Ti^{4+} in Ba$_{4.2}$Sm$_{9.2}$Ti$_{18}$O$_{54}$ ceramics by doping with MgO, Al$_2$O$_3$ and MnO$_2$. Ceram Int 38: 3011-3016.

10. Thatikonda SK, Goswami D, Dobbidi P (2014) Effects of CeO$_2$ nanoparticles and annealing temperature on the microwave dielectric properties of MgTiO$_3$ ceramics. Ceram Int 40: 1125-1131.

11. Bari M, Taheri-Nassaj E, Taghipour-Armaki H, Chen XM (2013) Role of nano- and micron-sized particles of TiO$_2$ additive on microwave dielectric properties of Li$_2$ZnTi$_3$O$_8$- 4wt% TiO$_2$ ceramics. J Am Ceram Soc 96: 3737-3741.

12. Tartaj P, Morales MP, González-Carreño T, Veintemillas-Verdaguer S, Serna CJ (2005) Advances in magnetic nanoparticles for biotechnology applications. J Magn Magn Mater 290-291: 28-34.

13. Liao P, Qiu T, Yang J, Lu X (2014) Effect of Al$_2$O$_3$ addition on microwave dielectric properties of BaCo$_{0.194}$Zn$_{0.116}$Nb$_{0.69}$O$_3$ ceramics. Electron Mater Lett 10: 121-125.

14. Hyun Yoon K, Soo Kim E, Jeon JS (2003) Understanding the microwave dielectric properties of (Pb$_{0.45}$Ca$_{0.55}$)[Fe$_{0.5}$(Nb$_{1-x}$Ta$_x$)$_{0.5}$]O$_3$ ceramics via the bond valence. J Eur Ceram Soc 23: 2391-2396.

15. Moulson AJ, Herbert JM (1990) Electroceramics. Chapman & Hall, London.

16. Harrop PJ (1969) Temperature coefficients of capacitance of solids. J Mater Sci 4: 370-374.

T (°C)	Q×f (GHz)		εr		τf (ppm/°C)	
	1380	1420	1380	1420	1380	1420
0 wt%	36,500	37,291	44.69	44.93	4.38	2.12
0.05 wt%	36,766	38,215	44.71	45.1	6.78	3.65
0.1 wt%	37,054	37,964	44.7	45.01	4.96	2.71
0.2 wt%	37,346	37,110	44.51	44.85	4.87	0.98
0.4 wt%	37,144	36,127	44.56	44.76	0.36	-3.47
0.8 wt%	35,285	35,072	44.43	44.74	-4.7	-7.89

Table 1: The microwave dielectric properties of the CTLA ceramics doped with x wt% Al$_2$O$_3$ nanoparticles.

14

Deterioration of Stainless Steel Corrosion Resistance Due to Welding

Fandem QA*

Aramco Qatif, Eastern Region, Saudi Arabia

Abstract

The objective of this paper is to provide an overview of the main welding defects that frequently exist in piping and equipment and how to detect these defects without destroying the welds using Non Destructive Testing (NDT) methods. Then, the sensitization of stainless steel weld, characterization, processing and structure properties of HAZ as well as weld metal will be discussed in details. Furthermore, this paper will illustrate the treatment of the weld decay by recovery of passivation film after welding which will be followed by prevention of intergranular corrosion or weld decay of SS using surface mechanical attrition treatment and all of these will be described with help of characterization techniques XRD, SEM, TEM and EPMA.

Keywords: Welding; Welding decay; Stainless steel; Passivation; Sensitization; NDT; SMAT; Cr depletion

Introduction

Welding is an important technique of joining metals homogenously in industries due to its effectiveness with low overall cost. However, presence of welding imperfections is a challenge which may form due to poor workmanship such as cracks, porosity, lack of fusion, incomplete penetration and weld decay in stainless steel because of material properties issue. Furthermore, stainless steel material is one of the best choices in industries because of its high resistant to corrosion, however the issue with stainless steel material is the sensitization or weld decay after welding fabrication particularly in Heat Affected Zone (HAZ) [1-3].

Non-destructive testing (NDT) techniques

In order to check the soundness of the weld for the equipment without damaging or destroying it, the manufacturers or fabricators are using NDT methods. There are several types of these such as Penetrant Testing, Radiographic Testing (RT), Ultrasonic Testing (UT), Magnetic Testing (MT) and of course Visual Testing (VT). There are also advanced NDT techniques used for welding critical services such as Time of Flight Diffraction Ultrasonic (TOFD). Some of these methods will be discussed briefly in this report.

Penetrant Testing is simple and low cost technique used to detect open to surface defects such as crack by using three different sprayers i.e. penetrant, developer and cleaner sprayers. The applying procedure of this method is by cleaning the surface and applies the penetrant which is in red color, and then after five minutes the area cleaned off with use of the cleaner followed by applying the developer which is in white color to bleed out the penetrant and make a color contrast. The crack will be visible easily after application as in Figure 1. The disadvantages of this method are; used only for open to surface defects, temperature limit of 125 F maximum, cannot be used to measure flaw size [4,5].

Another NDT method is Ultrasonic Testing which is a manual and effective to measure the defect size as well as detect internal defects in the weld by sending a beam of sound waves with frequency range (>20,000 cycles per second). If there is any defect the waves will be reflected back to the probe and amplitude intensity displayed on the screen. The disadvantages are requiring certified operator, even with the certified personnel it is difficult to recognize the type of welding defect and no permanent record of the measured defects [6,7].

Radiographic Testing is one of the most effective techniques used to detect internal defects by sending X-rays from radioactive source (Ir 192 or Cobalt 60) placed at one side of the weld and on the other side an image film is placed with image quality indicator (IQI). The main disadvantages of this method are health hazardous, high cost and very sensitive to the defect orientation for example if there is a crack parallel to the X-rays as shown in Figure 2, it appears on the film as a dot and the interpreter will recognize it as porosity instead of the crack, that is why it is not recommended to use RT for crack detection or to be used at several different shooting angles at the same joints which means more cost [8].

Stainless steel's corrosion resistance deterioration due to welding

Stainless steel gains the property of corrosion resistant at room temperature from presence of Chromium Cr (at least 10.5%) and Ni elements. The Cr produces a passivation film layer (Chromium Oxide) over the surface when exposed to atmosphere which protects the

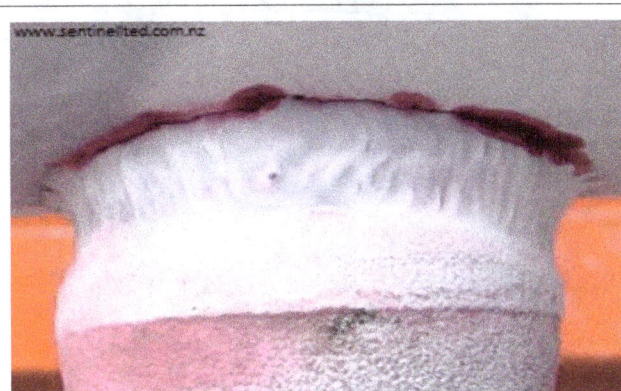

Figure 1: Crack is visible after applying the three sprayers.

*Corresponding author: Fandem QA, Aramco Qatif, Eastern Region, Saudi Arabia
E-mail: q_2010@hotmail.com

metal from corrosion. There are several type of SS based on formation temperature and Cr% i.e. Austenite has Face Centered Cubic structure (300 and 200 series), Ferrite has Body Centered Cubic structure (400 series, Martensitic has Body Centered Cubic structure... etc. [9].

Although the SS is highly corrosion resistant at room temperature, when it is exposed to high temperature such as welding process (500-800°C), the HAZ is susceptible to corrosion as intergranualr corrosion and Stress Corrosion Cracking (SSC) as shown in Figure 3. This is occurred due to chromium carbide precipitation at grain boundary which leads to chromium depletion, this is called sensitization [10].

Material and Methods

To understand the cause and characterization of the stainless steel corrosion as well as SCC, detailed experiments and results will be discussed for a material of ferritic SS 0Cr18Mo2Ti per [11]. The Manufacturing process sequence of 0Cr18Mo2Ti is vacuum melting technique (VMT), continues casting, plate mill, annealing and finally pickling. Moreover, the chemical composition and mechanical properties of the base metal material are listed in Table 1 [11].

A test plate of 0Cr18Mo2Ti with thickness 3 mm produced by VMT technique and heated for 10 min at temperature 850°C and then cooled by air. The plate was welded using SMAW process with lime titania super low carbon SS electrode. The chemical compositions of the weld in Wt.% are listed in Table 2.

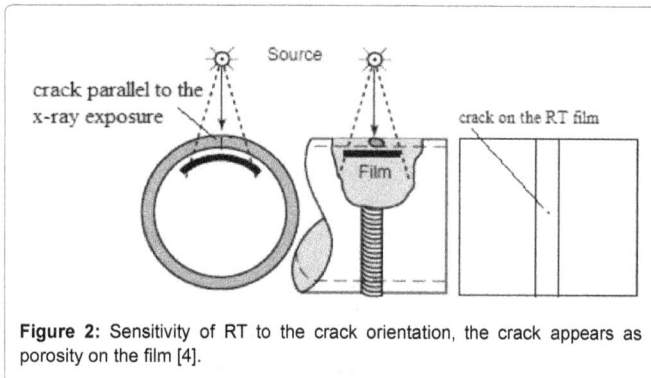

Figure 2: Sensitivity of RT to the crack orientation, the crack appears as porosity on the film [4].

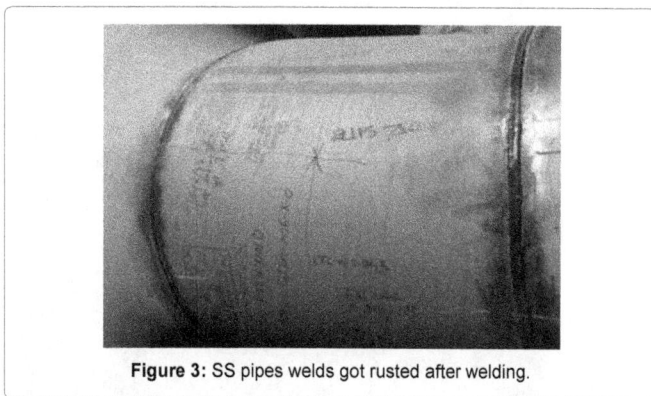

Figure 3: SS pipes welds got rusted after welding.

Also, the welding essential variables used as per welding procedure from ASME IX are listed in Table 3. The test specimens were prepared by cutting the weld from the weld and then an etchant (HNO_3 and HCl ratio 1:3) was used to display microstructures of the cut weld. The HAZ was analyzed by TEM and electron diffraction to identify the lattice structure as well as the microstructure. Moreover, the phase composition precipitation was identified by SEM, XRD and electron probe microanalysis (EPMA) [12].

Results and Discussion

The microstructures in the weld zone are austenite, ferrite and just a little of martensite, its morphology using SEM is shown in Figure 4a and 4b. Moreover, the weld was also analyzed by use of XRD with the following working parameters; voltage 40 KV, current 150 mA and scanning range 30-90 degrees. The result revealed that it is mainly austenite and ferrite just like in SEM.

For coarse grained heat affected zone (CGHAZ) which is only ferrite (single phase) and has al lower toughness due to heating up at high temperature during welding. Obviously, as we go far from the fusion line to the base metal, the grain size is reduced. Therefore, the welding heat input should be small enough as possible to prevent the microstructure coarse grain and hence the toughness reduction by following the proper welding procedure per ASME IX [11].

In order to get more detail analysis of the CGHAZ and the fusion line between the base metal-weld, thin films from these two zones were analyzed by TEM and electron diffraction. The TEM, electron diffraction pattern and indexing diagram of both zones were performed on [210] and [012] directions as shown in Figure 5 for CGHAZ and Figure 6 for fusion line.

It can be observed from the TEM Figures 5 and 6a, there are many dislocations and twin clusters exist inside grains due to the residual stress caused by welding. The diffraction pattern confirmed that the CGHAZ is a single phase ferrite with BCC structure and the lattice parameter a =0.2866 nm. Also for the fusion line, it is a single phase ferrite with a BCC structure. Furthermore, micro cracks were also observed at CGHAZ, this is due to the residual stress which is shown in Figure 7 [11].

Passivation treatment of stainless steel after welding

As mentioned previously the CGHAZ of stainless steel got affected after welding and the solution to recover the protection film is by performing passivation process. Passivation is the removal of impurities or iron from the surface and can be achieved by electropolishing, electrochemical cleaning or chemical passivation [12].

Electrochemical cleaning approach will be discussed in the report on 316L weld material. The surface was first polished with 1000 grit abrasive paper in Al_2O_3 (0.5 um) solution Figure 8. Then it has been cleaned chemically with 2% citric acid solution and 5% ammonia at 80°C. After that, the passivation was performed at temperature of 60°C with 6% HNO_3 solution that has $CuSO_4.5H_2O$ at 2%. The analysis of the surface was performed before and after polishing as well as after

Chemical Composition (wt%)										
C	N	Mn	Si	Cr	Ni		Mo	Ti	S	P
0.019	0.014	0.28	0.16	18.65	0.14		1.6	0.24	0.005	0.024
Mechanical properties										
Tensile Strength (Mpa)		Yield Strength (Mpa)			Elongation (%)			Impact energy (J)		
550-700 (560)		300-400 (360)			30-40 (34)			148-155 (152)		

Table 1: base metal's mechanical properties and chemical compositions [11].

C	Cr	Ni	Mo
<0.03	17-20	Nov-14	1.5-2.5

Table 2: Chemical composition of the used weld deposit [11].

Electrode Diameter (mm)	Welding Voltage (Volts)	Welding current (Amp)	Travel Speed (mm/s)
3.2	24-26	90-105	06-Aug

Table 3: Welding essential variables used for the plate [11].

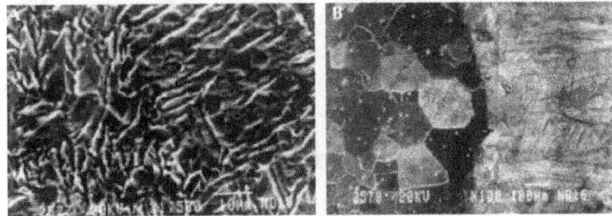

Figure 4: (a) Microstructures of weld metal SEM (X 1500), (b) weld metal & HAZ SEM(X100) [11].

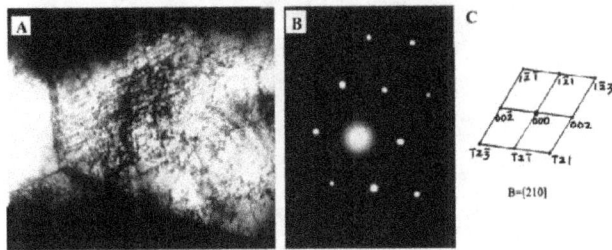

Figure 5: Structure characterization of CGHAZ (a) TEM morphology (X35000), (b) Electron Diffraction Pattern, (c) Schematic Diagram [11].

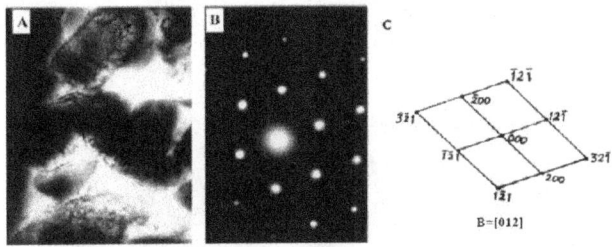

Figure 6: Structure characterization of welding fusion line (a) TEM morphology (X35000), (b) Electron Diffraction Pattern, (c) Schematic Diagram [11].

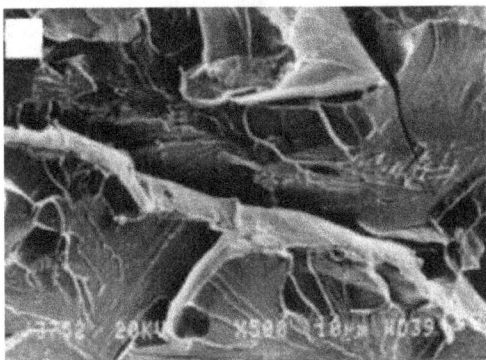

Figure 7: Micro crack in CGHAZ -SEM (X 500) [11].

passivation process. Figure 9 shows the SEM surfaces of base metal A and weld metal B, presence of cracks, grain boundaries and impurities are clear in both. On the other hand, the polished surfaces were characterized by smooth and uniform structure and the cracks as well as impurities were removed as shown in Figure 10.

After passivation SEM the base and weld metal were analyzed which presents an irregular distribution of indentations with a maximum diameter 5 um in the base metal as shown in Figure 11. Moreover, the quantitative analysis of the base metal has been performed by EDXS before and after passivation process as shown in Figure 12a and 2b. From this figure it is very clear that chromium carbide percentage dropped after passivation which enhances formation of the passivation film [12].

Prevention of SS corrosion using surface mechanical attrition treatment

The weld decay or the corrosion of SS material after getting exposed to high temperature such as welding process can be prevented by Surface Mechanical Attrition Treatment (SMAT) technique as per [10] in order to induce grain refinement as well as formation of twins.

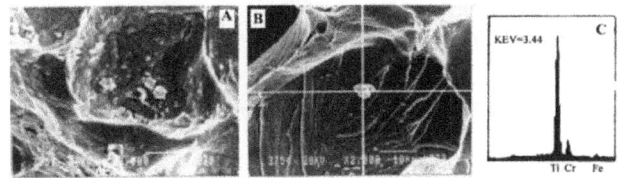

Figure 8: (a) Precipitates in CGHAZ SEM (X 2000), (b) Analysis location point for EPMA (X 2000), (c) EPMA spectra [11].

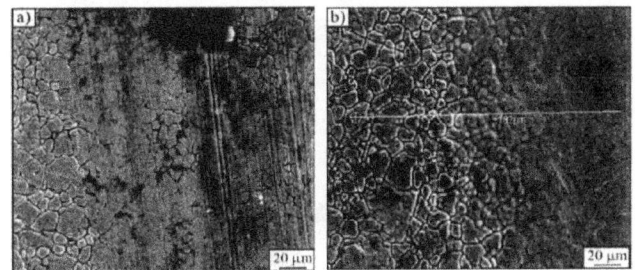

Figure 9: A (SEM) surface of base metal before polishing. B. Surface of weld before polishing [12].

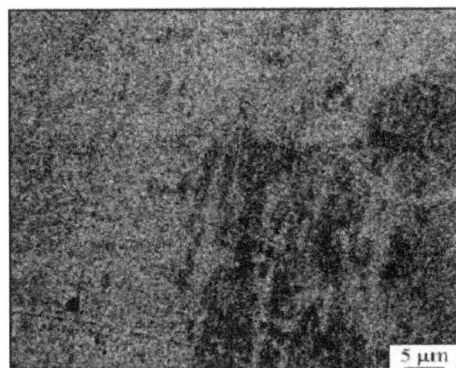

Figure 10: (SEM) the surface after polishing [12].

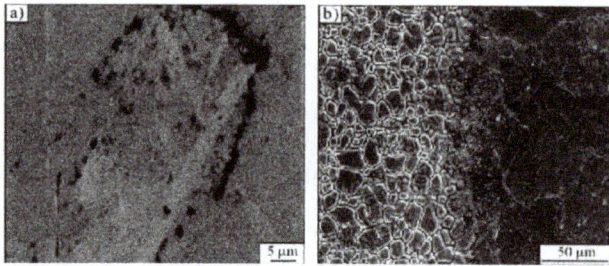

Figure 11: (a) (SEM) surface of base metal after passivation, (b) surface of weld after passivation [12].

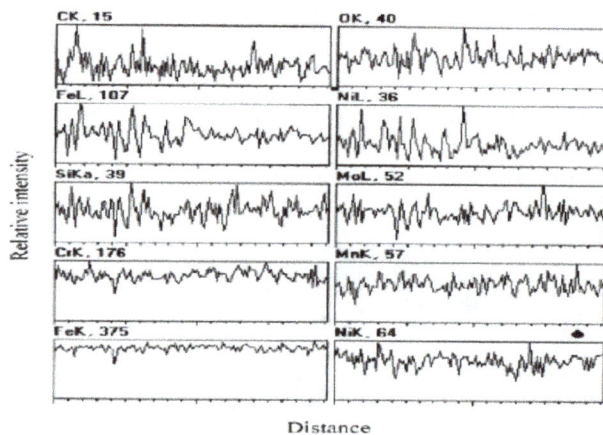

Figure 12a: (EDXS) Quantitative linear analysis before passivation [12].

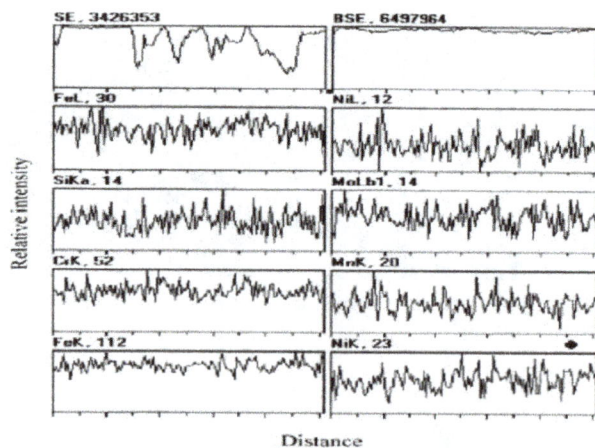

Figure 12b: (EDXS) Quantitative linear analysis after passivation [12].

SS 304 material has been used in this experiment with use of GTAW welding process. The samples were annealed at 1070°C for one hour and then quenched in water. The inducement of grain was refinement, the samples were put under vacuum at room temperature for 30 minutes with a vibrating frequency 20 kHz.

Optical micrograph of electroetching in 10% oxalic acid solution samples is shown in Figure 13. Grooved grain of the untreated sample are very clear, while for the SMATed ones are not. The single twins and their intersections can be seen with about 300 um thick below the surface. Moreover , TEM with magnification of 100 nm taken for the

SMATed top surfaces as shown in Figure 14, which characterized by ultrafine equiaxed grains with random crystallographic orientation and as can be seen the average grain size is about 10 nm.

Moreover, SEM micrographs were also taken for the samples before and after SAMT as shown in Figure 15 and it is clearly observed the deep groove along the boundaries due to weld decay or the sensitization as Figure 15a and 15b. On the other hand, after treatment

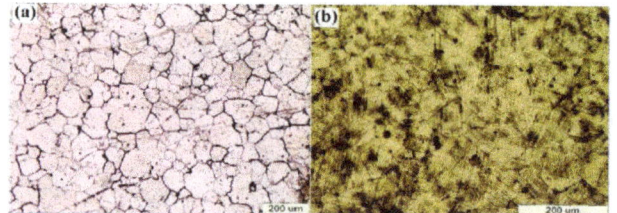

Figure 13: (a) Optical photos of untreated HAZ, (b) SMATed HAZ after electroetching [10].

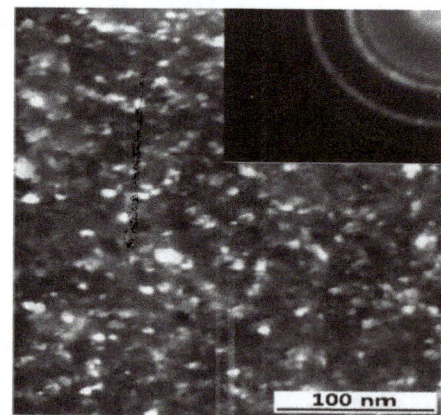

Figure 14: Dark field TEM of the top surface SMATed HAZ.

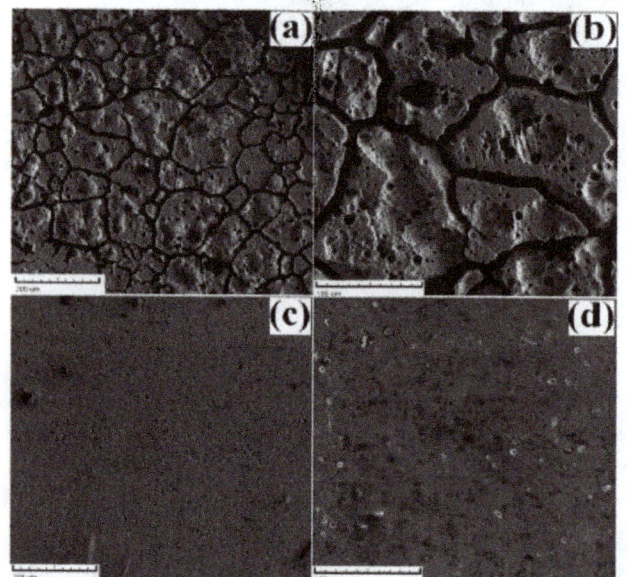

Figure 15: (a and b) SEM photos for untreated surface while, (c and d) for SAMTed surfaces [10].

the deep grooves in the boundaries become shallow and there is no sign for intergranular corrosion or sensitization because of formation high density twins as well as grain refinement. Thus, the SMAT improve the SS material to overcome the sensitization after welding by about 50 times than the untreated one [10].

Conclusions

It has been observed that the stainless steel material is corroded (weld decay) after welding due to sensitization or Cr depletion in the grain boundaries in form of chromium carbide. Thus, to enhance the recovery of the protection film, passivation treatment is used to clean the impurities or iron from the surface and hence the iron is reduced with recovery of Cr% atomic mass. Furthermore, the prevention technique with use of Surface Mechanical Attrition Treatment (SMAT) can be utilized before welding to enhance the sensitization for about 50 times. Also, it has been explained for the importance of the characterization techniques in failure investigation over the NDT methods with the example of SS welding.

References

1. Hayes B (1996) Classic brittle failures in large welded structures. Engineering Failure Analysis 3: 115-127.

2. Zerbst U, Ainsworth RA, Beier HT, Pisarski H, Zhang ZL, et al. (2014) Review on fracture and crack propagation in weldments-A fracture mechanics perspective. Engineering fracture mechanics 132: 200-276.

3. http://www.twi-global.com/technical-knowledge

4. API Standard (2004) Welding Inspection and Metallurgy API 577. American Petroleum Institute Washington.

5. ASME Standard (2007) Nondestructive Examination ASME V. The American Society of Mechanical Engineering. New York.

6. http://met-tech.com/crane-weldment-failure.htm

7. ISO 5817 (2003) Arc welded joints in steel-Guidance on quality levels for imperfections. Geneva: International Organization for Standardization.

8. Harrison JD (1972) Basis for a Proposed Acceptance-standard for Weld defects, Part. 1: Porosity. Metal Constr Br Weld J 4: 99-107.

9. https://en.wikipedia.org/wiki/Stainless_steel

10. Laleh M, Kargar F, Rouhaghdam AS (2012) Prevention of weld-decay in austenitic stainless steel by using surface mechanical attrition treatment. International Nano Letters 2: 37.

11. Yajiang LI, Yonglan Z, Bin S, Juan W (2002) Tem observation and fracture morphology in the cghaz of a new 0cr18mo2ti ferritic stainless steel. Bulletin of Materials Science 25: 361-366.

12. Gojić M, Marijan D, TuDja M, Kožuh S (2008) Passivation of welded AISI 316L stainless steel Pasivacija varjenega nerjavnega jekla AISI 316L. Original Scientific Papers-Izvirni znanstveni članki Passivation of welded AISI 316L stainless steel 55: 408-419.

15

Electrochemical Synthesis and Characterization of Cu_2ZnSnS_4 Thin Films

Lakhe MG, Bhand GR, Londhe PU, Rohom AB and Chaure NB*

Electrochemical Laboratory, Department of Physics, Savitribai Phule Pune University, India

Abstract

Cu_2ZnSnS_4 (CZTS) thin films have been electrochemically deposited from aqueous electrolyte containing $CuCl_2$, $ZnCl_2$, $SnCl_4$ and $Na_2S_2O_3$ onto fluorine doped tin oxide (FTO) coated glass substrates. A conventional three-electrode geometry consisting working, counter and reference electrodes was used to perform the electrochemical experiments. The films were deposited at - 1.1 V with respect to Ag/AgCl reference, which was optimized by cyclic voltammetry. CZTS layers were annealed in tubular furnace at 400°C for 15 minutes in vacuum. As-deposited and annealed CZTS films were characterized using range of characterization techniques to study the structural, optical, morphological, and compositional and optoelectronic properties. Annealed sample revealed (112), (220) and (312) planes corresponds to tetragonal kesterite CZTS structure and secondary peaks of CuZn alloy. The optical study shows that the band gap of the as-deposited CZTS film was found to be 1.68 eV. Upon annealing the optical band gap ~ 1.49 eV corresponds to CZTS were estimated from UV-Visible Spectroscopy and photoluminescence. Densely packed, void free and relatively uniform thin films were deposited by electrodeposition technique. The grain size has been increased upon the heat treatment. Copper and zinc rich off-stoichiometric films were deposited at -1.1 V. Current density-Voltage (J-V) measurements showed Schottkey behavior. The flat band potential and carrier concentration estimated by C-V measurement for annealed CZTS sample was 0.30 V and ~ 2.4×10^{16} cm^{-3} respectively.

Keywords: CZTS thin films; Kesterite structure; Cyclic voltammetry; Characterization

Introduction

It is important to fabricate thin film solar cells with high efficiency from earth abundant, non-toxic and environmentally friendly elements/materials. In this scenario $Cu_2ZnSnSe/S_4$ (CZTSe/S) brings new hopes. It is I_2–II–IV–VI_4 promising quaternary kesterite non-toxic semiconducting compound with p-type conductivity [1-5] and large absorption coefficient, 10^4 cm^{-1} [2]. Its optical band gap varies from 1.0 eV to 1.5 eV [1-5] by replacing selenium with sulphur. The highest reported efficiency is 12.6% [6]. It can be synthesized by number of techniques such as hydrazine based solution process [7-9], nanoparticles from nontoxic solutions [10], thermal evaporation [11-13], chemical vapor deposition [14], sputtering [15-18], e-beam evaporation [19], electrodeposition using ionic liquids [20-25], spray pyrolysis [26,27] etc. Electrodeposition is one of the easy, scalable, cost-effective and found to be very successful technique. Either single-step or multistep approach has been accepted for the electrodeposition of CZTS absorber layer. Slupska et al. [28] has grown Sn-Zn-Cu alloy from aqueous bath containing $CuSO_4$, $SnSO_4$ and $ZnSO_4$ precursors by electrodeposition technique. Tri-sodium citrate was used as complexing agent/supportive electrolyte. Khalil et al. [29] has deposited Cu-Zn-Sn metal alloys on molybdenum substrate from electrolyte containing $CuSO_4$, $ZnSO_4$, Na_2SnO_3 and $K_4P_2O_7$. CZT alloy thin films were subsequently annealed in elemental sulphur ambient for the formation of CZTS. Valdes et al. [30] has electrodeposited CZTS thin films by electrochemical atomic layer deposition and conventional one-step electrodeposition. The deposition of non-stoichiometric CZTS films is reported by one-step electrodeposition. Here we report the synthesis of Cu_2ZnSnS_4 (CZTS) onto FTO substrate by single-step electrodeposition from aqueous bath. The preliminary results obtained from the pristine and annealed thin films are discussed.

Experimental Details

The CZTS films have been synthesized on FTO substrate by cathodic potentiostatic electrodeposition technique at pH 4.5 and bath temperature 50°C with moderate stirring. $CuCl_2$, $ZnCl_2$, $SnCl_2$

and $Na_2S_2O_3$ were used as precursors for the co-deposition of Cu, Zn, Sn and S, respectively. Tri-sodium citrate is used as complexing agent for the stoichiometric co-deposition of precursors [23,28,30-32]. A standard three-electrode system consisting working, counter and reference electrodes was employed for the electrodeposition of CZTS films. Commercially available fluorine doped tin oxide (FTO) coated soda lime glass substrates of resistivity 10-15 Ω/sq, platinum sheet and Ag/AgCl were used as working, counter and reference electrode, respectively. Potentiostat/galvanostat Model, Biologic SP 300 was used to perform the cyclic voltammetry (CV) and electrodeposition of CZTS thin films. Prior to the experimentation, all substrates were thoroughly cleaned using double distilled boiled water, acetone and iso-propanol followed by few minutes ultra-sonication with iso-propanol. The CV experiments were performed for various bath temperatures and stirring rate to optimize the suitable deposition potential for co-deposition of Cu, Zn, Sn and S. Deposition potential -1.1 V was optimized by using cyclic voltammetry experiments. The samples were annealed in vacuum at 400°C for 15 minutes. The pristine and annealed samples were studied by range of characterization techniques to study structural, optical, morphological, compositional and electrical properties. X-ray diffractometer (Model Bruker D8 Advance, Germany) of Cu Kα radiation, with λ=0.154 nm was used to study the structural properties. Optical measurements were performed using JASCO, UV-VIS-NIR Spectrophotometer model V-670. Photoluminescence was studied by

*****Corresponding author:** Chaure NB, Electrochemical Laboratory, Department of Physics, Savitribai Phule Pune University, India
E-mail: n.chaure@physics.unipune.ac.in

Perkin Elmer LS-55 Spectrophotometer. JEOL JSM-6360 A SEM/EDAX at accelerating voltage 20 kV and probe current 1 nA was used to study the surface morphology and elemental composition. Current density-Voltage (J-V) measurements were performed using the Potentiostat/galvanostat Model, Biologic SP 300. Frequency response analyzer (FRA) facility available in above mentioned potentiostat was used to study the C-V measurements at frequency 100 KHz.

Results and Discussion

Cyclic voltammetry

The cyclic voltammogram recorded with respect to Ag/AgCl reference electrode in the bath consisting precursors of Cu, Zn, Sn and S is shown in Figure 1. The temperature of the bath was maintained at 50°C with continuous stirring with 180 rpm throughout the experimentation. The redox potentials of Cu, Zn, Sn and S are different; therefore the co-deposition of these elements is difficult. However with the help of complexing agent the stoichiometric deposition can be obtained. The complexing agent can slow down the rate of reaction by forming complex or ligands with the nobel ionic species. The cathodic and anodic curves are marked by forward (black) and reverse (red) arrows. During the cathodic scan, upto ~ - 0.6 V, current was nearly steady indicates the applied growth potential was not sufficienct to deposite the precursors. The current found to be increased beyond -0.6 V to -0.9 V could be due to the metallic deposition of copper and zinc. The small plateau region revealed around – 1.0 V to -1.1 V ('A') could be suitable for the co-deposition of CZTS. The sharp linear rise in current beyond -1.1 V is proposed due to the rapid growth of metallic Cu_xZn_y alloy along with hydrogen evolution. The peak attributed during the anodic scan about -0.30 V, -0.15 V and + 0.30 V are associated to the stripping of Zn, Sn and Cu respectively. Indeed, we observed that the layer was completely stripped out after complition of CV measurement.

The flow of the complexed ionic species towards the electrode was maintained with continuous stirring. The number of samples were scanned during cyclic voltammetry experiment to optimize the deposition potential and it was found to be ~ - 1.1 V with respect to Ag/AgCl reference electrode. The reduction of copper, zinc, tin and sulphur occurs by the following reaction mechanism [33];

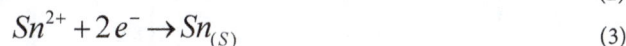

$$Cu^{2+} + 2e^- \rightarrow Cu_{(S)} \tag{1}$$

$$Zn^{2+} + 2e^- \rightarrow Zn_{(S)} \tag{2}$$

$$Sn^{2+} + 2e^- \rightarrow Sn_{(S)} \tag{3}$$

The deposition of sulphur is not straight forward. In the present bath $SnCl_2$ and $Na_2S_2O_3$ are used as sources of Sn and S respectively. Both the sources are strong reducing agents which can reduce the copper and zinc; therefore copper and zinc rich layers can be deposited. Upon the application of deposition potential $Na_2S_2O_3$ may decomposes by the following reaction mechanism,

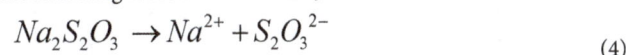

$$Na_2S_2O_3 \rightarrow Na^{2+} + S_2O_3^{2-} \tag{4}$$

The $S_2O_3^{2-}$ reacts with the ligands of the Cu and Zn and directly get deposited on the substrate in the form of CuS or ZnS. Another possible mechanism is, $S_2O_3^{2-}$ reacts with the CuZn metallic species which was already deposited on substrate and sulfurization occurs on the upper surface only which leads to the metal rich graded deposition of CZTS on the CuZn surface. The reported reduction potential for Cu, Zn and Sn [33] with respect to normal hydrogen are + 0.34 V, -0.763V and -0.136 respectively.

X- ray diffraction

The XRD pattern of (a) as-deposited and (b) annealed CZTS sample is shown in Figure 2. FTO peaks are marked by dark solid circles (•). In as-deposited sample, the two peaks present at 43.26° and 50.15° corresponds to (210) and (020) planes of Cu_5Zn_8, respectively [JCPDF file No. 14-1435]. The broad hump ranging from 20-28° corresponds the mixed amorphous and crystalline nature having short range periodicity of mixed ternary and quaternary phases of CuZnS or CZTS materials.

Upon heat treatment the broad hump is disappeared and the peaks exhibited at 28.58°, 47.61° and 56.71° corresponds to (112), (220) and (312) planes of CZTS of tetragonal kesterite structure [JCPDF file 26-0575]. The transfer of one phase to another phase depends on the formation of enthalpy of the particular phase. An important issue regarding the phase formation with quaternary semiconductors is whether homogeneous samples can be synthesized experimentally, or some secondary phases are also unintentionally formed. People have grown CZTS samples successfully using a variety of techniques (vacuum and non-vacuum; solution and solid-state) reported above, and significant variation has been achieved in the Cu: Zn: Sn atomic ratio, depending on the growth environment. If secondary phases are formed during the synthesis, these phases can be removed upon annealing in inert atmosphere. To describe the phase stability of CZTS as compared to the secondary phases Walsh et al. [34] has calculated the chemical stability region in the atomic chemical potential. To maintain a stable Cu_2ZnSnS_4 crystal, the chemical potentials of Cu, Zn, Sn, and S must satisfy the following equation:

$$2\mu_{Cu} + \mu_{Zn} + \mu_{Sn} + 4\mu_S = \Delta H_f(Cu_2ZnSnS_4) = -4.21\ eV \tag{5}$$

where ΔH_f represents the formation of enthalpy for CZTS from their respective elements. Both factors i.e., variation of Cu:Zn:Sn:S ratios and annealing treatment controls the chemical potential which shows the stable phase of CZTS within a narrow thermodynamic window.

Secondary phases observed at 42.75° and 49.83° corresponds to CuZn metallic alloy. It is also noticed that depending upon the enthalpy

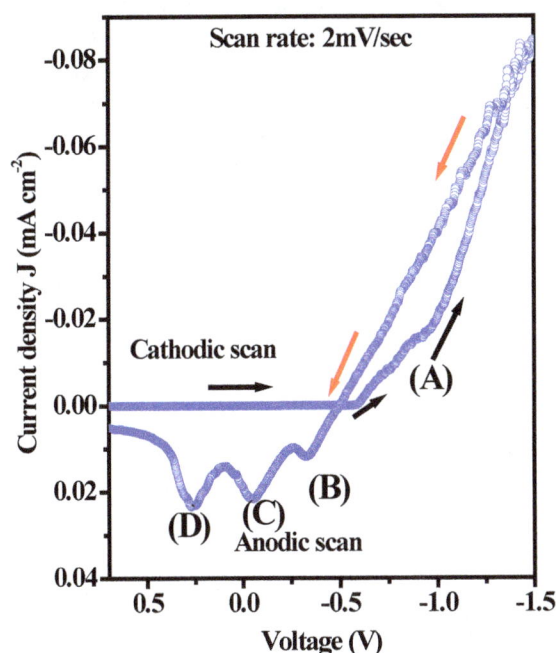

Figure 1: Cyclic voltammogram recorded at 50°C on FTO substrate in electrolyte containing Cu, Zn, Sn and S ionic species at pH 4.5.

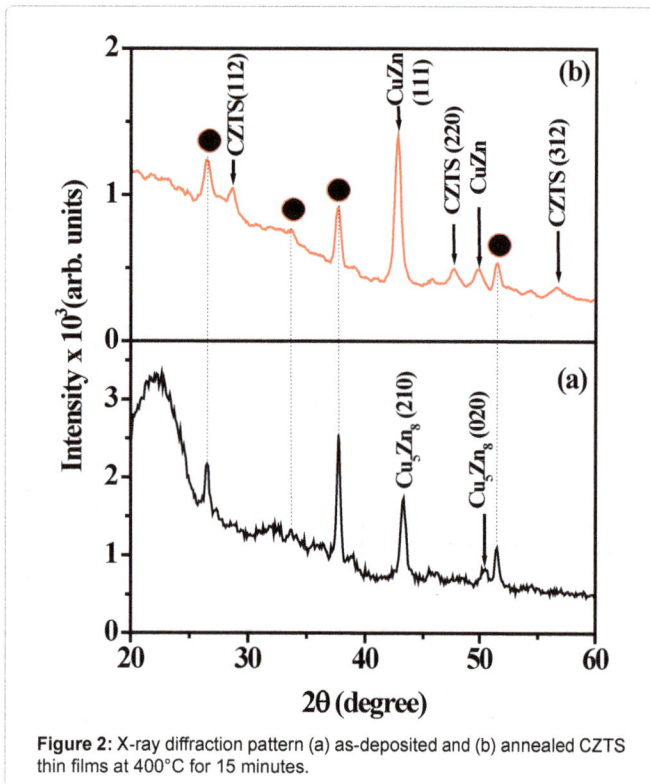

Figure 2: X-ray diffraction pattern (a) as-deposited and (b) annealed CZTS thin films at 400°C for 15 minutes.

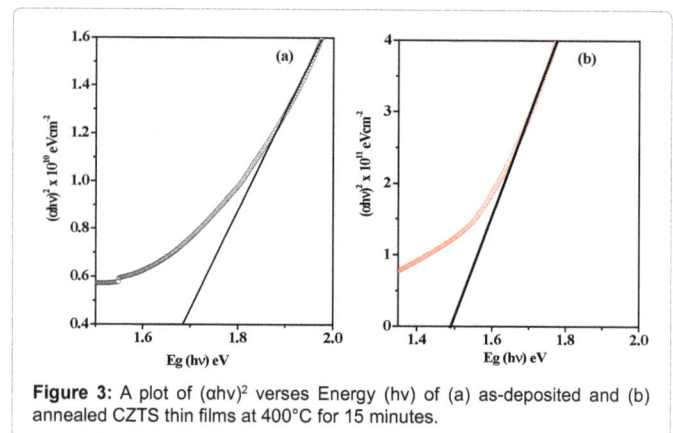

Figure 3: A plot of $(\alpha h\nu)^2$ verses Energy (hν) of (a) as-deposited and (b) annealed CZTS thin films at 400°C for 15 minutes.

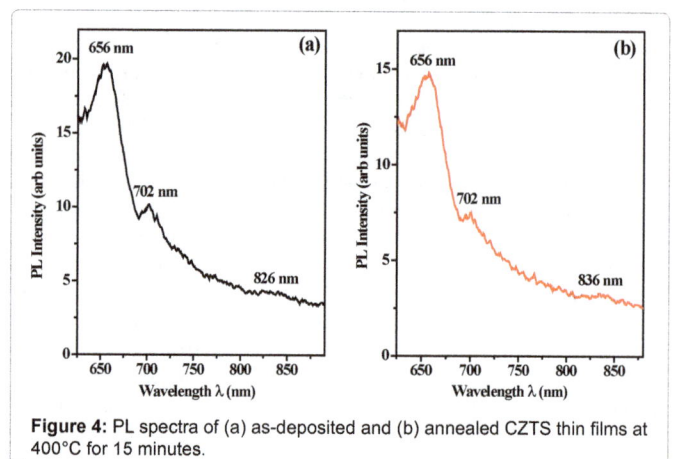

Figure 4: PL spectra of (a) as-deposited and (b) annealed CZTS thin films at 400°C for 15 minutes.

or free energy formation of Cu_5Zn_8 phase observed in as-deposited sample has been disappeared upon annealing and transformed into CuZn alloy due to the desired condition of free formation of CuZn phase.

UV-Visible spectroscopy

The optical study of as-deposited and annealed CZTS sample deposited at -1.1 V is carried out by UV- Visible –IR spectroscopy. Figure 3 depicts the $(\alpha h\nu)^2$ verses energy (hν) of (a) as-deposited and (b) annealed CZTS thin films. The optical band gap of as-deposited CZTS sample is found to be 1.68 eV. The large band gap value of as-deposited sample could be due to the short range crystallinity, presence of mixed secondary and ternary phases along with metallic phases which agrees well with XRD results. The reported band gap for Cu_2ZnSnS_4 is ~1.5 eV [1-5], however, due to variation in the chemical composition of the precursors and/or mixed surface morphology and grains size may affect on the band gap of CZTS, which can be varied in the range 1.4 eV to 1.6 eV [35,36]. Upon annealing the band gap estimated ~ 1.49 eV corresponds to CZTS. It is also noticed that after annealing the sample, the absorption intensity was found to be increased by one order of magnitude as compared to as-deposited sample.

Photoluminescence

Photoluminescence of (a) as-deposited and (b) annealed CZTS samples were studied by Perkin Elmer LS-55 Spectrophotometer and shown in Figure 4. The small shoulders observed in both as-deposited and annealed samples at 826 nm and 836 nm, respectively are associated with CZTS [1-5]. The peak exhibited around 702 nm could be associated with the ternary and quaternary alloys of CZTS. The highest intensity peak exhibited about 656 nm (1.89 eV) is associated to the formation of ternary alloys of CuZnS [37,38].

Scanning electron microscopy (SEM)

The morphology of CZTS samples were studied by scanning electron microscopy (SEM). Figures 5a and 5b depicts the SEM images of as-prepared and annealed CZTS thin films respectively. The compact and well adherent CZTS layers were deposited by electrodeposition technique at -1.1 V. It can be clearly seen from Figure 5 that the small grains are agglomerated in both as-prepared and annealed sample to form the clusters. The enhancement in the size of cluster upon annealing could be clearly seen. As both SEM images were obtained for same magnification, therefore the scale bar given in each SEM image can be used to estimate the cluster size. The particle size was found to be spherical because of the higher concentration of metallic copper and zinc. The grain size in the as-deposited film was ~ 1μm whereas after annealing it is found to be increased ~ 3-4 μm. A cauliflower like morphology has been observed in both as-deposited and annealed samples.

Energy dispersive spectroscopy (EDS)

The elemental composition of (a) as-deposited and (b) annealed CZTS films was studied by energy dispersive spectroscopy and summarized in Table 1. The elemental composition of as-deposited film is, Cu=48.28, Zn=21.88, Sn=9.43 and S=20.41, whereas after annealing it is Cu=48.03, Zn=27.89, Sn=8.82 and S=15.26. CZTS films were found to be copper and Zn rich. The sample close to the stoichiometry could be deposited by optimizing the concentration of complexing agent and/ or pH and temperature of the bath (Figure 6).

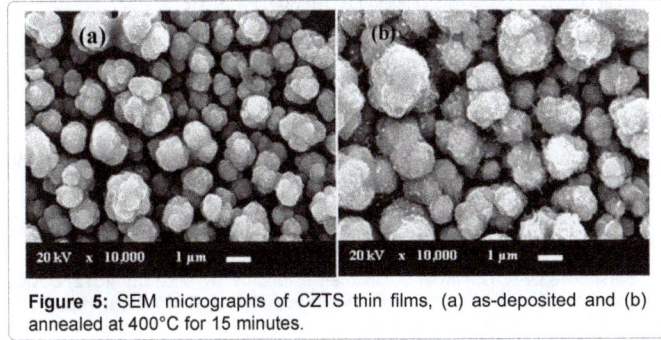

Figure 5: SEM micrographs of CZTS thin films, (a) as-deposited and (b) annealed at 400°C for 15 minutes.

Elemental Composition in At. %					
Deposition potential (V)	Sample condition	Cu	Zn	Sn	S
-1.1 V	As-deposited	48.28	21.88	9.43	20.41
	Annealed	48.03	27.89	8.82	15.26

Table 1: A summary of elemental compositions obtained by EDAX for as-deposited and annealed Cu_2ZnSnS_4 sample.

Current density-Voltage (J-V) measurement

The current density–voltage (J-V) measurement of as-deposited and annealed sample studied under dark condition is depicted in Figure 7. Both as-prepared and annealed CZTS sample shows Schottky behaviour under dark condition. The potential barrier, 'φ_b' was calculated by the following equation [39];

$$\varphi_b = -\frac{kT}{q}\ln\left(\frac{A^{**}T^2}{J_s}\right) \quad (6)$$

where, 'φ_b' is the barrier height, 'k' is the Boltzmann's constant, 'T' is the temperature, 'q' is the charge on electron, 'A*' is the effective Richardson's constant for CZTS (63.6 A/cm² K²) [40] and 'J_s' is the reverse saturation current density. The barrier height 'φ_b' was found to ~ 0.29 eV and 0.26 eV for as-deposited and annealed CZTS sample, respectively. The barrier height φ_b is found to be decreased upon annealing the sample. This is associated with several parameters viz. the crystallinity of the layer, the formation of homogeneous mixture of ternary/quaternary alloy, presence of secondary phases, grain boundaries and nature of metal semiconductor contact.

Capacitance – Voltage (C-V) measurement

The capacitance – voltage measurement of (a) as-deposited and (b) annealed CZTS sample was performed under dark condition with frequency 100 kHz and plots are shown in Figure 8. The observed Mott-Schottky plots of both as-deposited and annealed samples were nearly similar except small change in the flat band potential. The inversion, depletion and accumulation region are observed in both as-deposited and annealed samples however, inversion region is very small. The inversion and depletion region are related to the depletion of the charge carriers whereas accumulation is related to diffusion of the charge carriers. The values of flat band potential are found to be 0.36 V and 0.30 V for as-deposited and annealed CZTS sample respectively. The carrier concentration is calculated by using the following relation [39,40].

$$N_A = \frac{2}{\varepsilon_s q}\left[-\frac{1}{d\left(1/C^2\right)/dV}\right] \quad (7)$$

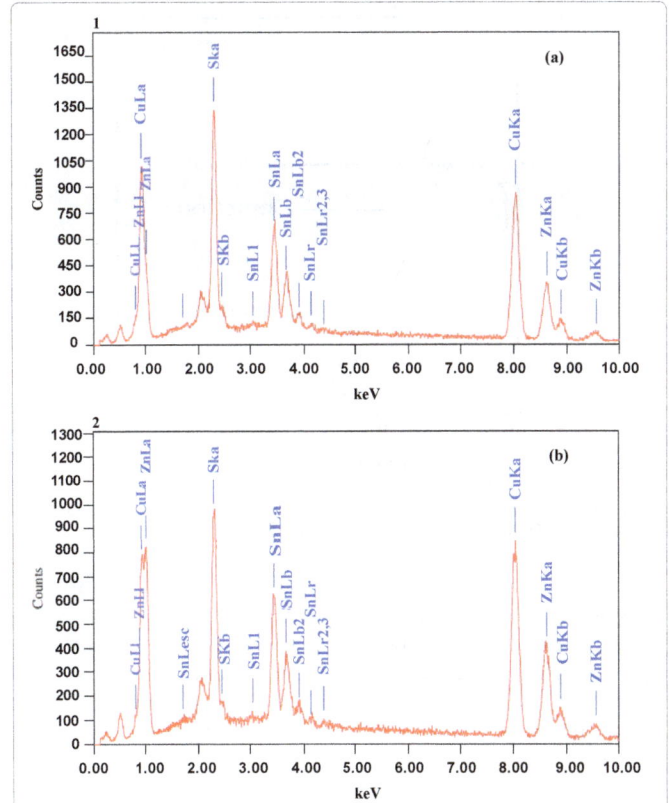

Figure 6: EDS spectra of CZTS thin films, (a) as-deposited and (b) annealed at 400°C for 15 minutes.

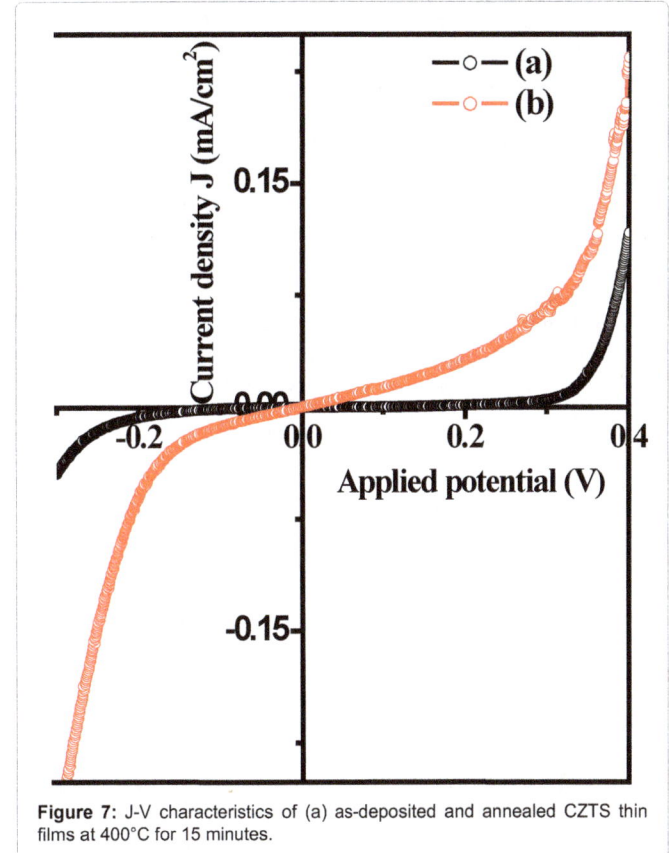

Figure 7: J-V characteristics of (a) as-deposited and annealed CZTS thin films at 400°C for 15 minutes.

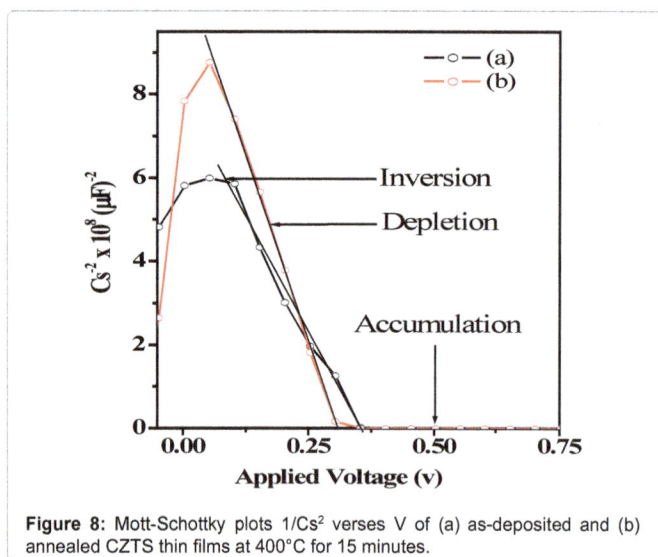

Figure 8: Mott-Schottky plots $1/Cs^2$ verses V of (a) as-deposited and (b) annealed CZTS thin films at 400°C for 15 minutes.

The value of static relative dielectric constant of CZTS material is taken 4.27 [10] for calculation of carrier concentration. The carrier concentration of the as-deposited film was found to be 1.4×10^{16} cm^{-3}. Upon annealing the carrier concentration was calculated ~ 2.4×10^{16} cm^{-3} which is very similar to that of as-deposited sample. Further optimization in annealing condition is under progress to obtain the photoactive semiconductor layers.

Conclusions

In conclusion, CZTS thin films can be deposited by low-cost electrodeposition technique. XRD results revealed (112), (220) and (312) planes corresponds to tetragonal kesterite CZTS structure. The optical band gap of the as-deposited CZTS film was found to be 1.68 eV. Upon annealing the band gap was estimated ~ 1.49 eV corresponds to CZTS which was further confirmed by photoluminescence study. Cauliflower like morphology of grain size 3 - 4 μm was observed upon annealing the CZTS sample. EDS data confirms the deposition of off stoichiometric CZTS films. The elemental composition of as-deposited film was Cu=48.28, Zn=21.88, Sn=9.43 and S=20.41, whereas, upon annealed the composition were Cu=48.03, Zn=27.89, Sn=8.82 and S=15.26. Both as-prepared and annealed CZTS sample showed Schottky behaviour under dark condition. The carrier concentration calculated from Mott-Schottky plot was found to be in the order of ~10^{16} cm^{-3}. The deposition of highly crystalline CZTS layers with desired composition close to the ideal is under progress.

Acknowledgement

One of the authors ML is thankful to University Grant Commission (UGC) for UGC-BSR fellowship.

References

1. Walsh A, Chen S, Wei S, Gong X (2012) Kesterite thin-film solar cells: advances in materials modelling of Cu$_2$ZnSnS$_4$. Adv Energy Mater 2: 400-409.

2. Grossberg M, Krustok J, Raudoja J, Timmo K, Altosaar M, et al. (2011) Photoluminescence and Raman study of Cu$_2$ZnSn(Se$_x$S$_{1-x}$)$_4$ monograins for photovoltaic applications. Thin Solid Films 519: 7403-7406.

3. Kahraman S, Çetinkaya S, Çetinkara HA, Guder HS (2014) A comparative study of Cu$_2$ZnSnS$_4$ thin films growth by successive ionic layer adsorption–reaction and sol-gel methods. Thin Solid Films 550: 36-39.

4. Peng C, Dhakal TP, Garner S, Cimo P, Lu S, et al. (2014) Fabrication of Cu$_2$ZnSnS$_4$

5. Salome PMP, Malaquias J, Fernandes PA, Ferreira MS, daCunha AF, et al. (2012) Growth and characterization of Cu$_2$ZnSn(S,Se)$_4$ thin films for solar cells. Solar Energy Materials & Solar Cells 101: 147-153.

6. http://www.solar-frontier.com/eng/news/2013/C026764.html

7. Bag S, Gunawan O, Gokmen T, Zhu Y, Mitzi DB (2012) Hydrazine-processed Ge-substituted CZTSe solar cells. Chem Mater 24: 4588-4593.

8. Todorov TK, Reuter KB, Mitzi DB (2010) Photovoltaic devices: high-efficiency solar cell with earth-abundant liquid-processed absorber. Adv Mater 22.

9. Barkhouse DAR, Gunawan O, Gokmen T, Todorov TK, Mitzi DB (2012) Device characteristics of a 10.1% hydrazine-processed Cu$_2$ZnSn(Se,S)$_4$ solar cell. Prog Photovolt: Res Appl 20: 6-11.

10. Saha SK, Guchhait A, Pal AJ (2012) Cu$_2$ZnSnS$_4$ (CZTS) nanoparticle based nontoxic and earth-abundant hybrid pn-junction solar cells. Phys Chem Chem Phys 14: 8090-8096.

11. Zhang J, Long B, Cheng S, Zhang W (2013) Effects of sulfurization temperature on properties of CZTS films by vacuum evaporation and sulfurization method. International Journal of Photoenergy 2013: 1-6.

12. Xinkun W, Wei L, Shuying C, Shuying C, Hongjie J (2012) Photoelectric properties of Cu$_2$ZnSnS$_4$ thin films deposited by thermal evaporation. J Semicond 33.

13. Shin B, Gunawan O, Zhu Y, Bojarczuk NA, Jay Chey S, et al. (2011) Thin film solar cell with 8.4% power conversion efficiency using an earth-abundant Cu2ZnSnS4 absorber. Prog Photovolt: Res Appl 21: 72-76.

14. Washio T, Shinji T, Tajima S, Fukano T, Motohiro T, et al. (2012) 6% efficiency Cu$_2$ZnSnS$_4$-based thin film solar cells using oxide precursors by open atmosphere type CVD. J Mater Chem 22: 4021-4024.

15. Katagiri H, Jimbo K, Maw WS, Oishi K, Yamazaki M, et al. (2009) Development of CZTS-based thin film solar cells. Thin Solid Films 517: 2455-2460.

16. Fernandes PA, Salome PMP, da Cunha AF, Schubert B (2010) Cu$_2$ZnSnS$_4$ solar cells prepared with sulphurized dc-sputtered stacked metallic precursors. Thin Solid Films 519: 7382-7385.

17. Hartman K, Johnson JL, Bertoni MI, Recht D, Aziz MJ, et al. (2011) SnS thin-films by RF sputtering at room temperature. Thin Solid Films 519: 7421-7424.

18. Han J, Shin SW, Gang MG, Kim JH, Lee JY (2013) Crystallization behaviour of co-sputtered Cu$_2$ZnSnS$_2$ precursor prepared by sequential sulfurization processes. Nanotechnology 24: 095706.

19. Katagiri H, Sasaguchi N, Hando S, Hoshino S, Ohoshi J, et al. (1997) Preparation and evaluation of Cu$_2$ZnSnS$_4$ thin films by sulfurization of E•B evaporated precursors. Sol Energy Mater Sol Cells 49: 407-414.

20. He X, Shen H, Wang W, Zhang B, Dai Y, et al. (2012) Effect of donor concentration on the PTCR behavior of Y-doped BaTiO$_3$–(Bi$_{1/2}$Na1/2)TiO$_3$ ceramics. J Mater Sci: Mater Electron 24: 431.

21. Sarswat PK, Free ML (2012) A comparative study of co-electrodeposited Cu$_2$ZnSnS$_4$ absorber material on fluorinated tin oxide and molybdenum substrates. Journal of Electronic Materials 41: 2210-2215.

22. Yang K, Ichimura M (2012) Preparation, characterization, and activity evaluation of CuO/F-TiO$_2$ photocatalyst. International Journal of Photoenergy 2012: 1-9.

23. Pawar SM, Pawar BS, Moholkar AV, Choi DS, Yun JH, et al. (2010) Single step electrosynthesis of Cu$_2$ZnSnS$_4$ (CZTS) thin films for solar cell application. Electrochimica Acta 55: 4057-4061.

24. Zhang X, Shi X, Ye W, Ma C, Wang C (2009) Electrochemical deposition of quaternary Cu$_2$ZnSnS$_4$ thin films as potential solar cell material. Appl Phys A 94: 381-386.

25. Bhattacharya RN, Kim JY (2012) Cu-Zn-Sn-S thin films from electrodeposited metallic precursor layers. The Open Surface Science Journal 4: 19-24.

26. Kamoun N, Bouzouita H, Rezig B (2007) Fabrication and characterization of Cu$_2$ZnSnS$_4$ thin films deposited by spray pyrolysis technique. Thin Solid Films 515: 5949-5952.

27. Kumar YBK, Babu GS, Bhaskar PU, Raja VS (2009) Preparation and characterization of spray-deposited Cu$_2$ZnSnS$_4$ thin films. Solar Energy Materials and Solar Cells 93: 1230-1237.

28. Slupska M, Ozga P (2014) Electrodeposition of Sn-Zn-Cu alloys from citrate solutions. Electrochimica Acta 141: 149-160.

29. Khalil MI, Bernasconi R, Magagnin L (2014) CZTS layers for solar cells by an electrodeposition-annealing route. Electrochimica Acta 145: 154-158.

30. Valdes M, Modibedi M, Mathe M, Hillie T, Vazquez M (2014) Electrodeposited Cu_2ZnSnS_4 thin films. Electrochimica Acta 128: 393-399.

31. Ge J, Jiang J, Yang P, Peng C, Huang Z, et al. (2014) A 5.5% efficient co-electrodeposited $ZnO/CdS/Cu_2ZnSnS_4/Mo$ thin film solar cell. Solar Energy Materials & Solar Cells 125: 20-26.

32. Lee SG, Kim J, Woo HS, Jo Y, Inamdar AI, et al. (2014) Structural, morphological, compositional, and optical properties of single step electrodeposited Cu_2ZnSnS_4 (CZTS) thin films for solar cell application. Current Applied Physics 14: 254-258.

33. Pandey RK, Sahu SN, Chandra S (1996) Handbook of semiconductor electrodeposition. CRC Press, USA.

34. Walsh A, Chen S, Wei S, Gong X (2012) Kesterite thin-film solar cells: advances in materials modelling of Cu_2ZnSnS_4. Adv Energy Mater 2: 400-409.

35. Han J, Shin SW, Gang MG, Kim JH, Lee JY (2013) Crystallization behaviour of co-sputtered Cu_2ZnSnS_2 precursor prepared by sequential sulfurization processes. Nanotechnology 24: 095706.

36. Zaberca O, Oftinger F, Chane-Ching JY, Datas L, Lafond A, et al. (2012) Surfactant-free CZTS nanoparticles as building blocks for low-cost solar cell absorbers. Nanotechnology 23:185402.

37. Kitagawa N, Ito S, Nguyen D, Nishino H (2013) Copper zinc sulfur compound solar cells fabricated by spray pyrolysis deposition for solar cells. Natural Resources 4: 142-145.

38. Yildirim MA, Ates A, Astam A (2009) Annealing and light effect on structural, optical and electrical properties of CuS, CuZnS and ZnS thin films grown by the SILAR method. Physica E 41: 1365-1372.

39. Sze SM, Poplai HS (1983) Physics of semiconductor devices. 3rd Edn., Wiley Eastern Limited, New Delhi, India.

40. Rakhshani AE, Thomas S (2015) Cu2ZnSnS4 films grown on flexible substrates by dip coating using a methanol-based solution: electronic properties and devices. Journal of Electronic Materials 44: 4760-4768.

16

Determination of Optimum Post Weld Heat Treatment Processes on the Microstructure and Mechanical Properties of IS2062 Steel Weldments

Chennaiah MB[1]*, Kumar PN[2] and Rao KP[3]

[1]*Assistant Professor in ME Department, V.R.Siddhartha Engineering College, Vijayawada, India*
[2]*Professor in ME Department, N.B.K.R Institute of Science and Technology, Vidyanagar, India*
[3]*Professor in ME Department, J.N.T.University College of Engineering, Ananthapuram, India*

Abstract

This study investigates the effect of post heat treatment on the microstructure and mechanical properties of IS2062 steels. Similar metal joints of IS2062 weldments are prepared by using MIG welding process. This melting is occurring at edges of the plates because of sufficient amount of heat energy is passing over the plate per unit time and density of energy is supplied to the wire. In this connection, heating and cooling of weldment some of the disturbances in metallurgical and mechanical point of view. To overcome we will choose suitable post weld heat treatment to avoid the disturbances and improve industrial requirement irrespective of the mechanical, microstructure of the weldment. The objective is to determine the optimum post weld heat treatment method for the IS2062 steel. After welding, the effects of post weld heat treatment on weld metal microstructure and mechanical properties including weldment tensile strength, impact and hardness over the room temperature range 32°C are investigated. In particular this study the effect of these properties to understand estimate heat treatments on the tensile impact, Hardness materials are considered as weldments before heat treatment and after post weld heat treatment.

Keywords: Heat input; Mechanical properties; Post weld heat treatment; Welding zone; HAZ

Introduction

In any welding technique can be heating and cooling of the parent materials, strength setup of the entire materials can be obtained overmatched depending upon the temperature or heat energy is transferred in to the parent materials. This overmatched strength of the welded metals often restricts some of the welding processes are controlled, but metallurgical point of view it is not suitable for entire material to maintain uniformity. So researcher chooses the post weld heat treatment processes to improve the tensile, impact and hardness of the parent metals in order to choose optimal post weld heat treatment processes increasing the mechanical and metallurgical results, which derive the industrial requirement and literature [1,2]. To improve mechanical properties in pressure vessels and boilers of thicker sections, considerable research efforts have been directed and many papers have described the methods of improving mechanical properties by conducting tests and results. The literature survey provides on post weld heat treatment, shows the effect of post weld heat treatment on certain mechanical properties and the problems arising during post weld heat treatment. The toughness of the CGHAZ recovers the slowest as a function of increasing post weld heat treatment temperature and remains low until a 730°C heat treatment. To guarantee an adequate HAZ toughness, a minimum post weld heat treatment temperature of 730°C for 2 h is recommended. This recommendation agrees with the ASME code required 732°C minimum tempering temperature for the base metal. Ahmad studied the effect of a post-weld heat treatment (post weld heat treatment) on the mechanical and microstructure properties of an AA6061 sample welded using the gas metal arc welding (GMAW). The welded samples were divided into as-welded and post weld heat treatment samples. The post weld heat treatments used on the samples were solution heat treatment, water quenching and artificial aging. Both welded samples were cut according to the ASTM E8M-04 standard to obtain the tensile strength and the elongation of the joints. A Vickers micro hardness testing machine was used to measure the hardness across the joints. By implementing post weld heat treatment, a 3.8% increase was recorded for tensile strength, hardness strength was increased by 25.6% and a 21.5% higher elongated was achieved. The results proved that post weld heat treatment was able to enhance the hardness strength and tensile properties of AA6061 welded joints using GMAW. The higher values of hardness, tensile strength and elongation are due to the fact that post weld heat treatment produces a fine and uniform distribution of precipitates at the weld joints.

Sample Preparation

The experiment was performed on samples which were made with specific dimensions of approximately 250 x 100 x 10 mm single V-type grooved samples are prepared from IS2062 as main test plate. The chemical composition of IS2062 and filler material is shown in Table 1. Filler metal as MIG wire (Copper Coated Mild Steel) with diameter 1.2 mm is taken. The importance of copper coating on Mild steel is used to prevent rust and also current to pass current easily. The steel has a 0.22% of carbon content; as a result weldability and toughness are improved. These plates having the same type groove are welded together using MIG welding processes. As we have considered 10 mm thickness plate, before doing Welding we have to preheat (100°-150°C) the materials in order to prevent the moisture in the metal, distortion control and also for cracks rectification. 16 samples are prepared by using welding parameters of current 160-180 amp, voltage 26-30 V, welding speed 3.3-3.5 mm/sec. 4 samples are before post weld heat treatment and remaining after post weld heat treatment processes is conducting as

---

*Corresponding author: Chennaiah MB, Assistant Professor in ME Department, V.R.Siddhartha Engineering College, Vijayawada, India
E-mail: chennai303.mech@vrsiddhartha.ac.in

S.No	Name of Material	Composition in % of Weight						
		C	Mn	Si	Cr	S	P	Mg
1.	Mild steel (IS 2062)	0.22	1.5	0.40	-	0.045	0.045	-
2.	Copper Coated Mild Steel	0.1	1.86	0.73	0.2	0.30	0.03	-

Table 1: Chemical composition of IS2062 and filler material.

follows, namely Annealing, Normalizing, Tempering conditions [3]. The MIL-STD welding procedure is followed in preparation of IS2062 weldment [4].

Tensile test samples

The tensile test samples are prepared as per procedure is used to cut from the weldment. While conducting tensile test firstly, to measure the initial length and the diameter of the test piece and noted and then test piece is mounted on the testing machine and then apply load and after sometime test piece is elongated and necking is formed and then loading process continued until fracture. This test is conducted for test pieces of 4 joints. Heat treatment is as stage in the fabrication of structures and is often forgotten; but it has perhaps more wide-reaching and important ramifications than many of the other stages in the fabrications of structures or components. Especially at the welded zone careful observation is required, the test values are tabulated in the Table 1.

Impact test samples

This test samples are prepared and tested as per ASTM-IS2062 procedure [5,6]. The samples are cut longitudinally to the weld axis with notch in the middle of the welded area as shown in Figure 1. This work is carried out at room temperature.

Hardness

The sample preparation and evolution is carried based on ASTM procedure [7]. The hardness measure is done at different zones of weldment namely base metal, HAZ and WZ, and their values are tabulated.

Microstructure- a process methodology

Welding processes and the associated heating and cooling of edges of the similar plates leading to the development of different microstructures in different zones have been observed. For metallographic observation before post weld heat treatment and after post weld heat treatment specimens were etched with 4% nital for 20 s and consequently the microstructures of the base, weld and the heat affected zone were defined. Specimens were prepared for Electron Back Scattered Diffraction (EBSD) analysis using standard sample preparation method. A Zeiss 940SEM with a tungsten filament was used. The SEM device is coupled with automatic OIMTM (Orientation Imaging Microscopy) software [8-10].

Results and Discussion

Tensile test

The tensile test is conducted on before post weld heat treatment and after post weld heat treatment on test specimens. One of the most significance observation at annealing conditioned test sample given higher value than the other samples (Figure 2 and Table 2).

Impact strength

While conducting Impact test firstly we have to prepare the test

piece as per dimensions by using milling machine and also by using V cutter in order to make a notch up to 2 mm thickness. The arrangement of test setup consist of big size of pendulum is used to hit the prepared samples at an standard height, at the same time the amount of energy is absorbed by the samples from the pendulum to fracture the test samples, the results obtained from the testing machine. Here the minimum impact energy in before heat treatment and maximum at annealing condition test sample (Figure 3 and Table 3).

Hardness

Hardness was found to be a reliable method of estimating the yield and tensile strength of the different zones of the weldment before post weld heat treatment and after post weld heat treatment. The tensile data obtained from the sections of B.M, W.Z and HAZ could be useful to

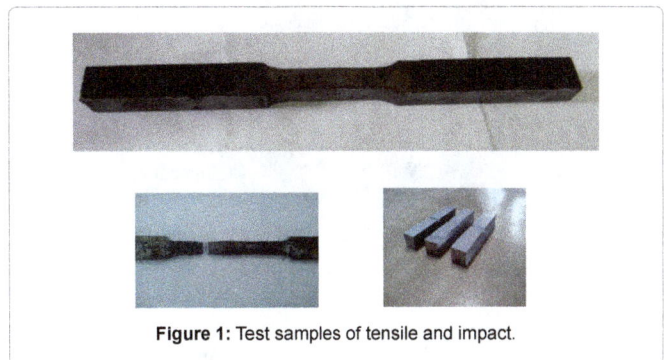

Figure 1: Test samples of tensile and impact.

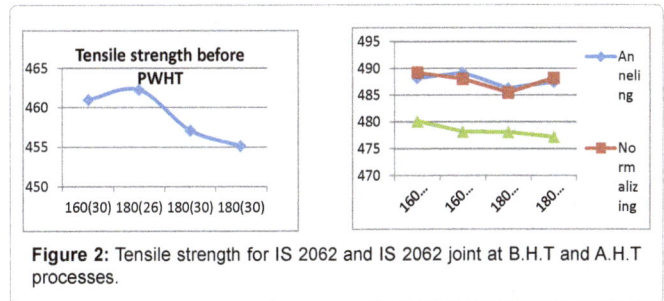

Figure 2: Tensile strength for IS 2062 and IS 2062 joint at B.H.T and A.H.T processes.

	Current (Amps)	Voltage (Volts)	Tensile strength (N)			
			Before post weld heat treatment	After Annealing	After Normalizing	After Tempering
1	160	26	461.01	488.17	489.23	480.12
2	160	30	462.24	489.23	488.01	478.21
3	180	26	457.112	486.27	485.49	478.11
4	180	30	455.16	487.48	488.23	477.22

Table 2: Before post weld heat treatment& after post weld heat treatment processes of IS 2062 welded joints tensile strength.

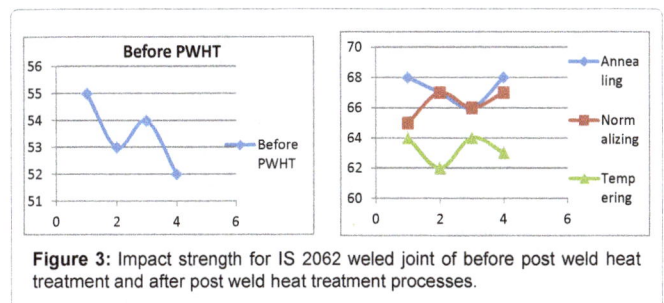

Figure 3: Impact strength for IS 2062 weled joint of before post weld heat treatment and after post weld heat treatment processes.

	Current (Amps)	Voltage (Volts)	Impact strength (J)			
			Before post weld heat treatment	After Annealing	After Normalizing	After Tempering
1	160	26	55	68	65	64
2	160	30	53	67	67	62
3	180	26	54	66	66	64
4	180	30	52	68	67	63

Table 3: Impact strength results for IS 2062 welded joint for before and after post weld heat treatment process.

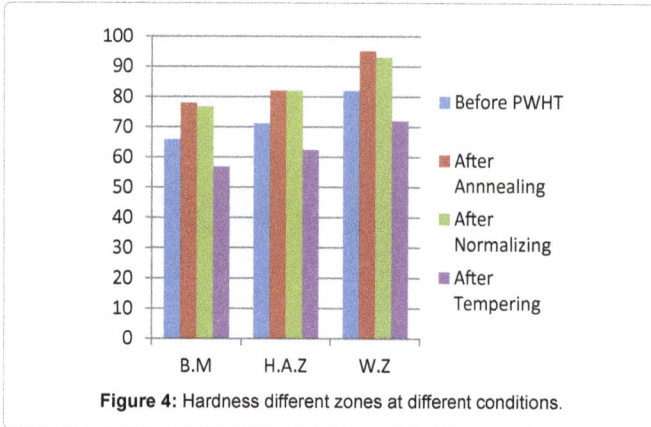

Figure 4: Hardness different zones at different conditions.

Hardness	Before post weld heat treatment	After Annealing	After Normalizing	After Tempering
B.M (IS 2062)	66	78	76.75	57
H.A.Z (IS 2062)	72	82	82	62.5
W.Z (IS 2062)	82	95	93	72

Table 4: Hardness at different locations in IS 2062 and IS 2062.

identify for industrial requirements. A practical standpoint, hardness measurements are quicker and more straightforward than numerical calculation of the HAZ. Especially at the welded zone of similar material is 15.85% is increases for the before heat treatment condition test samples (Figure 4 and Table 4).

Microstructure- a process methodology

Welding processes and the associated heating and cooling of edges of the similar plates leads to the development of different microstructures in different zones have been observed. For metallographic observation before post weld heat treatment and after post weld heat treatment specimens were etched with 4% nital for 20 s and consequently the microstructures of the base, weld and the heat affected zone were defined. Specimens were prepared for Electron Back Scattered Diffraction (EBSD) analysis using standard sample preparation method. A Zeiss 940SEM with a tungsten filament was used. The SEM device is coupled with automatic OIMTM (Orientation Imaging Microscopy) software. In this before post weld heat treatment similar metals are to be joined by using MIG processes. It may be seriously after the size and shape of the grains of which materials is composed with filler metals. Depending upon the welding processes the grains of the welded zone grow to larger size and sometimes destroyed stresses setup during the welding and cooling (Table 5).

Annealing process: In this Heat treatment processes the weldment is heated to above the recritilization temperature (>750°C) with the weldment the carbon content is varied with respect to the temperatures. The carbon can be dissolved in it to form a solid solution ferrite (C = 0.006%) at room tempeture, this can be increased to the heating temperatures limits. The micro structure of IS2062 steel obtained with carbon content of 0.83% or less is normally grains of pearlite and ferrite in the base metals, HAZ and welded zone (Table 6).

Tempering: At low tempering temperatures (Approx. 80-200°C) a hexagonal close-packed carbide (called epsilon carbide) begins to form, and with this rejection of carbon the crystal structure of martensite changes ultimately from tetragonal to the body-centered cubic characteristic of ferrite. The second stage at about 200 to 300°C, depending upon the steel is characterized by transformation of the retained austenite to bainite. During third stage from 300 to 475°C (Approx) there is formation of Fe_3C (cementite) from epsilon carbide and Change from low-carbon martensite to cubic ferrite. From 450 to 705°C (Approx.) the cementite (Fe_3C) agglomerates and coalesces. The structure becomes an aggregate of ferrite with cementite in quite fine spheres, referred to as tempered martensite and tempered bainite.

Record of Microstructure		Specimen name	Material	Condition
		ISO 2062 Gr.E250B	MS Plate	Before Heat Treatment
Heat Treatment	Description	Before Heat Treatment		
	Temperature (Max)	-		
	Temperature cooling (Min)	-		
	Soaking Time	-		
	Cooling Time	-		
Etchant Used		4% Nital		
Magnification		100X		
Parent Metal		HAZ	Wedment	
Result		The microstructure in parent metal consists of pearlite and ferrite, while the weld consists of coarse martensite.		

Table 5: Represents microstructure welding processing of grains of materials.

Record of Microstructure	Specimen name	Material	Condition	
	ISO 2062 Gr.E250B	MS Plate	Annealing	
Heat Treatment	Description	Annealing Heat Treatment		
	Temperature (Max)	750°C		
	Temperature cooling (Min)	200°C		
	Soaking Time	1 hr		
	Cooling Time	3 hrs		
Etchant Used	4% Nital			
Magnification	100X			
Parent Metal	HAZ	Weld Zone		
Result	The microstructure in parent metal consists of pearlite and ferrite, while the weld consists of martensite.			

Table 6: Represents microstructure of annealing processes.

Record of Microstructure	Specimen name	Material	Condition	
	ISO 2062 Gr.E250B	MS Plate	Tempered	
Heat Treatment	Description	Tempered Heat Treatment		
	Temperature (Max)	705°C		
	Temperature cooling (Min)	200°C		
	Soaking Time	1 hour		
	Cooling Time	2 hrs		
Etchant Used	4% Nital			
Magnification	100X			
Parent Metal	HAZ	Weld Zone		
Result	The microstructure in parent metal consists of pearlite and ferrite, while the weld consists of fine tempered martensite.			

Table 7: Represents microstructure of tempering processes.

+Record of Microstructure	Specimen name	Material	Condition	
	ISO 2062 Gr.E250B	MS Plate	Normalising	
Heat Treatment	Description	Normalised Heat Treatment		
	Temperature (Max)	1100°C		
	Temperature cooling (Min)	200°C		
	Soaking Time	1 hour		
	Cooling Time	4 hrs		
Etchant Used	4% Nital			
Magnification	500X			
Base Metal	HAZ	Weldment		
Result	The microstructure in parent metal consists of pearlite and ferrite, while the weld consist of martensite.			

Table 8: Represents microstructure of normalizing processes.

The structure may become more or less uniformly spheroidized from prolonged heating at the upper end of the range (Table 7).

Normalizing: Normalizing or air quenching consists in heating steel to about 40-50°C above its upper critical temperature (i.e., A_3 and A_{cm} line) and, if necessary, holding it at that temperature for a short time and then cooling in still air at room temperature. The purpose of structure is obtained by normalizing largely depends on the thickness of cross section as this will affect the rate of cooling. Thin sections will give a much finer grain than thick sections. Normalizing produces microstructures consisting of ferrite (white network) and pearlite (dark areas) for hypoeutectoid (i.e., up to about 0.8% C) steels. For eutectoid steels, the microstructure is only pearlite and it is pearlite and cementite for hypereutectoid steels (Table 8).

Conclusions

In this work IS2062 of 10 mm thickness of the plate is used as single V-type grooved weldment. Optimal Post Weld Heat Treatment processes are used for the following annealing, normalizing and tempering condition. Tensile test, harpy-V test and hardness samples are evaluated from the IS2062 as compared to the before post weld heat treatment conditions. This work is effectively used to improve the properties of weldment by using optimal post weld heat treatment processes. Particularly in between HAZ and WZ some cracks are initiated, finaly it's propagated during the preparation of the samples. These are observed in micro hardness examinations at different zones of weldment.

References

1. Ritter JC, Dixon BF (1987) Improved properties in welded HY-80 steel for Australian warship. Weld J 66: 33-44.

2. Brosilow R (1991) High-strength steels: a progress report. Weld Design Met Fabr 64:40-44.

3. Sampath K, Civis DA, Kleinosky MJ (1993) Effects of GMA welding conditions on high strength steel weld metal properties for ship structures. Proceedings of international symposium on low-carbon steels for the 90's, Warrendale, PA, USA.

4. MIL-STD-1688 fabrication, welding and inspection of HY 80/100 submarine application (Replacement Document Navsea Pub T9074-ad-GIB-010/1688).

5. American Society for Testing and Materials (2008) ASTM E8M-04 standard test methods for tension testing of metallic materials.

6. American Society for Testing and Materials (1982) ASTM E 23 standard test methods for notched bar impact testing of metallic materials.

7. ASTM E92-82(2003) Standard test methods for vickers hardness of metallic materials. ASTM International, West Conshohocken, PA.

8. MIL-STD-16216G Steel plate, alloy, structural High Yield strength (HY-80 AND HY-100).

9. Cakici B (2002) Investigation of mechanical properties of HY 80 steel joints, welded by using arc welding methods. Kocaeli University.

10. Gungor ON (1996) The effects of welding processes on the mechanical properties of the welded joint and HAZ for the quenched and tempered HY 80 high strength low alloy steel. Kocaeli University.

17

Fabrication and Testing of Pure and NaOH Treated Sisal Fibre Reinforced Bio-Composite

Nandan Kumar GM and Kingsly Jasper M*

VIT University, Vellore, Tamil Nadu, India

Abstract

Natural fibres are now widely used for reinforcements of materials as they are a reliable resource and the cost is also less. One of the significant reasons of using natural fibres as reinforcements instead of polymers or other plastics is that they are degradable and don't cause any harm to the environment. This article is particularly centered on polymer matrix composites made of Sisal fibre which was accessible locally and finding the composite's mechanical property (tensile strength). A number of test samples of sisal fibre composite are fabricated by ASTM standards and made to undergo tensile tests in order to obtain the desired properties. In the later part of the work a comparison between the mechanical properties of a non-treated pure Sisal fibre composite and a NaOH treated Sisal fibre composite is carried out to check for any enhancements in the properties.

Keywords: Bio-Composite; Fabrication; Materials; Matrix

Introduction

Sisal fibre the second hardest natural fibre is obtained from the leaves of the plant Agave Sisalana. It is widely cultivated in India, Brazil, East Africa and Indonesia [1,2]. It has very good durability and strength. It is also one of the most extensively cultivated hard fibre which makes it easily available. The fibers are extracted from the leaves of the plant. The fibre mass in each leaf is about 4% [3]. The diameter of the fibre varies between 0.1 mm to 0.3 mm and is extracted by retting or scrapping [4].

Paramasivam in their article investigated the feasibility of using sisal fibre composites due to their amenability to laminating and low cost of production. On experimenting, they also found that the unidirectional modulus of the composite was found to be 8.5 Gpa.

Chandramohan stated in his paper all the various applications where the Sisal fibre is constantly utilized. It was also stated that the fibre has a good strength, durability and a resistance to deterioration in salt water. The wall covers made up of Sisal were also found to meet the ASTM and national fire protection association standards. Also, many a lot focus is also placed in trying to put it into utilization in the automobile industry [5].

J M L Reis in his paper studied the effect of fibre surface treatment on fracture properties for natural sisal fibre reinforced polymers. The surface treatment was done by soaking the fibres in NaOH solution and Acetic acid. Epoxy and unsaturated resin were used as the matrix for the composite. For the comparison a study with untreated fibre was also made. One observing the results it was given that the untreated fibre had better fracture properties and the 10% NaOH solution treated fibre had the least properties [6].

Maries Idicula investigated the filler concentration for several fibre surface treatments. The thermophysical behavior was analysed and evaluated for a constant load along the fibres. From the results it was inferred that the thermal contact resistance is reduced due to the surface treatments. It was also suggested that a hybrid composite would result in better heat transportation [7].

P K Bajpai in their article studied the possibilities of forming a tribo material by combining a bio polymer with a locally available plant fibre. They combined three different plant fibres including sisal in PLA polymer and developed laminate composites. Thermal, wear and frictional characteristics were analysed. On experimentation they found

that due to this incorporation the wear resistance is increased extensively and a reduction in friction co-efficient was also observed [8].

A Karthikeyan presented a study on the effect of fibre length and NaOH treatment on composites. Various NaOH solution concentrations were taken from the range of 2-10%. The treated fibre was then made into a composite by combining it with Epoxy matrix and fabricated by hand layup technique [9].

On observing the results it was seen that the treated fibres showed an enhancement of about 15% in impact strength when compared with an untreated fibre. As indicated by the Cambridge advanced Dictionary, Epoxy resins also known as the poly epoxides are reactive polymers containing epoxide groups. It is a strong adhesive used for gluing things together and covering the surfaces. The playing point of using Epoxy is that the curing can be either attached or eased off by adding additives called as Hardeners or Curatives (Figures 1 and 2) [10].

Lucas Sobczak in their article have analysed the mechanical properties of the natural fibre composites. The focus is also placed on the mechanical filler as they play a majorrole in composites. The performance characteristics data were taken and compiled and comparison was made in so called Ashby plots. Based on the Ashby plots a full examination was done on the natural fibre composites (Tables 1 and 2) [11].

Methodology

In this work, Sisal fibre composite laminates were made in the dimensions 300 × 300 × 10 mm. The laminate is formed by Bi-directional arrangement of the fibres by orienting fibre in two angles (0 and 90 degree). The laminate comprises of totally four layers of the fibre spread over for a final thickness of 10 mm.

The fibres were taken and combed and then cut into 300 mm length. The composite was made by hand layup technique. So a mould was made

*****Corresponding author:** Kingsly Jasper M, VIT University, Vellore, Tamil Nadu 632014, India, E-mail: kingslyjasper@yahoo.co.in

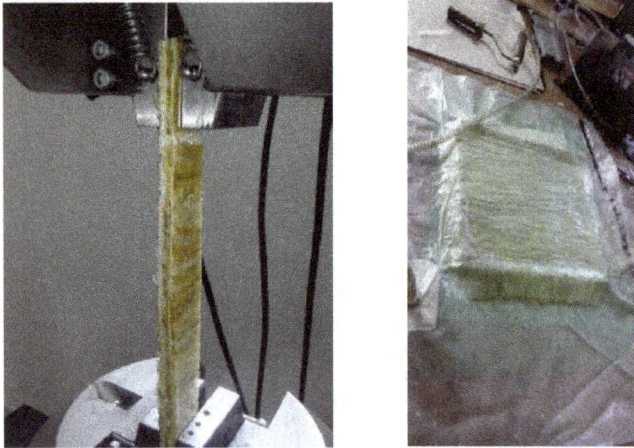

Figure 1: Specimens taken according to the ASTM standards.

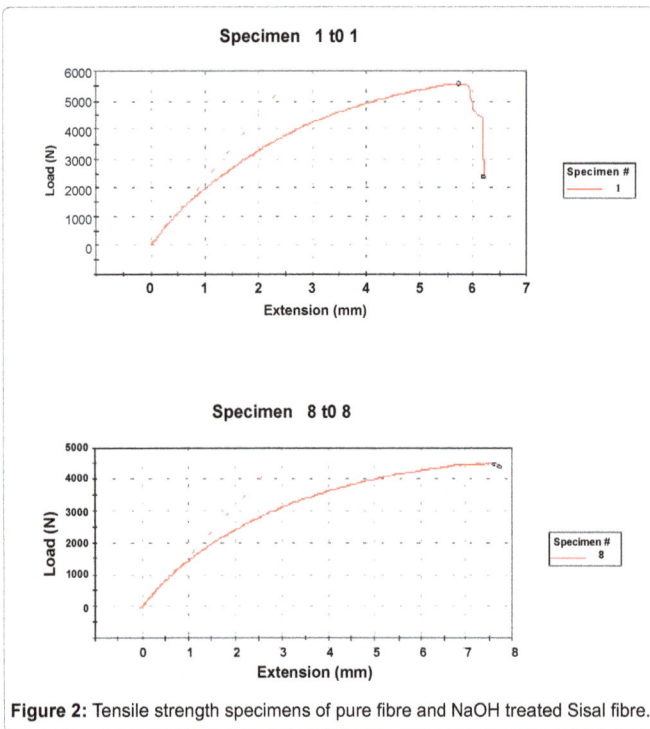

Figure 2: Tensile strength specimens of pure fibre and NaOH treated Sisal fibre.

S.no	Specimen Label	Tensile Strain at Maximum Load (%)	Load at Break (Standard) (kN)	Tensile Stress at Break (Standard) (MPa)
1	001	3.82332	2.37	7.32
2	002	3.54214	2.86	8.83
3	003	4.1733	4.92	15.19
4	004	3.83078	3.42	10.57
5	005	3.32244	1.2	3.71

Table 1: The tensile tests were conducted with specimens of pure fibre.

out of wood. The inner dimension of the mould was the same as that of the laminate. The top of the mould box was kept open to arrange the fibres. The matrix used is Epoxy GY257 and the hardener used for accelerating the curing process is HY951. The Epoxy and the hardener

S.no	Specimen Label	Tensile Strain at Maximum Load (%)	Load at Break (Standard) (kN)	Tensile Stress at Break (Standard) (MPa)
1	006	5.0921	3.6	9.18
2	001	6.7762	4.17	10.64
3	001	5.4253	4.41	11.24
4	009	4.8514	3.87	9.88
5	010	6.1374	3.16	8.05

Table 2: The tensile tests were conducted with NaOH treated Sisal fibre.

were mixed in a 3:1 ratio and the matrix for combining the matrix and fibre was made. The mould box was taken and Silicon spray is sprayed at the bottom to avoid the gluing of the mould with the composite. A layer of sisal fibre is laid and the epoxy matrix is spread evenly over them till there are no dry places on the layer.

The next layer of sisal is kept perpendicular to the previous layer and rolled using a roller to spread the fibre evenly all over the layer. The same is repeated for four layers. After laying up of all the layers another sheet sprayed with Silicon spray is kept on top of the mould and covered. This entire set up is kept inside vacuum bag and 760 mm of Hg pressure is applied for half hour and then weight is kept on top of them to make sure the fibre glues with the matrix and within itself properly. The setup is left for curing for about 48-60 hours.

The laminates are then cut into required sizes as per the ASTM standards for tensile testing using jig saw and the specimens are tested in the Universal Testing Machine and the results are obtained.

Calculation

Density of epoxy GY257: 1.2 gm/cc

Density of sisal fibre: 1.33 gm/cc

Areal weight of fibre: 57.5 gm

Volume fraction= $(\frac{W_f}{\delta_f})/(((\frac{W_f}{\delta_f})+(100-W_f))/\delta_r)$

δf : Density of sisal fibre

δr: Density of epoxy resin

Therefore volume fraction = 0.55

Volume of sisal fibre: 57.5/1.33 = 43.2 cc

Volume of one layer: 43.2/0.5 = 86.46 cc

Volume of resin required for one layer of sisal fibre and epoxy mixture: 86.46 × 0.45 = 38.90 cc

Mass of resin: 38.9×1.2= 46.68 gms

Mass of resin required for 12 layers:

46.68 × (4+1) = 234 gms

Resin includes hardener and epoxy in 1: 3 ratio, therefore 58 gm of hardener and 176 gm of epoxy.

Results and Discussions

The specimens taken according to the ASTM standards are tested in the Universal Testing Machine and the following values were obtained. The image during the testing of the specimen in the Universal testing Machine is given below. The results obtained from the Universal testing Machine are represented graphically and the overall results are also tabulated.

Conclusion

The tensile tests were conducted with respect to the ASTM standards on comparison it was found that the two specimens of pure fibre and NaOH treated Sisal fibre. The following conclusions were made; when sisal was treated with the NaoH, it was found that the surface characteristic of sisal changes and strength gets increased. It also makes the fabrication easier by making fibres into individual strands. Therefore, from the above study it can be concluded that Tensile strength of sisal fibres can be increased by increasing the duration of alkali treatment to certain extent.

References

1. Ramesh S, Sakthivei M (2013) Mechanical properties of natural fibres (Coir, Banana, Sisal) polymer composite. Science park 1: 2321-8045.

2. Bouzakis KD, Tsifis I (2008) Determination of epoxy resins mechanical property by experimental-computational procedures in tension 1224-5615.

3. Bose NR, Dipa R, Sarkar BK, Rana AK (2001) Effect of alkaline treated jute fibres on composite material. Bull Mater Sci 24: 129-135.

4. Tara S, Jagananatha RHN (2011) Application of coir, bamboo, sisal and jute natural composites instructural up-gradations. IJIMT 2: 186-191.

5. Wang H, Pattarachaiyakoop N, Trada M (2011) A review on the tensile properties of natural fibre reinforced polymer 42: 856-873.

6. Rao DRS, Madhusudan S, Madhukiran J (2013) Fabrication and testing of natural fibre reinforced hybrid composites banana/pineapple. IJMER 3: 2239-2243.

7. Srinath LS (2013) Advanced mechanics of solids. (3rd edn.), Tata McGraw-Hill Education, USA.

8. William D, Callister Jr (2012) Fundamental of material science and engineering. (7th edn.), Wiley India Pvt Ltd, USA.

9. Mallick PK (2007) Fibre-reinforced composites: materials, manufacturing and design. (3rd edn.), CRC Press, USA.

10. Raman S, Shattaki C, SalilKumar CS, Chandran SM (2013) Fabrication and testing of jute epoxy bio-composite. IJAER 8: 2525.

11. Lucas JK (1999) Review on sisal fibre reinforced polymer composites. Revista Brasileira de Engenharia Agrícola e Ambiental 3: 367-379.

Reduced Graphene Oxide Thin Films with Very Large Charge Carrier Mobility Using Pulsed Laser Deposition

A Bhaumik[1,2], A Haque[1,2*], MFN Taufique[1,3], P Karnati[1,4], R Patel[1], M Nath5 and K Ghosh[1]

[1]Department of Physics, Astronomy and Materials Science and Center for Applied Science and Engineering, Missouri State University, Springfield, MO 65897, USA
[2]Department of Materials Science and Engineering, North Carolina State University, Raleigh, NC 27695, USA
[3]School of Mechanical and Materials Engineering, Washington State University, Pullman, WA 99164-2920, USA
[4]Department of Materials Science and Engineering, Ohio State University, Columbus, OH, 43210, USA
5Department of Chemistry, Missouri University of Science and Technology, Rolla, MO 65409, USA

Abstract

Large area reduced graphene oxide (RGO) thin films have been grown using pulsed laser deposition (PLD) technique. A very large carrier mobility of 372 cm^2 V^{-1}s^{-1} has been observed in a PLD grown RGO thin film with a large sp^2 carbon fraction of 87% along with narrow Raman 2D peak profile. The fraction of sp^2 carbon and carbon/oxygen ratios are tuned through PLD growth parameters, and these are estimated from X-ray photoelectron spectroscopy (XPS) data. The electrical properties of the RGO thin films are comprehended by the intensity ratios between different optical phonon vibrational modes of Raman Spectra. The photoluminescence spectra also indicate a less intense and broader blue fluorescence spectrum detecting the presence of miniature sized sp^2 domains in the near vicinity of π^* electronic states which favor the variable range hopping transport phenomena. This study on large area RGO thin films with very large carrier mobility fabricated by PLD process will be very useful for high mobility electronic device applications and could open a roadmap for further extensive research in functionalized 2D materials.

Keywords: Reduced graphene oxide; Raman spectroscopy; Variable Range Hopping; 2D-materials; Pulsed laser deposition

Introduction

Chemical functionalization of graphene has enticed significant research interests due to its potential in obtaining an optical band gap and subsequent tuning of its optoelectronic properties for device applications [1,2]. The solution processed route for producing RGO flakes has received significant attention because of its high throughput manufacturing, and tunable electrical and optical properties by controlling the ratio of sp^2 and sp^3 hybridized carbon clusters [3,4]. Functionalization of graphene that creates disorders and electron localization-hopping phenomena plays a significant role in determining its physical properties [5,6]. However, the electronic parameters for device applications are not quite good, for example, the carrier mobility of solution processed RGO is not high enough for many device applications [6]. The change in physical properties of functionalized graphene is due to (i) the transformation of hybridization state in carbon atoms, (ii) formation of a barrier (scattering entity) at the functionalization site and within the electron potential sequence, (iii) distortion of its two-dimensional planar lattice due to functionalization or thermal energy, and (iv) introduction of sp^2 clusters and defects thereby introducing different energy levels [7]. The alterations in the chemical, structural, and electronic properties of RGO can be originated by the molecular level interaction arising by a number of phenomena i.e. (i) covalent bonding, (2) π-π^* interfacing, (3) lattice substitution, or (4) physisorption [8]. In graphene the direct covalent functionalization of sp^2 hybridized carbon transforms it into a tetrahedral sp^3 hybridized carbon. This conversion of the hybridization state of C atoms causes a loss of the free, sp^2 associated π electrons. On the contrary, the substitution of carbon atoms with other elements (for instance N or B) retains the sp^2 character disrupting the π-cloud continuum [9]. Additionally, it is imperative to remark that the non-covalent functionalization of graphene via π-π^* interfacing or physisorption does not distort the sp^2 network. However this phenomena changes the doping concentration, raises the density of electron-hole puddles substantially, and also may create scattering sites for the free charge carriers [9].

The most studied approach of graphene functionalization is covalent functionalization which has been conducted by following different routes. This process causes for the significant changes to the electrical properties in graphene structure. Reducing GO produces RGO; however, RGO contains unreduced, covalently bonded oxy groups [10]. The removal of the π electron from the carbon atom reduces the carrier density and can introduce an optical band gap. The functionalizing molecule can also introduce energy levels (edge states and functionalization states) in the band structure of covalently functionalized graphene (CFG) to make it an n-type or p-type semiconductor [9]. Therefore, a combination of the carrier deficiency at the sp^3 site, the associated disruption of the electron-potential continuum, and the distorted planar lattice cause a drastic reduction in charge carrier mobility and a change in charge polarity in graphene. Compare to graphene (mobility of 10,000-50,000 cm^2 V^{-1}s^{-1} and intrinsic mobility limit 200,000 cm^2 V^{-1}s^{-1} at room temperature), RGO, with few covalently bonded residual functional groups, exhibits a drastically reduced carrier mobility (0.05-200 cm^2 V^{-1}s^{-1}), p-type conductivity, and a finite effective optical band gap of 0.2 to 2 eV [5]. However a recent report indicates that the FET mobility in RGO can be increased up to ~1000 cm^2 V^{-1} s^{-1} by microwave reduction technique [11]. Till date, the most common approach to synthesize RGO involved multistep processing i.e. (i) synthesizing graphene oxide (GO) by oxidative exfoliation of graphite (Hummers' method) and (ii) reducing GO by

*****Corresponding author:** Ariful Haque, Department of Physics, Astronomy and Materials Science and Center for Applied Science and Engineering, Missouri State University, Springfield, MO 65897, USA, E-mail: ahaque@ncsu.edu

thermal and chemical processes. To remove additional oxidative agents in the GO structure, centrifugation, filtration, etc. are required, which increases the production cost of large area GO. Successive reductions of the cleaned GO using additional treatments are also required to produce high quality RGO which usually involves more processing steps such as thermal reduction, photocatalytic reduction, chemical reduction, etc. [12,13]. These processing steps increase complexity in RGO synthesis.

PLD has been established as a very efficient technique to produce thin films of various materials with unique properties. PLD has a number of advantages over other physical vapor deposition techniques. For example, PLD can transfer the composition of the target to the substrate. Since the plume temperature in PLD is very high (order of 10^4 K) it gains more energy and increases the surface mobility which leads to improved quality films. Many materials can be deposited using a wide variety of reactive gases such as oxygen, nitrogen, and other gas mixture. This is possible because there are no ions or evaporation sources that contain hot filament in the vacuum. Because of the many advantages, PLD has been extensively used for the growth of High TC oxides, ferroelectrics, ferrites, semiconductors, polymers, metals, alloys, carbon nanotubes, and many others [11-13]. However, to our knowledge there is no report available on the growth of high-quality graphene and RGO films using PLD [14]. This study attempts to synthesize high-quality large area RGO thin films with tunable electronic and optical properties using PLD.

By employing PLD we have fabricated RGO thin films with controlled variation in sp^2/sp^3 ratios calculated from XPS analysis, and FWHM of *2D* peak along with *I2D/IG* ratios obtained from Raman spectroscopy. These factors interplay significant roles to influence the electronic properties of thin films of RGO. The ratios of sp^2/sp^3, hence, structural and electronic properties can be controlled through growth parameters of the PLD and introducing a seed layer of single layer of graphene on a substrate before the growth of RGO. Here we present results of some representative samples termed as sample A, B, C, and D on the basis of the gradual increment in the intensity and sharpness of the *2D* peak. Sample D shows significantly high concentration of sp^2/sp^3 ratio, has sharpest 2D peak compared to the other samples and exhibits best electronic properties. To better understand the improved electronic transport properties of PLD grown RGO samples detailed temperature dependent electronic transport properties have also been investigated and correlated with structural and optical properties.

Experimental Section

Thin films of RGO were deposited on SiO2 (300 nm)/Si substrates purchased from MTI corporation by pulsed laser deposition (Excel Instrument, PLD-STD-18) using a dense and high-quality graphite target (99.9%) purchased from Kurt J. Leskar. A KrF excimer laser (Lambda Physik, COMPEX 201) with energy density of 2 J cm^{-2}, λ=248 nm, pulsed duration of 20 ns, was used at a pulse rate of 10 Hz for the deposition. Thin films were grown by varying the number of PLD shots (sample A: 5000 shots, sample B: 100 shots, sample C: 10,000 shots, and sample D: 500 shots) maintaining a constant oxygen pressure (1 × 10^{-5} Torr) and growth temperature (700°C) in the PLD chamber. Before the growth of sample D, a seed layer of a single layer graphene was introduced to improve the mobility of the samples. All the deposited films were cooled in forming gas, a gas mixture of Ar (96%) and H2 (4%) at pressure of 10^{-4} Torr. The number of laser shots were critically determined for the best growth of reduced graphene oxide thin films.

To calculate the fraction of GO and RGO and lattice parameters of

these two phases, X-ray diffraction (XRD) measurement was carried out. XRD data of all thin film samples were recorded using a X-ray diffractometer (Bruker, D8 Discover) using the theta-2theta scan with CuKα (λ=1.5405 Å). XRD data were analyzed using Gaussian-Lorentzian fitting profile using Origin Pro 8.5.1 software. The vibrational phonon modes were determined by Raman spectroscopy using 532nm green laser (Horiba, Labram Raman-PL). Photoluminescence (PL) measurements were carried out employing 325 nm UV laser using the above-mentioned instrument. The Raman spectra and PL spectra was analyzed using Gaussian and Lorentzian peak fittings in NSG Lab spec software. The ratio of sp^2/sp^3 carbon and carbon/oxygen ratio were estimated from X-ray photoelectron spectroscopy (XPS) performed on a Kratos Axis 165 photoelectron spectrometer with a hemispherical 8 channel analyzer where photoemission was stimulated by a monochromatic Al source. The electrical conductance and Hall coefficient measurements were carried out by a standard four-probe technique. The Ohmic contacts were made using low-temperature In-Ag solder. The sample voltage was measured with a nanovoltmeter (Keithley, 182) with a current of 1 to 100 mA using a 20 ppm stable current source (Keithley, 220). The magnetic field dependence of the Hall effect was measured with the field applied perpendicular to film surface in the Van der Pauw configuration. The low-temperature resistance measurements were carried out using a SQUID magnetometer coupled with Labview programming interface. The temperature was stabilized within 0.1 % of that temperature and measured using a calibrated sensor. The morphology of all RGO thin film samples was characterized using atomic force microscopy (AFM). AFM imaging was performed under ambient conditions using a Digital Instruments (Veeco) Dimension-3100 unit with Nanoscope® III controller, operated in tapping mode. The etched silicon tapping tips employed in the study had a nominal apex radius of curvature of ~ 10-20 nm and a cone angle of ~34°. The spring constant of the cantilever was ~42 N/m. The cantilevered tip was oscillated close to the mechanical resonance frequency of the cantilever (typically, 200-300 kHz) with amplitudes ranging from ~10 to 30 nm.

Results and Discussion

Raman scattering

Figure 1 represents the room temperature unpolarized Raman spectroscopy data of RGO thin films deposited onto SiO2 (300 nm)/Si. Raman spectra of graphene-based materials show almost all the significant features in the range between 1000 cm^{-1} to 3000 cm^{-1}, and the corresponding Raman modes are known as *D*, *G*, *2D* and *D+G* band [15]. To ensure the best fit an extra peak was found in between *D* and *G* band which is known as *f* band [16]. The A1g symmetry phonon near K zone boundary involved with the breathing modes of sp^2 carbon atoms present in graphene and GO introduces *D*-peak around 1355 cm^{-1} which arises only due to the presence of defects and asymmetry in the lattice [17]. The *D* peak position of the as-synthesized samples are in between 1345 cm^{-1} to 1350 cm^{-1}. The E2g phonon mode in the Brillouin zone center correspond to the *G* peak is positioned in between 1500 cm^{-1}-1630 cm^{-1}. The origin of the *G* peak is due to the bond stretching of sp^2 C atoms both for rings and chains [17]. It has been observed that the *G* peak of RGO is shifted to lower frequencies due to the removal of functional groups by reduction reaction [18]. Stiffening of the *G* mode also indicates the increase in carrier density of either sign [19]. The *G* peak position of the four samples observed in between 1597-1600 cm^{-1} indicates the presence of RGO structure on the thin films.

The ratio between the peak intensities of *D* band and *G* band is an

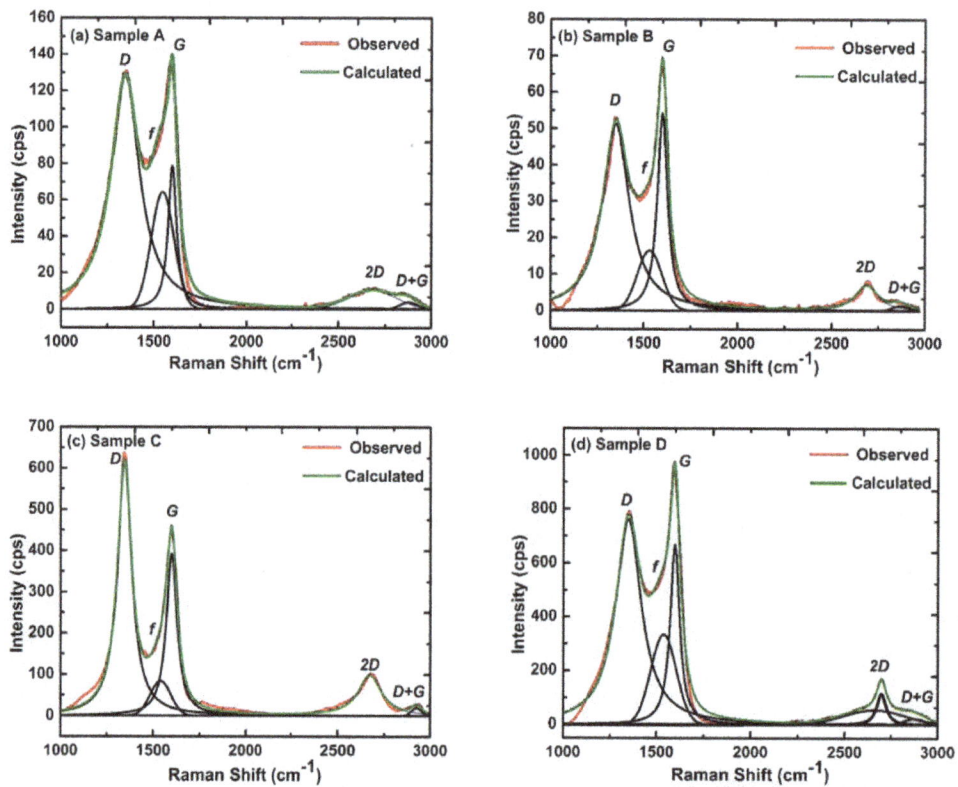

Figure 1: Raman spectra of RGO thin film samples synthesized by PLD technique, (a) Sample A, (b) sample B, (c) sample C, and (d) sample D, showing the variations in FWHM of 2D peaks and I2D/IG, I2D/ID+G, and ID/IG, ratios to illustrate the restoration of sp2 carbon clusters and the degree of reduction.

indication of the degree of reduction [20]. Researchers suggested that according to Tuinstra and Koenig (TK) relationship with increasing the disorder in the structure, *ID/IG* increases. However, further reduction reaction creates more disorder and distortion on the sp^2 structure [17]. As a result, *ID* decreases compared to *IG* and TK relationship no longer exists. The corresponding *ID/IG* ratio of sample A, B, C and D is 1.653, 0.973, 1.565, and 1.14, respectively. The higher *ID/IG* ratio signifies that higher growth temperature and reaction environment assist in reduction of the GO thin films.

The second order overtone of the *D* peak around 2710 cm^{-1} is known as *2D* peak. This peak is the characteristic to the double resonance transition resulting from the generation of two phonons with opposite momentum to each other [21,22]. It has been reported that the *2D* peak profile is sharp in the Raman spectrum of Graphene, while small *2D* peak intensity compared to *D* and *G* peaks is an indication of disorder in GO [23]. For sample C and D it is observed that the *2D* peak is sharper compared to sample A and B. The intense and narrow *2D* peak of sample D indicates that the number of defects is less compared to the other samples. The FWHM of *2D* peak for sample A, B, C and D are 360 cm^{-1}, 155 cm^{-1}, 152 cm^{-1} and 59 cm^{-1}, respectively. The reduced FWHM of 2D peak of sample D suggests the quality of reduction i.e. reduced structure with less amount of defects. The high *I2D/IG* ratio can be used to predict the charge carrier mobility [24]. This *I2D/IG* ratio for sample A, B, C and D is found to be 0.133, 0.121, 0.251, and 0.171, respectively. From this data, we can predict that sample C and D can show better electrical property.

The *D+G* peak around 2890 cm^{-1} is a combination band and depends on the defect concentration [10,25]. In our case peak position

of the *D+G* band lies in the range of 2880 cm^{-1} to 2925 cm^{-1}. Scientists also reported that *I2D/ID+G* ratio increases due to the restoration of sp^2 hybridized structure [16]. This ratio for sample A, B, C, and D is 2.6, 4.5, 5.5, and 5, respectively. The higher value of *I2D/ID+G* indicates that there is a possibility for high restoration of sp^2 clusters in sample C and D. Therefore, the large value of *I2D/IG* and *I2D/ID+G* is also responsible for improved electrical transport properties.

X-ray diffraction

Figure 2 represents the XRD pattern of the RGO thin film samples (A, B, C, and D) deposited onto SiO2 (300 nm)/Si. The curve fittings for the XRD patterns of the RGO samples were done with the help of Gaussian-Lorentzian fitting profile using Origin Pro 8.5.1 software. The corresponding 2θ value for GO in the RGO samples varies from 9.770 to 12.380 which is in accordance to the previous results for GO synthesized by chemical route [26,27]. Reports have shown that GO has the largest interlayer distance (5Å-9Å) because of the presence of intercalated water molecules and functional groups such as hydroxyl, epoxy, and carboxyl [28].The interlayer distance for GO has been calculated, using Bragg's relation which varies from 7.1-9 Å [29]. As the films were deposited at relatively higher temperature (700°C), we can see the peaks of RGO, corresponding 2θ value ranging from 15.780 to 19.370 and interlayer distance varying from 5.58-4.57 Å. The shift of the 2θ peak from GO (7.1-9 Å) to RGO (5.58-4.57 Å) suggests that the GO is reduced. The shift of the GO peak towards the right is mainly dependent on the vaporization of the water molecules present between the stable GO layers which results in decrease of the interlayer distance [30]. The crystallite size for the RGO films were calculated using the Debye-Scherer formula which is shown by eqn. (1) [31].

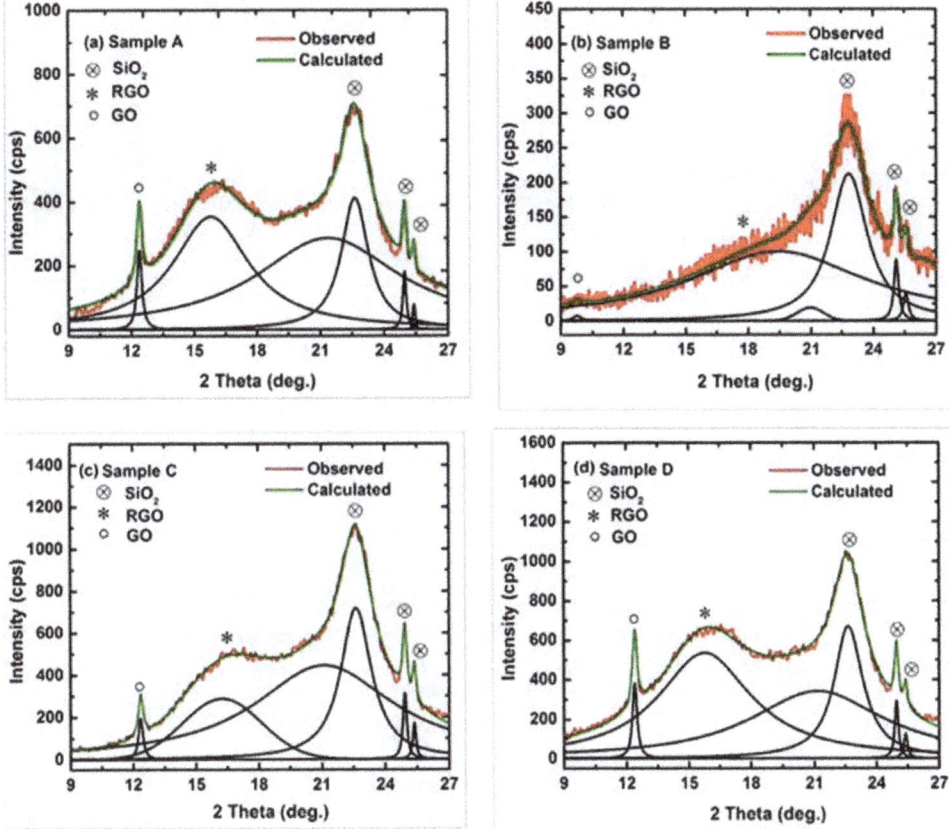

Figure 2: XRD patterns of the (a) sample A, (b) sample B, (c) sample C, and (d) sample D, interpreting interplanar separation and confirming the presence of multiphase structures of GO and RGO.

$$D_{(002)} = \frac{K\lambda}{B\cos\theta} \tag{1}$$

Where $_{(002)}$ represents the crystallite size, K (0.91) is a constant dependent on the crystallite shape, λ is the X-ray wavelength (1.5406 Å), B is the corrected FWHM, and θ is the scattering angle. The corrected FWHM for the corresponding GO and RGO peaks has been calculated by utilizing a strain correction, subtracting the strain from the substrate used [32].

The strain correction is represented by the eqn. (2),

$$B = \sqrt{|B_S^2 - B_{Su}^2|} \tag{2}$$

Where, B_S is the FWHM of the corresponding peak sample, B_{Su} is the FWHM of the corresponding substrate peak.

For sample A and C there is no appreciable difference in the calculated crystallite size for RGO peaks. But for samples B and D we could see an appreciable difference in the crystallite size (8.25-15.5 Å). The change in the crystallite size explains about the growth of the RGO thin films. With the help of Debye-Scherer equation, researchers have calculated the number of layers present in GO (N_{GO}) using the following eqn. (3), [33]

$$N_{GO} = \frac{D_{(002)}}{d_{(002)}} \tag{3}$$

Where $d_{(002)}$ is the interlayer distance of RGO thin films. The least number of layers (2) is found to be in sample B which is mainly dependent on the number of shots given. From the XRD pattern

of sample B a broader RGO peak is visible resulting from a smaller crystallite size (8.25Å) which refers the amorphous nature of the sample B when compared with the remaining samples (A, C and D). Detailed structural parameters of GO and RGO of four samples obtained from XRD data are given in Table 1.

X-ray photoelectron spectroscopy

Figure 3 depicts C1s XPS data of the PLD grown RGO samples A, B, C, and D fabricated under different growth conditions. The curve fitting of the C1s spectra was performed using a Gaussian-Lorentzian peak profile after performing a Shirley background correction using Origin Pro 8.5.1 data plotting and acquisition software. The binding energy of the C-C and C-H bonding are assigned at 284.5-285 eV and chemical shifts to 288.0 eV, 286.7 eV are typically assigned for the C=O, and -O-CH3 functional groups, respectively [34]. A peak at 283.6 eV after deconvolution of C1s spectra for the sample D originates, which represents the sp^2 binding energy peak of C-C molecular bond [34]. Recent research reports indicate that graphite oxides have an epoxide group (C-O-C), which has C1s binding energy similar to that of C-OH [34]. It is also possible that there might occur a larger chemical energy shift in the C-O-C emission into the binding energy range of the C=O emission, which can be the cause of disappearance of the epoxide group peak. The C1s spectra in Figure 3a and 3c predominantly have peaks of sp^3 C-C atom, C-H, C=O, while sample B has an extra -O-CH3 instead of the C=O ketone peak. The sample D has broader C1s XPS peaks with an additional peak at 283.6 eV corresponding to the π bonds activation energy in the graphitic network, which generally has ~ 1eV less binding energy than the σ bond. An attempt was made to fit the

	Sample A		Sample B		Sample C		Sample D	
	RGO	GO	RGO	GO	RGO	GO	RGO	GO
Peak Position (deg)	15.86	12.38	19.37	9.77	16.22	12.36	15.78	12.38
d (Å)	5.587	7.143	4.57	9.045	5.45	7.15	5.61	7.14
FWHM(deg)	4.182	1.51	9.87	2.03	4.183	1.77	5.23	1.67
Crystallite size (Å)	19.4	53.5	8.25	39.71	19.39	45.64	15.5	48.37
Number of layers	3.47	7.48	1.8	4.39	3.55	6.37	2.76	6.77

Table 1: Structural parameters of GO and RGO samples obtained from XRD data.

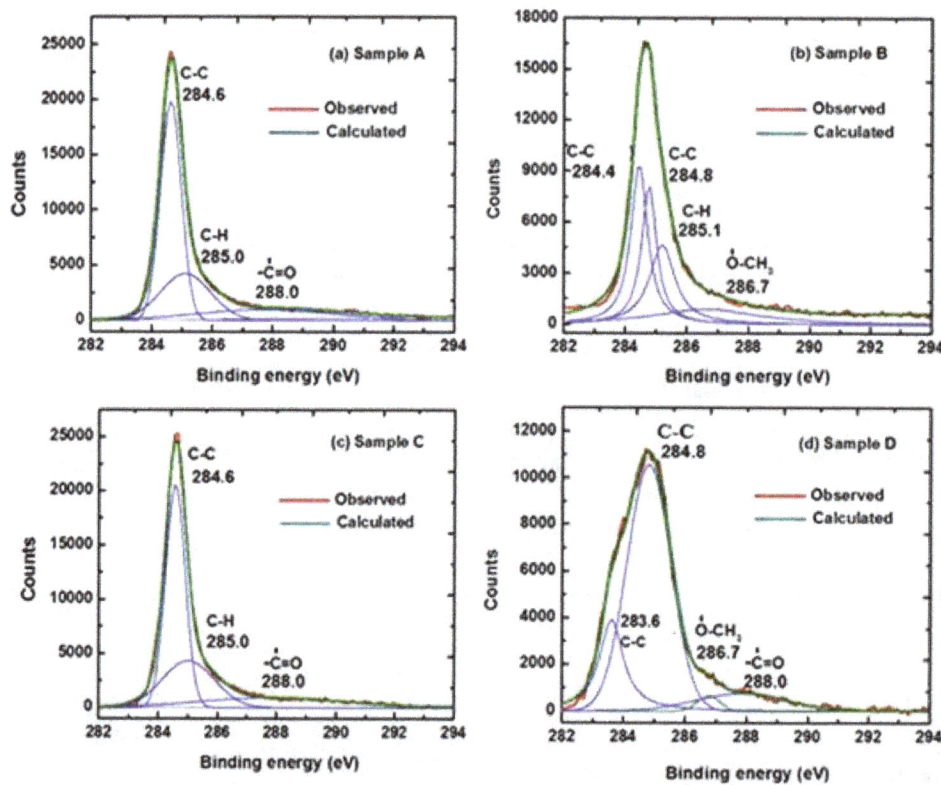

Figure 3: Deconvolution of the C1's peaks of XPS spectra of (a) sample A, (b) sample B, (c) sample C, and (d) sample D confirms the presence of C-C, C-H, =C=O, -CH$_3$, and -O-CH$_3$ functional groups. From the fitted data the calculated percentage of sp^2/sp^3 is 54%, 57%, 54%, and 87% for sample A, B, C, and D, respectively.

expected peaks for graphene oxide at 286.6, 287.6, and 289.1 eV but the mean square error for such a fit was much larger. The graphitic network formed by PLD technique was thereby fitted for the best goodness of fit and the peak positions and FWHM area were thus extracted. Higher values of FWHM and a broader tail towards higher binding energy is an indicative of contribution from a variety of carbon bonding configurations in the thin films. The similarity of the C1s peaks profile and position in sample A and C suggest no considerable change in the surface as well as in the depth of the sample with increasing the number of shots as the photoelectron kinetic energies of C1s is larger than O1s, and so the sampling depth is larger. This proves that relatively thicker thin films of RGO grown by the PLD technique are having quite similar surface properties. All the RGO samples show a marked reduction in the intensities of all of the related oxygen peaks were indicating that the delocalized π conjugation was restored in the samples.

The C/O ratio was calculated for all the samples. Samples A, B, C, and D show a C/O ratio of 7.2, 6.25, 4.28, and 5.95, respectively. This ratio is an indicative of the reduction in the graphene oxide structure. The above-mentioned values are appreciably lower than the reported

values in a recent article, [35] suggesting that more reduction should be undertaken for complete formation of RGO. However, it should be pointed out that during thermal reduction extreme high pressure is created in the stacked layers, thereby occurring disruption of the graphitic network by loss of carbon. A pressure of 40 MPa is generated at 300 C as predicted by state equations, while higher pressures (130 MPa) are generated at elevated temperatures (1000°C) The evaluation of the Hamaker constant indicate that 2.5 MPa pressure is needed to delaminate layers of stacked graphite oxide platelets [34]. An immense amount of structural defects are created during this pressure induced thermal exfoliation which also leads to the removal of carbon dioxide from graphene oxide structures [36]. Theoretical and experimental calculations have predicted a 30% loss of mass of GO during this process. This process leaves behind vacancies and various others topological defects (Stone Wales defect) throughout the basal plane in RGO [34]. The presence of these defects drastically affect the optoelectronic properties of RGO thin films by reducing the ballistic path lengths and introducing electronic charge carrier scattering moieties.

It will be discussed in electrical measurements section that the sample D is having high Hall mobility than the rest of the PLD-prepared RGO samples. Though the thermal reduction is less in the sample B, but due to the increased defect density there occurs more scattering centers for the charge carriers thereby reducing the charge carrier mobility. Carbon-oxygen signatures were detected in the XPS spectrum of the samples, but these are far diminished in the case of sample B and sample D than sample A or sample C, and were also dwarfed by the C-C and C=C signals. The absence of C=O in the sample B indicates that the number of pulsed laser shots is a crucial parameter for fabricating functionalized graphene oxide. With the increase in the number of shots from 100 (in sample B) to 500 (in sample D) evolution of C=O peak occurs. Further increase in the number of PLD shots completely replaces the -O-CH3 to C=O functional group. The amount of sp^2 content is generally indicated by the C-C peak, while the oxygen-containing functional groups give rise to sp^3 hybridized states [5]. The presence of sp^3 defect sites cause a distortion in the π continuum cloud of the intrinsic π electronic states of the sp^2 sites. Therefore the sp^2 fraction in RGO thin films provide a valuable insight to its optoelectronic properties and is commonly referred to as the reduction efficiency [5]. The sp^2 fraction is calculated from the XPS high resolution data. This is done by considering the ratio of the integrated area of the peak corresponding to the C-C bond to the total area under the C1s XPS spectra. The percentage of the carbon sp^2 fraction is determined by the following equation [35]:

$$carbon\,sp^2\;fraction = \frac{Area(C\text{-}C)}{Area(C-O)+Area(C-H)+Area(C-C)}\times100 \quad (4)$$

The percentage of sp^2 fraction calculated from XPS is 54 %, 57%, 54%, and 87% for sample A, B, C, and D, respectively. Increased sp^2 fraction in sample D is an indicative of the restoration of the π network

in the RGO structure, which will be the determining factor for increased electrical mobility observations.

Electrical transport properties

Figure 4a represents the resistance- temperature (R-T) characteristics of the RGO thin film samples. Electrical measurements indicate good electronic properties of the as-synthesized RGO thin films. Within the voltage range of -100 to 100 mV, the I-V curves are Ohmic which allows to get more reliable resistance measurements of the samples at different temperatures. Four probe measurement setup also helps avoid the contact resistance and lead resistance. To better understand the electronic properties R-T data have been collected over a broad range of temperature (5 K to 400 K). In all the samples the value of resistance increases nonlinearly with decrease in temperature which implies the semiconducting nature of the thin films. With decreasing temperature from 400 K to 5 K, we have noticed two completely different types of rate of change of resistance i.e. at higher temperature region the resistance increases slowly and at lower temperature region it increases rapidly. So the transport mechanism is thought to be following two different kind of models. Researchers have experimentally found that at high and low-temperature regimes mode of electrical transport mechanism is dominated by band gap and variable range hopping processes, respectively [37]. Resistance vs. temperature data of all the samples in temperature range 5 K to 190 K elucidate that the transport mechanism is dominated by Efros-Shklovskii variable range hopping (ES VRH) and from range 190 K to 400 K the data best fit with Arrhenius equation. As at lower temperature activated type of conduction is not possible in the RGO thin films, the variable range hopping mechanism plays a significant role in the conduction process. The sp^3 hybridization of C atoms and discontinuous sp^2 clusters in sp^3

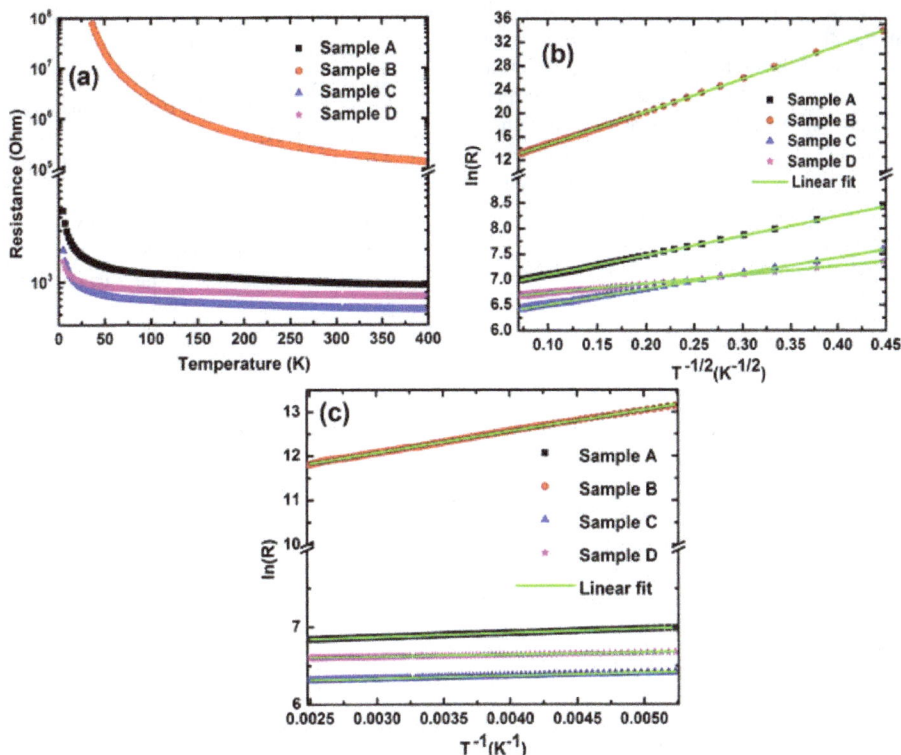

Figure 4: (a) Temperature (T) dependence of electrical resistance of samples A, B, C, and D in the range 5 K<T<400 K. (b) ES VRH region is identified by plotting ln(R) as a function of $T^{-1/2}$ in the temperature range 5 K<T<190 K, (c) ln(R) as a function of T^{-1} in the temperature range 190 K<T<400 K to illustrate the band-gap dominated Arrhenius-like temperature dependence transport mechanism.

matrix play a major role for the insulating property of GO. Scientists suggested that new domains of sp^2 clusters are formed by the removal of oxygen groups due to the reduction of GO. This gives the percolation path between domains already present. Initial sp^2 clusters of the GO do not take part in conduction [38].

The new sp^2 clusters, which were formed by the subsequent reduction reaction of the thin film at high temperature (700°C), help in electron transport process. Newly formed sp^2 clusters can be considered electronically as isolated states that aid in hopping of the free electrons. As the number of sp^2 clusters is finite so the conductance in the as grown RGO film is measurable. To determine the conduction mechanism in the RGO samples we fitted our results with various conductivity mechanism models. At low temperature regime we found that the data best fit to a straight line with ES-VRH model, shown in Figure 4b, compared to Mott-VRH or 3D models [39]. ES model was considered from the specific form of single particle density of states in Coulomb gap to include the effects of Coulomb interactions and modified Mott's argument of variable range hopping [40]. From this model a relationship between the conductivity and temperature is formed. This relationship can be expressed in terms of resistance, R and temperature, T by eqn. (5) [5].

$$R = R_0 \, e^{\left(\frac{T_0}{T}\right)^{1/2}} \qquad (5)$$

Where T_0 is the characteristic temperature and the exponent ½ is independent of dimensionality of the sample. This characteristics temperature T_0 can be used to find out the localization length, ξ using eqn. (6) [41].

$$T_0 = \frac{\beta e^2}{4\pi\varepsilon\varepsilon_0 k_B \xi} \qquad (6)$$

Where $\beta = 2.8$ is a constant, k_B is the Boltzmann constant, e is the charge of an electron, $_0$ is the vacuum permittivity, and ε is the effective dielectric constant of RGO thin films on SiO_2. From different studies we can estimate the value of ε to be 3.5 for our RGO samples [28,42,43]. From the fitted data demonstrated in Figure 4b we obtained characteristic temperature which allows to calculate the localization lengths of corresponding samples. From the values of localization length reported in Table 2 we can infer that the longest localization length has been found in the sample with larger conductivity i.e. sample D. Minimum localization length is obtained in the least conductive sample i.e. sample B.

Room temperature Hall mobility measurement was performed in all the samples. Only sample D has shown a considerable amount (372 cm^2 V^{-1} s^{-1}) and sample C has shown a meager amount of Hall mobility. The Hall mobility of sample D calculated from the Magnetic field vs Hall voltage data collected from four probe measurement setup. It has been experimentally shown that the fraction of sp^2 sites in RGO increases with increasing localization length [42]. According to Raman analysis sample D has less defect states and as it is having highest sp^2 fraction among all the samples, noticeable mobility was observed in this sample. Another possible explanation for improved mobility could

be a longer localization length which is the outcome of the shortened barrier heights for electron transportation among the sp^2 clusters. Hopping energy (Eh) is an important parameter in ES VRH model of 2D materials. The hopping energies for all the samples are calculated by using the following formula represented by eqn. (7) and given in Table 2 [30].

$$\sigma\sqrt{T} = A e^{-\frac{E_h}{k_B T}} \qquad (7)$$

Where A is a constant, and σ is the conductivity. As lower hopping energy always facilitates the transportation of electrons in the sample, we found the maximum conductivity in sample D where minimum hopping energy is obtained. On the other hand, as sample B demonstrates maximum hopping energy, maximum resistivity is observed in this sample.

At higher temperature regime (190-400 K) the data is analyzed based on the relationship between temperature and resistance which is represented in eqn. (8) [17].

$$R = R_0 \, e^{\frac{E_g}{k_B T}} \qquad (8)$$

Where Eg is the activation energy. The plotted data of natural logarithm of resistance $ln(R)$ versus inverse of temperature T^{-1} best fit to a straight line in Figure 4c. This implies that at higher temperature regime band-gap dominated Arrhenius-like temperature dependence plays the major role in transport mechanism. Using the slope of the fitted lines the calculated 15 activation energy is found to be in a range from 2.35 meV to 37.54 meV which is comparable with the values reported previously [44,45].

The activation energy is minimum in case of sample D. So, at room temperature it is easier for the electrons in the donor atoms lying in the crystal to reach to the conduction band. Completely opposite phenomenon occurs in the case of sample B with higher activation energy and this scenario is portrayed in Figure 5. The work of Pearson and Bardeen also showed a decrease in activation energy which results an increase in carrier concentration [46]. For each kind of defect/impurity, there should be a specific characteristic ionization energy also termed as activation energy. This energy is fixed under normal operational conditions [47]. So, the value of the activation energy is dependent only on the nature of the intrinsic and extrinsic defects/impurities in the samples. The variance in the activation energy of

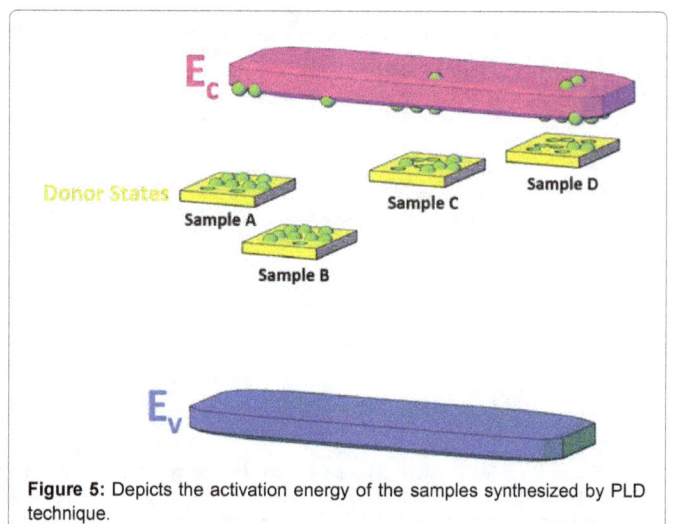

Figure 5: Depicts the activation energy of the samples synthesized by PLD technique.

Sample Name	Localization Length (nm)	Activation Energy (meV)	Hopping Energy (meV)
Sample A	916	4.7	4.2
Sample B	4	42.1	17.7
Sample C	1401	3.4	3.9
Sample D	4227	2.3	3.7

Table 2: Electrical parameters of RGO thin film samples A, B, C, and D.

our samples implies that the nature and the level of the defects and impurities are dependent on different growth parameters.

The decrease in carbon sp^2 fraction leads to an increase in R. Among all the samples maximum value of R is found in sample B ($\sim 10^8$ Ω) while the minimum is found in sample A ($\sim 10^3$ Ω) demonstrating that the value of R can be tuned by more than 5 orders of magnitude by varying the carbon sp^2 fraction which supports the information illustrated from Raman spectroscopy. The decrease in resistance with increasing sp^2 fraction demonstrates that the restoration of π-π* bond improves charge percolation pathways in the RGO sheet. It has also been observed that such improvement occurs at the expense of increasing topological defects in the samples [48,49]. The salient features of improved mobility can be accounted for (i) less hopping energy, (ii) longer localization length, (iii) minimal defect density, and (iv) smaller activation energy. Better conductivity of our samples suggests the formation of numerous sp^2 clusters by the removal of oxygen groups in the sp^3 matrix [38]. The presence of the huge number of small sp^2 clusters can also be proven by the blue shifting of widened PL peak.

Photoluminescence spectroscopy

Figure 6 represents the PL spectra of the RGO samples A, B, C, and D synthesized by PLD technique. We have represented the PL after subtracting from the background PL signal of the substrate measured under identical conditions. RGO is a graphene sheet functionalized with oxygen groups on the basal plane and at the edges, thereby exhibiting interesting steady-state PL properties. The PL spectra show a broad PL region between 1.25 eV to 3.00 eV (410 nm to 990 nm) and is also reported to be a characteristic spectrum for oxygen plasma-treated, mechanically exfoliated, single-layer graphene sheet [38]. The prominent PL peak seen in sample D is observed at 2.3 eV which moves towards lower energy values for all other synthesized samples. Increased sp^2 fraction (as calculated by XPS) in sample D decreases the PL intensity and exhibits blue fluorescence (as the peaks move towards right). The increased PL intensity in sample C is due to less reduction (C/O ratio) as observed by XPS studies. The sharper peak profile for the 2D peak also influences the PL spectra. A decrease in FWHM of the 2D peak also corresponds to blue fluorescence in the PL spectra.

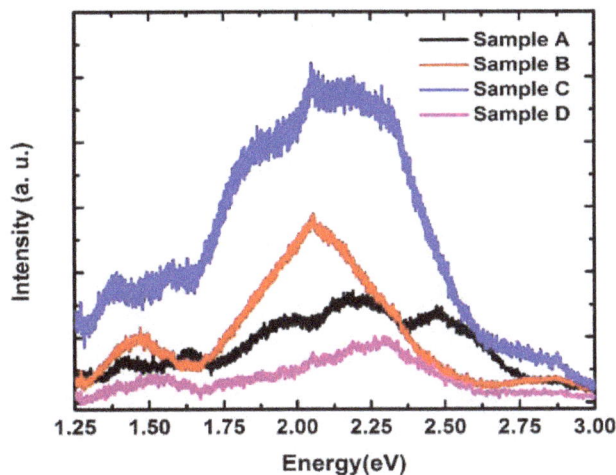

Figure 6: PL spectra of all the RGO samples after subtracting from the background signal of the substrate. The reduced intensity and the higher energy position of the PL peak of sample D imply relatively high C/O ratio and high sp^2 hybridized carbon fraction compared to the other samples.

This can be due to the reason that a sharper 2D peak is an indicative of reduced defect regions in the layered structure of GO. First-principle calculation suggests a large fraction of the C atoms in the hydroxyl and epoxy units are bonded to each other to form strips of sp^2 carbon atoms in GO compounds [46]. In GO compounds, containing a mixture of sp^2 and sp^3 moieties, the optoelectronic properties are mainly determined by the π and π* electronic states of the sp^2 sites, which lie within the σ-σ* gap [38]. The π bonding is weaker and has a lower formation energy than the σ bond, Therefore, there occurs the presence of a large density of disorder-induced localized states in the two-dimensional structure of as synthesized GO and RGO. There are electronic states associated with the functionalized carbon atoms situated in the basal plane and edges of the structure which gives rise to interesting PL features. It is well studied that the interactions of the π states are strongly dependent on their projected dihedral angles. There also occurs the formation of these structurally disordered localized electronic states in the band tail or deep states in the π-π* electronic gap [38]. The optical transitions resulting from these states result in a broader absorption or emission band and is evident from the PL spectra. The deoxygenation process which leads to reduction (formation of RGO) decreases the density of these disorder- induced localized states and forms well defined sp^2 clusters. This causes a decrease in the intensity of the PL emission with reduction. The creation of RGO gives rise to formation of numerous smaller sized sp^2 clusters which plays an important role in the variable hopping range electronic conduction. The formation of new sp^2 clusters with reduction also broadens and reduces the intensity of PL spectra. The formation of these miniature sized sp^2 clusters in RGO provides percolation pathways between sp^2 clusters already present. Thus, it can be realized that the reduction of GO to RGO gives rise to zero-band gap regions. There are still some oxygen functional groups which remains unreduced during this process, which acts as scattering sites. The ratio of the zero gap sp^2 clusters to the sp^3 clusters in the RGO is quite high which results in the quenching of the PL signal due to the phenomena of weak carrier confinement [38]. The original GO consists of numerous disorder induced defect states within the π-π* gap and exhibits a broad prominent PL spectrum centered at 500-600 nm. After deoxygenation, the number of disorder-induced states within the π-π* gap decreases, and an increased number of cluster-like states from the newly formed small and isolated sp^2 domains are formed. The electron-hole recombination among these sp^2 cluster-like states exhibits blue fluorescence at shorter wavelengths with a narrower bandwidth. The heterogeneous electronic structures of GO and RGO with variable sp^2 and sp^3 hybridizations through reduction gives rise to broader and interesting featured PL spectra in the PLD synthesized samples.

Discussion

The typical broadening of D peaks and then their gradually overlapping with G peaks in the RGO thin films indicate a successful hybridization phenomena. Varying the number of pulsed laser shots altered the physical properties of the fabricated RGO films. Kinetic energy of the atoms, molecules or ions ablated form the target are mainly dependent on the laser parameters and target characteristics (graphite target). Graphitic entities were being formed onto the SiO2 (300 nm)/Si substrates after laser ablation of graphite target. When the plume comes in contact to another surface (substrate), it condenses to form thin film which mainly depends on the surface energy of the substrate. Depending on the thermodynamics and surface energies between the film and substrate and also film-interface energy it follows three different growth mechanisms [50]. Increased surface energy of the substrate facilitated random movement of sp^2 clusters on the substrate thereby forming layered GO. GO sheets could be thought as graphene that was decorated with oxygen-containing functional groups, these functional groups could be gradually removed and sp^2 conjugated graphene network would partially be restored during the thermal reduction process, which would result in the transition from insulator to semimetal in RGO [51]. Recently, we have also reported formation of graphene, RGO, and GO using pulsed laser annealing technique [51]. This transition would change the manners of charge transportation and generate tunable electronic properties within the thin films as presented in this research article. Calculation of inter planar distances from XRD plot confirms the presence of multi-phase structures of GO and RGO in the thin films. The broadened XRD peak in the case of sample B indicates the formation of amorphous sp^2 clusters, thereby having an enormous resistance. Raman spectroscopy was carefully analyzed to extract interesting structural properties in the samples pertaining to defect states. The sharper $2D$ peak profile was an indicative of better electrical mobility in the sample D. A layer of graphene is introduced to enhance the charge carrier mobility of the PLD deposited RGO layer. Pulsed laser deposition (unlike CVD) is a non-equilibrium processing technique, where the energies of the ablated species are ~100 kT. Increasing the thickness of the deposited layer will introduce defects (Stone-Wales and functionalization defects) in the structure of the RGO layer which will impede the charge carrier mobility. So, the thickness of the deposited thin film is an important factor governing the electronic properties of RGO. In this report, the highest value of Hall mobility was achieved in the 500 pulsed laser shots sample (sample D). Detailed experiment of thickness dependent mobility in RGO thin films is under investigation and will be reported elsewhere. Again, the calculated defect densities indicate an increase with increasing number of PLD shots. FWHM of the $2D$ peak is an indicative of the degree of disorderness in the graphene (or GO and RGO structures). This peak is the characteristic of the double resonance transition arising from the generation of two phonons with wave vectors $+k$ and $-k$ (travelling in opposite direction). The FWHM of $2D$ peak for sample A (5000 shots), B (100 shots), C (10,000 shots) and D (500 shots on graphene) are 360 cm^{-1}, 155 cm^{-1}, 152 cm^{-1} and 59 cm^{-1}, respectively. This clearly indicates that the sample A has the highest degree of disorderness as compared to the other synthesized RGO thin films. The partial pressure of oxygen gas maintained throughout the pulsed laser deposition time was 10^{-5} torr. With the increase in the number of PLD shots there is an increase in the size of the sp^3 moieties formed after functionalization of the edges and basal plane of RGO. This increases the FWHM of the 2D peak and also the ID/IG ratio and is evident from the Raman spectroscopy of samples A and B. With the further increase in the number of the PLD shots there occurs an increased concentration of the sp^2 entities, thereby decreasing the FWHM of the $2D$ Raman peak. This is completely in accordance with PLD surface kinematics. Though the defect density was least in sample B, the resistance was appreciably larger due to the presence of isolated sp^2 clusters. In addition to that, the measured electrical parameters such as localization length, hopping energy, and activation energy is not favorable to produce a large charge carrier mobility. In case of sample D, the incorporation of graphene as a seed layer helps attain high sp^2 carbon fraction as well as large $I2D/IG$ ratio along with reduced FWHM of the $2D$ peak of Raman spectrum. The combination of these favorable parameters contribute to the enhancement in charge carrier mobility and we observed the maximum amount of charge carrier mobility in this sample. Furthermore, an increased number of miniature sp^2 moieties favors charge carrier variable range hopping mechanism. Other important factors that played a crucial role in the exuberant increase in charge carrier mobility for sample D were: increased localization length (4227 nm), decrease in hopping energy (3.7 meV), and decrease in activation

energy (2.3 meV). Although the ratio of sp^2/sp^3 carbon fraction is less in sample C, but due to the favorable values of localization length, hopping energy, activation energy, and $I2D/IG$ ratio it showed a moderate value of charge carrier mobility. In sample A though the measured values of electrical characterization parameters are close to those of sample C but the large FWHM of $2D$ Raman peak and small value of $I2D/IG$ result no measureable mobility in this sample. Surface defects usually act as scattering centers as they pin the electronic charge flow. The C/O ratio as calculated from XPS data suggests reduced surface defects due to moderate evolution of CO2 from the layered RGO structure. The low temperature resistance measurements suggests that the density of states (DOS) near the Fermi level (EF) is variable and it disappears linearly with energy for a two dimensional (2D) system (ES-VRH) [51]. When an electron hops from one site to another leaving behind a hole, the system must have enough energy to overcome this electron-hole Coulomb interaction for reduced resistance [5]. The non-constancy of the DOS predicts presence of disorder in the layered samples, is also presumably observed in Raman spectroscopy, which in particular is sensitive to sp^2 electronic states for the green laser used. The decreased activation energy as calculated in sample D by Arrhenius equation is also an indicative of increased concentration of sp^2 clusters near the π^* electronic state. Qualitatively it can also be inferred from the broader blue fluorescence in PL spectra that the sample D is having more sp^2% than the rest of the PLD grown samples. A sp^2 fraction of 87% was also being calculated from XPS analysis which reinforces the above mentioned fact. Detailed structural and electronic correlation parameters found through this study are given in Table 3.

The dark gray balls in the structural drawing represents C, the yellow ones H, and the red ones O [51].

Figure 7 is a schematic representation of the structural, optical, and electrical properties of the sample D along with a snapshot of the sample grown in the research facility. The pictorial representation of PL spectra represented in this figure indicates blue florescence due to the presence of innumerous sp^2 clusters, which also favors VRH mechanism. During synthesis of RGO samples the as-deposited RGO sheets could cover grain boundaries and wrinkles uniformly. This could facilitate easy movement of charge carriers underneath the film to cross over the grain boundaries with the assistance of RGO sheet via variable-range hopping process through the highly conductive graphene islands. In addition, the thermal reduction resulted in conductivity enhancement of RGO which facilitated the hopping process of charge carrier, thus leading to increased value of charge carrier Hall mobility.

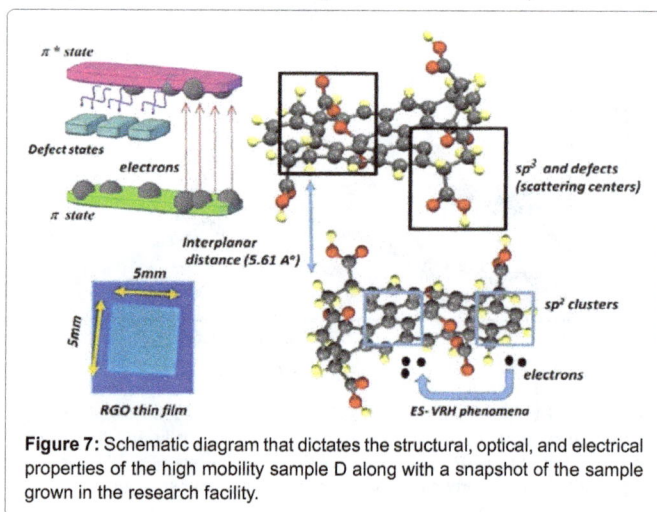

Figure 7: Schematic diagram that dictates the structural, optical, and electrical properties of the high mobility sample D along with a snapshot of the sample grown in the research facility.

Sample Name	ID/IG	I2D/IG	I2D/ID+G	FWHM of 2D (cm⁻¹)	sp²/sp³ (%)	Hall mobility (cm² V⁻¹s⁻¹)
Sample A	1.65	0.133	2.6	360	54	Small
Sample B	0.97	0.121	4.5	155	57	Small
Sample C	1.57	0.251	5.5	152	54	34
Sample D	1.14	0.171	5	59	87	372

Table 3: Structural and electronic properties of samples A, B, C, and D.

Conclusions

Large area RGO thin films have been grown using PLD having a very large carrier mobility of 372 cm² V⁻¹ s⁻¹. Larger carrier mobility has been observed with a larger fraction of sp^2 carbon fraction and a narrower Raman $2D$ peak profile in PLD grown RGO thin films. The ratio of sp^2 and sp^3 hybridized carbon atoms in functionalized graphene plays a major role in controlling the electronic properties. At lower temperature the charge carriers are having limited thermal energy to move in the plane, which can be drastically reduced with the increase in defect density with reduction. Increased sp^2 fraction helps ES-VRH mechanism thereby introducing exuberant electrical mobility. The fabrication of large area RGO thin films with very large carrier mobility fabricated by PLD method and understanding the correlation with the structural and electrical properties in reduced graphene oxide thin film will be very useful for high mobility electronic devices and could open a roadmap for further extensive research in functionalized 2D materials.

Acknowledgements

The authors would like to acknowledge Dr. William Mitchel, Air Force Research Laboratory for providing CVD grown graphene. The authors 1 through 4 contributed equally to this work. The authors would also like to acknowledge the NSF Grant (DMR-08211593) supporting this research.

References

1. Novoselov KS, Geim AK, Morozov SV, Jiang D, Zhang Y, et al. (2004) Electric field effect in atomically thin carbon films. Science 306: 666-669.

2. Lee JW, Hall AS, Kim JD, Mallouk TE (2012) A facile and template-free hydrothermal synthesis of Mn3O4 nanorods on graphene sheets for supercapacitor electrodes with long cycle stability. Chemistry of Materials 24: 1158-1164.

3. Li D, Kaner RB (2008) Graphene-based materials. Nat Nanotechnol 3: 101.

4. Zhu J, Zhang H, Kotov NA (2013) Thermodynamic and structural insights into nanocomposites engineering by comparing two materials assembly techniques for graphene. ACS nano 7: 4818-4829.

5. Joung D, Khondaker SI (2012) Efros-Shklovskii variable-range hopping in reduced graphene oxide sheets of varying carbon s p 2 fraction. Physical Review B 86: 235423.

6. Zhang Y, Tan YW, Stormer HL, Kim P (2005) Experimental observation of the quantum Hall effect and Berry's phase in graphene. Nature 438: 201-4.

7. Robertson J (2006) Amorphous carbon. Advances in Physics 35:317-374.

8. Wang ZF, Li Q, Zheng H, Ren H, Su H, et al. (2007) Tuning the electronic structure of graphene nanoribbons through chemical edge modification: A theoretical study. Physical Review B 75: 113406.

9. Ohta T, Bostwick A, Seyller T, Horn K, Rotenberg E (2006) Controlling the electronic structure of bilayer graphene. Science 313: 951-954.

10. Stankovich S, Dikin DA, Piner RD, Kohlhaas KA, Kleinhammes A, et al. (2007) Synthesis of graphene-based nanosheets via chemical reduction of exfoliated graphite oxide. Carbon 45: 1558-1565.

11. Tello PG, Castano FJ, O'Handley RC, Allen SM, Esteve M, et al. (2002) Ni–Mn–Ga thin films produced by pulsed laser deposition. Journal of applied physics 91: 8234-8236.

12. Bonaccorso F, Bongiorno C, Fazio B, Gucciardi PG, Marago OM, et al. (2007) Pulsed laser deposition of multiwalled carbon nanotubes thin films. Applied Surface Science 254: 1260-1263.

13. Habermeier HU (1993) Pulsed laser deposition-a versatile technique only for high-temperature superconductor thin-film deposition? Applied surface science 69: 204-211.

14. Kumar I, Khare A (2014) Multi-and few-layer graphene on insulating substrate via pulsed laser deposition technique. Applied Surface Science 317: 1004-1009.

15. Trusovas R, Ratautas K, Račiukaitis G, Barkauskas J, Stankevičienė I, et al. (2013) Reduction of graphite oxide to graphene with laser irradiation. Carbon 52: 574-582.

16. Mathew S, Chan TK, Zhan D, Gopinadhan K, Barman AR, et al. (2011) The effect of layer number and substrate on the stability of graphene under MeV proton beam irradiation. Carbon 49: 1720-1726.

17. Ferrari AC, Robertson J (2000) Interpretation of Raman spectra of disordered and amorphous carbon. Physical review B 61: 14095.

18. Krishnamoorthy K, Veerapandian M, Yun K, Kim SJ (2013) The chemical and structural analysis of graphene oxide with different degrees of oxidation. Carbon 53: 38-49.

19. Yan J, Zhang Y, Kim P, Pinczuk A (2007) Electric field effect tuning of electron-phonon coupling in graphene. Physical review letters 98: 166802.

20. Wang L, Park Y, Cui P, Bak S, Lee H, et al. (2014) Facile preparation of an n-type reduced graphene oxide field effect transistor at room temperature. Chemical Communications 50: 1224-1226.

21. Al-Jishi R, Dresselhaus G (1982) Lattice-dynamical model for graphite. Physical Review B 26: 4514-4522.

22. Gupta A, Chen G, Joshi P, Tadigadapa S, Eklund PC (2006) Raman scattering from high-frequency phonons in supported n-graphene layer films. Nano letters 6: 2667-2673.

23. Malard LM, Pimenta MAA, Dresselhaus G, Dresselhaus MS (2009) Raman spectroscopy in graphene. Physics Reports 473: 51-87.

24. Su CY, Xu Y, Zhang W, Zhao J, Tang X, et al. (2009) Electrical and Spectroscopic Characterizations of Ultra-Large Reduced Graphene Oxide Monolayers. Chemistry of Materials 21: 5674-5680.

25. Zhan D, Ni Z, Chen W, Sun L, Luo Z, et al. (2011) Electronic structure of graphite oxide and thermally reduced graphite oxide. Carbon 49: 1362-1366.

26. Some S, Kim Y, Yoon Y, Yoo H, Lee S, et al. (2013) High-quality reduced graphene oxide by a dual-function chemical reduction and healing process. Scientific reports 3: 1929.

27. Marcano DC, Kosynkin DV, Berlin JM, Sinitskii A, Sun Z, et al. (2010) Improved synthesis of graphene oxide. ACS nano 4: 4806-4814.

28. Jung I, Vaupel M, Pelton M, Piner R, Dikin DA, et al. (2008) Characterization of thermally reduced graphene oxide by imaging ellipsometry. The Journal of Physical Chemistry C 112: 8499-8506.

29. Mikhailov S ed (2011) Physics and Applications of Graphene - Experiments. InTech.

30. Cullity BD, Weymouth JW (1957) Elements of X-ray Diffraction. American Journal of Physics 25: 394-395.

31. McKeehan M, Warren BE (1953) X-Ray Study of Cold Work in Thoriated Tungsten. Journal of Applied Physics 24: 52.

32. Ju HM, Huh SH, Choi SH, Lee HL (2010) Structures of thermally and chemically reduced graphene. Materials Letters 64: 357-360.

33. Eda G, Fanchini G, Chhowalla M (2008) Large-area ultrathin films of reduced graphene oxide as a transparent and flexible electronic material. Nature nanotechnology 3: 270-274.

34. Becerril HA, Mao J, Liu Z, Stoltenberg RM, Bao Z, et al. (2008) Evaluation of solution-processed reduced graphene oxide films as transparent conductors. ACS nano 2: 463-70.

35. Kudin KN, Ozbas B, Schniepp HC, Prud'homme RK, Aksay IA, et al. (2008) Raman spectra of graphite oxide and functionalized graphene sheets. Nano letters 8: 36-41.

36. Muchharla B, Narayanan TN, Balakrishnan K, Ajayan PM, Talapatra S, et al. (2014) Temperature dependent electrical transport of disordered reduced graphene oxide. 2D Materials 1: 11008.

37. Eda G, Lin YY, Mattevi C, Yamaguchi H, Chen HA, et al. (2010) Blue Photoluminescence from Chemically Derived Graphene Oxide. Advanced Materials 22: 505-509.

38. Lee SJ, Ketterson JB, Trivedi N (1992) Metal-insulator transition in quasi-two-dimensional Mo-C films. Physical Review B 46: 12695-12700.

39. Efros AL, Shklovskii BI (1975) Coulomb gap and low temperature conductivity of disordered systems. Journal of Physics C: Solid State Physics 8: 49-51.

40. Peters EC, Giesbers AJM, Burghard M (2012) Variable range hopping in graphene antidot lattices. physica status solidi B 249: 2522-2525.

41. Joung D, Zhai L, Khondaker SI (2011) Coulomb blockade and hopping conduction in graphene quantum dots array. Physical Review B 83: 115323.

42. Joung D, Chunder A, Zhai L, Khondaker SI (2010) Space charge limited conduction with exponential trap distribution in reduced graphene oxide sheets. Applied Physics Letters 97: 93105.

43. Eda G, Mattevi C, Yamaguchi H, Kim H, Chhowalla M (2009) Insulator to Semimetal Transition in Graphene Oxide. The Journal of Physical Chemistry C 113: 15768-15771.

44. Ci L, Song L, Jin C, Jariwala D, Wu D, et al. (2010) Atomic layers of hybridized boron nitride and graphene domains. Nature Materials 9: 430-435.

45. Pearson GL, Bardeen J (1949) Electrical Properties of Pure Silicon and Silicon Alloys Containing Boron and Phosphorus. Physical Review 75: 865-883.

46. Tsang WT, Schubert EF, Cunningham JE (1992) Doping in semiconductors with variable activation energy. Applied Physics Letters 60: 115.

47. Erickson K, Erni R, Lee Z, Alem N, Gannett W, et al. (2010) Determination of the local chemical structure of graphene oxide and reduced graphene oxide. Advanced materials (Deerfield Beach, Fla.) 22: 4467-4472.

48. Gómez-Navarro C, Meyer JC, Sundaram RS, Chuvilin A, Kurasch S, et al. (2010) Atomic Structure of Reduced Graphene Oxide. Nano Letters 10: 1144-1148.

49. Adschiri T, Hakuta Y, Sue K, Arai K (2001) Hydrothermal Synthesis of Metal Oxide Nanoparticles at Supercritical Conditions. Journal of Nanoparticle Research 3: 227-235.

50. An X, Simmons T, Shah R, Wolfe C, Lewis KM, et al. (2010) Stable Aqueous Dispersions of Noncovalently Functionalized Graphene from Graphite and their Multifunctional High-Performance Applications. Nano Letters 10: 4295-4301.

51. Bhaumik A, Narayan J (2016) Wafer scale integration of reduced graphene oxide by novel laser processing at room temperature in air. Journal of Applied Physics 120: 105304.

19

Laser Induced Shock Wave Studies of Para and Ferro Magnetic Materials

Walid K Hamoudi, Dayah N Raouf and Narges Zamil*

Applied Sciences Department, University of Technology, Baghdad, Iraq

Abstract

In the present work, laser induced shock wave on different materials was studied, the used materials is [Paramagnetic metal (Stainless steel 304) and Ferro magnetic metals (Iron)]; immersed in some media (air and water) to study the media effect on its properties. These materials are irradiated by using different laser intensities. Laser induced shock wave was achieved by employing Nd-YAG laser at wavelength 1064 nm. X-ray diffraction revealed shifting in diffracted angles of Fe toward higher angles due to phase transformation where in air transition from Austenite to Martensite while in water from Martensite to Martensite has smaller grain size because high generated temperature followed by quenched, the diffracted angle of S.S.304 metal shifted toward very little value due to twining effects where high stress and strain are produced.

Keywords: Laser induced shock wave; Hardening; XRD of shocked and un-shocked metal

Introduction

Laser Shock Peening or Laser Shockwave Processing (LSP) is one of three processes of laser-induced shock waves in deformation processing. It is similar to Shot Peening (SP), but laser pulses replace the shots [1]. LSP is a cold work process in which a high-intensity laser pulse hits the selected surface, and then shock waves are generated [2]. LSP applies a force to the selected part of a surface, resulting in a mechanical energy applied to the surface [3]. In solid alloys; like aluminum, the surface depresses little, but depression becomes deep when increasing the treating laser intensity [4]. LSP is an effective life extension technology of fatigue life (very important property of aircraft machinery) because of the deeper residual stresses and a surface finish improves relative to conventional SP. This enhances fatigue life by modifying microstructure and/or compressive residual stresses in the surfaces/sub-surfaces [1-4].

Experimental Section

This section describes all of the tools and equipment used in the generation of the shock wave as well as the techniques used to study and explain the impact of the shock wave on the properties of the materials under study. The Laser system consisted from nanosecond Q-S Nd: YAG laser system; made by (DELIXI) company was utilized in this work. This laser system is working at 1064 nm wavelength and giving laser energy ranging between (1-500) mJ. This laser system provides short 7 ns pulses at 10 repetition frequency. The laser is facilitated with closed system cooling and has the ability of second harmonic generation, but this was not used in the present work. To achieve a high level of laser intensities, this laser is equipped with some focusing lenses made of high optical quality and high damage threshold ED2 glass. Its front panel accommodates a digital readout for the selected laser energy and controlling knobs for pulse repletion frequency and the number of pulses required per one firing touch. A straight forward set-up was adopted in this study. The set-up consists of Nd: YAG laser source, high quality, 10 cm focal length lens, and samples; cut with relevant dimensions to fit the experiment requirement. The laser source was vertically mounted in a holder above the sample, aligned and focused by 10 cm lens to achieve a circular spot with a suitable intensity. Figure 1 show the setup used in our experiments.

Sample Preparation

In this work, the laser-induced shock wave effect on different materials in different media (air, water, and ethanol) was tested. The metals were classified into; Paramagnetic metal (S.S. 304), Ferromagnetic metals (Fe). Some of the used materials were thin plates (Iron, Stainless steel). Each material was prepared in specific dimension to fit the measurement's type required. For structures and morphology measurement, the samples were cut in (1×2) cm^2. This was followed by polishing them by using graded Amery papers in the range (220-2000), then immersed in diluted HF acid in water (1:50) ml to remove dirt and the native oxide layer. The preparation process was finalized by putting them in an ultrasonic ethanol bath for about 30 min. The used materials can be summarized in the present work as follow: Fe and stainless steel have thick native oxide layer. Getting rid of oxide layer necessitated the use of a combination of Amery paper with benzene, in addition to an ultrasonic bath for 30 min. To complete the cleaning process, these samples also required etching for about 10 seconds in diluted HF: Water (1:10) percent followed by thorough rinsing in water and then in ethanol ultrasonic bath with ethanol.

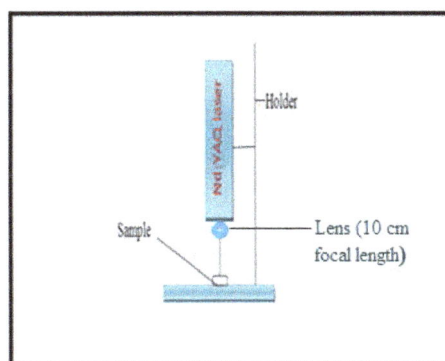

Figure 1: Experimental set-up.

***Corresponding author:** Narges Zamil, Applied Sciences Department, University of Technology, Baghdad, Iraq, E-mail: ner_ner2@yahoo.com

Laser Induced Shock Wave Procedure

The working setup was based on a Nd: YAG laser operating at a wavelength of 1.06 μm and guided by red diode laser pointer. The emitted laser power can produce high laser intensity (10^{13} W/m^2) formed a plasma and subsequently shock wave generation; as shown in Figure 2. This is usually associated with a rapidly expanding gas; associated with a strong pressure wave, a shock wave, emanating from the breakdown volume [1]. When reaching the free surface, the pressure wave creates surface waves on the substrate surface. The spatial and temporal profile of the surface waves contains information on the location and shape of the breakdown volume [2].

The material was irradiated by single and multi-pulses (2 to 5) for different laser energy (50 to 400) mJ on different materials.

XRD need long length and large irradiated area, to cover large areas, the laser irradiates pulses were made to In many rows with 50% overlapping ratio the set-up of overlapping ,overlapping system is shown in Figure 3.

The laser intensity (power density) was measured by using equations (2.20). The spot area was measured by using carbon material because it absorption is high and has small diffusivity rather than other materials. The laser spot on carbon is shown in Figure 4.

The used Nd: Yag laser given energy in voltage the calibration and converted from voltage to joule achieved by using Joulemeter Genetic type. Table 1 show the laser energy and intensity values.

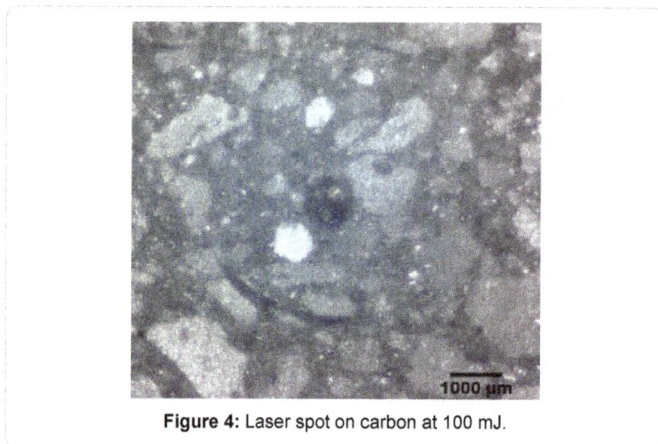

Figure 2: The experimental procedure set up, (a) sketch and (b) photographic image.

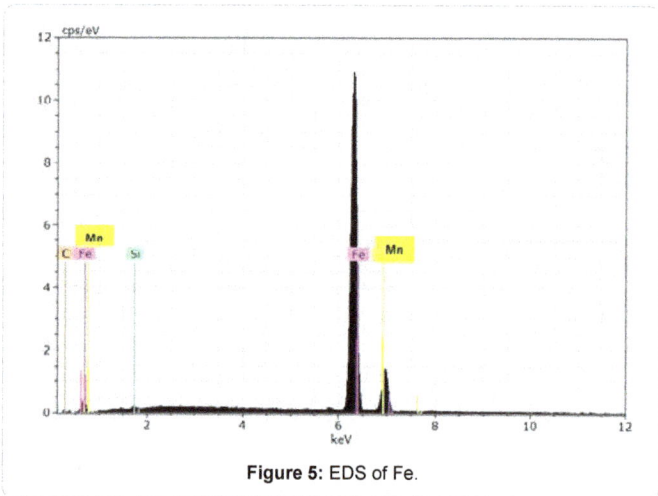

Figure 3: LSP overlapping scheme diagram.

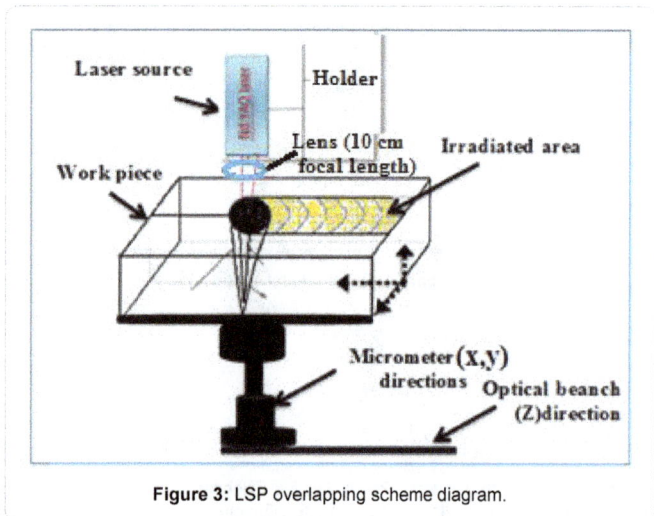

Figure 4: Laser spot on carbon at 100 mJ.

Laser operation voltages	Laser energy (mJ)	I=×10^{13} W/m^2
550	50	1.5
600	100	3.1
650	150	4.6
700	200	6.1
750	250	7.6
800	300	9.2
850	350	10.7
900	400	12.2

Table 1: Laser intensity calculations using 10 cm focal length lens. The averaged spot diameter was 0.8 mm and its cross-sectional area is 0.005 cm^2.

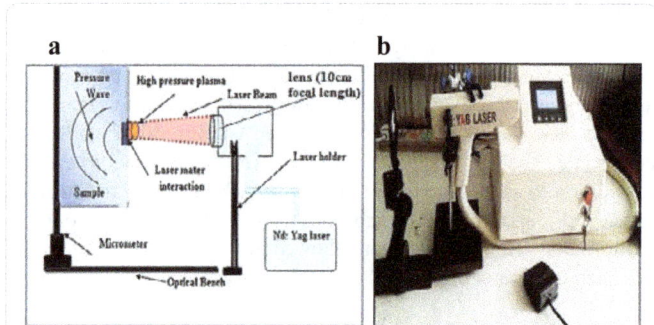

Figure 5: EDS of Fe.

X-ray diffraction analyses (XRD) were achieved in Nanotechnology and Advance Material Research Center in Al-Technology University.

Results

The effect of LSP on different material in a different media like air and water. This study includes (structure properties by using x-ray diffraction, shock wave parameters like (shock wave pressure, particle velocity, and shockwave velocity) where calculated.

Energy dispersive spectroscopy and X-ray diffraction analysis of Iron before and after LSP

The energy dispersive spectroscopy (EDS) of Fe is shown in Figure 5. The concentration of elements in the used sample is shown in Table 2. Figure 6 shows the XRD of Fe before and after LSP in different media by using different laser intensities for the unshocked sample

El	Fe	Mn	C	Si
Norm. wt.%	79.51	17.17	2. 76	0.57

Table 2: The used Iron samples elements concentration.

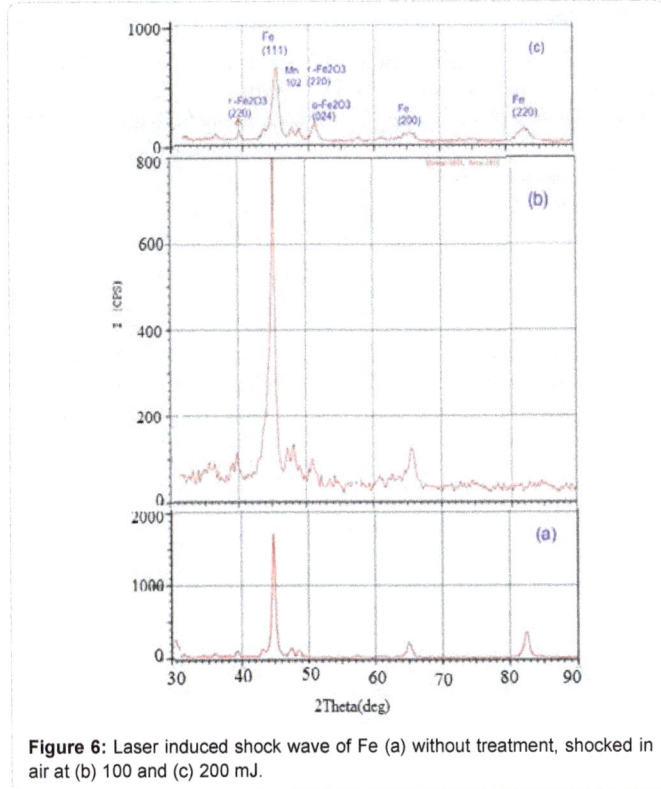

Figure 6: Laser induced shock wave of Fe (a) without treatment, shocked in air at (b) 100 and (c) 200 mJ.

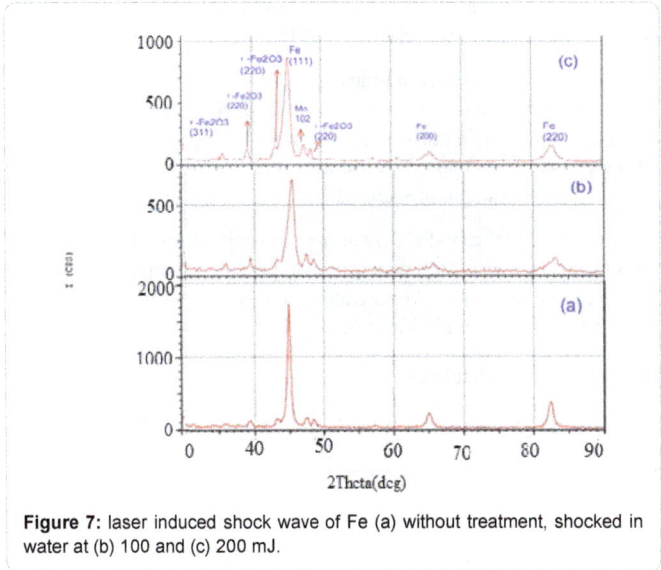

Figure 7: laser induced shock wave of Fe (a) without treatment, shocked in water at (b) 100 and (c) 200 mJ.

	2 Theta θ1	2 Theta θ2	2 Theta θ3	d(A) θ1	d(A) θ2	d(A) θ3	FWHM θ1	FWHM θ2	FWHM θ3
Without shock	44.83	65	82.37	2.02	1.434	1.17	0.621	0.754	0.784
100 mJ air	45.13	65.64	-	2.01	1.424	-	0.806	1.017	-
200 mJ air	45.02	-	82.52	2.01	-	1.168	1.076	-	1.976
100 mJ water	45.21	-	83.02	2	-	1.162	1.079	-	1.017
200 mJ water	45.01	-	82.77	2.013	-	1.165	1.002	-	1.448

Table 3: Iron diffraction angle, inter planer spacing and fall width half maximum before and after LSP.

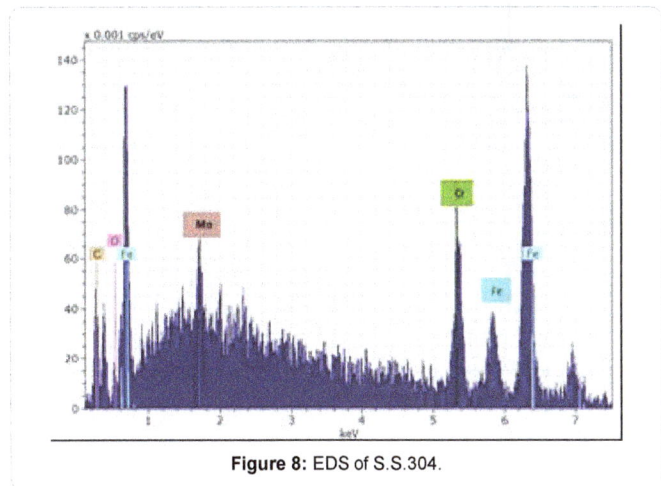

Figure 8: EDS of S.S.304.

the FeO group is a presenced with high intensity and decreased after laser treatment, it means the laser cleaned Fe layer from native oxygen. Also, that means the change in atomic mass is happened where the particles with high atomic mass give more intense XRD in this case the (FeO has higher atomic mass than Fe) [3]. The Fe shocked in air with different laser intensity, the new peaks appeared in 100 mJ laser energy component due to the formation of Fe_2O_3 and disappeared at 200 mJ laser energy due to high laser energy evaporate or removed this component [4], smaller shifting in diffraction angle due from increasing in lattice parameter. The Fe peaks were decreased with increased laser energy because the phase transition of Fe will happen from Martensite to Austenitic (Martensite has a particle size greater than Austenitic) [5].

Figures 7 represents the material shocked under water at 100 and 200 mJ respectively, the oxide has lower intensity than fresh sample, in Table 3 the diffraction angle little shift toward higher diffracted rather than fresh samples and constant with small broadening in 200 mJ due to (the strain and ductility effect) [6].

Energy dispersive spectroscopy and X-ray diffraction of Stainless steel 304 before and after laser-induced shock wave process

The energy dispersive spectroscopy (EDS) of S.S.304 is shown in Figure 8 and Table 4.

Figure 9 represents S.S.304 shocked in air by 100 and 200 mJ. At low temperatures (550-600) K the Ferrite (α-Fe) is presence in S.S.304, by increasing laser energy Austenite peaks appeared with high intensity due to phase transition because of the high generated temperature [7].

At 900°C typical low-carbon steel is composed entirely of Austenite, a high-temperature phase. At a temperature lower than 700°C, the austenite is thermodynamically unstable and, under equilibrium conditions, it will (undergo a eutectoid reaction and form pearlite) [8].

A shifting in diffraction angle happened toward small angle because of the substantial residual stress due to volume change on reaction. A compression happened to inner direction lead to decrease the lattice constant, also because of a phase transformation is taken place, a twining is a dominant deformation mechanism by serve a plastic deformation of the grain refinement [9].

Figure 10 represents stainless steel is shocked by 100 mJ, and 200 mJ in water medium respectively, the diffraction angle little shift towards smaller angle rather than unshocked sample due to the smaller lattice constant, and phase transition the stress that generated due to higher generated shock pressure [10].

Under water the (γ-Fe_2O_3) has high intensity due to quenching and phase transition also high generated temperature changed magnetism from paramagnetic to ferromagnetic type that effect to XRD shifting and intensity as shown in Table 5.

Structure Parameters

From X-ray diffraction, we obtain the results given in Tables 4-7.

Grain size

Since the size of grains directly affect the on structure and hence

properties of materials. Crystals are regular arrays of atoms, and X-rays can be considered waves of electromagnetic radiation. Atoms scatter X-ray waves, primarily through the atoms' electrons. A few specific directions, determined by Bragg's law:

$$2d \sin \theta = n\lambda \tag{1}$$

Here d is the spacing between diffracting planes, θ is the incident angle, n is any integer, and λ is the wavelength of the beam. The relation between the inter planer spacing and lattice constant (a) is given by [11]:

$$d = \frac{a}{\sqrt{h^2 + k^2 + l^2}} \tag{2}$$

The grain size was calculated from Sherrie eq. given by [12]:

$$D_{hkl} = \frac{K\lambda}{B\cos\theta} \tag{3}$$

Where B is the integral half width, K is a constant equal to 0.94, λ is the wavelength of the incident X-ray (λ=0.15406 nm), D is the crystallite size, and θ is the Bragg angle. Its value is shown in Table 6 for used materials as follow:

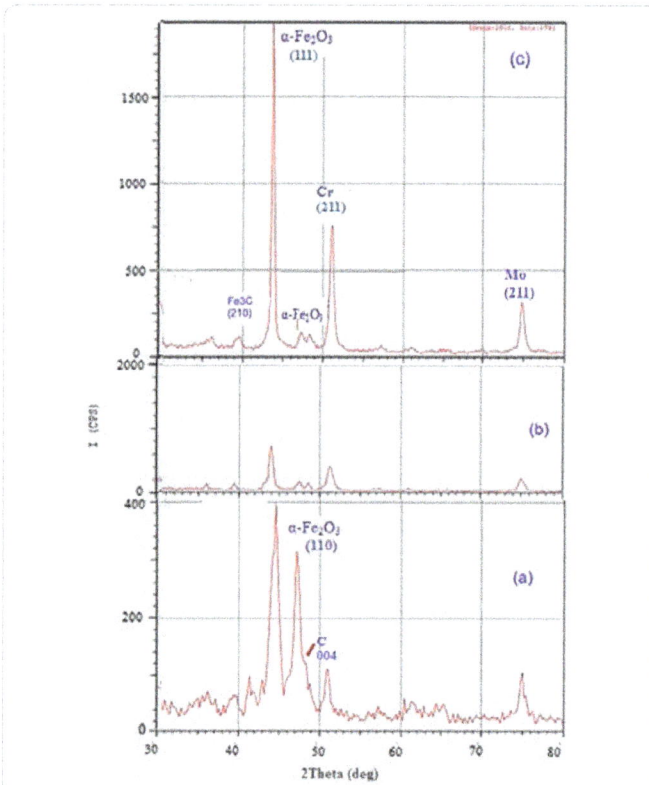

Figure 9: Laser-induced shockwave of S.S.304 (a) without treatment, shocked in air at (b) 100 and (c) 200 mJ.

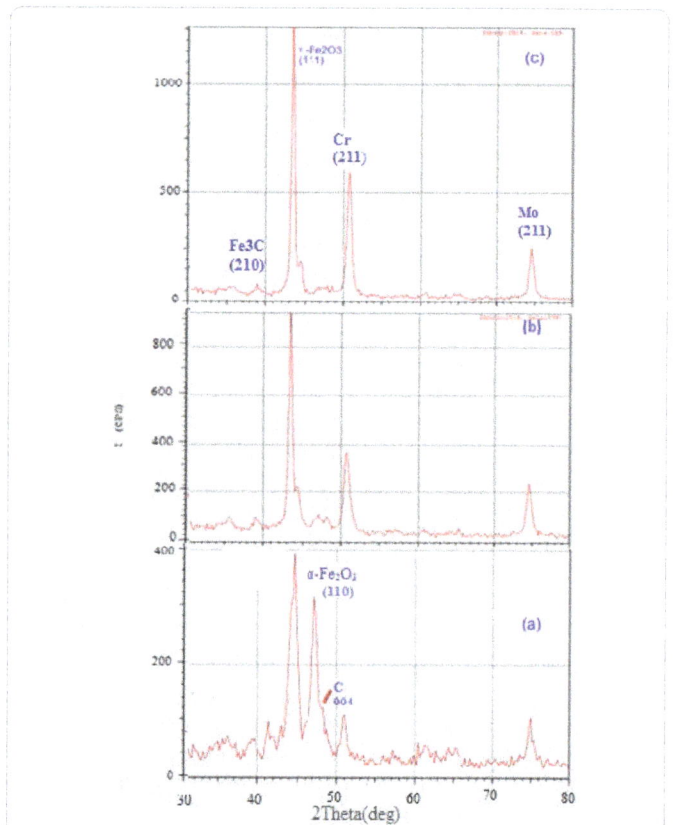

Figure 10: Laser-induced shockwave of S.S.304 (a) without treatment, shocked in water medium at (b) 100 mJ and (c) 200 mJ.

El	Fe	Cr	Mo	C	O
Norm. wt.%]	68.83	12.98	11.3	0	0.58

Table 4: The used S.S. 304 samples elements concentration.

	2 Theta θ1	2 Theta θ2	2 Theta θ3	d(A) θ1	d(A) θ2	d(A) θ3	FWHM θ1	FWHM θ2	FWHM θ3
Unshock	44.45	50.94	74.92	2.04	1.79	1.267	1.16	0.62	0.72
100 mJ air	44	51.26	75.01	2.06	1.78	1.265	0.65	0.78	0.77
200 mJ air	43.8	51.07	74.83	2.07	1.79	1.268	0.54	0.65	0.66
100 mJ water	43.44	50.72	74.57	2.08	1.8	1.27	0.61	0.84	0.79
200 mJ water	43.45	50.74	74.58	2.08	1.8	1.27	0.598	0.77	0.74

Table 5: S.S.304 diffraction angle, inter planer spacing and fall width half maximum before and after LSP.

Sample condition	Fe	S.S.304
Unshocked	13.86	12.33
100 mJ air	10.43	13.16
200 mJ air	6.97	15.48
100 mJ water	9.61	12.89
200 mJ water	8.3	13.47

Table 6: Grain size (nm) of used samples.

Sample condition	Fe	S.S.304
Unshocked	0.0052	0.0074
100 mJ air	0.0077	0.0061
200 mJ air	0.0106	0.0052
100 mJ water	0.0082	0.0063
200 mJ water	0.0089	0.0068

Table 7: Lattice strain of used samples.

Material	constant (a) (Km/s)	Constant (b)	Impedance (kgm^{-2}s^{-1} × 10^6)
Fe	3.574	1.920	46.4
Stainless steel 304	4.580	1.49	45.7

Table 8: Acoustic impedance, and (a) and (b) constants value of different metals.

Grain size of Fe: The grain size of Fe in air is decreased after LSP due to phase transition from Martensite to Austenite that has smaller grain size. In water the decreased of grain size due to quenching led to phase transition to Martensite differs from old one by has smaller grain size.

Grain size of S.S.304: The grain size in air increased due to both multidirectional mechanical twins and multidirectional Martensite bands led to grain refinement mechanism this consisted with [13]. In water the shock pressure is greater than in air, the distortion becomes stronger and stronger, finally the accumulated distortion make the original twin into intercrossed.

Lattice strain

The strain-induced broadening due to crystal imperfection and distortion was calculated using the formula:

$$\varepsilon = \frac{B_{hkl}}{4tan\theta} \tag{4}$$

Fe lattice strain: in air lattice strain increased with increasing laser energy due to cold working that induced by plastic deformation of materials. The shocked Iron in water medium, is exposed to the quickly quenching led to increasing the ductility between the particles rather than a fresh sample, but shocked material under water is smaller than shocked it in the air due to increased shock pressure when samples shocked in water [1,5].

S.S.304 lattice strain: in air, lattice strain decreased with increased laser energy due to Martensite transitions, In water the decreasing of lattice strain due to increased heating led to increased diffusion enhanced dislocation motion, decreased dislocation density by annihilation, formation of low energy dislocation configurations, relieve of the internal strain energy (Table 7).

Dislocation densities

The dislocation density (δ), the dislocation toughly influences many of the properties of materials, which represents a number of defects in the sample is defined as the length of dislocation lines per unit volume of the crystal and is calculated using the equation,

$$\delta = \frac{1}{D_{hkl}^2} \tag{5}$$

Where D_{hkl} is the crystallite size.

Fe dislocation density: After shocked Fe the dislocation is increased with increasing laser intensity in different media (air and water), due to temperature effect also because of cold working that induced by deformation of materials.

S.S.304 dislocation density: In air and water decreasing dislocation density with increasing laser energy due to twining effects.

Laser Shock Wave Properties

This includes shock wave pressure particle velocity, and shock wave velocity.

Shockwave pressure

The pressure value; due shock wave effect could be measured when the size of the Gaussian laser beam is relatively great. The shock model makes modifications to Fabbro's model after assuming that the laser beam spot size in the order of microns. The 1-D assumption is followed, but 2-D equivalence is considered to account for the small laser spot size [1].

$$P = (AZI)^{1/2} \tag{6}$$

$$Z = \rho * V_l \tag{7}$$

Where, Z: acoustic impedance of material, ρ: density V_i: velocity of acoustic waves in the material. When using a confining medium, the acoustic impedance $z = \dfrac{2}{\left(\dfrac{1}{z1}+\dfrac{1}{z2}\right)}$ is expressed in terms of those of the confining medium (z_1) and the target material (z_2).

and illustrated in Table 8. The shock pressure is increased with increasing the laser intensity due to increase surface temperature so increased the number of ions that emerged from the irradiated surface [1], Shock pressure of ferrite metals (Fe) is greater than S.S.304). That means increased pressure with increased magnetism of metals due to increased acoustic impedance. A shock pressure for all metals is great in water than in air medium due to the shock wave that generated by laser is wave that is transferred by particles science particles are spread so far a part in gas. Water is incompressible; particles are together and give impedance to applied force greater than air consequently increased shock pressure. The shock pressure depending on absorption [1] (Figure 11).

Shock velocity and particle velocity

When a laser pulse of high energy interacts with the material, the hot plasma is produced which is expanding at very high speed in the opposite direction of the laser. Externally expanded plasma utilizes high pressure in the direction of the interior, to form a wave of severe shock towards the interior. P is shocked pressure, V_L is shocked velocity, and V_p is the partial velocity. The relation between particle velocity and shock pressure related with acoustic impedance (Z), and given as flow:

$$V_p = \frac{P}{Z} \tag{8}$$

The Shockwave velocity and particle velocity related to each other tor as:

$$V_l = a + bV_p \tag{9}$$

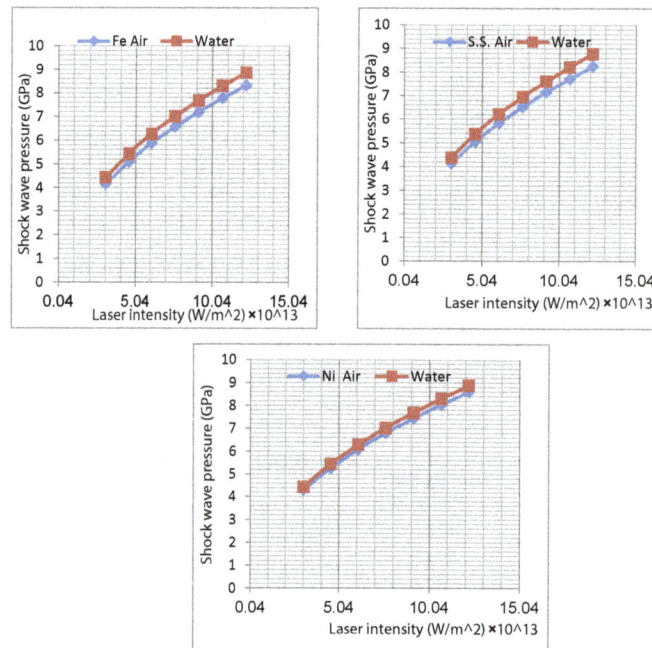

Figure 11: Laser-induced shockwave pressure vs. laser intensity at different metals at air, water, and ethanol media.

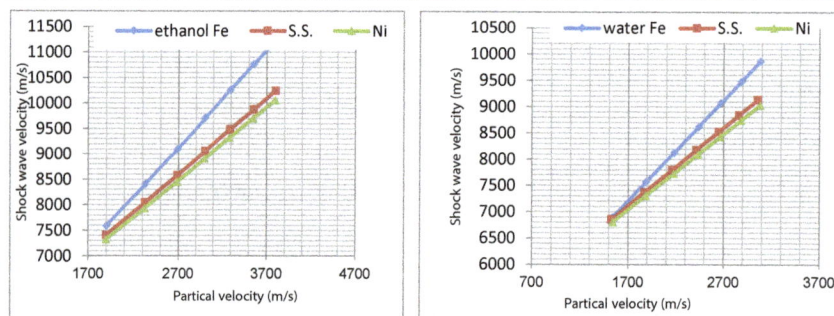

Figure 12: Laser-induced shockwave velocity vs. particle velocity.

Where *a* and *b* is constant given in Stanley *P*. Marsh data book and shown in the Table 8. Figure 12 shows particle velocity.

A linear relation between shock velocity and particle velocity for all metals in different media is observed because of the dependence on each other. Shock velocity is great in water and ethanol than in air, because (the acoustic impedance of air is smaller than water and ethanol), the shock and particle velocity is affected by temperature when laser intensity is increased the temperature increased so shock wave tend to travel faster at height temperatures.

Conclusions

Laser induced shock wave field is a very interesting area of study. It turns out to be a very useful non-destructive technique to study the material properties especially with very small area. We can summarize the main conclusions of our work as follow; LSP is a successful method to increase the hardness of some metals. LSP is a simple, low cost and controllable. X-ray diffraction showed a reduction in the grain size after the shocking process. The deformation increased after LSP process and this means increased ductility of metals. This effect led to increased dislocation density and enhanced stress.

References

1. Gujba AK, Medraj M (2014) Laser peening process and its impact on materials properties in comparison with shot peening and ultrasonic impact peening. Materials 7: 7925-7974.

2. Gornikowska MR, Kusinski J, Blicharski M (2011) The Influence of The Laser Treatment on Microstructure of the Surface Layer of an X5CrNi18-10 Austenitic Stainless Steel. Archives of Metal Allergy and Materials 56: 717-721.

3. Clauer A (1997) Laser Shock Peening For Fatigue Resistance. J Metal Society 18: 217-230.

4. Jay FT, Alexander GP (2007) Low Speed Laser Welding of Aluminum Alloys Using Single-Mode Fiber Lasers. J. Welding 86: 179-186.

5. Callisster WD (2006) Materials Science And Engineering. An Introduction. John Wiley and Sons, Inc., 200-252.

6. Bo H, Ma Z, Jing X, Ting YY (2011) Structural Electrical and Optical Properties of Azo/Sio2/P-Si SIS Heterojunction Prepared by Magnetron sputtering. J Optical Applicant 11: 15-24.

7. Ameen HA, Hassan KS, Mubarak EM (2011) Effect Of Loads, Sliding Speeds and Times on the Wear Rate for Different Materials. American J Science and Industrial Research 2: 99-106.

8. Young DA (1975) Phase Diagrams of the Element. Lam/Lam/Hencehunermofe Laboratory 20: 21.

9. Bindu P, Thomas S (2014) Estimation of Lattice Strain in Zno Nanoparticles: X-Ray Peak Profile Analysis. J Theor Appl Phys 8: 123-134.

10. Cheng YT, Cheng CM (1998) Scaling Approach to Conical Indentation in Elastic-Plastic Solids with Work Hardening. J Appl Phys 84: 1284-1291.

11. Vogel A, Busch S (1996) Shock Wave Emission And Cavitation Bubble Generation by picosecond and nanosecond optical breakdown in water. J. Acoust. Soc. Am 100: 148-165.

12. Ariga S, Sigel R (1976) Picosecond Microphotography as a Diagnostic Method for Laser Produced Plasmas. 31: 697-706.

13. Marsh P (1980) Lasl Shock Hugoniot Data. The Regents of the University of California, USA 89: 110-212.

20

Nanoparticle Shape and Thermal Radiation on Marangoni Water, Ethylene Glycol and Engine Oil Based Cu, Al$_2$O$_3$ and SWCNTs

Nur Atikah Adnan, Kandasamy R* and Mohammad R

Research Centre for Computational Fluid Dynamics, FSTPi, Universiti Tun Hussein Onn Malaysia, Malaysia, 86400, Batu Pahat, Johor, Malaysia

Abstract

The aim of this paper is to investigate the relationship between particle shape and radiation effects on Marangoni boundary layer flow and heat transfer of water, ethylene glycol and engine oil based Cu, Al$_2$O$_3$ and SWCNTs. There are three types of nanoparticle shapes are considered in this research such as sphere, cylinder and lamina. The governing nonlinear partial differential equations are reduced into a set of nonlinear ordinary differential equations by applying similarity transformation which is solved using shooting technique in conjunction with Newton's method and Runge Kutta algorithm. Temperature profiles are graphically and tabularly provided for the effects of solid volume fraction parameter, radiation parameter and empirical shape factor. The result shows that solid volume fraction and radiation energy gives a good impact on thermal boundary layer. Sphere nanoparticle shape predicts a better result on heat transfer rather than other nanoparticle shapes.

Keywords: Nanoparticle shape; Thermal radiation; Marangoni water; Ethylene glycol; Engine oil; Exponential temperature

Introduction

Nanotechnology is one of widely technology rapidly progress in various fields such as chemistry, physics, materials science, biotechnology and other applications. It is due to their structures that are determined on the nanometer scale. It seems that Choi [1] is the first introduce term "nanofluids" for reference to base fluids suspended nanoparticles. Nanofluid is a fluid contains nanometer particles known as nanoparticles and made of metals, oxides, carbines or carbon nanotubes. The fluids are engineered colloidal suspension of nanoparticles in a base fluid. Water, ethylene glycol and oil are commonly example base fluid. Studies have shown that adding nanoparticles such as metal particles, metal oxides, metalloid oxides and carbon nanotubes, in the base fluids can effectively improve the thermal conductivity of the base fluids and enhance heat transfer performance of the liquid.

Studies have shown that nanofluid exhibit heat transfer characteristic compare to conventional fluid. There are several numerical and experimental studies on heat transfer in nanofluids: conductive, convective and radiative. Sidik et al. [2] presented an inclusive review on preparation methods and challenges of nanofluids. In addition, Pang et al. [3] presented the recent development and research effort of heat and mass transfer in nanofluid. Based on the research, most of the researchers are focusing on thermal conductivity and the heat transfer performance affected by the following parameters: nanoparticle material, nanoparticle size, nanoparticle shape, temperature and additives.

Marangoni boundary layer is the dissipative layer which may occur along the liquid-gas or liquid-liquid interfaces. Marangoni flow, occur at surface temperature gradient or the surface concentration gradient, appears in many practical projects such as chemical reaction process [4], aerospace engineering, crystal growth [5] and silicon melt [6]. There are two types of Marangoni which is thermal Marangoni effect (EMT) and solute Marangoni effect (EMS). Pearson [7] was introduced the mechanism of EMT.

The mechanism of thermal Marangoni effect (EMT) occur when a thin layer of fluid is heated from below and the temperature gradient is such that small variations in the surface temperature lead to surface tractions which cause the fluid to flow. Then, it tends to maintain the original temperature variations. Scriven and Sternling [8] are researchers who introduced the mechanism of EMS. Recently, there have been several papers published on the mechanism of Marangoni boundary layer flow transport. Christopher and Wang [9] present the effects of Prandtl number and Marangoni number on the Marangoni boundary layer around a vapor bubble during nucleation and growth by the shooting method.

In addition, Zheng et al. [10] examined the analytical result for Marangoni convection over a liquid-vapor surface due to an imposed temperature gradient by the Adomian decomposition technique couple with the Padé approximant technique. Then, Chamkha et al. [11] obtained a set of exact analytical results for the MHD thermosolutal Marangoni boundary layers over a flat surface. Later on, Chen [12] investigated the influence of Marangoni boundary layer on the flow and heat transfer of power-law fluids in a finite thin film over an unsteady stretching sheet.

Thus a research is carry out to investigate the relationship between particle shape and radiation effects on Marangoni boundary layer flow and heat transfer of copper, alumina and SWCNTs- water nanofluid, ethylene glycol and engine oil. There are three types of nanoparticle shapes are considered in this research such as sphere, cylinder and lamina. The governing nonlinear partial differential equations was reduced into a set of nonlinear ordinary differential equations by applying similarity transformation which are solved using shooting technique in conjunction with Newton's method and Runge Kutta algorithm. Temperature profiles are graphically and tabularly presented

*Corresponding author: Ramasamy Kandasamy, Research Centre for Computational Fluid Dynamics, FSTPi, Universiti Tun Hussein Onn Malaysia, Malaysia, 86400, Batu Pahat, Johor, Malaysia, E-mail: ramasamy@uthm.edu.my

for the effects of solid volume fraction parameter, radiation parameter and empirical shape factor.

Mathematical Analysis

The problem is consider as two dimensional steady Marangoni boundary layer flow and heat transfer of copper, alumina and SWCNTs in the presence of water, ethylene glycol and engine oil above a flat interface with surface tension gradient due to an exponential temperature. The physical model is assumed as incompressible, the base fluid and nanoparticle are in thermal equilibrium, no slippage and the flow is laminar.

Schematic of the physical system is shown in Figure 1 where Y is the coordinate measured normal to the surface whereas X axis pointing toward the porous medium. In addition, the thermo physical properties of the base fluid and nanoparticles copper, alumina and SWCNTs are given in Table 1 [13,14]. Marangoni effect acts as a boundary condition on the governing equations for the flow and it is unlike the Boussinesq effect in buoyancy-induced flow. The basic equations can be written in Cartesian coordinates X and Y as:

$$\frac{\partial U}{\partial X} + \frac{\partial V}{\partial Y} = 0 \tag{1}$$

$$U\frac{\partial U}{\partial X} + V\frac{\partial U}{\partial Y} = \frac{\mu_{nf}}{\rho_{nf}}\frac{\partial^2 U}{\partial Y^2} \tag{2}$$

$$U\frac{\partial T}{\partial X} + V\frac{\partial T}{\partial Y} = \alpha_{nf}\frac{\partial^2 T}{\partial Y^2} - \frac{1}{(\rho C_p)_{nf}}\frac{\partial q_r}{\partial Y} \tag{3}$$

The boundary conditions are

$$Y = 0 : \mu_{nf}\frac{\partial U}{\partial Y}\Big|_{Y=0} = \frac{\partial \sigma}{\partial X}\Big|_{Y=0} = 0, T\Big|_{Y=0} = T_w = T_\infty + T_{const}e^{-X/L_0} \tag{4}$$

$$Y \to \infty : U\Big|_{Y \to \infty} = 0, T\Big|_{Y \to \infty} = T_\infty \tag{5}$$

Where U and V are the velocity components along the X and Y directions respectively. M_{nf} is the viscosity of nanoparticle-nanofluid and ρ_{nf} is the density of the nanofluid. T is temperature, α_{nf} is thermal

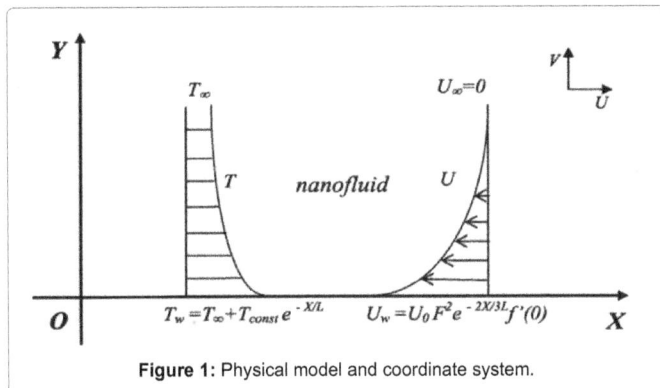

Figure 1: Physical model and coordinate system.

	C$_p$ (J/kgK)	ρ (kg/m³)	k (W/Mk)
Cu	385	8933	400
Al$_2$O$_3$	765	3970	40
SWCNTs	425	2600	6600
Water	4179	997.1	0.613
Ethylene glycol	2430	1115	0.253
Engine oil	1910	884	0.144

Table 1: Thermo physical properties of Copper, Alumina and SWCNTs – water, ethylene glycol and engine oil.

diffusivity of the nanofluid, c_p is specific heat at constant pressure, $(\rho c_p)_{nf}$ is heat capacity of the nanofluid and q_r is the radiative heat flux. T_w is surface temperature and it is assumed to be an exponential function with X, T_∞ is the temperature of nanofluid far from the interface and T_{const} is a reference temperature.

The others physical characteristics of the nanofluid are given by refs. [15,16].

$$\mu_{nf} = \frac{\mu_f}{(1-\varphi)^{2.5}} \tag{6}$$

$$\rho_{nf} = (1-\varphi)\rho_f + \varphi\rho_s \tag{7}$$

$$\alpha_{nf} = \frac{k_{nf}}{(\rho C_p)_{nf}}, \quad \gamma_{nf} = \frac{\mu_{nf}}{\rho_{nf}} \tag{8}$$

$$(\rho C_p)_{nf} = (1-\varphi)(\rho C_p)_f + \varphi(\rho C_p)_s \tag{9}$$

ρ_{nf} is the density of water, ρ_s is density of solid nanoparticles, $(\rho c_p)_f$ is the heat capacity of fluid $(\rho c_p)_s$ is the heat capacity of solid nanoparticles. γ_{nf} is kinematic viscosity of nanofluid, k_{nf} is thermal conductivity of nanofluid. In this research, the nanoparticles shapes are taken into account by using Hamilton and Crosser model [17].

$$\frac{K_{nf}}{K_f} = \frac{[k_s + (m-1)k_f] - (m-1)\varphi(k_f - k_s)}{[k_s + (m-1)k_f] + \varphi(k_f - k_s)} \tag{10}$$

where k_f is thermal conductivity of nanofluid, k_s is thermal conductivity of solid nanoparticles, $m=3/\varphi$ is empirical shape factor where φ is sphericity. By using Rosseland approximation, the radiative of heat flux is become to:

$$q_r = -\frac{4\delta^*}{3k^*}\frac{\partial T^4}{\partial Y} \tag{11}$$

where k* is the mean absorption and δ* is Stefan Boltzman. The temperature on the surface is an exponential function with X. Moreover, there are temperature differences within the flow; T⁴ is expressed as a linear function of temperature. This accomplished by expanding T⁴ in a Taylor series about T∞. The higher-order terms are neglecting, thus it become:

$$T^4 \approx 4T_\infty^3 T - 3T_\infty^4 \tag{12}$$

According to the boundary condition 4, σ is defined as surface tension. The temperature gradient occur by interfacial surface tension gradient at the interface induced flow as

$$\frac{\partial \sigma}{\partial X} = \frac{\partial \sigma}{\partial T} \cdot \frac{\partial T}{\partial X} \tag{13}$$

In addition, it is assumed that the surface tension is linear with the temperature such that

$$\sigma = \sigma_0 - \gamma_T(T - T_\infty), \gamma_T = -\frac{\partial \sigma}{\partial T} \tag{14}$$

σ_0 is a positive constant and it represents the surface tension when $T=T_\infty$. γ_T is the temperature coefficient of the surface tension. We introduced the similar dimensionless variables (U_0 is velocity unit, L_0 is length unit):

$$u = \frac{U}{U_0}, v = \frac{V}{U_0}\left(\frac{U_0 L_0}{\gamma_f}\right)^{\frac{1}{2}} = \frac{v}{U_0}\text{Re}^{\frac{1}{2}} \tag{15}$$

$$x = \frac{X}{L_0}, y = \frac{Y}{L_0}\left(\frac{U_0 L_0}{\gamma_f}\right)^{\frac{1}{2}} = \frac{Y}{L_0}\text{Re}^{\frac{1}{2}} \tag{16}$$

$$t = \frac{T}{T_\infty} \tag{17}$$

$$a = \left(1-\varphi\right)^{2.5}\left[\left(1-\varphi\right)+\varphi\frac{\rho_s}{\rho_f}\right] \tag{18a}$$

$$b = \left[\left(1-\varphi\right)+\varphi\frac{\left(\rho C_p\right)_s}{\left(\rho C_p\right)_f}\right]\frac{[k_s+(m-1)k_f]-(m-1)\varphi(k_f-k_s)}{[k_s+(m-1)k_f]+\varphi(k_f-k_s)} \tag{18b}$$

$$= \left(\ -\ \right)^{2.5} \tag{18c}$$

$$\mathrm{Re} = \frac{U_0 L_0}{\gamma_f}, \mathrm{Pr} = \frac{\gamma_f}{\alpha_f}, \gamma_f = \frac{\mu_f}{\rho_f}, Ma = \frac{\gamma_T T_{const} L_0}{\mu_f \alpha_f}, Nr = \frac{16\delta^* T_\infty^3}{3k_{nf}k^*} \tag{19}$$

Where R_e is Reynolds number, P_r is Prandtl number. In this paper, we consider P_r=7.8 and Ma is Marangoni number and Nr is the radiation parameter. So, the governing eqns. (1)-(3) can be transformed to:

$$\frac{\partial u}{\partial x} + \frac{\partial v}{\partial y} = 0 \tag{20}$$

$$u\frac{\partial u}{\partial x} + v\frac{\partial u}{\partial y} = \frac{1}{a}\frac{\partial^2 u}{\partial y^2} \tag{21}$$

$$u\frac{\partial t}{\partial x} + v\frac{\partial t}{\partial y} = \frac{1+Nr}{b\,\mathrm{Pr}}\frac{\partial^2 t}{\partial y^2} \tag{22}$$

Boundary conditions (4)-(5) become:

$$y=0: \frac{T_{const}}{T_\infty}\frac{\partial u}{\partial y}\Big|_{y=0}=-\frac{Ma}{\mathrm{Pr}}\left(\frac{1}{\mathrm{Re}}\right)^{\frac{3}{2}}c\frac{\partial t}{\partial x}\Big|_{y=0}, v=0, t\big|_{y=0}=1+\frac{T_{const}}{T_\infty}e^{-x} \tag{23}$$

$$y\to\infty: u\big|_{y\to\infty}=0, t\big|_{y\to\infty}=1 \tag{24}$$

Using the boundary layer approximation and introducing the stream function ψ defined as $u=\frac{\partial\psi}{\partial y}$ and $v=\frac{\partial\psi}{\partial y}$, the following transformation variables are:

$$\psi(x,y)=Fe^{\frac{-x}{3}}f(\eta), \eta=Fe^{\frac{-x}{3}}y, F=\left(\frac{Ma}{\mathrm{Pr}}\right)^{\frac{1}{3}}\left(\frac{1}{\mathrm{Re}}\right)^{\frac{1}{2}} \tag{25}$$

$$t(x,y)=1+\frac{T_{const}}{T_\infty}e^{-x}\theta(\eta) \tag{26}$$

The partial differential eqns. (20)-(22) are transformed to the following ordinary equations:

$$f'''(\eta)=a\left[\frac{1}{3}f(\eta)f''(\eta)-\frac{2}{3}f'(\eta)^2\right] \tag{27}$$

$$\theta''(\eta)=\frac{b\,\mathrm{Pr}}{1+Nr}\left[\frac{1}{3}f(\eta)\theta'(\eta)-f'(\eta)\theta(\eta)\right] \tag{28}$$

Boundary conditions (23)-(24) can be expressed as

$$f(0)=0, f''(0)=c, f'(\infty)=0 \tag{29}$$

$$\theta(0)=1, \theta(\infty)=0 \tag{30}$$

The X and Y component of the velocity and temperature are:

$$U=U_0\left(\frac{Ma}{\mathrm{Pr}}\right)^{\frac{2}{3}}\frac{1}{\mathrm{Re}}e^{\frac{-2x}{3L_0}}f'(\eta)=\left(\frac{\gamma_T T_{const}\gamma_f}{\rho_f L_0^2}\right)^{\frac{2}{3}}\frac{\mathrm{Re}}{U_0}e^{\frac{-2x}{3L_0}}f'(\eta) \tag{31}$$

$$V=\frac{1}{3}\left(\frac{\gamma_T T_{const}\gamma_f}{\rho_f L_0^2}\right)^{\frac{1}{3}}e^{\frac{-x}{3L_0}}\left[f(\eta)+\eta f'(\eta)\right] \tag{32}$$

$$T=T_\infty+T_{const}e^{\frac{-x}{L_0}}\theta(\eta) \tag{33}$$

Local Nusselt number Nu_x defined as:

$$Nu_x=\frac{xq_w(x)}{k(T)\left[T(X,0)-T(X,\infty)\right]} \tag{34}$$

$q_w(X)$ is heat flux of nanofluid as $q_w(X)=-k(T)\left(\frac{\partial T}{\partial Y}\right)\Big|_{y=0}$. By using eqns. (25), (26), (33) and (34) (Table 2 and Figure 2),

$$Nu_x=-\frac{X}{L_0}e^{\frac{-x}{3L_0}}\left(\frac{\gamma_T^2 T_{const}^2}{\rho_f^2 U_0^3 L_0\gamma_f}\right)^{\frac{1}{6}}\theta'(0) \tag{35}$$

Numerical Techniques

The set of nonlinear ordinary differential eqns. (27) and (28) with boundary condition (29-30) are reduced to the first order differential equations and have been solved by using Runge Kutta scheme in conjunction with modified Newton- Raphson shooting method .The terms in eqns. (27) and (29) are expressed as a set of first order equations in terms of three variables u, v and r. Introducing the shooting parameter t as $f'(0)=t$ and $f(\eta)$, $f'(\eta)$ and $f''(\eta)$ are denoting by using variables u, v and r yields:

$$f=f(\eta)$$
$$f'=\frac{d}{d\eta}f(\eta)=u(\eta)$$
$$f''=\frac{d^2}{d\eta^2}f(\eta)=\frac{d}{d\eta}u(\eta)=v(\eta) \tag{36}$$
$$f'''=\frac{d^3}{d\eta^3}f(\eta)=a(f(\eta).v(\eta)-2u(\eta)^2)/3$$

$$f(0)=0, \quad u(0)=t, \quad v(0)=c \tag{37}$$

Then the eqns. (27) and (28) can be converted into

$$diff(v(\eta),\eta)-a.\left(\left(\frac{1}{3}\right).f(\eta).v(\eta)-\left(\frac{2}{3}\right)u(\eta)^2\right) \tag{38}$$

$$diff(r(\eta),\eta)-\left(b.\frac{(\mathrm{Pr})}{(1+q)}\right).\left(\left(\frac{1}{3}\right).f(\eta).r(\eta)-u(\eta).\theta(\eta)\right) \tag{39}$$

The asymptotic boundary conditions ∞ given in eqns. (29) and (30) were replaced by a finite value of 8 for similarity variable η maximum as follows:

$$u(8)=0 \quad v(8)=1 \quad \theta(8)=0 \tag{40}$$

	Sphere	Cylinder	Lamina
φ	1	0.4710	0.1857
m	3	6.3698	16.1576

Table 2: Values of the sphericity and the empirical shape factor for different particles shapes.

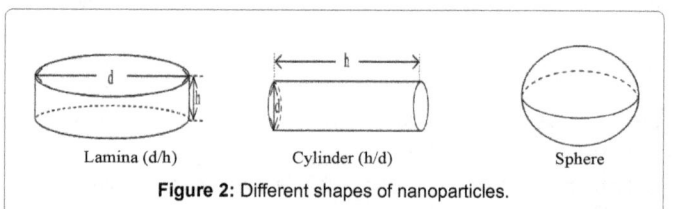

Lamina (d/h) Cylinder (h/d) Sphere

Figure 2: Different shapes of nanoparticles.

MAPLE 18 software for fourth fifth order Runge Kutta method is using to find the values of heat transfer and velocity.

Results and Discussion

In this research, Marangoni boundary layer flow and heat transfer of nanoparticle shapes (sphere, cylinder and lamina) in the presence of water; ethylene glycol and engine oil based on copper, alumina and SWCNTs are investigated by exponential temperature. Organization the rate of heat transfer and temperature within the nanofluid with different nanoparticle shapes are observed in terms of figures and tables where the influences of the solid volume fraction, radiation parameter and empirical shape factor are considered.

Analysis of nanoparticle shape and volume fraction on temperature profiles

The solid volume fraction is an important component in the physical parameter for nanofluids and plays a key role in Marangoni boundary layer flow and heat transfer. The academic literature [18,19] had revealed that the solid volume fraction on copper-water nanofluid in the range $0.05\% \leq \phi \leq 6.00\%$. If the concentration is exceeds 6.00%, the sendimation would take place. In this study, the solid volume fraction parameter consider as ϕ=0.0%, 0.1% and 0.2% while the others physical parameters are fixed as Pr=7.8, Nr=1 and m=3 (sphere particle), m=6.3698 (cylinder particle) and m=16.1576 (lamina particle). Figure 3

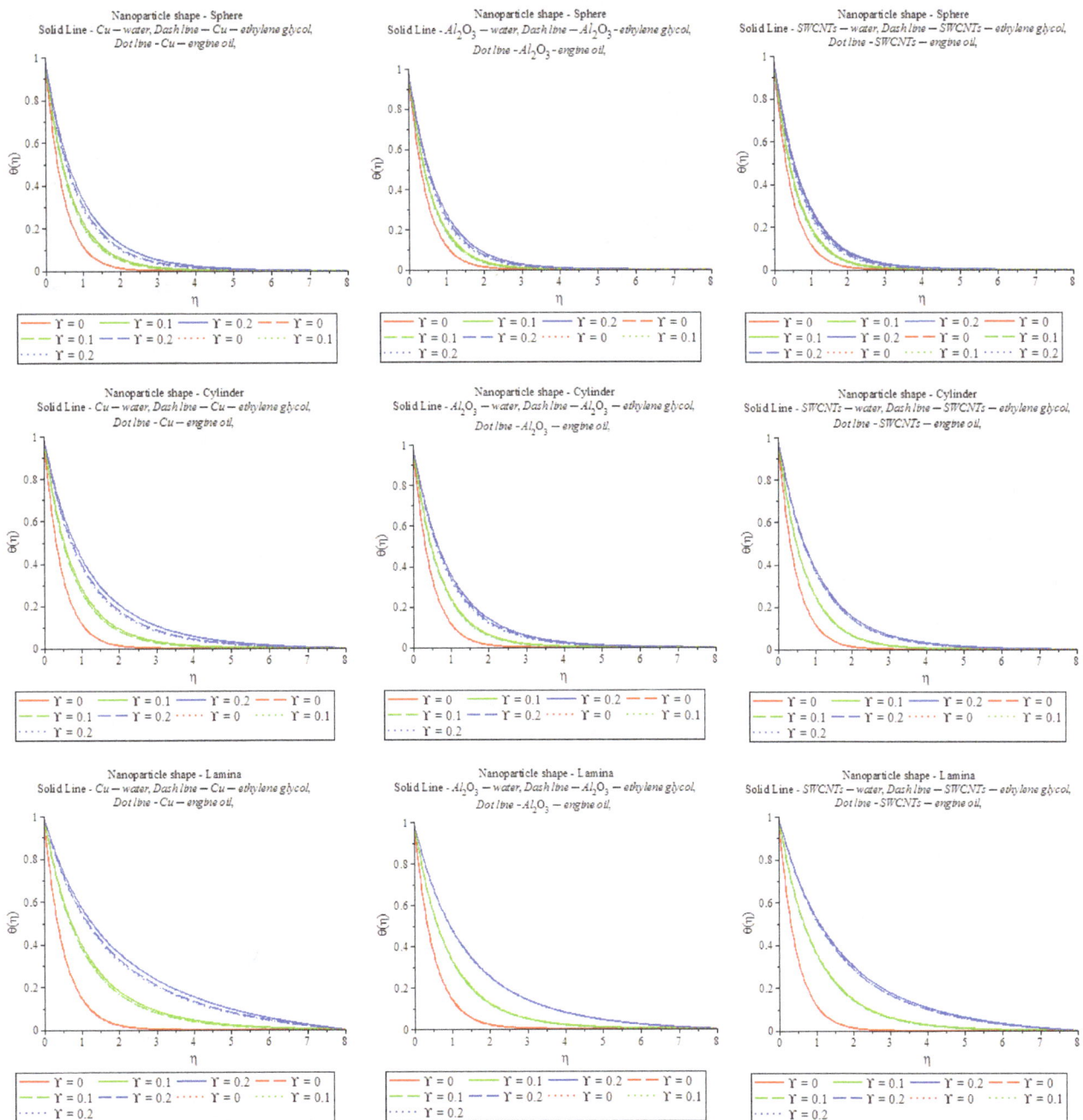

Figure 3: Effect of the nanoparticle volume fraction and shapes on temperature profiles.

displays the effects of solid volume fraction parameter on heat transfer of the water, ethylene glycol and engine oil Cu, Al_2O_3 and SWCNTs for sphere, cylinder and lamina particles.

It is already known that Marangoni effect on fluid flow is fluid moves from a region with low surface tension to a region with high surface tension. It is observed that the temperature of the water, ethylene glycol and engine oil based Cu, Al_2O_3 and SWCNTs increases with increase of nanoparticle volume fraction. The outcomes of the investigation shows that the thermal boundary layer thickness of sphere shape copper nanoparticles in Cu-water is stronger than that of all the other mixtures in the flow regime. This is due to the combined effects of the density and

thermal conductivity of the Cu-water is more significant as compared to the other mixtures in the flow regime.

Analysis of the radiation and nanoparticle shape on temperature profiles

Figures displays the results obtained from the influence of the thermal radiation energy on heat transfer characteristic of the water ethylene glycol and engine oil based Cu, Al_2O_3 and SWCNTs on dimensionless temperature $\theta(\eta)$. As shown in the Figure 4, the dimensionless temperature of the water ethylene glycol and engine oil based Cu, Al_2O_3 and SWCNTs increases with increase of thermal

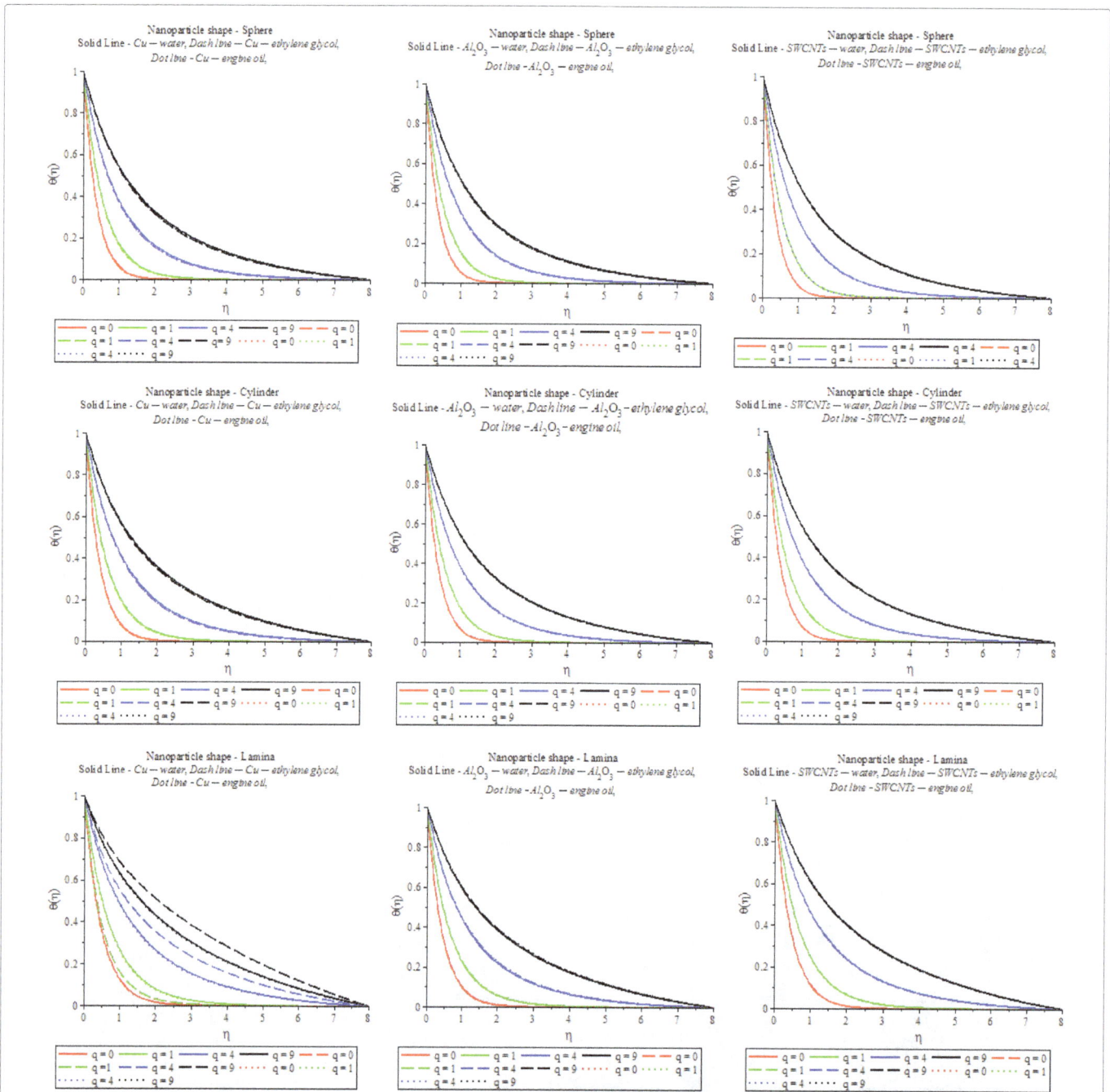

Figure 4: The thermal radiation energy and nanoparticle shape on temperature profiles.

radiation energy. It is because presence of radiation to put off the repairing effect of thermal Marangoni boundary layer. Besides that the temperature distribution within the thermal boundary layer invert to T_∞ slowly. It is interesting to notice that the thermal boundary layer thickness of the lamina shape Cu - ethylene glycol is more powerful as compared to the other mixtures in the flow regime with increase of thermal radiation energy because of the combined effect of thermal conductivity and the specific heat of the Cu - ethylene glycol along the flow. What stand out in the Figure 4 is lamina shape copper nanoparticles in ethylene glycol leads an efficient hits on temperature distribution with increase of thermal radiation energy.

Effect of the empirical shape factor (m)

Nanoparticle shape is another significant aspect on flow and heat transfer of nanofluid. Most of the researchers investigating nanofluid have utilized spherical shape because it's effective thermal conductivity. Thermal conductivity of nanofluids can be approximated by Hamilton and Crosser model as $\frac{K_{nf}}{K_f} = \frac{[k_s+(m-1)k_f]-(m-1)\varphi(k_f-k_s)}{[k_s+(m-1)k_f]+\varphi(k_f-k_s)}$ where the sphericity and empirical shape factor (m) are considered. In this study, three types of nanoparticle shapes are taken into account such as sphere, cylinder and lamina.

Different shapes of copper, alumina and SWCNTs particles are presented in Figure 2 and empirical shape factor are derived in Table 1. According to Figure 2, N is height and diameter ratio of cylinder defined as $N=h/d$ where h is height of the cylinder and d is the diameter. The empirical shape factor and the sphericity of a cylinder can be expressed by:

$$m(N)=\frac{2N+1}{2N}\sqrt[3]{12N}, \phi(N)=\frac{2N}{2N+1}\sqrt[3]{\frac{9}{4N}} \quad (41)$$

There is a differentiation between cylinder and lamina. Cylinder is defined as a column for $N\geq 10$ whereas lamina is defined for $N\leq 0.1$.

Figure 5 displays an overview influence of different type's nanoparticle shapes (sphere, cylinder lamina) on dimensionless temperature of the water, ethylene glycol and engine oil based Cu, Al_2O_3 and SWCNTs. The results show that for different nanoparticle shapes with the same base fluid and other parameters, the dimensionless temperature is sphere<cylinder<lamina. Meanwhile, the rate of temperature is decreases sphere>cylinder>lamina. According to the result, it is indicated that the sphere shape nanoparticles in the water

Cu, Al_2O_3 and SWCNTs plays a dominant role as compared with the other mixtures in the flow regime.

The results of this investigation show that nanoparticle volume fraction and particle shape give a significant impact on thermal conductivity of the copper, alumina and SWCNTs. The ratio thermal of conductivity is increases as solid volume fraction increases for the same particle shape (sphere, cylinder and lamina) in the presence of water, ethylene glycol and engine oil based on Cu, Al_2O_3 and SWCNTs. In addition, thermal conductivity increases when sphericity $\varphi(N)$ increases or empirical shape factor $m(N)$ decreases in the same solid volume fraction. In sequence order, the ratio thermal conductivity is sphere<cylinder<lamina. These results are similar to Jiao et al. [19].

Conclusion

The present study was designed to determine the effect of particle shapes and radiation parameter on Marangoni boundary layer flow and heat transfer of copper, alumina and SWCNTs in the presence of water, ethylene glycol and engine oil driven by exponential temperature. The following results are obtained:

1. There is a positive significant correlation between nanoparticle volume fraction parameter and particle shape on temperature and thermal conductivity.

2. The sphere shape nanoparticles in the presence of water Cu, Al_2O_3 and SWCNTs play a dominant role as compared with the other mixtures in the flow regime.

3. Sphere shape nanoparticles in the nanofluid is a more assurance in terms of enhancing the heat transfer compared to other nanoparticles shapes.

4. Lamina shape copper nanoparticles in ethylene glycol leads an efficient hit on temperature distribution with increase of thermal radiation energy.

5. The thermal boundary layer thickness of sphere shape copper nanoparticles in Cu-water is stronger than that of all the other mixtures in the flow regime because of the thermal conductivity of Cu - water.

It is registered that the lamina shape nanoparticle in the existence of Cu-ethylene glycol is researched in this work can be gainful in the solar radiation energy systems. Resultantly, the lamina shape in the cu-ethylene glycol is a more affirmation in terms of complementing

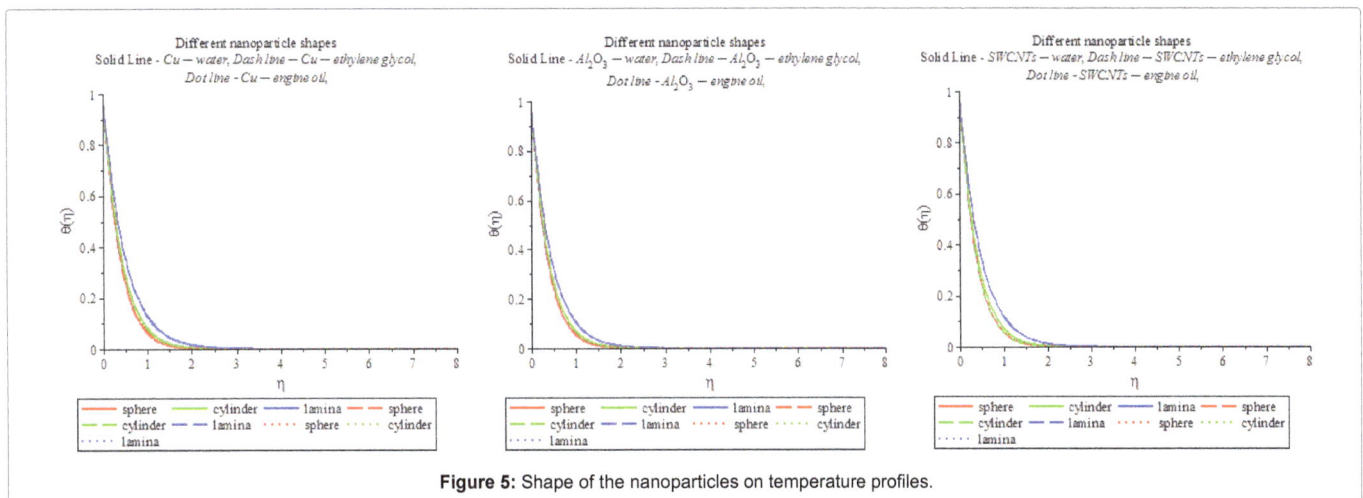

Figure 5: Shape of the nanoparticles on temperature profiles.

the heat transfer reinforcement of the Marangoni boundary layer flow system.

References

1. Chol SUS (1995) Enhancing thermal conductivity of fluids with nanoparticles. ASME-Publications-Fed 231: 99-106.

2. Sidik NAC, Mohammed HA, Alawi OA, Samion S (2014) A review on preparation methods and challenges of nanofluids. International Communications in Heat and Mass Transfer 54: 115-125.

3. Pang C, Lee JW, Kang YT (2015) Review on combined heat and mass transfer characteristics in nanofluids. International Journal of Thermal Sciences 87: 49-67.

4. Zhang Y, Zheng L (2012) Analysis of MHD thermosolutal Marangoni convection with the heat generation and a first-order chemical reaction. Chemical engineering science 69: 449-455.

5. Arafune K, Hirata A (1999) Thermal and solutal Marangoni convection in In–Ga–Sb system. Journal of Crystal Growth 197: 811-817.

6. Li J, Li M, Hu W, Zeng D (2003) Suppression of Marangoni convection of silicon melt by a non-contaminating method. International journal of heat and mass transfer 46: 4969-4973.

7. Pearson JRA (1958) On convection cells induced by surface tension. Journal of fluid mechanics 4: 489-500.

8. Scriven LE, Sternling CV (1960) The marangoni effects. Nature 187: 186-188.

9. Christopher DM, Wang B (2001) Prandtl number effects for Marangoni convection over a flat surface. International Journal of Thermal Sciences 40: 564-570.

10. Zheng L, Zhang X, Gao Y (2008) Analytical solution for Marangoni convection over a liquid–vapor surface due to an imposed temperature gradient. Mathematical and Computer Modelling 48: 1787-1795.

11. Al-Mudhaf A, Chamkha AJ (2005) Similarity solutions for MHD thermosolutal Marangoni convection over a flat surface in the presence of heat generation or absorption effects. Heat and Mass Transfer 42: 112-121.

12. Chen CH (2007) Marangoni effects on forced convection of power-law liquids in a thin film over a stretching surface. Physics Letters A 370: 51-57.

13. Malvandi A, Hedayati F, Ganji DD (2014) Slip effects on unsteady stagnation point flow of a nanofluid over a stretching sheet. Powder Technology 253: 377-384.

14. Sheikholeslami M, Ganji DD, Ashorynejad HR (2013) Investigation of squeezing unsteady nanofluid flow using ADM. Powder Technology 239: 259-265.

15. Sheikholeslami M, Gorji-Bandpy M, Ganji DD, Soleimani S (2013) Effect of a magnetic field on natural convection in an inclined half-annulus enclosure filled with Cu–water nanofluid using CVFEM. Advanced Powder Technology 24: 980-991.

16. Kang HU, Kim SH, Oh JM (2006) Estimation of thermal conductivity of nanofluid using experimental effective particle volume. Experimental Heat Transfer 19: 181-191.

17. Sourtiji E, Gorji-Bandpy M, Ganji DD, Hosseinizadeh SF (2014) Numerical analysis of mixed convection heat transfer of Al2O3-water nanofluid in a ventilated cavity considering different positions of the outlet port. Powder Technology 262: 71-81.

18. Sourtiji E, Ganji DD, Seyyedi SM (2015) Free convection heat transfer and fluid flow of Cu–water nanofluids inside a triangular–cylindrical annulus. Powder Technology 277: 1-10.

19. Jiao C, Zheng L, Lin Y, Ma L, Chen G (2016) Marangoni abnormal convection heat transfer of power-law fluid driven by temperature gradient in porous medium with heat generation. International Journal of Heat and Mass Transfer 92: 700-707.

21

Pitch-Based Carbon Fibre-Reinforced PEEK Composites: Optimization of Interphase Properties by Water-Based Treatments and Self-Assembly

Martin A*, Addiego F, Mertz G, Bardon J, Ruch D and Dubois P

Luxembourg Institute of Science and Technology, Materials Research and Technology, Luxembourg

Abstract

This work addresses the challenging fibre-matrix compatibilization and interface adhesion improvement of poly(etheretherketone) (PEEK) composites reinforced with pitch-based carbon fibre. An innovative and environmentally friendly method, inspired both by supramolecular "layer-by-layer" (LBL) assembly and by the composition of adhesive proteins in mussels was designed to modify the carbon fibre surface and improve the composites transverse properties by supramolecular interactions. The results proved that few sensitive carbon surfaces can be selectively modified by stable polyelectrolyte complexes and catechol amine polymer partners dispersed in water in such a way that a sizing treatment can be applied by techniques as simple as immersion or spraying procedure. It was shown that the combination of these solutions self-organized to form thin deposits containing compatibilization and/or crystallization promoter partners, thought transcrystallinity, onto carbon surface. This approach is an innovative and environmentally-friendly method which improves fibre-matrix interface quality in terms of compatibilization, adhesion and mechanical properties.

Keywords: Composite interphase; Fibre sizing; Interfacial adhesion; Interfacial microstructure; Polyelectrolyte; Polydopamine; PEEK/pitch-based carbon fibre composites; Surface modification

Introduction

The advanced carbon fibre reinforced polymer (CFRP) composite market is dominated by materials based on thermoset polymer (TS) matrices such as epoxy, polyimides, bismaleimides and cyanate esters which have found many applications in aeronautic, space and military usages whose extreme environments demand high-performing materials. However, TS have limitations in terms of storage, hydrothermal ageing, insufficient toughness and have constraints in processing since long and strict multi-step processing by autoclave are needed [1]. This is why the replacement of TS matrices by thermoplastic polymers (TP) is an area of intense research with primarily use in short fibres reinforced composites. Currently, TP-based matrix composites associated with continuous fibres are growing rapidly and the TP used as matrix potentially present major competitive advantages over thermosets because of their ease of implementation and their intrinsic characteristics such as an extremely long shelf life as well as their recycling possibilities. Thermoplastics are usually high molecular weight, linear polymers which form bulk material by non-covalent bonds like hydrogen bonding, dipole-dipole, van der Waals and π-π interactions, and linear structures with network possibility [2]. When high performance applications are targeted, thermoplastics with high thermal stability are required. One can cite space applications such as orbiting satellites, wherein investigations must not only be focused in space junk and orbital debris but also in material improvement with new multi-functionality and cost reduction in order to increase the application range.

The materials studied in this work include a continuous high modulus pitch-based carbon fibre with high thermal conductivity reinforcing a poly(etheretherketone) (PEEK) matrix. PEEK is one of the most promising high performance thermoplastic polymers and arouses a strong interest, not only for medical applications [3] but also as an alternative polymer for thermoset matrices replacement to the extent where it has sufficiently good properties to be compatible with severe conditions. Applications of PEEK-based composites are increasingly popular and lead to the manufacture of primary and secondary structures such as PEEK/glass fibre laminates, [4] carbon nanofibre composites [5] and carbon nanotube composites [6]. However, carbon fibre-reinforced PEEK composites (CFRPEEK) only show sufficient performances for few CFRP developments and do not fully meet the requirements for high performance applications, in particular when the material is reinforced with continuous high modulus pitch-based carbon fibre because of issues linked to fibre-matrix lack of compatibility, viscosity, thermal stability and transcrystallinity. Indeed, the manufacturing process of PEEK-based composites, which has to be performed at elevated temperatures, i.e., between 380°C and 400°C, is challenging due to the following points: (i) the high melt viscosity is an obstacle to the lay-up of fibres combined with matrix into either a flat consolidated and oriented laminate or a three-dimensional shaped component, (ii) the fibre commercial sizings have insufficient thermal stability and, more generally, sizings have to be developed with a coating chemistry showing sufficient stability at such high temperatures, and (iii) the transcrystallinity of PEEK at the fibre-matrix interface depends on the pitch-based carbon fibre surface properties and has to be taken into account to explain the interphase properties. Furthermore, to the best of our knowledge, there is a limited understanding about the fibre-matrix compatibilization and adhesion mechanisms in such high performance TP composites [7]. More precisely it is assumed that fibre-matrix interface is actually the key to improve CFRP composites [8]. Indeed, one of the challenges linked to these materials relate to the fibre-matrix interface compatibilization, adhesion promotion, ageing and improvement of damage tolerance.

*Corresponding author: Martin A, Luxembourg Institute of Science and Technology, Materials Research and Technology, 5, rue Bommel, Z.A.E. Robert Steichen, L-4940 Hautcharage, Luxembourg; E-mail: arnaud.martin@list.lu

In this context, the mechanical properties of high performance composites are dependent on several important microstructural properties [9,10]. It is well known that in carbon material reinforced polymer composites, local interactions between the fibre surface and the matrix at the vicinity of this fibre, can be transferred over longer distances thanks to the creation of an interphase area. In the specific case of semi-crystalline polymer-based matrix composites, this interphase, wherein the matrix properties are different than in the bulk matrix, is made of transcrystalline layers, also called transcrystallinity, and transition regions of decreasing or increasing segmental mobility within the amorphous phase [11]. Thus, the mechanical properties of CFRPEEK depend on the crystallinity of the matrix, the size of the spherulites but also on the microstructure transcrystallinity-related properties. Indeed, tensile modulus and tensile strength increase with increasing crystallinity up to a certain point after which the properties decrease because of crack propagation through the spherulites [12]. The formation and growth mechanisms of transcrystallinity have been published in a recent review and discussed for several semi-crystalline polymers [13]. It is important to note that it is well known that some fibres may act as heterogeneous nucleating agents and promote the nucleation of crystallization along the interface with a sufficiently high density of nuclei; where the crystals grow in perpendicularly direction to the fibre surface. Furthermore, many studies have shown that the composite properties, hence the transcrystallinity, can be greatly affected by the (melt) processing cycle [14]. Thus, interphase properties such as transcrystallinity are affected by the fibre-matrix interface quality that is directly connected to the fibre-matrix compatibility.

A very limited number of sizing formulations compatible with thermoplastic-based matrices and particularly with PEEK matrix has been reported so far [15]. However, sizing agents proved necessary for making fibre handling and prevent fibre degradation but they also are very interesting as a way to fibre-matrix compatibilization, to adhesion promotion and to enhance interphase properties [16,17]. Based on our understanding of sizing chemistry and processes and on the compatibility between fibre surfaces with PEEK matrix, it seems suitable to develop sizing formulations from thermostable thermoplastics [18]. Different criteria must be considered, both the compatibility of coating oligomers and main constituents with PEEK matrix and constraints linked to sizing implementation and composite material manufacturing. Indeed, polymers compatible with a PEEK matrix [19,20] present a poor adhesion with pitch-based carbon fibres. The fibre-matrix interface chemical engineering methodology presented here is expected to improve the interface by reactive compatibilization due to adhesion promotion of the carbon fibre surface with the PEEK matrix through growth of crystallinity directly onto the fibre, which is essential to obtain high performance materials [21-24]. In order to develop an effective solution to improve the fibre-matrix compatibility, particular attention was paid to the reinforcement material and to the methodology for coating deposition of a material able to promote the compatibilization and/or the adhesion with the matrix.

Assuming that the surface of the as-received and unsized carbon fibre is electrochemically-oxidized, we expected that the surface contains some oxide functional groups conferring to the surface an electronegative potential, which shall allow the physisorption of cationic polyelectrolyte materials. To this end, an innovative and environmentally friendly method, inspired by supramolecular layer by layer (LBL) assembly, [25] is used to modify the carbon fibre surface. The LBL technique relies upon the deposition of an oppositely charged macromolecule to the previously adsorbed polyelectrolyte, hence modifying the surface properties. Recent relevant reviews have

been published detailing the advantages of LBL technique [26-29]. This surface modification technique is suited for the deposition of multilayered films covering a broad range of applications. Nevertheless, to our knowledge this method can only be applied to small surface areas. Another important aspect concerns the fact that LBL films can not only be deposited from polyelectrolytes (due to electrostatic interactions) but also from all kinds of molecules or particles owing to mutually complementary interactions, such as hydrogen donor-acceptors, [30,31] covalent coupling through click chemistry, [32] complementary stereoregular polymers, [33] and nanomaterials [34] justifying the use of a mussel-based adhesive promoter namely polydopamine (PDA) due to its exceptional adhesive performance in wet environment and its role in interfacial adhesion[35,36]. A major challenge needs to be addressed to make this technique a key-pillar and an applicable technology for high performance applications: the adsorption of these two partners (polyelectrolyte and PDA) and particularly their complexation at the surface of a high modulus pitch-based carbon fibre.

Thus, this contribution aims at focusing on (i) the carbon fibre surface modification investigation by sessile drop testing, scanning electron microscope (SEM) observations, X-ray photoelectron spectroscopy (XPS) and multi-mode atomic force microscopy (AFM) characterization to identify and quantify the chemical modifications prior and after surface treatments, (ii) the influence of the fibre-matrix interface chemical engineering on the interphase properties of the composites reinforced with the so-modified carbon fibres by X-ray diffraction measurements to assess any crystallinity change, and (iii) the investigation of the impact of the reactive compatibilization on the CFRP composite mechanical properties by transversal tensile strength testing. We assume that this environmentally friendly methodology is an innovative challenge that allows the possibility to improve fibre-matrix interface quality and CFRP composite mechanical properties.

Experimental Section

Materials

Carbon fibres: Two carbon fibres were used in this study: high performance GRANOC yarn pitch-based carbon fibres YSH-50A-60Z and YSH-50A-60S produced by Nippon Graphite Fiber Corporation. Based on the supplier datasheet, these are high modulus and high strength 6 k tow count fibres (6000 monofilaments) with a tensile modulus of 630 GPa and a tensile strength of 3.9 GPa. One fibre was surface oxidized and not sized (CF-Z), the other was surface oxidized and thermoset-sized (CF-TS) with a sizing chemistry based on epoxy polymers.

Thermoplastic polymer: One thermoplastic polymer was used in this study. An unreinforced and uncharged poly(etheretherketone) (PEEK) was selected and used as a matrix of the composites. The PEEK materials were supplied in the form of a fine powder (PEEK 150XF) and thick films (300 μm) (APTIVTM 2000) by Victrex. The powder material was specifically used for the crystallinity investigation of the composites by x-ray diffraction, while the film material was used for the rest of the study. The remolding agent was CIREX 041WB from SICOMIN.

Chemicals: All chemicals (detailed in the section 2.2) were supplied by Sigma-Aldrich and used as received.

Processes

Preparation of the water-based surface treatment solutions: Water (W/Milli-Q or ultrapure water) used in all experiments (with the exception of fibre rinse cycles) was prepared in a three-stage Millipore

Milli-Q purification system and was air-equilibrated before use. All glassware was submitted to cleaning steps using ethanol and acetone and rinsed with ultrapure water (W/Milli-Q). Poly(ethyleneimine) (PEtI) was used without further purification as the polyelectrolyte for physisorbed deposition and was added (1 mg/mL) to a buffer solution pH=7.4 containing 10 mM tris(hydroxymethyl)aminomethane (THAM), 15 mM NaCl and ultrapure water. The solution was stirred at room temperature. Fresh dopamine hydrochloride (PDA) solutions were prepared before use because the mixture reacted immediately upon addition to a 10 mM THAM-HCl (pH=8.5) buffer solution.

Carbon fibre surface treatments: Continuous carbon fibres (substrates) were immersed in polyelectrolyte solution (PEtI) during times ranging from 10 s to several hours and rinsed with water before any further use. Similarly, substrates (with and without PEtI surface treatments) were immersed in polydopamine mixtures (PDA) (during times ranging from 10 s to 24 h) and rinsed with water.

PEEK-based composites implementation: We prepared 50 mm × 50 mm × 0.8 mm plates of PEEK composites reinforced with 2 unidirectional (UD) plies of as received or modified carbon fibres by compression-molding. To this end, a hydraulic manual press Carver equipped with heating plates and a Carver tile mold with an insert for 50 mm × 50 mm specimens were used. 50 mm × 50 mm UD plies of as received and modified carbon fibres were produced by the winding methodology. To fabricate a composite, two UD plies of fibres were subsequently sandwiched between three films of PEEK, then placed in the mold and pressed at 400°C for 30 min. To remove trapped air, the material was first sequentially pressed to 2 Tons and released and was finally subjected to 8 Tons for 30 s. The cooling step is performed by switching off the heating and leaving the material under pressure until the temperature reached room temperature.

Characterizations

Scanning electron microscopy: A pressure-controlled Quanta FEG 200 environmental scanning electron microscope (SEM) from FEI Company Europe BV was used in secondary electron detection mode to get information about the carbon substrate surface topography, the fibre diameter, the surface treatment deposition quality and the fibre-matrix interface quality after transverse tensile strength testing of the composites. The environmental mode, with gaseous secondary electron detector (GSED) enables wet samples to be observed through the use of partial vapour pressure in the microscope specimen chamber to analyze the drop growth and geometry at the surface of the treated fibres. All observations were directly performed on specimens without any particular preparation procedure (no conductive coating).

Sessile drop testing: The wettability of the fibres prior and after surface treatment was investigated by static contact angle measurements (apparatus OCA 15 from Dataphysics) using the sessile drop method and distilled water as a liquid. Droplet volume is 2 µL. Elliptical model [37] was used to accurately estimate contact angle values, while 5 images were recorded to increase statistics.

Atomic force microscopy: A new generation of Bruker Atomic force microscopy (AFM) (Multi Mode 8 AFM with nanoscope 5 controller) was used to measure profiles of the surface of some samples at ambient conditions in order to investigate the electrostatic, structural and mechanical properties. Surface potential was measured in Peak Force (PF) KPFM in amplitude and frequency modulation modes (in conjunction with peak force tapping), topography in PF tapping mode, Young's modulus and adhesion in PF quantitative nano-mechanical mapping (QNM) mode.

X-ray photoelectron spectroscopy: X-ray photoelectron spectroscopy (XPS) analyses of the carbon substrates were performed with a Kratos AXIS Ultra DLD instrument using a hemispherical energy analyzer and a monochromatic Al Kα X-ray source (1486.6 eV) as the incident radiation. For each samples, at least two measurements were performed at different locations. Scans were collected from 0 to 1300 eV with a power of 225 W and an anode voltage of 12 kV. The pressure in the analysis chamber was about 5.10^{-8} Pa and the pass energies were set to 160 eV and 40 eV for survey and higher resolution scans, respectively. The binding energy scale was referenced from the carbon contamination using the C (1s) peak signal at 284.6 eV. Core level peaks were analyzed using a nonlinear Shirley-type background. Concerning the analysis of the C (1s) high resolution spectra, the peak positions and areas were optimized by a weighted least-square fitting method using a GL function (product of a Lorentzian by a Gaussian) by fixing the full width-at-half-maximum (FWHM) using XPSCASA software, except for the C-C sp^2 for which the spectrum was fitted allowing some variations of the FWHM. It is assumed that this method has a scanned depth approximately between 5 nm and 10 nm.

X-ray diffraction: X-ray diffraction (XRD) experiments were conducted with a Panalytical X'Pert Pro MPD in reflection configuration, equipped with a temperature chamber (reference TTK 450). The X-ray beam corresponding to the Kα copper radiation (1.54 Å) was generated at 40 kV and 45 mA. As primary optics, a programmable divergent slit was used with a constant sample irradiated length of 5 mm, a 0.04 rad Soller slit, and a mask of 5 mm. As secondary optics, a Pixcel detector was used with a constant programmable antiscatter slit of 5 mm, a 0.04 rad Soller slit, and a reception slit of 0.1 mm. The samples, consisting of 10 mm-continuous fibre on which was deposited a thin layer of PEEK powder, were positioned on the TTK 450 sample holder and subjected to a temperature program. The latter consisted of an heating step from 25°C to 400°C at 10°C/min to erase the thermal history of the PEEK matrix and enable the composite consolidation, and then to a cooling step from 400°C to 25°C with a cooling rate of -30°C/min. The intensity (I) - 2 theta (2θ) diffractogram of the composite was first measured at 400°C to verify that the melting was totally achieved and then at the end of the program to calculate the material crystallinity. The crystallinity was calculated as the ratio between the area of crystalline peaks to the total area of crystalline and amorphous peaks. As crystalline peaks, we considered the (110), (113), (200), (213), (216) reflections positioned at 18.7°, 20.8°, 22.7°, 28.8°, 32.9° and 38.8°, respectively, corresponding to the orthorhombic phase of PEEK [38]. As amorphous peaks, we considered the bump centered at 20° and 28.6°. The mathematical deconvolution was done with the software PeakFit by using a Gaussian equation for each peak. The influence of the fibre/matrix compatibilization on PEEK crystallinity was evaluated.

Transverse tensile testing: A miniature tensile/compression testing machine from Kammrath and Weiss was utilized to characterize the tensile behavior of PEEK/carbon fiber composites in the transverse direction. Tensile bars of dimensions 50 mm × 10 mm × 1 mm were machined perpendicular to fibre direction from the initial molded plates with a precision saw (Struers). The mechanical testing was performed at room temperature (20°C) and at a displacement rate of 10 µm.s^{-1}, corresponding to a strain rate of 10 µm/28000 µm = 0.00036 s^{-1}. The tensile strength, corresponding to the maximum stress at failure, was measured and discussed as a function of the fibre-matrix compatibilization parameters while at least 5 tastings were performed to increase statistics.

Results and Discussion

In this work, influence of water-based surface treatments on carbon fibre, particularly an environmentally friendly methodology consisting of water-based solutions of polyelectrolytes, of mussel-inspired adhesive promoter and thermostable thermoplastic polymer, was investigated in the field of high performance materials. To this end, the following fibres were studied: i) the as received unsized carbon fibre (CF-Z), ii) the as received carbon fibre with commercial thermoset sizing (CF-TS), iii) the poly(ethyleneimine) treated carbon fibre (CF-PEtI), iv) the carbon fibre treated with hydrochloride polydopamine solutions (CF-PDA) and v) the carbon fibres modified with a combination of the two treatments (CF-PEtI-PDA). The interphase properties and interface mechanical performances of the composites have been investigated using PEEK as TP matrix.

Carbon fibre surface properties were first investigated by measuring the water contact angle (WCA, sessile drop tests) before and after surface treatments. This surface characterization allowed us to identify the influence of surface treatments on carbon roving wettability (Figure 1) and to illustrate the wetting difference between the untreated fibre (CF-Z) with the commercial-sized fibre (CF-TS) (Figure 1).

The as-received unsized carbon fibre (CF-Z) is hydrophobic with a contact angle of about 120°. As expected, the thermoset-sized CF (CF-TS), with a sizing chemistry based on epoxy polymers, is more hydrophilic with a contact angle close to 80°. The results obtained on the treated CF (Figure 1) show a decrease of the hydrophobic character with all the surface treatments whatever the treatment duration. Regarding the fibres treated with PDA solutions (CF-PDA), there is a decrease of the hydrophobic character from 115° to 103° as a function of the treatment duration indicating CF surface property modifications through self-assembly of the polydopamine onto the surface. With the polyelectrolyte treatments (CF-PEtI) the contact angle was about 90° whatever the treatment duration, which proved the existence of electrostatic interactions between the polyelectrolyte chains and the CF surface. Moreover, this result seems to prove that physisorption of the polyethyleneimine is effective even after rinses cycle and occurs very rapidly comparing to some industrial processes, which could even require thermal treatments.

The steady value of contact angle with treatment duration shows that the covering of the fibre is immediate and complete and that a prolonged time treatment is not necessary to further modify the surface properties. Indeed, the overcompensation of the surface charge, which will be demonstrated further down this contribution, avoids the addition of the polyelectrolyte and an oppositely charged partner can be added in order to tune the CF surface modification. Thus, with a combination of the two treatments (CF-PEtI_PDA), the hydrophobic character decreases with treatment duration as expected since the PDA lies at the top of the surface. Moreover, with aged PDA (PEtI_PDA$_{24h}$), contact angle faster decreases than with fresh PDA obviously because self-assembly of the polydopamine occurred in solutions yielding larger amount of macromolecules physisorbed at the surface. Therefore, at short treatment time, the amount of polydopamine trapped in the coating is certainly higher than in the case of aged solutions. Nevertheless, with longer treatment the properties became almost the same indicating that self-assembly occurred with fresh solution, which improves PDA concentration and homogeneity of deposition at the surface. However, surface topography should be different and will be discussed (vide infra). These results indicate that the fibre water-based surface modification is due: i) to the electrostatic interaction of the fibre surface with the polyelectrolyte (PEtI) certainly resulting into charge overcompensation at the surface, and ii) to the self-assembly of the polydopamine at the carbon fibre surface. Moreover, it seems that PDA is more compatible with the CF surface after PEtI surface treatment. These measurements were completed with environmental SEM observations of the carbon fibres in wet conditions to get results on monofilaments of the carbon fibres (Figure 2).

To this end, drop growth and geometry at the micrometer scale were obtained by SEM observations, using partial vapour pressure in the sample chamber. Results of the CF-Z fibres are presented on Figures 2a and 2b. One can specify that these are qualitative observations and no quantitative measurements because of the difficulty to control the drop volume when the images were observed and the difficulty to calculate the contact angle when substrate is curved. Thus CF-Z monofilament wetting was clearly hydrophobic confirming results obtained on roving. The drop geometry was almost spherical (Figure 2b) and the contact angle (Figure 2a) appeared to be >90°. Regarding fibres treated with the polyelectrolyte solutions (CF-PEtI), Figures 2c and 2d confirmed a decrease of the hydrophobic character with drops more homogeneously spread onto monofilament and particularly, Figure 2c exhibited a good reproducibility with almost the same geometry for all droplets confirming WCA results. Figures 2e and 2f allowed observation of the PDA concentration influence for the same duration of fibre treatment. The higher was the PDA concentration (2f), the less spherical was the drop shape highlight that the surface wetting was clearly influenced by the PDA concentration and PDA surface treatment duration (self-assembly). Figure 2g showed that a PEtI treatment prior to PDA self-assembly was relevant and enabled a more suitable coating of the droplet onto the monofilament than without polyelectrolyte physisorption confirming that, for such carbon fibres, a surface priming (pretreatment) of the substrate with a polyelectrolyte complex-based film triggers a more efficient subsequent deposition and self-assembly of hydrogen-bonded film [39]. Influence of aged PDA (24 h) is presented on the Figures 2h and 2i and it seems that a large quantity of PDA effectively covers the monofilaments after 24h duration, making the surface hydrophilic. These observations confirmed the WCA results obtained on roving.

The main difficulty concerning the fibre-matrix interface compatibilization by fibre dipping or spraying is the understanding of

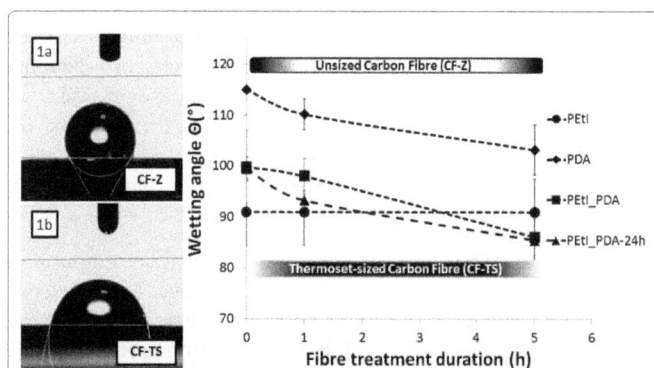

Figure 1: Contact angle values (averaged from 5 measurements with standard deviation) of pitch-based carbon fibre (CF) as a function of the nature and treatment duration (from 10 s to 5 h) with CF-Z: unsized CF, CF-TS: thermoset-based sized CF, PEtI: CF treated with PEtI solution [1mg/mL], PDA: CF treated with PDA solution [0.05 mg/mL], PEtI_PDA: CF treated with solutions of PEtI [1 mg/mL] then fresh PDA [0.05 mg/mL] and PEtI_PDA-24h: CF treated with solutions of PEtI [1 mg/mL] then 24h aged PDA [0.05 mg/mL]. Sessile drops from experimental set (image 1a): as received unsized CF; (image 1b): as received thermoset-sized CF. Black dashed lines are given as guides to the eye.

the carbon surface chemistry and its influence after surface treatments. Indeed, these treatments trigger surface modifications which may introduce functional groups onto the surface able of interact and/or react with the matrix. Thus, in order to go deeper into the characterization of surface properties, X-ray photoelectron spectroscopy (XPS) testing was performed on the fibres to determine the atomic composition of the extreme surface by measuring the elemental composition and calculating the O/C and N/C ratios. The elemental compositions and the calculated surface atomic O/C and N/C ratios prior and after water-based treatments are presented in Figure 3. Assuming that the carbon was homogeneously distributed in the mass of the carbon fibre and that the XPS sampling depth was approximately 10 nm, the results confirmed the success of the water-based surface modifications by chemical modification onto the fibre surface. Indeed, the untreated carbon fibre (CF-Z) contained about 2.4%-At of oxygen (O/C=0.024) and no nitrogen as expected with a pitch precursor. For all treated samples, the oxygen concentration is different from the case of bare fibres and nitrogen was detected, which was obviously a consequence of the surface modifications by the polyelectrolyte and polydopamine solutions. Indeed, PEtI treatment induced an increase of oxygen atom amount to about 8.3%-At (O/C=0.095) and identification of nitrogen functional groups (4.1%-At; N/C = 0.047). Moreover, making the assumption that the film thickness was bigger than the depth of analysis of XPS, [25] the ratios allowed to calculate the polyelectrolyte complexes (PEtI and THAM) ratio in the covering through equations (1) and (2):

$$[O/C]_{CF-PEtI}=a.[O/C]_{PEtI}+b.[O/C]_{THAM} \tag{1}$$

$$[N/C]_{CF-PEtI}=a.[N/C]_{PEtI}+b.[N/C]_{THAM} \tag{2}$$

Where $[O/C]_{CF-PEtI}$ and $[N/C]_{CF-PEtI}$ are the polyelectrolyte complexes covering measured by XPS (corresponding to respectively 0.095 and 0.047), $[O/C]_{PEtI}$ and $[N/C]_{PEtI}$ are the carbon and nitrogen atom concentration in pure PEtI (corresponding to respectively 0

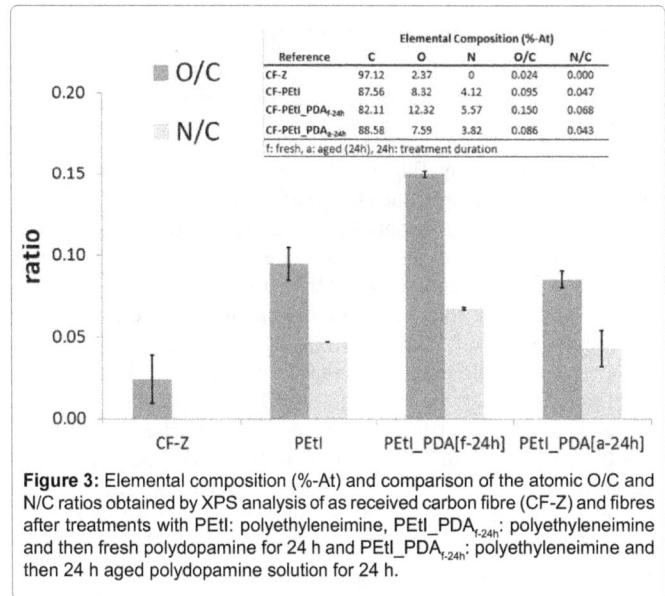

Figure 3: Elemental composition (%-At) and comparison of the atomic O/C and N/C ratios obtained by XPS analysis of as received carbon fibre (CF-Z) and fibres after treatments with PEtI: polyethyleneimine, PEtI_PDA$_{f-24h}$: polyethyleneimine and then fresh polydopamine for 24 h and PEtI_PDA$_{f-24h}$: polyethyleneimine and then 24 h aged polydopamine solution for 24 h.

and 0.5), $[O/C]_{THAM}$ and $[N/C]_{THAM}$ are the carbon and nitrogen atom concentration in pure THAM (corresponding to respectively 0.75 and 0.25), a the total amount of PEtI measured by XPS and b the total amount of THAM measured by XPS.

It derived from (1) and (2) that the polyelectrolyte complex-based film contained a THAM/PEtI ratio of approximately equal. Considering that we used a poly(ethyleneimine) solution with a relative molar mass, based on the supplier documentation, M_r of about 600000-1000000, it is not surprising that the deposited film had such composition by adsorption of complexes formed from polyanion interacting and/or trapped in the branched polycation. This result paves the way to the development of tailored fibre-matrix interfaces by using other polyanionic species and suitable (nano) fillers.

Regarding the combination of PEtI and PDA treatments, we used fresh and 24-h aged polydopamine solutions to modify the CF-PEtI fibres. XPS analysis showed that significant improvements in O/C and N/C ratios were obtained with fresh polydopamine solution treatment and tended to confirm the success of the surface modification and self-assembly onto the carbon fibre surface and thus the presence of PDA. Concerning 24-hour aged PDA treatment, the O/C and N/C ratios determined by the technique are respectively 0.086 and 0.043. It can be observed that these ratios are lower than that of PEtI surface priming and of PDA reference in solution (0.25 and 0.125), which means that the excess of carbon measured could come from fibres which are not fully covered. Indeed, PDA treatment is performed in solution with higher pH and lower salt concentration increasing the negative surface charge of the fibre and inducing higher interaction between the surface and the PEtI, which could modify the film structure and thickness. Moreover for aged-PDA, self-assembly occurred in solution. The PDA deposited onto the fibre is built on larger macromolecules assembled in solution and not self-assembled directly onto the fibre. The topography between both samples and their surface chemistry are thus different as confirmed by XPS, WCA and wet-ESEM results.

After the successful surface treatments and chemical property modification of the carbon fibres with PEtI and PEtI_PDA solutions, multi-mode Atomic Force Microscopy (AFM) characterizations were performed to confirm and finely identify the physico-chemical surface modifications and particularly the charge overcompensation of the

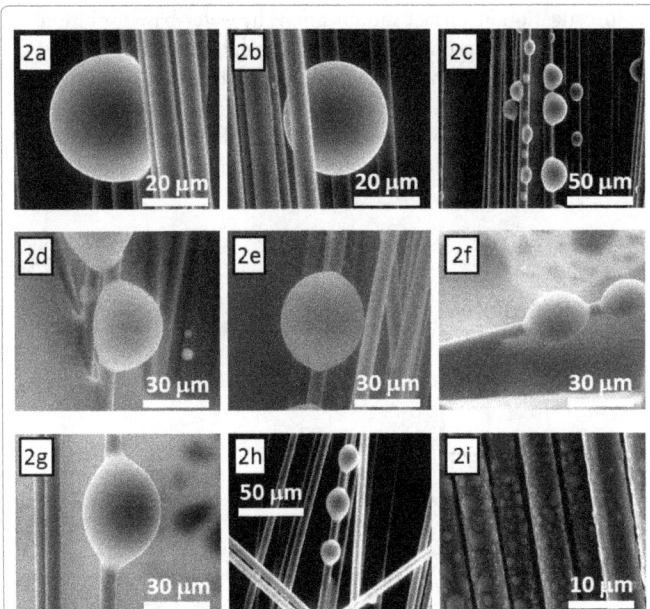

Figure 2: Wet environmental SEM observations of carbon fibres monofilaments with 2a and 2b: as received unsized CF, 2c and 2d: CF-PEtI, 2e and 2f: CF-PDA[0.1]$_{f-24h}$ and CF-PDA[1.0]$_{f-24h}$, 2g: CF-PEtI_PDA$_{f-24h}$, 2h and 2i: CF-PEtI_PDA$_{24h-s}$ and $_{24h-24h}$ ([x]$_{y-z}$, x: PDA concentration [mg/mL] , y: fresh "f" or 24h aged PDA solutions and z: treatment duration several seconds "s" or 24 hours "24h").

surfaces after treatment. Peak force AFM was used to measure surface profiles of the CF-Z, CF-PEtI, CF-PEtI_PDA$_{f-24h}$ and CF-PEtI_PDA$_{a-24h}$ carbon fibres at ambient conditions in order to investigate the electrostatic, structural and mechanical properties of the treated fibre surfaces.

Surface topography (Figures 4a and 4b) was measured in PF tapping mode. Seemingly, the change of surface roughness indicated the change of the carbon fibre surface morphology with respect to the as-received unsized carbon fibre. The polyelectrolyte complex-based film and the combination of the polyelectrolyte complex with fresh polydopamine solution induced a subtle decrease of the fibre roughness (respectively Ra=2.1 nm and 1.9 nm against 4.6 nm for the reference). This could indicate a suitable coating of a polyelectrolytes complexes film onto the fibre and a PDA self-assembly onto the nanoscale roughness of the physisorbed film. On the contrary, the aged PDA solution induced deposition of large PDA molecules (self-assembled in solution) improving surface roughness. These results demonstrated that the PDA self-assembly was distinctly different if it occurred directly onto the fibre surface or in the solution. Surface Young's modulus and adhesion properties were measured in PF QNM mode (Figure 4c). Regarding PF QNM measurements, modulus and adhesion were only compared one with the others in a qualitative manner. Interestingly, the Young's modulus remained almost constant for all fibres (CF-Z: 0.13 Arb.; PEtI: 0.16Arb.; PEtI_PDA$_{f-24h}$: 0.132Arb. and PEtI_PDA$_{a-24h}$: 0.114Arb.). The slight improvement measured with the PEtI treatment must be triggered by the polyelectrolyte complex-based physisorbed film, which was obviously of a different nature than the self-assembled PDA deposit. This suggests that all over changes of the PDA deposits were extreme surface related. Moreover, the decrease

in adhesion of PDA-treated substrates indicates a change of the surface functional groups of these fibres (CF-Z: 0.12V; PEtI: 0.186V; PEtI_PDA$_{f-24h}$: 0.04V and PEtI_PDA$_{a-24h}$: 0.06V). Finally, the surface potential (Figure 4d) was measured in Peak Force (PF) KPFM in frequency modulation mode (in conjunction with peak force tapping). The consistent changes of surface potential (CF-Z: -41mV; PEtI: 69mV; PEtI_PDA$_{f-24h}$: -204mV and PEtI_PDA$_{a-24h}$: -229mV) showed a modification of the fibre surface electrostatic properties after treatment (with respect to the reference). Moreover, the charge modification with PEtI treatment demonstrated the charge overcompensation induced by the physisorption of the polyelectrolyte film onto the carbon fibre. These results clearly demonstrated the efficiency of our methodology inspired by the LBL technique. Finally, the charge modification and surface potential improvement with PEtI_PDA treatment seemed to prove that the hydroxyl functional groups of the adhesive promoter are exposed at the top surface of the coating. The interaction mechanism and the polyelectrolyte and adhesive promoter conformations are schematically explained in Figure 5.

The influence of the water-based surface treatments onto the carbon fibre surface properties has been investigated and detailed. Surface modification, particularly in terms of wettability, chemical composition at the extreme surface, roughness and surface potential have been measured and discussed. It has been proven that polyelectrolyte complex-based film, in combination or not with the adhesive promoter self-assembly directly onto the carbon surface by dipping in water-based solutions containing these chemicals, modified the physico-chemical properties of carbon fibre surface. After these successful surface modifications of carbon fibres with PEtI and PEtI_PDA, XRD analysis has been approached to study the crystallization behavior of PEEK with constant processing parameters, as a function of fibre surface treatments. The composite heating treatment and cooling steps were directly performed in the XRD apparatus. Specific cooling step parameters have been designed in such a way that crystallinity could not grow at the interface. This procedure allowed the investigation of surface modifications by water-based solutions on the fibre-matrix interphase properties through the monitoring of the crystallinity behavior. Indeed, it is well known that to erase the thermal history of PEEK matrix, a thermal treatment at 400°C was required [40] and we verified that cooling step affected PEEK crystallization (results not presented here). A cooling rate of 30°C/min was chosen which was

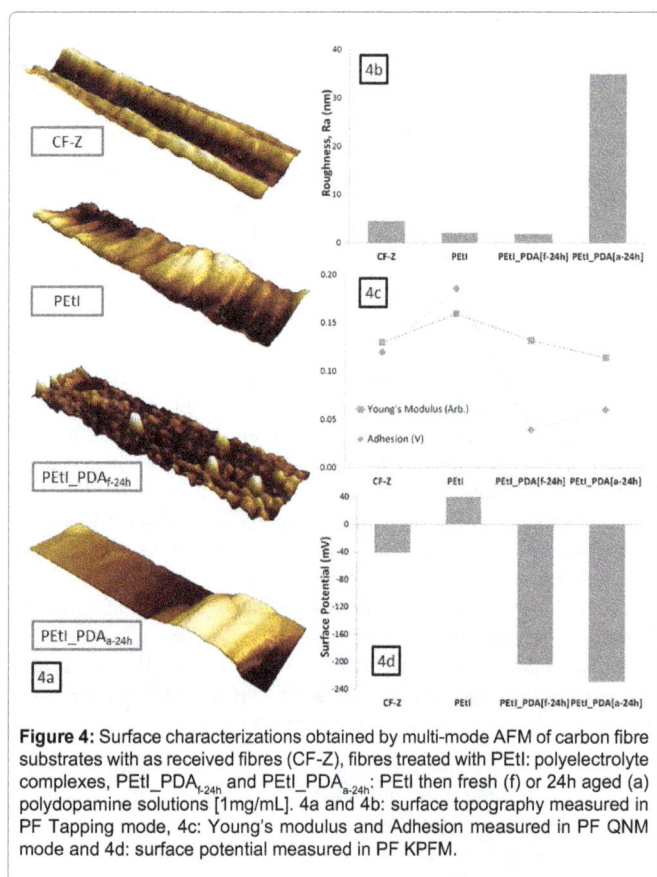

Figure 4: Surface characterizations obtained by multi-mode AFM of carbon fibre substrates with as received fibres (CF-Z), fibres treated with PEtI: polyelectrolyte complexes, PEtI_PDA$_{f-24h}$ and PEtI_PDA$_{a-24h}$: PEtI then fresh (f) or 24h aged (a) polydopamine solutions [1mg/mL]. 4a and 4b: surface topography measured in PF Tapping mode, 4c: Young's modulus and Adhesion measured in PF QNM mode and 4d: surface potential measured in PF KPFM.

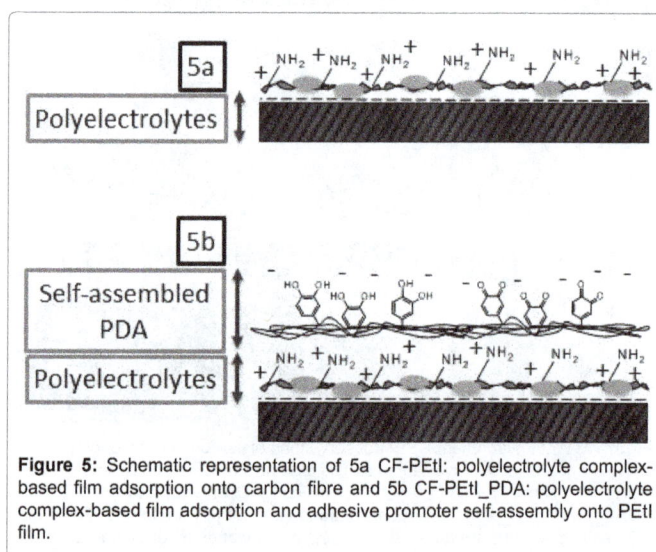

Figure 5: Schematic representation of 5a CF-PEtI: polyelectrolyte complex-based film adsorption onto carbon fibre and 5b CF-PEtI_PDA: polyelectrolyte complex-based film adsorption and adhesive promoter self-assembly onto PEtI film.

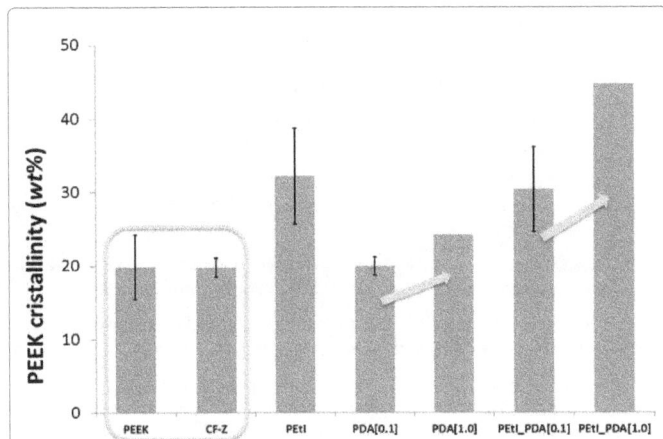

Figure 6: PEEK crystallinity (*wt*%) obtained by XRD analysis of a PEEK polymer and of CFRPEEK composites reinforced with as received fibres (CF-Z) and surface treated fibres with PEtI: polyelectrolyte complexes, PDA[0.1] and PDA[1.0]: fresh PDA solution at 0.1 and 1.0 mg/mL, PEtI_PDA[0.1] and PEtI_PDA[1.0]: PEtI then PDA solutions.

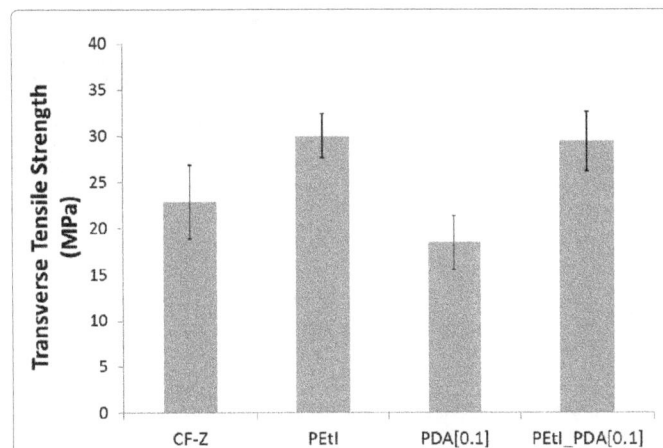

Figure 7: Transverse tensile strength of PEEK composites reinforced with 2 unidirectional plies of pitch-based carbon fibres prior (CF-Z: as received) and after surface modifications by dipping in water-based solutions of PEtI, PDA [0.1] and of PEtI_PDA [0.1].

the faster controlled process representative of an air cooling because of the mold temperature gradients. Crystallinity assessments presented in Figure 6 showed that, as expected, unreinforced PEEK polymer and CFRPEEK composites reinforced with as received carbon fibres (CF-Z) had almost the same crystallinity, i.e. about 20 *wt*%. XRD results confirmed that with typical industrial scale parameters, there is a lack of compatibility and interaction between the matrix and the surface of the CF-Z fibres since crystallinity improvement trough transcrystallinity is not detected. Indeed, as far authors know, transcrystallinity has only been observed by microscopy with very slow cooling parameters [41] with pitch-based carbon fibre-reinforced PEEK composites.

Concerning the fibres treated with fresh PDA solutions at 0.1 mg/mL and 1.0 mg/mL, the lower PDA concentration samples did not show any influence on the composites crystallinity. This experiment did not provide the anticipated results because even at low concentration and without pre-surface treatment, PDA treatment induced chemical and physical properties modification onto the fibre surface. But, with the improvement increase of the PDA concentration (to 1 mg/mL), the PEEK crystallinity increased from 20 to about 24 *wt*% highlighting that

the PDA self-assembly onto the fibre influenced the total crystallinity of the PEEK. Furthermore, the fibres treated with the polyelectrolyte complex solution (PEtI) proved to increase the matrix crystallinity in the composites up to ca. 32 *wt*%. This significant improvement seemed to demonstrate that transcrystallinity occurred at the fibre-matrix interface creating an interphase area. This interphase should be an area of strong interaction between PEEK polymer and carbon fibre surface. Interestingly, as far as the fibres treated with both solutions, i.e. PEtI then PDA, a substantial some improvement to about 30 *wt*% and even 45 *wt*% was determined. These improvements of PEEK crystallinity demonstrated that PEtI permitted a higher interaction between PEEK and the fibre surface and that PDA did not interfere with matrix crystallinity improvement and the very likely creation of an interphase area. These results could demonstrate an enhancement of fibre-matrix interface compatibilization and improvement of adhesion of the PEEK onto the fibre through the formation of a transcrystalline interphase which was influenced by the fibre surface properties. Indeed, we assume that the presence of the polyelectrolyte complex-based film forces the PEEK to get very close to the carbon fibre surface owing to electrostatic interactions and hydrogen bonding. During the cooling phase, the PEEK macromolecules (containing aromatic rings) close to the fibre surface (also made of aromatic rings due to the high density graphitic sheet-like microstructure of the fibre and high preferred orientation in terms of crystallite size [42,43] create Π-Π stacking and electrostatic interactions making the fibre surface an effective nucleating site and a large number of nuclei can be induced directly onto the surface. The crystal structure similarity, the high thermal conductivity of the fibre increasing the nucleation, and the high degree of molecular orientation inducing by the polyelectrolyte complex-based coating, allowed a nucleation growth at the interphase, which is thus referred to as transcrystallinity. Moreover, this Π-Π stacking interaction effect was not obstructed by the presence of PDA and could be even more pronounced by the self-assembly of PDA resulting from Π-Π stacking assembly. Actually, fibres treated with the higher PDA concentration [1.0 mg/mL] proved to increase the crystallinity of the composites to around 24 *wt*% and the higher crystallinity to about 45 *wt*%, was obtained with PEtI_PDA [1.0 mg/mL] samples, which tended to support our proposed mechanism. These results seem to confirm that PDA improved PEEK crystallinity by improving the Π-Π stacking interaction concentration induced by it self-assembly especially as when it optimized (by the presence of the polyelectrolyte complex-based film, for example).

Assuming these promising results, the water-based surface treatments were proposed as a convenient method for the sizing treatment of the high modulus pitch-based carbon fibres. Indeed, sizing process was considered as the simplest procedure to modify the carbon fibre surface [44-46]. The use of these modified carbon fibres was expected to improve the composite mechanical properties, particularly because of better adhesion between the fibre and the matrix as a result of interface compatibilization by interphase area formation at the vicinity of the fibres. The identified matrix crystallinity improvement was assumed to be transcrystallinity and was obtained by fibre surface modifications, was expected to generate interaction responsible for the PEEK adhesion on the fibre surface.

The transverse tensile strength at failure obtained with as received carbon fibres (CF-Z) and surface treated fibres were reported in Figure 7. Enhancements were actually obtained on the transverse tensile strength at failure, as measured via transverse tensile testing of unidirectional composites, by applying the PEtI and PEtI_PDA surface treatments. Indeed, 30 MPa and 29.4 MPa were respectively measured, whereas transverse tensile stress of 22.9 MPa was recorded for the composites reinforced with the reference (CF-Z). The effect of the influence of the transcrystallinity (inducing fibre-matrix compatibilization and adhesion promotion at the interface) on the composite mechanical behavior is confirmed. The influence of the crystallinity improvement, which is certainly a transcrystalline interphase, on the mechanical behavior is strongly assumed. The poor mechanical properties (18.5 MPa) obtained with the composites reinforced with mussel-inspired adhesive promoter-treated fibres confirmed the poor fibre-matrix interface quality and the lack of crystallinity with low concentration of adhesive promoter. Encouraging results allow the use of PEEK composites reinforced with high modulus pitch-based carbon fibres as structural materials for high performance applications by the use of fibres sized applying this environmentally friendly methodology.

Summary and Conclusions

This study showed an innovative and environmentally friendly methodology for fibre-matrix compatibilization and adhesion promotion in the case of PEEK composites reinforced with pitch-based carbon fibres. The composite interphase properties were improved by the carbon fibre surface modification using water-based treatments. Transcrystallinity at the interphase, which was successfully improved by means of incorporating polyelectrolyte complex-based film and self-assembled adhesive promoter coating by water-based sizing procedure, allowed the improvement of the composite transverse mechanical properties. Indeed, carbon fibre surfaces were modified thanks to the physisorption of polyelectrolyte complex-based film. This film adsorption, based on electrostatic interaction, had a positive effect on the adhesive promoter self-assembly onto the fibre. The influence of these water-based surface treatments onto the carbon fibre surface properties (tow and monofilament) have been investigated and detailed. Surface modification particularly in terms of wettability, chemical composition at the extreme surface, roughness and surface potential have been measured and discussed and an interaction mechanism was proposed. Interestingly enough, this is the first time that PEtI_PDA/pitch-based carbon fibre were used in a real continuous pitch-based carbon fibre reinforced PEEK composite. Finally, this environmentally friendly methodology appeared to be very interesting and efficient ways to the development of reactive compatibilization and high performance eco-sizing. This work paves the way to the formulation of cost efficient thermoplastic sizing easily processed by water-based solutions/process.

Acknowledgements

The authors thank Jean-Luc Biagi (LIST) for his skillful characterization (SEM) and Dr. Khaled Kaja from Bruker Company for the AFM measurements. Ph. Dubois thanks the FNR of Luxembourg for his PEARL chair "SUSMAT".

References

1. Reyne M (2006) Solutions composites : Thermodurcissables et thermoplastiques. JEC, Paris.

2. Campbell FC (2010) Matrix resin systems. ASM International.

3. Zhao Y, Wong HM, Lui SC, Chong EYW, Wu G, et al. (2016) Plasma Surface Functionalized Polyetheretherketone for Enhanced Osseo-Integration at Bone-Implant Interface. ACS Appl Mater Interfaces 8: 3901-3911.

4. Díez-Pascual A, Ashrafi M, Nakkakh B, González-Domínguez M, Johnston JM, et al. (2011) Influence of carbon nanotubes on the thermal, electrical and mechanical properties of poly(ether ether ketone)/glass fiber laminates. Carbon 49: 2817-2833.

5. Bartolucci SF, Mago G, Fisher FT, Troiano E, Kalyon DM (2012) Unusual fracture surface morphology of fatigued carbon nanofiber/poly(ether ether ketone) composites. Carbon 50: 2359-2361.

6. Zhai T, Lizhi D, Yang D (2013) Study on the Pretreatment of Poly(ether ether ketone)/Multiwalled Carbon Nanotubes Composites through Environmentally Friendly Chemical Etching and Electrical Properties of the Chemically Metallized Composites. ACS Appl Mater Interfaces 5: 12499-12509.

7. Flöck J, Friedrich K, Yuan Q (1999) On the friction and wear behavior of PAN- and pitch-carbon fiber reinforced PEEK composites. Wear 225-229: 304-311.

8. Zhang X, Fan X, Yan C, Li H, Zhu Y, et al. (2012) Interfacial Microstructure and Properties of Carbon Fiber Composites Modified with Graphene Oxide. ACS Appl Mater Interfaces 4: 1543-1552.

9. Sharma M, Gao S, Mäder E, Sharma H, Wei LY, et al. (2014) Carbon fiber surfaces and composite interphases. Compos Sci Technol 102: 35-50.

10. Qin W, Vautard F, Drzal LT, Yu J (2014) Modifying the carbon fiber-epoxy matrix interphase with graphite nanoplatelets. Polym Compos 37: 1549-1556.

11. Jonas A, Legras R (1993) Crystallization and chain adsorption of poly(etheretherketone) in discontinuous pitch-derived carbon fiber composites. Polym Compos 14: 491-502.

12. Park JM, Kim DS (2000) The influence of crystallinity on interfacial properties of carbon and SiC two-fiber/polyetheretherketone (PEEK) composites. Polym Compos 21: 789-797.

13. Quan H, Li ZM, Yang MB, Huang R (2005) On transcrystallinity in semi-crystalline polymer composites. Compos Sci Technol 65: 999-1021.

14. Denault J, Vu-Khanh T (1992) Crystallization and fiber/matrix interaction during the molding of PEEK/carbon composites. Polym Compos 13: 361-371.

15. Giraud I, Franceschi-Messant S, Perez E, Lacabanne C, Dantras E (2013) Preparation of aqueous dispersion of thermoplastic sizing agent for carbon fiber by emulsion/solvent evaporation. Appl Surf Sci 266: 94-99.

16. Martin A, Pietras-Ozga D, Ponsaud P, Kowandy C, Barczak M, et al. (2014) Radiation-curing of acrylate composites including carbon fibres: A customized surface modification for improving mechanical performances. Radiat Phys Chem 102: 63-68.

17. Martin A, Defoort B, Coqueret X (2015) Sizing composition for reinforcing fibres and application thereof. Patent WO 2015121274.

18. Torrecillas R, Baudry A, Dufay J, Mortaigne B (1996) Thermal degradation of high performance polymers-influence of structure on polyimide thermostability. Polym Degrad Stab 54: 267-274.

19. Crevecoeur G, Groeninckx G (1991) Binary blends of poly(etheretherketone) and poly(etherimide). Miscibility, crystallization behavior, and semicrystalline morphology. Macromolecules 24: 1190-1195.

20. Harris JE, Robeson LM (1988) Miscible blends of poly(aryletherketone)s and polyetherimides. J Appl Sci 35: 1877-1891.

21. Drzal LT (1986) The interphase in epoxy composites. Adv Polym Sci 75: 1-32.

22. Gérard JF (1988) Characterization and role of an elastomeric interphase on carbon fibers reinforcing an epoxy matrix. Polym Eng Sci 28: 568-577.

23. Guigon M (1991) Microtexture and mechanical properties of carbon fibers: Relationship with the fiber-matrix adhesion in a carbon-epoxy composite. Polym Eng Sci 31: 1264-1270.

24. Kuttner C, Hanisch A, Schmalz H, Eder M, Schlaad H, et al. (2013) Influence of the Polymeric Interphase Design on the Interfacial Properties of (Fiber-Reinforced) Composites. ACS Appl Mater Interfaces 5: 2469-2478.

25. Kharlampieva E, Kozlovskaya V, Sukhishvili SA (2009) Layer-by-Layer hydrogen-bonded polymer films: from fundamentals to applications. Adv Mater 21: 2053-2065.

26. Lavalle P, Voegel JC, Vautier D, Senger B, Schaaf P, et al. (2011)Dynamic aspects of films prepared by a sequential deposition of species: perspectives for smart and responsive materials. Adv Mater 23: 1191-1221.

27. Ariga K, Hill JP, Ji Q (2007) Layer-by-layer assembly as a versatile bottom-up nanofabrication technique for exploratory research and realistic application. Phys Chem Chem Phys 9: 2319-2340.

28. Klitzing RV (2006) Internal structure of polyelectrolyte multilayer assemblies. Phys Chem Chem Phys 8: 5012-5033.

29. Schönhoff M, Ball V, Bausch A, Déjugnat C, Delorme N, et al. (2007) Hydration and internal properties of polyelectrolyte multilayers. Colloids Surf A Physicochem Eng Asp 303: 14-29.

30. Cheung JH, Stockton WB, Rubner MF (1997) Molecular-Level Processing of Conjugated Polymers. 3. Layer-by-Layer Manipulation of Polyaniline via Electrostatic Interactions. Macromolecules 30: 2712-2716.

31. Sukhishvili SA, Granick S (2000) Layered, Erasable, Ultrathin Polymer Films. J Amer Chem Soc 122: 9550-9551.

32. Rydzek G, Thomann JS, Ben Ameur N, Jierry L, Mesini P, et al. (2010) Polymer Multilayer Films Obtained by Electrochemically Catalyzed Click Chemistry. Langmuir 26: 2816-2824.

33. Serizawa T, Hamada K, Kitayama T, Fujimoto N, Hatada K, et al. (2000) Stepwise Stereocomplex Assembly of Stereoregular Poly(methyl methacrylate) s on a Substrate. J Amer Chem Soc 122: 1891-1899.

34. Jiang H, Zhao X, Shelton AH, Lee SH, Reynolds JR, et al. (2009) Variable-Band-Gap Poly(arylene ethynylene) Conjugated Polyelectrolytes Adsorbed on Nanocrystalline TiO_2: Photocurrent Efficiency as a Function of the Band Gap. ACS Appl Mater Interfaces 1: 381-387.

35. Lee H, Dellatore SM, Miller WM, Messersmith PB (2007) Mussel-inspired surface chemistry for multifunctional coatings. Science 318: 426-430.

36. Guvendiren M, Messersmith PB, Shull KR (2008) Self-assembly and adhesion of DOPA-modified methacrylic triblock hydrogels. Biomacromolecules 9:122-128.

37. Bortolotti M, Brugnara M, Della Volpe C, Siboni S (2009) Numerical models for the evaluation of the contact angle from axisymmetric drop profiles: A statistical comparison. J Colloid Interface Sci 336: 285-297.

38. Franiti AV, Cross EM, Whitaker RB, Adams WW (1986) Refinement of the structure of PEEK fibre in an orthorhombic unit cell. Polymer 27: 861-865.

39. Kozlovskaya V, Yakovlev S, Libera M, Sukhishvili SA (2005) Surface Priming and the Self-Assembly of Hydrogen-Bonded Multilayer Capsules and Films. Macromolecules 38: 4828-4836.

40. Gao SL, Kim JK (2000) Cooling rate influences in carbon fibre/PEEK composites. Part 1. Crystalization and interface adhesion. Composites: Part A 31: 517-530.

41. Ogata N, Yasumoto H, Yamasaki K, Yu H, Ogihara T, et al. (1992) Evaluation of interfacial properties between carbon fibres and semicrystalline thermoplastic matrices in single-fibre composites. J Mater Sci 27: 5108-5112.

42. Naito K, Yang JM, Xu Y, Kagawa Y (2010) Enhancing the thermal conductivity of polyacrylonitrile- and pitch-based carbon fibers by grafting carbon nanotubes. Carbon 48: 1849-1857.

43. Bennet SC, Johnson DJ, Murray R (1976) Structural characterization of a high-modulus carbon fibre by high-resolution electron microscopy and electron diffraction. Carbon 14: 117-22.

44. Martin A, Pietras-Ozga D, Kowandy C. Defoort B, Coqueret X (2012) Optimization of carbon fibres by treatment and modification at the surface and study of the influence on the radiation-initiated polymerization of acrylate-based composites. Proceedings of the world conference on Carbon 2:1145.

45. Martin A, Bardon J, Michel M, Addiego F, Ruch D (2015) Compatibilization of pitch-based carbon fibre with high performance thermoplastic by supramolecular assembly for adhesion promotion. Proceedings of the world conference on Carbon.

46. Sharma M, Gao S, Mäder E, Sharma H, Wei LY, et al. (2014) Carbon fiber surfaces and composite interphases. Compos Sci Technol 102: 35-50.

22

High Temperature Oxidation Resistance of Ni$_{22}$Cr$_{11}$Al Bond Coat Produced by Spark Plasma Sintering as Thermal Barrier Coatings

Omoniyi FIS*, Olubambi PA and Sadiku ER

Department of Chemical, Metallurgical and Materials Engineering, Tshwane University of Technology, Pretoria, South Africa

Abstract

Thermal barrier coating (TBC) system is used in both aero engines and other gas turbines offer oxidation protection to super alloy substrate component. In the present work, it shows the ability of a new fabrication technique to develop rapidly new coating composition and microstructure. The compact powder were prepared by powder metallurgy method involving powder mixing and the bond coat was synthesized through the application of spark plasma sintering (SPS) at 1100°C, 1050°C and 1100°C to produce a fully dense 94%) Ni$_{22}$Cr$_{11}$Al bulk samples.

The influence of sintering temperature on hardness of Ni$_{22}$Cr$_{11}$Al done by micro vickers hardness tester was investigated. And oxidation test were carried out at 1100°C for 20 hr, 40 hr and 100 hr. The resulting coat was characterised with Optical microscopy, Scanning electron microscopy (SEM) and X-ray diffraction (XRD). Micro XRD analysis after the oxidation test revealed the formation of protective oxides and non-protective oxides.

Keywords: High temperature; Oxidation; Ni$_{22}$Cr$_{11}$Al; Spark plasma sintering; Thermal barrier coating

Introduction

In order to meet the challenging rising cost of high performance materials used for components operating at elevated temperature in aeroengines and other gas turbines, surface modification and techniques have attracted the attension of investigators globally [1,2]. Thermal barrier coatings are mostly used for this application. The TBCs are primarily made up of super alloy substrate, bond coat (BC) and YSZ topcoat, in which the bond coat layer is strongly linked to the oxidation behaviour of the TBC which affects its durability [3-5]. The bond coat is typically a platinum aluminide or NiAl based which provides oxidation and adhesion in the TBC System [6]. Application of TBC at elevated temperature gives rise to transfer of oxygen from the topcoat into the bond coat, so that an oxidized scale can be formed on the bond coat which is referred to as the thermally grown oxide (TGO) typically alfa-alumina [7,8].

However, this scale protects the super alloy components from further oxidation, the growth of TGO during thermal exposure leads to failure of the TBC System. This failure mechanism is generated at the TGO, the TGO/BC interface and the TBC. Studies shows that oxygen is transferred through ionic diffusion from the crystalline structure of ZrO$_2$ at the topcoat and penetration of gas through micro-cracks and porosities [9,10]. It has been proposed that these mechanism is facilitated mainly by the stress state arising from the residual compression in the TGO formed in service on bond coat which is the Al reservoir promoting α- alumina to form in preference to oxides [11-13]. In this study a Ni$_{22}$Cr$_{11}$Al bond coat were consolidated using spark plasma sintering technique and the high temperature oxidation resistance of the coating has been investigated.

Materials and Methods

Feed stock powder and characterization

Commercial starting powders for coating were Ni (flow master metal powder 28/0.5-3.0 micronparticle size 99.5% Ni), Cr (flow master metal powder<10 micron particle size, 99.2% Cr) and Al (flow master metal powder 99.7%) powders as bond coat. The three powders were mixed using the tubular shaker mixer T2F with the ball to powder ratio of 10:1 using stainless steel vial and five balls as milling media. Mixing speed of 72 rpm was chosen and a mixing time of 8 hr was allowed.

The Mixed powders were weighed according to Ni$_{22}$Cr$_{11}$Al molar composition as shown in Table 1. For the microstructures observations of the mixed powders, a small amount of the powders was mounted. The Cross section of the powders particles were prepared by conventional metallographic techniques. Microstructures and chemical composition of different phases in the powders were studied by employing Scanning electron microscope (SEM) with an energy dispersive spectrometer (EDS). The XRD analysis was done to identify the phases present in the samples in Bruker D$_2$. Advanced diffractometer using cobalt as anode material at 30 Kv and 10 MA. The powders were scanned at a step scan mode of 0.02.

Consolidation of the composites powders

In this work, the powders were consolidated by spark plasma sintering unit (H-HPD25-FCT Systeme GmbH Germany). The powders were loaded into a graphite die 20 mm in diameter and poured into packs of thickness between 3 to 8 mm in which 5 mm was aimed at. A Pressure of 30 MPa is applied to die through hydraulic rams fixed to a press. The current pulse of 3.3 ms fixed is generated by the power supply in which each pulse sequence contain 12 pulse was used. Temperature was measured by infra-red pyrometer to obtain a true value of temperature during consolidation. The densification process was carried out using a multistep heating method with heating rate of 100°C/min from room temperature to 1050°C. A holding time of 1 min

*Corresponding author: Omoniyi FIS, Department of Chemical, Metallurgical and Materials Engineering, Tshwane University of Technology, Pretoria, South Africa
E-mail: OmoniyiF@tut.ac.za, folorunsoomoniyi@gmail.com

Material	Ni	Cr	Al
NiCrAl	67%	22%	11%

Table 1: Chemical composition (in wt%) of $Ni_{22}Cr_{11}Al$ mixed feedstock powder.

at 600°C and 3.5 min at 1050°C.

Characterization of the sintered samples

The Sintered bulk surface were cleaned by grinding the surface using (Saphir 520 grinding machine made in Germany) with SiC P320 paper at speed 150 micron/min and a force 25 N for 30 min. Polished using both Aka-largan with dia max 9 micronmeter poly and Aka-napal with double 1 micronmeter poly to remove diffused carbon. The density of the sintered polished samples were performed by water displacement method (Archimedes principle process) using OHAUS density scale weighing balance of 0.001 mg accuracy. The density of the final sintered samples was 94% of the theoretical density.

Micro-hardness was measured using Vickers micro hardness tester (Future-Tech Corporation Tokyo, Japan) under loads of 25 gf for a dwell time of 10 secs. For the load five indents were made of which the average values were calculated. Fully dense SPsed samples were subjected to oxidation test in an electric furnace with an air atmosphere at 1100°C for 20, 40 and 100 hrs. Micro-XRD was carried out on the cross section of the sample after oxidation to determine the crystalline structure of oxide scales. SEM to investigate the microstructure and chemical composition.

Results and Discussion

Feedstock powder characterization

In Figure 1, XRD pattern for the mixed powders shows that there was mechanical activation of powders and also reactant phases. SEM micrographs of the mixed powder particles are represented in Figure 2, Mechanical mixing of the new material occurs, while there were still some regions which still consisted of one phase. EDS analysis showed that white areas are composed of aluminium and other areas encapsulating of Cr and Ni in softer Al particles which is visible extensively. Based on the XRD results of the mixed powders it can be expected that a reaction would occur during spark plasma sintering.

Sintering behaviours of the mixed powders during spark plasma sintering

Spark plasma sintering involves (relative piston travel as an indirect measure of densification and changes in the thickness of a sample die to movement of punches against the die with time is indicated by displacement [14-16]. The expansion or shrinkage is as a result of displacement towards the negative or positive direction [17].

The densification behaviour of the mixed powder was explained through the consideration of the shrinkage displacement and sintering time at different holding temperature during consolidation process [18,19].

$Ni_{22}C_{11}Al$ bulk samples were successfully densely sintered at 1050°C for holding time of 1 min at 600°C and finally 3 min under a pressure of 30 MPa. From Figures 3a-3c, the curve of displacement against temperature, it was observed that there was a preheating stage from room temperature to 585°C, thereafter shrinkage occurred at 970°C before thermal expansion of the dense sample at temperature greater or equal to 1050°C. This changes in displacement occurred due to sintering behaviour of the thermal expansion of the dense sample and the green body. And these behaviour could further be explained based

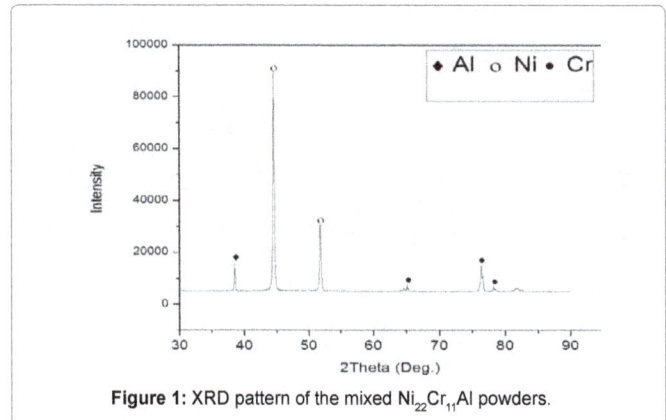

Figure 1: XRD pattern of the mixed $Ni_{22}Cr_{11}Al$ powders.

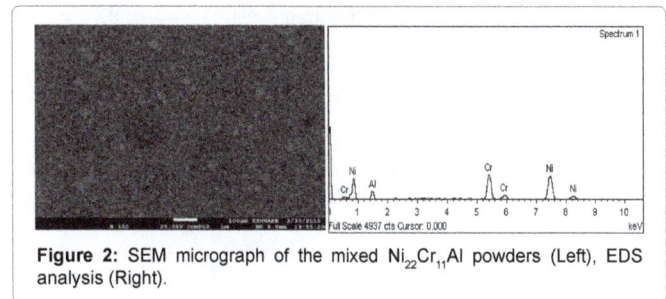

Figure 2: SEM micrograph of the mixed $Ni_{22}Cr_{11}Al$ powders (Left), EDS analysis (Right).

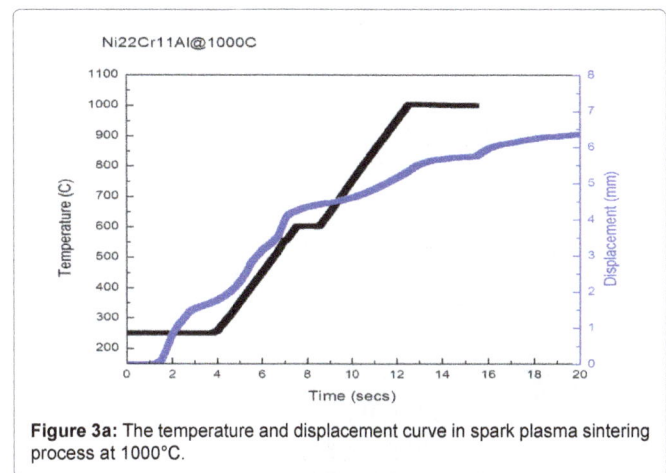

Figure 3a: The temperature and displacement curve in spark plasma sintering process at 1000°C.

on the fact that temperature was of great influence and important than the current during sintering process [20-23].

Characterization of microstructure, density and micro hardness

Figures 4 and 5 show the SEM images of sintered $Ni_{22}Cr_{11}Al$ samples at 1050°C and 1100°C. It can be observed that grain coarsening occurs with sintering temperature increase and grain growth was seen with sintering temperature rise. XRD diffractograms of the sintered $Ni_{22}Cr_{11}Al$ at 1050°C and 1100°C were shown in Figures 6 and 7 along XRD results revealed diffraction peaks of all the elements . no phase transformation or decomposition is observed comparison with the starting powders.

The density values of the sintered $Ni_{22}Cr_{11}Al$ shows that the material is fully dense which indicates low porosity and voids. From

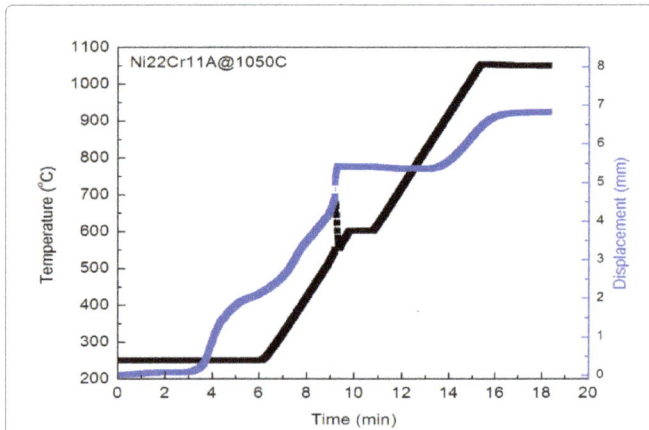

Figure 3b: The temperature and displacement curve in spark plasma sintering process at 1050°C.

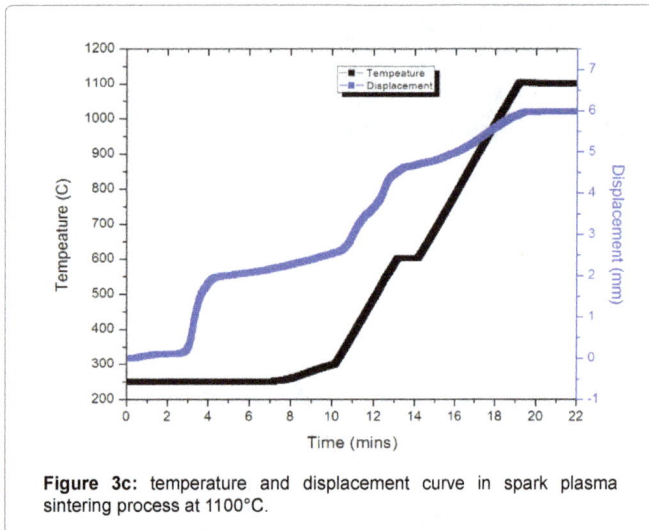

Figure 3c: temperature and displacement curve in spark plasma sintering process at 1100°C.

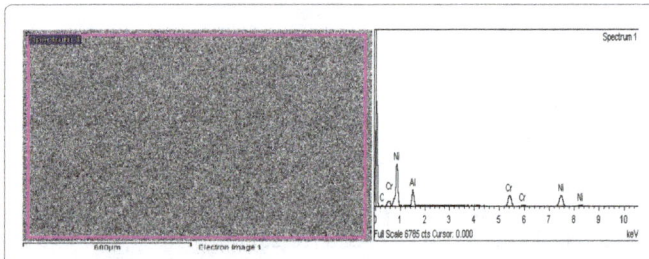

Figure 4: SEM micrograph of $Ni_{22}Cr_{11}Al$ powder (Left), EDS analysis (Right) sintered at 1050°C.

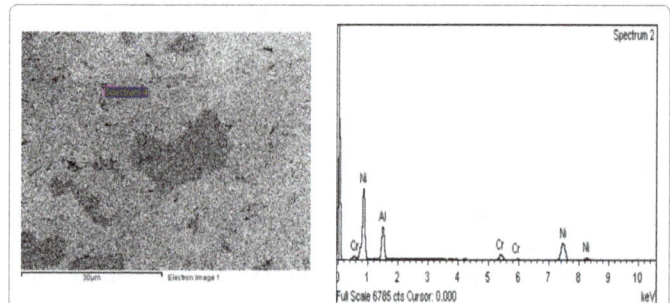

Figure 5: SEM micrograph of $Ni_{22}Cr_{11}Al$ powder (Left), EDS analysis (Right) sintered at 1100°C.

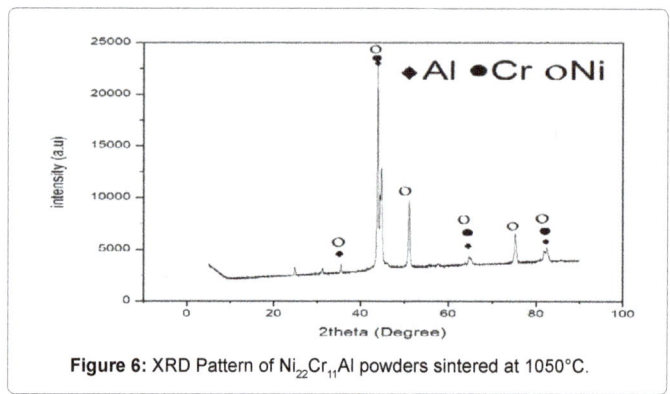

Figure 6: XRD Pattern of $Ni_{22}Cr_{11}Al$ powders sintered at 1050°C.

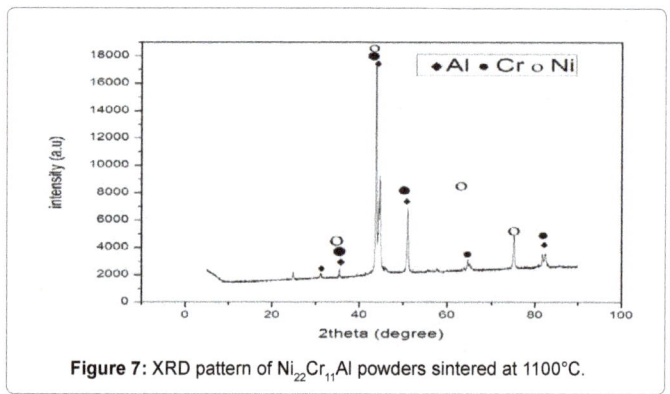

Figure 7: XRD pattern of $Ni_{22}Cr_{11}Al$ powders sintered at 1100°C.

Powder ($Ni_{22}Cr_{11}Al$)	Sintering Temperatures	Relative Density
	1000°C	92.2
	1050°C	93
	1100°C	94

Table 2: Sintering temperature as compared with relative density.

Table 2 and Figure 8 where the relative density values increases with the sintering temperature because the grain size is getting higher at higher sintering temperature. The material may shrink slightly at higher sintering temperature ;therefore the density increase slightly. Density is mass per unit volume. Density of Ni = 8.908 g/cm³, Cr = 7.19 g/cm³, Al = 2.7 g/cm³. The density of the sintered grinded and polished bodies was determined by Archimedes principle using the density determination kit of OHAUS density scale. The Principle states that every solid body immersed in a fluid loss weight by an amount equal to that of the fluid it displaces.

Density (Q) = A / (A-B) * Q0.

Where, A is weight of solid in air.

B is weight of solid in water.

Q0 is the density of water = 1.006 g/cm³

Micro hardness values of the sintered $Ni_{22}Cr_{11}Al$ are shown in Table 3 and in Figure 9, hardness value increases with increasing sintering temperature from 1000°C to 1100°C, this is due to more developed sintering necks , rounder pores and change in microstructure. Consequently, the micrographs of the hardness was studied with SEM as shown in Figure 10, it was observed that extent of damages at the load 25 gf is governed by mechanism of intergranules and transgranular

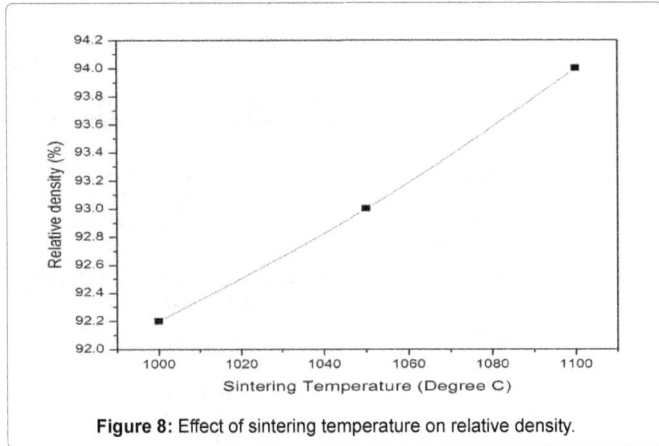

Figure 8: Effect of sintering temperature on relative density.

Powder Ni$_{22}$Cr$_{11}$Al	Sintering Temperatures	Av. Vickers Hardness 25gf
	1000°C	261.15
	1050°C	270.34
	1100°C	290.46

Table 3: Vickers micro hardness Hv at various sintering temperature.

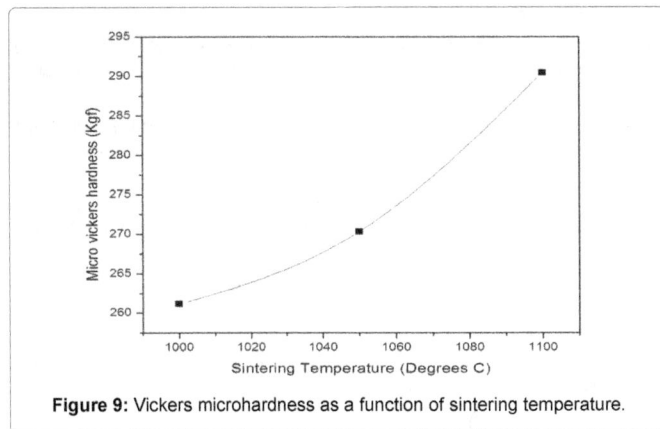

Figure 9: Vickers microhardness as a function of sintering temperature.

fracture, grain crushing and pulling effects.

Oxidation behaviour of sintered samples

NiCrAl alloy SPSed specimens were exposed for 20, 40, 100 hr at 1100°C.

SPSed: A thickness of 2-4 micro meters oxide layer is formed on the sample surface. According to the linear scan map, it is seen by EDS analysis that the layer is rich in Al as such corresponds to Al$_2$O$_3$ [22]. However, as reported in Figure 11, XRD analysis performed on the surface of oxidized samples reveals the presence of alumina and other oxides such as Cr$_2$O$_3$, spinel-like compounds like NiO, NiCr$_2$O$_4$ which are non-protective oxides [24-26]. It is well established that while the formation of slow-growing oxides alumina has to be maintained to prevent oxidation of the super alloy substrate, the presence of mixed oxides like spinels should be avoided as they grow too fast to produce protective scale [26-29].

When NiCrAl sintered samples are initially exposed to high temperature oxidizing environment, Al is the first element that tends to be preferentially oxidized [30-33]. From thermodynamic point of view Al has high affinity to react with 0$_2$, which brings about O$_2$ penetration and Al diffusion gives rise to Al$_2$O$_3$ formation [34-36]. EDS investigation

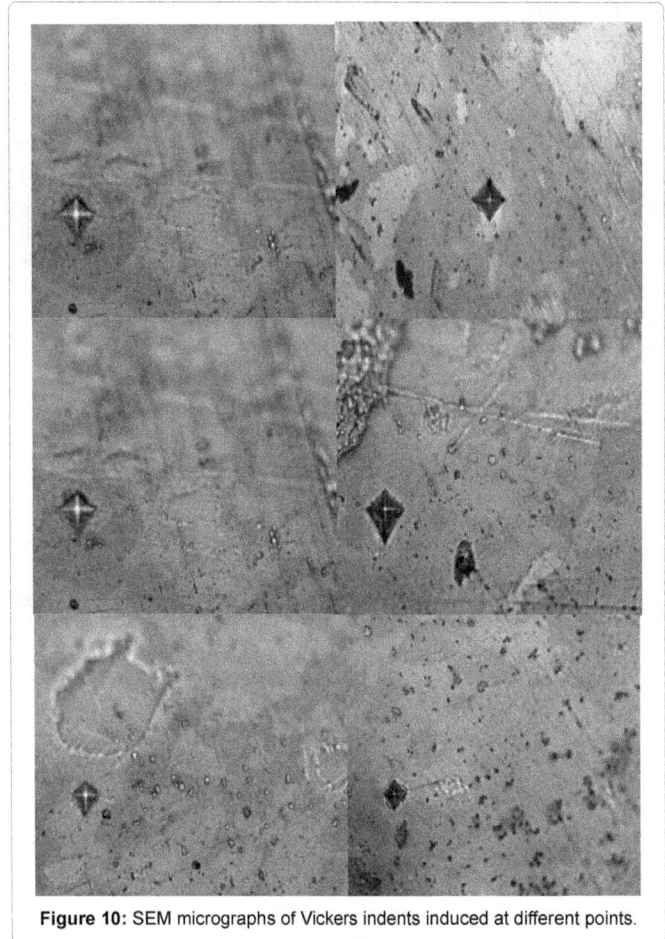

Figure 10: SEM micrographs of Vickers indents induced at different points.

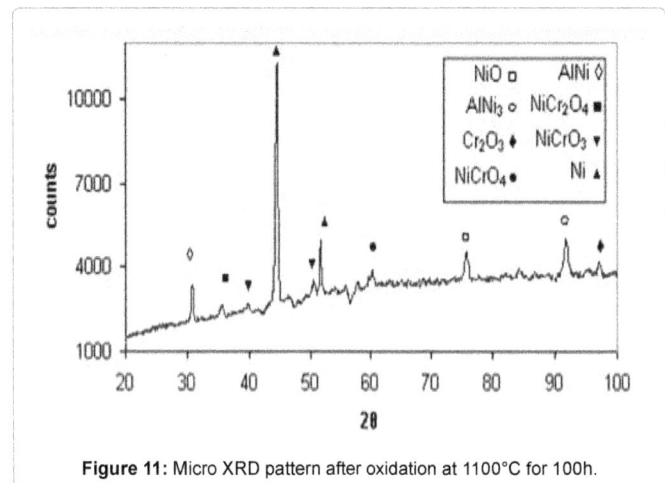

Figure 11: Micro XRD pattern after oxidation at 1100°C for 100h.

of the TGO reveals the high percentage of O$_2$ and Al and less of Cr and Ni element. Hence the main component of the TGO is Al$_2$O$_3$.

XRD patterns through the cross section of the bond coat after oxidation at 1100°C for 100 hr, revealed the oxides phases in the bond coat are NiCr$_2$O$_4$, NiCrO$_3$, NiCrO4, and NiO [37]. The degradation of the bondcoat leading to overall TBC failure is caused by these oxides formation accompanied with rapid volume increase [38,39].

Conclusion

$Ni_{22}Cr_{11}Al$ powder sintered at 1000°C, 1050°C and 1100°C was produced. And the relative density of sintered sample was as high as 94%. Microstructure observation showed that the achieved coating was well dense with attractive mechanical properties at room temperature. The grain growth of the samples was seen with interesting sintering temperature. These features described above are attributed to the oxidation behaviour displayed by $Ni_{22}Cr_{11}Al$ when exposed to air at high temperature. The homogeneity of the final sintered product coupled with microstructure refinement favour the selective formation of a continuous and dense layer of alumina while the presence of less protective mixed oxides (spinels) is hindered. SPS process indicates to be a promising technique due to shorter time process which gives rise to many advantages such as rapid turnover and low cost.

Reference

1. Chai Y, Chan P, Fu Y, Chang Y, Liu C (2008) Copper/Carbon nanotube composite interconnect for enhanced electromigration resistance. Electronic Components and Technology Conference, 2008, Lake Buena Vista, FL.

2. Bukhari MZ, Brabazon D, Hashmi MSJ (2011) Application of metal matrix composites cusic and alsic as electronics packaging materials. 28th International Manufacturing Conference.

3. Schubert T, Trindade B, Weißgärber T, Kieback B (2008) Interfacial design of cu-based composites prepared by powder metallurgy for heat sink applications. Mater Sci Eng A 475: 39-44.

4. Datsyuk V, Firkowska I, Gharagozloo-Hubmann K, Lisunova M, Vogt A, et al. (2011) Carbon nanotubes based engineering materials for thermal management applications. Semiconductor Thermal Measurement and Management Symposium (SEMI-THERM). 2011 27th Annual IEEE, San Jose, CA.

5. Cocke DL, Schennach R, Hossain MA, Mencer DE, McWhinney H, et al. (2005) The low-temperature thermal oxidation of copper, cu < sub > 3 < /sub > o < sub > 2 < /sub >, and its influence on past and future studies. Vacuum 79: 71-83.

6. Orth JE, Wheat HG (1997) Corrosion behavior of high energy-high rate consolidated graphite/copper metal matrix composites in chloride media. applied composite materials 4: 305-320.

7. Silvain JF, Vincent C, Heintz JM, Chandra N (2009) Novel processing and characterization of Cu/Cnf nanocomposite for high thermal conductivity applications. Composites Science and Technology 69: 2474-2484.

8. Sule R, Olubambi PA, Abe BT, Johnson OT (2012) Synthesis and characterization of sub-micron sized copper–ruthenium–tantalum composites for interconnection application. Microelectronics Reliability 52: 1690-1698.

9. Akbarpour MR, Salahi E, Alikhani Hesari F, Simchi A, Kim HS (2013) Fabrication, characterization and mechanical properties of hybrid composites of copper using the nanoparticulates of sic and carbon nanotubes. Materials Science and Engineering: A 572: 83-90.

10. O'reilly M, Jiang X, Beechinor JT, Lynch S, NiDheasuna C, et al. (1995) Investigation of the oxidation behaviour of thin film and bulk copper. Applied surface science 91: 152-156.

11. Deng CF, Ma YX, Zhang P, Zhang XX, Wang DZ (2008) Thermal expansion behaviors of aluminum composite reinforced with carbon nanotubes. Materials Letters 62: 2301-2303.

12. Thompson CV (2008) Carbon nanotubes as interconnects: Emerging technology and potential reliability issues. reliability physics symposium, 2008. IRPS 2008. IEEE International, Phoenix, AZ.

13. Nie J, Jia C, Jia X, Zhang Y, Shi N, et al. (2011) Fabrication, microstructures, and properties of copper matrix composites reinforced by molybdenum-coated carbon nanotubes. Rare Metals 30: 401-407.

14. Firkowska I, Boden A,Vogt AM, Reich S (2011) Effect of carbon nanotube surface modification on thermal properties of copper–cnt composites. Journal of Materials Chemistry 21: 17541-17546.

15. Damayantia M, Sritharan T, Mhaisalkar SG, Phoon E, Chan L (2007) Study of Ru Barrier failure in the Cu/Ru/Si system. J Mater Res Soc 22: 2505-2511.

16. Yang CC, Spooner T, Ponoth S, Chanda K, Simon A, et al. (2006) Physical, electrical, and reliability characterization of Ru for Cu interconnects.

Interconnect Technology Conference, 2006 International, Burlingame, CA.

17. Wang Y, Cao F, Zhang ML, Liu YT (2011) Comparative study of Cu-Zr and Cu-Ru alloy films for barrier-free Cu metallization. Thin Solid Film 519: 3169-3172.

18. Ghosh D, Chen S (2008) Solid-state electronic conductivity of ruthenium nanoparticles passivated by metal-carbon covalent bonds. Chemical Physics Letters 465: 115-119.

19. Ding SF, Xie Q, Müeller S, Waechtler T, Lu HS, et al. (2011) The inhibition of enhanced Cu oxidation on ruthenium/diffusion barrier layers for Cu interconnects by carbon alloying into Ru. Journal of the Electrochemical Society 158: H1228-H1232.

20. Carreno-Morelli E, Yang J, Couter E, Hernadi K, Seo JW, et al. (2004) Carbon Nanotube/Magnesium Composites. Physical status solid 201: R53-R55.

21. Esawi A, Morsi K (2007) Dispersion of Carbon nanotubes (Cnts) in Aluminum powder. Composites Part A: Applied Science and Manufacturing 38: 646-650.

22. Yamanaka S, Gonda R, Kawasaki A, Sakamoto H, Mekuchi Y, et al. (2007) Fabrication and thermal properties of carbon nanotube/nickel composite by spark plasma sintering method. Materials Transactions 48: 2506-2512.

23. Kim C, Lim B, Kim B, Shim U, Oh S, et al. (2009) Strengthening of copper matrix composites by nickel-coated single-walled carbon nanotube reinforcements. Synthetic Metals 159: 424-429.

24. Cho S, Kikuchi K, Miyazaki T, Takagi K, Kawasaki A, et al. (2010) Multiwalled Carbon nanotubes as a contributing reinforcement phase for the improvement of thermal conductivity in copper matrix composites. Scripta Materialia 63: 375-378.

25. Bhat BN, Ellis D, Smelyanskiy V, Foygel M, Rape A, et al. (2013) copper-multiwall carbon nanotubes and copper-diamond composites for advanced rocket engines.

26. Platzman I, Brener R, Haick H, Tannenbaum R (2008) Oxidation of Polycrystalline Copper Thin Films at Ambient Conditions. The Journal of Physical Chemistry C 112: 1101-1108.

27. Sule R, Olubambi PA, Sigalas I, Asante JKO, Garrett JC (2014) Effect of SPS consolidation parameters on submicron Cu and Cu–CNT composites for thermal management. Powder Technology 258: 198-205.

28. Hu Q, Luo P, Yan Y (2008) Influence of spark plasma sintering temperature on sintering behavior and microstructures of dense bulk mosi< sub> 2</sub>. Journal of Alloys and Compounds 459: 163-168.

29. Saheb N, Iqbal Z, Khalil A, Hakeem AS, Al-Aqeeli N, et al. (2012) Spark plasma sintering of metals and metal matrix nanocomposites: a review. Journal of Nanomaterials 2012: 18.

30. Pan MY, Gupta M, Tay AAO, Vaidyanathan K (2006) Development of bulk nanostructured copper with superior hardness for use as an interconnect material in electronic packaging. Microelectronics Reliability 46: 763-767.

31. Sule R, Olubambi PA, Sigalas I, Asante JKO, Garrett JC (2014) Densification and fracture characteristics of spark plasma sintered Copper- CNT- Ruthenium composites. Proceedings Euro PM2014 - PM for current and Future Applications (2014) Austria.

32. Xue ZW, Wang LD, Zhao PT, Xu SC, Qi JL, Fei WD (2012) Microstructures and tensile properties of carbon nanotubes reinforced Cu matrix composites with molecular-level dispersion. Materials and Design 34: 298-301.

33. Gupta M, Tay AAO, Vaidyanathan K, Srivatsan TS (2007) An investigation of the synthesis and characterization of copper sample for use in interconnect application. Mater Sci A 454-455: 690-694.

34. Lee KM, Oh DK, Choi WS, Weissgärber T, Kieback B (2007) Thermomechanical properties of Aln-Cu composite materials prepared by solid state processing. Journal of Alloys and Compounds 434-435: 375-377.

35. Tong XC (2011) Advanced materials for thermal management of electronic packaging. Springer.

36. James JD, Spittle JA, Brown SGR, Evans RW (2001) A review of measurement techniques for the thermal expansion coefficient of metals and alloys at elevated temperatures. Measurement Science and Technology 12: R1.

37. Pollack GL (1969) Kapitza resistance. Review of modern physics 41: 48-81.

38. Chu K, Guo H, Jia C, Yin F, Zhang X, et al. (2010) Thermal properties of carbon nanotube-copper composites for thermal management applications. Nanoscale research letters 5: 868-874.

39. Wagner CD, Naumkin AV, Kraut-Vass A, Allison JW, Powell CJ, et al. (2003) X-Ray photoelectron spectroscopy database. Nist Standard Reference Database 20, Version 3.4.

23

Surface Excess Free Energy: An Elaboration with Particular Insight for Use as a Predictor of Solvophilicity in Molecular Simulation

Mongelli GF*

Department of Chemical Engineering, Case Western Reserve University, Cleveland, USA

Abstract

This manuscript details the method to determine the surface excess from readily derivable ensemble properties, namely the pressure tensor, via computational molecular dynamics. It will then expand upon the theoretical and practical uses of quantities in Gibbs-Duhem like relationships for the surface excess and molecular concentration at the interface. Furthermore, it details several limitations of computational molecular dynamics, mainly to determine force field parameters natively and also to determine criteria for switching the bond order at certain temperatures. The goal in predicting surface presence is in inter-relating the relative surface excess free energies of each species with respect to the total system relative to the free energy of hydration of that system.

Keywords: Thermodynamic; Hydration; Polymers

Introduction

While the fundamental thermodynamic equations of state have been well documented, particularly relations for entropy and free energy of hydration for various systems, such relationships have not been explored with the use computational molecular dynamics. This is especially the case for parameters of unique interest to polymeric systems within mixed solvents. Such systems are a growing and multi-billion dollar global annual industry [1].

Two methods will be discussed to determine the free energy of hydration or solvation, and their relative complexity will be compared. The equations of relevance for calculation of the surface excess free energy will be explored in detail within this manuscript. Which equations are best suited for experimental or theoretical determination will be discussed. The surface excess free energy will be defined. How to compute this important quantity with computational molecular dynamics software will be elucidated. Additionally, some potential explanations for which aspects of molecular structure lead to reduced surface tension will be discussed.

Potentially Simpler Free Energy of Hydration Computation

An alternative method will employ the pull code within molecular dynamics simulations to obtain the free energy landscape for the surfactant molecule with respect to both the vacuum gas state, the bulk solubilized, and the interface. The pulling constant can be altered to perturb the polymer or target molecule of interest to various z-space values, where the interfaces to the various states are at z-axis parallels. In such a manner, the intermolecular forces present in different z-slabs can be determined [2]. As molecular structuring around the interface has been known to be different from the bulk, the density profile can be mapped onto the free energy of hydration in cases where the pulling constant is such that it allows the polymer or target molecule to swing throughout all z-space. Such density profiles can be proportional probability of a polymer or target molecule taking a certain position z in with respect to the interface is given by the following relation, and therefore indicates the relative free energies of the system:

$$Prob(z)=C*exp[-G(z)/kT] \tag{1}$$

where C is a constant and Prob[z] comes from umbrella sampled

systems. The free energy landscape, G(z), is thus obtained by inverting this equation,

$$G(z)= - k_B T \ln[Prob(z)]+C' \tag{2}$$

The values of Prob(z) are obtained as histograms constructed from a molecular dynamics trajectory. With the energy present at k_{BT}, umbrella sampling may be used to force the system into relatively unstable states. The free energy can then be related to the un-perturbed systems via Li-Makarov analysis [3]. This is required at low alcohol contents in co-solvated systems, where the intermolecular forces and free energies of solvation are of energies with magnitude greater than kB*T. To obtain better statistics, umbrella sampling will be used to bias the surfactant to certain positions with respect to the liquid surface; this bias is removed in the free energy calculation using the self-consistent histogram method [4,5]. Several simulations may need to be performed to start the system in the high energy state, i.e., solvated for insoluble polymers and in a surface state for soluble polymers. Soluble polymers might have even higher energy configurations when placed more than the cut-off radius away from the solvent interface in the vacuum region.

By establishing the differences between the relative free energy relationships, it becomes very computationally cheap to predict the solubility and surface activity properties of new molecules and new solvents. A single simulation of a polymer studied in this work with a new solvent allows for the prediction of the surface parameter of all of the molecules within such work in the studied solvents. Correspondingly, a simulation of a new solute in a previously studied solvent allows for the prediction of the surface parameter in each of the studied solvents. This is determinable because each solvent has a one-to-one mapping of free energy of hydration relative to surface parameter.

*****Corresponding author:** Mongelli GF, Department of Chemical Engineering, Case Western Reserve University10900 Euclid Ave, AW Smith 116, Cleveland, OH 44120, USA, E-mail: Gfm12@case.edu

After mapping out the free energy changes between solvents, and between polymers, we can do a single simulation and predict the surface parameter in all other contents for a particular solvents and indeed different solvents as well. The argument for exploring $\Delta G[z]$ beyond the cutoff radius in the 100% alcohol case allows us to estimate when polymers will go into the bulk solution that have less repulsion from the solvents than the least repulsive solute in a series. Additionally, we will only get all of the $\Delta G[z]$ information we need when the bulk state parameter goes to one and not non-zero --as stated earlier. This would imply that the surface parameter would need to go to zero in the 100% alcohol limit, which we have not seen for many systems. That being said, we need the free energy increases as a function of z beyond those estimated by these materials in the low alcohol content limit to predict when more soluble materials will go in and at what critical free energy values.

The free energy of hydration may be more simply determined by treatment of its time averaged quantity as a probability distribution in NVT simulations. The following functions are written for the isothermal-isobaric ensemble, but translate readily into the NVT/Canonical ensemble by replacing Gibb's Free energy, G[z], with Helmholtz Free Energy, A[z]. Then:

$$\text{Prob}[z] = C_0 \text{Exp}[-\Delta A[z]/RT] \tag{3}$$

Inverting to free energy space:

$$\Delta A[z] \propto \text{Ln}[\text{Prob}[z]] = \text{Ln}[\rho[z]] \tag{4}$$

Therefore, relative free energy relationships can be determined between NVT systems. It is not immediately clear how to compare the relative free energy of solvation between isothermic-isobaric simulations and NVS/Canonical simulations. In considering entropy changes from the simulation results, derivations are readily possible from general statistical mechanics theory. In the NPT and NVT systems, for bulk non-interface systems without a vacuum:

$$\Delta S = e^{\Delta A / k_B T} \tag{5}$$

In the case of NPT systems, the Helmholtz Free Energy is the negative of the Legendre transform with respect to entropy.

$$\Delta S = e^{pV - \Delta G / k_B T} \tag{6}$$

This entropy change implies a change in the number of configurations available to the system. The number of configurations available to the system can be compared in ideal gas states, interface active, bulk solvated states. The simple relation required to go from entropy, which can be established from (5) or (6) via an ensemble potential, is simply Boltzmann's Entropy formula:

$$\Delta S = k_B * T * Ln(\Delta \Omega) \tag{7}$$

Surface Excess

One quantity of interest which is available from these results via a derivation and many computational molecular dynamicists have not computed is the surface excess and associated Gibb's isotherms of hydrophobic and hydrophilic molecules. Such a surface excess [6] tells the area at the interface per molecule, α, as described by:

$$\alpha = \frac{1 * 10^{20}}{N * \Gamma} \tag{8}$$

N is Avogadro's number. Γ is the surface excess in units of moles/m². This quantity offers key information regarding how the surface tension, surface parameter, and chemical potentials of the solvent and the polymer vary together. Maximizing surface tension changes with minimum chemical potential changes represents an improved surfactantability, since the same surface tension can be achieved with reduced polymer concentrations. That is what should be observed from computing the other information within this section.

The following equations describe surface phenomena in the Gibb's surface formalism [7]:

$$-d\gamma = \Gamma_1 d\mu 1 + \Gamma_2 d\mu 2 \tag{9}$$

This tells how the surface tension will be impacted as a function of the surface excess quantities of various molecules and their respective chemical potentials. It is useful in estimation of whether molecules will increase or decrease the surface tension.

$$\Gamma_i = \frac{n_i^{TOTAL} - n_i^{\alpha} - n_i^{\beta}}{A} \tag{10}$$

The above relation [8] indicates how the surface excess is calculated as a function of the number of molecules of a specific type per unit surface area. It is simple to study more complex, multi-component systems by increasing the number of types of molecules detracting from n_i^{TOTAL} in the numerator. It is most useful for determining the surface excess via a single computational molecular dynamics simulation. Although the author is not aware of any codes presently which do this computation natively in one step, the number of molecules of each species within ca. one nanometer – or the cutoff length of non-bonded computations -- of the interface should be determined, and the relative number of molecules will allow for the computation of the surface excess via the above described method.

$$\Gamma_S = -\frac{1}{2RT} \left(\frac{\partial \gamma}{\partial C} \right)_{T,P} \tag{11}$$

The above relation is most useful for the determination of the surface excess from experimental methods. The alcohol or polymeric content is varied and the surface tension is measured. The surface tension decrease will be linear up to a breakdown point, at which additional concentration increases do not result in surface tension decreases. For immiscible materials, this means that the added molecules are no longer being added to the surface, but instead take bulk states in the preferred phase, and do not impact the surface tension. This holds true for polymers. For miscible materials, the free energy of solvation changes for each species as a function of the relative mass contents of the mixture. Therefore, adding more content of one species or another will change the free energies of solvation for either species and the energies of moving any individual species from the bulk to the interface. This relation also is useful for determining the surface excess from comparing the surface tensions of multiple simulations.

$$\Gamma_i^1 = \Gamma_i - \Gamma_1 \left(\frac{C_i^{\alpha} - C_i^{\beta}}{C_1^{\alpha} - C_1^{\beta}} \right) \tag{12}$$

If the surface excess of one species is known in a binary mixture, then the surface excess of the second member of the mixture may be determined from an experiment or simulation in a different content – assuming that the surface tension decrease has not occurred or is negligible.

$$dG = \sum_{\alpha, \beta, S} \left(\begin{array}{c} dU + PdV + VdP - TdS - SdT + \\ \sum_{i=1}^{k} \mu_i dn_i + \sum_{i=1}^{k} n_i d\mu_i \end{array} \right) + Ad\gamma + \gamma A \tag{13}$$

$$\sum_{i=1}^{k} n_i^S d\mu_i + Ad\gamma = 0 \tag{14}$$

The above equations also have implications in the activity coefficient of the system.

γ is the surface tension

Γ_i is the surface excess of component i

μ_i is the chemical potential of component i

Species 1/α is water.

The full NPT/isothermal-isobaric ensemble theory is written above, including terms for surface tension and area change effects. Of increased interest in this theory is the NVT/Canonical ensemble derivation, which utilizes the canonical partition function, Z, as defined earlier in an exponential of the Helmholtz free energy relative to the inverse temperature $k_B T$, rather than free energy.

For a system with a constant number of molecules and a constant interfacial area, as the alcohol content and alcohol structure varies, the area of the van der Walls surface varies as does the chemical potential in the vicinity of the interface. The surface excess refers to the energy per unit area that the molecule contributes to the interfacial chemical potential. The surface excess then tells how many molecules are required at an interface to alter the surface tension to a particular value and how many are required to be replaced to maintain a constant chemical potential. The solubility and intermolecular forces dictate whether such replacements actually occur.

The surface excess and its relationship to predicting the surface state presence of a molecule has an analogous relationship with fugacity. The surface excess tells of the surface presence of molecules spatial position in the system relative to the free energy of hydration whereas the fugacity predicts the vapor pressure and phase change fraction relative to the enthalpy of vaporization.

It is expected that there exists a relation similar and analogous to that of chemical potential, surface excess and surface tension which interrelates the volumetric hydrogen bond density and the free energy of hydration. Furthermore, a program should be written to will address the spatial correlation of the hydrogen bonds to classify them as within the cutoff radius of the interface and beyond the cutoff radius of the interface. These results can be related to the surface partitioning results discussed above, and will be used to assess the explanation of polyether solubility trends [5].

Possible Relation between Molecular Parameterizations and Surface Tension Reduction Capacity

The presence of a particular molecule in a mixture can only affect the surface tension if said solute is insoluble in the mixed solvent. Therefore, the foaming capability of insoluble polyethers is lost when adding alcohol co-solvents due to the loss of low surface tension-associated molecules from occupying the interface. The loss of foaming capability is not from alcohol molecules displacing polymer molecules from the interfacial surface caused by their own solvophobic repulsions. Since the surface tension of a mixture is dependent on the surface tensions of the individual components and what fraction of the surface they occupy, the addition of alcohols have low surface tension would decrease the overall mixture surface tension.

What is left to determine is which types of forces within the polymer or target molecule system result in the surface tension reduction relative to their non-polymer- or non-target-molecule- containing co-solvents. This phenomenon is likely the result of a needle or any other surface interacting body pushing against the dihedral interaction energies along the polymer backbone. Once a certain critical energy state is reached, the Lennard-Jones and Coulombic interactions combined with the physical external force of the body overcome the dihedral energy barrier. This is a sort of rotational energy problem. If you want to spin a water wheel so you can put grain through a mill or generate electricity but the wheel has some trash stuck in it and water isn't strong enough to fully rotate the wheel. If you give it a good push, then you will get the trash up to the top of the wheel and then the wheel will rotate down until it gets stuck again. In this case, the trash is sterically hindered groups on molecules, which actually increases the rotation of the wheel after it hits the top of the wheel, and the water flow is k_{BT}.

Conclusions

With new rigorous computational methods, such as those discussed in this work, it becomes easier to determine the free energies of solvation. We can receive results faster or more cheaply than with experiments in a large number case in the present force field. Faster, in the sense that doing many simulations scales fiscally much more readily than does doing many experiments. At a point beyond that, it may be determined that some presently non-parameterized atoms have such force field embodiments and whose associated to-be-determined-properties are desired. Though it is much more desired that molecules associated with certainly real force field parameterizations are studies via simulation when they cannot be synthesized above and before purely theoretically molecules which result from perturbed force field parameterizations of known molecules. Again, in the limit of more computational power than scholars can come up with short-term coding applications for – which may be the case given the acceleration of Moore's Law–and that complex coding and theory mathematics to run on supercomputers is limited by human derivations, perhaps the latter is desirable.

A possible explanation for the surface tension reducing capability of a molecule is suggested. One potential method to test for this explanation is to dial up the dihedral interaction constants for a surface active molecule and see if it will change the ensemble average of the pressure tensor and associated surface tension. Perhaps a more ambitious goal, then, is to determine which classes of structures could be surface active and maintain foaming capability from molecular simulation.

Future Work

Once polyethers, polysilicones and their polyalkane analogue materials can be simulated each of the quantities of interest can be determined via molecular simulation and from experimental studies. Any discrepancies in their derivations from these two methods may lead to changes in computational molecular dynamics theory for polymeric materials or, rather, to changes in the molecular parameterization constants for these materials within the Optimized Potentials for Liquid Simulations- All Atoms (OPLS-AA) standard.

Acknowledgements

I would like to acknowledge the National Science Foundation **Award Abstract #1159327**.

References

1. Specialty Surfactants Market - Global Scenario (2017), Raw Material And Consumption Trends, Industry Analysis, Size, Share & Forecast.

2. Dalvi VH, Rossky PJ (2010) Molecular origins of fluorocarbon hydrophobicity. Proc Natl Acad Sci USA 107: 13603-13607.

3. Li PC, Makarov DE (2004) Ubiquitin-like Protein Domains Show High Resistance to Mechanical Unfolding Similar to That of the I27 Domain in Titin: Evidence from Simulations. J Phys Chem B 108: 745-749.

4. Shah PP, Roberts CJ (2008) Solvation in mixed aqueous solvents from a thermodynamic cycle approach. J Phys Chem Lett 112: 1049-1052.

5. Mann EK, Langevin D(1991) Poly(dimethylsiloxane) molecular Layers at the Surface of Water and of Aqueous Surfactant Solutions. Langmuir 7: 1112-1117.

6. Mitropoulos SC (2008) What is a surface excess? Journal of Engineering Science and Technology Review 1: 1-3.

7. Geppert-Rybczynska M, Lehmann JK, Heintz A(2011) Surface Tensions and the Gibbs Excess Surface Concentration on Binary Mixtures of the Ionic Ethyl-3-methylimidazolium Bis[(tribluoromethyl)sulfonyl]imide with tetrahydrofuran and Acetonitrile. J Chem Eng Data 56: 1443-1448.

8. Butt HJ, Graf K, Kappl M (2004) Physics and Chemistry of Interfaces. Wiley, Belin.

24

Optimization of Sustainable Cutting Conditions in Turning Carbon Steel by CNC Turning Machine

Mohammad HM[1] and Ibrahim RH[2]*

[1]Materials Engineering Department, College of Engineering, University of Basrah, Iraq
[2]Mechanical Engineering Department, College of Engineering, University of Basrah, Iraq

Abstract

The current study aims to find the optimum cutting parameters in turning process without using cutting fluids (dry cutting condition) towards sustainable manufacturing. Where the power consumption and environmental pollution increase due to increase of the machining operations in manufacturing field, so to save energy and environment and reduce cost it is important to adopt sustainability in machining processes.

The experimental work in this study involves the preparation to a number of experiments on AISI 1045 carbon steel to collect the necessary data for implementing optimization process. The experiments were conducted by changing levels of cutting parameters (spindle speed, feed rate and cutting depth) in CNC turning machine. Surface roughness of the workpiece has been depended as a quality indicator. In addition, the temperature of cutting tool has been recorded during machining the work pieces in order to control the temperature of cutting process.

Theoretically, empirical equations for temperature of cutting tool and surface roughness of the work piece have been discovered. By using Genetic Algorithm technique these equations have been used to find the optimum of cutting parameters spindle speed, feed rate and depth of cut.

The optimum values that obtained by using Genetic Algorithm which achieve sustainable cutting were spindle speed 588.96 rpm, depth of cut 0.50 mm and feed rate 64.55 mm/min in order to have the optimum of surface roughness in low cutting temperature.

Keywords: Surface roughness; Optimization; Sustainable manufacturing and machining

Introduction

Metal cutting processes comprise the biggest part of manufacturing sector that in turn represents one of the largest energy consumers in the world. As the world is moving today towards sustainable industrialization to preserve the environment of manufacturing pollutants and protects non-renewable resources of energy and production goods with high quality and low cost, the improvement of operating conditions in the machining operations like turning, milling, boring ... etc., has become an urgent need.

The cutting conditions includes cutting parameters (spindle speed, feed rate and cutting depth) and other factors related to cutting process like cutting fluids used for cooling and lubrication, cutting temperature, the type of cutting tool and its geometry specification, …etc. Cutting parameters are the more effecting conditions on the cutting process and quality of the workpiece that is usually measured by surface roughness [1].

Surface roughness is one of the most important customer requirements where it measures the finer irregularities of the surface texture. Achieving the required surface roughness is critical for the functional behaviour of a part due to its impact on mechanical properties of the manufactured parts such as fatigue life, wearing, and light reflection, ability to distribute and hold a lubricant between contacting bodies, load bearing capacity and coatings [2,3]. Moreover, surface roughness plays a significant role on machining cost where it is related to the dimensions precision so contributes in reducing assembly time and avoiding the need for secondary operation, therefore cost will be reduced [3,4]. The value of surface roughness depends on numerous factors such as machining parameters, material of cutting tool and its geometry parameters, work piece material and its mechanical properties, generated temperature, machine vibrations and cutting conditions (wet or dry cutting). Even small changes in any of the mentioned factors may have a significant influence on the machined surface [4].

Sustainability has been applied to many fields, include engineering, manufacturing and design. In manufacturing sector, metal machining industry is under increasing pressure as a result of competition. Manufacturers are becoming increasingly concerned about the issue of sustainability because adopting sustainability in metal machining processes lead to improve their economic, environmental and social performance [5,6]. The U.S. Department of Commerce defines Sustainable manufacturing as the creation of manufactured products that use processes that minimize negative environmental impacts, conserve energy and natural resources, are safe for employees, communities, and consumers and are economically sound. Figure 1 shows Sustainability Elements of Manufacturing Processes [5,6]. One of the main problems in machining processes is the generated heat in the cutting zone where about 97% of the work that goes into cutting dissipated in the form of heat. The generated heat impacts on mechanical properties of the workpiece and wear rate of the cutting tool and consequently on surface

---

***Corresponding author:** Ibrahim RH, Mechanical Engineering Department, College of Engineering, University of Basrah, Iraq
E-mail: ceo_engg@yahoo.com

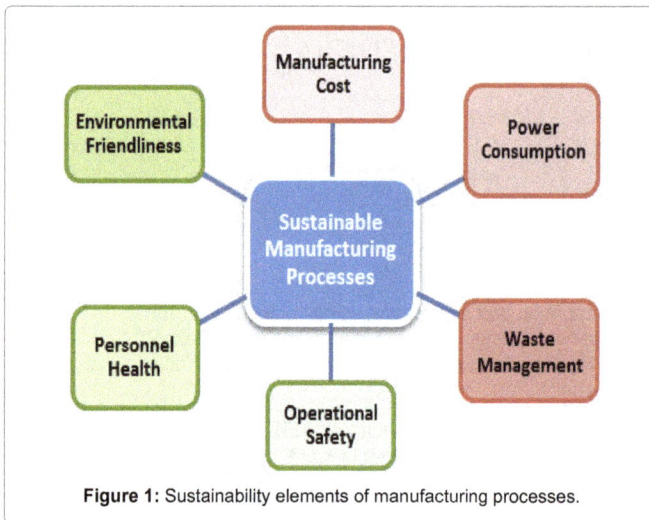

Figure 1: Sustainability elements of manufacturing processes.

roughness. To moderate the damaging effect of heat in machining processes, Cooling Lubricating Fluids (CLFs) were used. Economic and environmental troubles accompanied using conventional cutting fluids. Where cutting fluids are not naturally biodegradable, treatment of cutting fluids is required before disposal. In addition to maintenance and fluid purchase and preparation cost, this treatment estimated to be two or four times purchase price. Moreover, some of cutting fluids effect on employees health. For all that, conventional cutting fluids are a major non-sustainable factor in machining processes, which led to developed alternate cooling mechanisms [5,7].

To achieve sustainable manufacturing, alternatives to conventional fluid machining were developed such as cryogenic cooling, minimum quantity lubrication (MQL) and high pressure cooling. These new techniques contribute in reducing cost and reducing/avoiding health and environment problems that are usually caused when using oil-based CLF [1]. Towards sustainability, the complete omission of cutting fluids (dry cutting) was also used in machining materials that have relatively low hardness. Dry cutting condition was adopted in this study and measurement of the tool tip temperature was performed by infrared thermometer to control the temperatures range.

The growing demands of machining processes attract the attention of many researchers to explore the behaviour of factors on surface quality, tool wear, power consumption and ambient in machining operations especially in turning process. Yusuf Sahin [8], developed a mathematical models to predict surface roughness in terms of cutting speed, cutting depth and feed rate by using Response Surface Methodology (RSM). AISI 1040 steel was the workpiece material in the turning process with coated carbide cutting tools. The results showed that feed rate is the main influence factor. It has been seen that surface roughness increases as feed rate increases but decreases with increasing both depth of cut and cutting speed. Hasan Gokkaya [9] presented a study about the effects of various coating materials of cutting tool, coating method and cutting factors on surface roughness. Workpiece of AISI 1015 steel was turned without coolant using four types of cemented carbide cutting tools. Feed rate and cutting speed were selected as cutting parameters. According to the coating types, the results indicated that the best surface roughness was achieved when using cutting tools coated with TiN using the CVD technique. It is also noticed that surface roughness decreases when cutting speed increases while it increases when feed rate increases.

Yansong Guo [10] presented mathematical models in order to optimize energy consumption and surface quality in dry turning of steel (11SMnPb30) and aluminum (AlCuMgPb) with coated carbide cutting tool "SPUN120304". Cutting speed, feed rate and depth of cut were the input variables in the proposed models. It has been noticed that total specific energy (TSE) for both steel and aluminum decreases with increasing feed rate and depth of cut. The behaviour of total specific energy shows decreasing as cutting speed increases until a specific extent after that it begins in increasing with increasing speed. It is also obtained that surface roughness improves with increasing cutting speed and degrades with increasing feed rate and cutting depth. Ashvin J. Makadia [2] investigated the effect of cutting speed, feed rate, depth of cut and tool nose radius on the surface roughness in turning AISI 410 steel. A mathematical model was developed to predict surface roughness in terms of above parameters. Response Surface Methodology was used to find the optimum cutting parameters and study their effect on surface roughness. It has been seen that feed rate was the main factor that effects on the roughness, followed by the tool nose radius and cutting speed while depth of cut has no significant effect on the surface roughness. Murat Sarkaya [1] investigated the effect of cutting speed, feed rate, depth of cut and cooling condition on surface roughness in turning AISI 1050 steel by using CNC turning machine. The cooling condition involved dry cutting, conventional wet cooling and MQL. Taguchi design and response surface methodology were used to find optimal operating parameters and to create mathematical models for Ra and Rz. The results showed that feed rate is the most effective factor on the surface roughness and MQL is a good tool to improve surface roughness.

As there are many researchers studied the effect of cutting factors in turning, this work aims to investigate the influence of cutting parameters (spindle speed, feed rate and cutting depth) on both surface roughness and tool temperature and studies the behaviour of AISI 1045 steel under dry turning condition. The sustainable manufacture has been taken into account by identifying the optimal of these cutting parameters which achieve sustainability from economic perspective, Also, by specifying dry cutting condition sustainability can be achieved from economic, environmental and social aspects as stated in Figure 2.

Experimental Work

CNC turning machine of FANUC (Series oi Mate-TC) which is shown in Figure 3 was used to perform experiments, Table 1 shows machine specifications.

The choice of CNC machine is based on the high precision of the parts produced by it and as consequence, it reduces cost of machining and improves productivity. The used material in the present work was medium carbon steel of grade AISI 1045. Table 2 shows the chemical composition test that performed according to ASTM A751 standard [11]. Medium carbon steel of grade AISI 1045 has wide range applications because it machines readily. Typical applications of AISI 1045 steel are: various axles, bolts, connecting rods, hydraulic clamps and rams, various pins, various rolls, shafts, gears… etc. The specimens of carbon steel with length of 90 mm and diameter of 45 mm were cut by carbide cutting tool.

Five levels of cutting speed and three levels of both feed rate and depth of cut have been selected to conduct the experiments as shown in Table 3. As the turning process was conducted without cutting fluid, the generated heat was increasing especially at high speeds and form blue continuous chip which accumulated on tool. This type of chip which

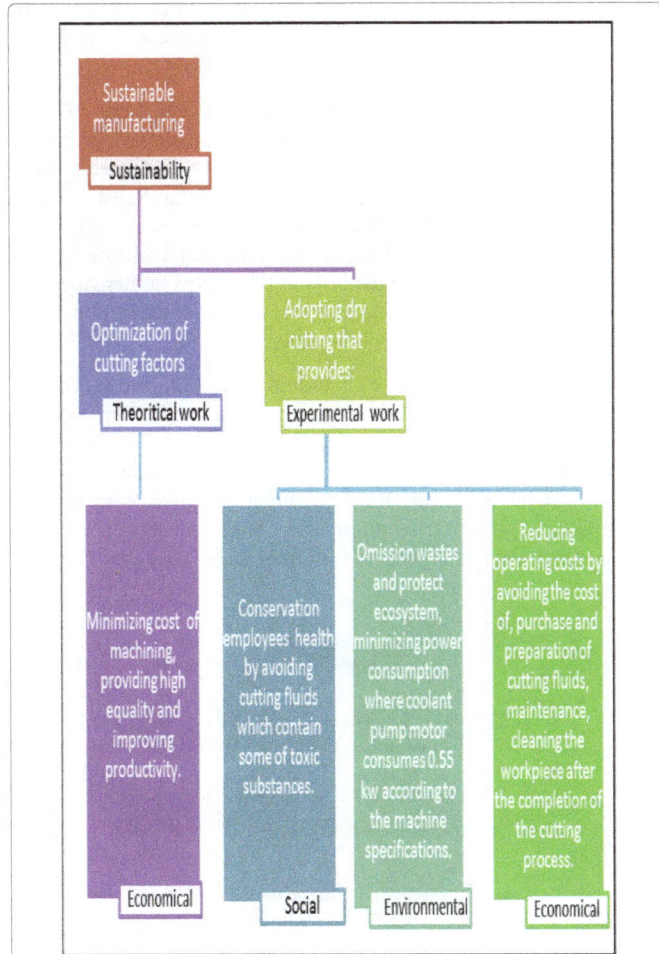

Figure 2: Dimensions of this study that lead to sustainable manufacturing system.

Figure 3: CNC turning machine of FANUC (Series Oi Mate-TC).

Element	Chemical composition Results %	ASTM results %[19]
Carbon	0.44	0.43-0.50
Manganese	0.52	0.60-0.90
Phosphorus	0.023	0.040 (max)
Sulfur	0.030	0.050(max)
Fe	Balance	Balance

Table 2: Chemical composition of AISI 1045 carbon steel in wt. % compared to ASTM.

Cutting parameters Units	Spindle speed (rev/min)	Feed rate (mm/min)	Depth of cut (mm)
Level-1	200	50	0.5
Level-2	300	100	1
Level-3	400	150	1.5
Level-4	500	—	—
Level-5	600	—	—

Table 3: Levels of cutting parameters that have been used in turning experiments.

Figure 4: Blue continuous chip which forms at high speeds.

Equipment	Unit	Range
Chuck diameter	mm	200
Max. cutting diameter	mm	200
Max. cutting length	mm	500
Spindle bore diameter	mm	55
Spindle rotating speed	rpm	45-4000
Main motor power (AC)	kW	11
Hydraulic unit power motor	kW	1.5
Chip conveyor motor	kW	0.2
Coolant pump motor	Kw	0.55
Weight	Kg	5000

Table 1: Specifications of FANUC Series oi Mate-TC machine.

shown in Figure 4 is dangerous and may be harmful for the worker and cutting tool so when this type was produced at any level of cutting parameters the higher levels of these parameters were not conducted.

Measurement of tool tip temperature was performed by infrared thermometer deviceof type UNI-Trend (UT303) as the operation of cutting progresses as shown in Figure 5 in order to control the temperature of cutting process. Three readings have been recorded during cutting the specified length of specimen and the maximum temperature has been chosen. Measurements of the surface roughness were taken in each of the mentioned levels in Table 3. The measurements were performed by a portable-type of surface roughness tester (Qualitest TR-110, US) as shown in Figure 6. Arithmetic average height (Ra µm) parameter was identified to measure the surface texture according to ISO 4287:1997 standard specifications. Five measurements have been recorded after each cutting process to minimize readings error. Calibration of tester device was done before each reading.

Optimization using genetic algorithm

Optimization is the act of obtaining the best result under given circumstances. In design, construction, and maintenance of any engineering system engineers need to specify the ultimate value of a particular function [13]. The goal of optimization is either minimization or maximization of an objective function within appropriate criteria. In this study, the Genetic Algorithm (GAs) has been used to find the

Figure 5: Measurement of tool tip temperature by Infrared Thermometer of type UNI-Trend (UT303) during cutting.

Figure 6: Measurement of the surface texture by surface roughness tester of type *Qualitest TR-110, US.*

optimum values of the specified cutting parameters. Genetic algorithms (GAs) are well suited for solving such problems; they can find the global optimum solution with a high probability [13].

Genetic Algorithms are considered good search tools to find the global maxima or minima of a practical application. The idea of genetic algorithm was inspired by biological system and Darwinian Theory of natural evolution depending on the principle "survival of the fittest". Genetic algorithm is a class of stochastic search and optimization methods where it works on three basic operators selection (reproduction), crossover (recombination) and mutation.

Genetic Algorithm initiates a random population of size N say P (t), and then the fitness function (objective function) of each individual in the population will be computed. After that if a convergence is achieved by a specific criterion the GA will stop and the best solution in P(t) will be chosen as an optimum solution, if the criteria has not met the operators (selection, crossover and mutation) are applied to form P(t+1) then the termination criterion is checked and so on. This process continues until a termination criterion has met. Figure 7 shows the cycle of the simple GA.

In engineering designs and problems that require finding optimal conditions there is usually more than one objective to be optimized. In single objective optimization (SOO), there is just one optimum solution while in (MOO) there is a set of optimal solution called *"Pareto optimal set"* [14]. Although multiple objectives optimization (MOO) are difficult but they are realistic problems, where the objectives are generally conflicting preventing simultaneous optimization of each objective. The Genetic Algorithm is considered one of the most popular metaheuristic approaches that are well suited for solving multi-

objective optimization problems, because that multi-objective Genetic Algorithm (MOGA) does not require user to prioritize, scale, or weight objectives [14].

Many methods were developed in order to solve multi-objective problems depending on the operators of the simple GA (selection, crossover and mutation) such as Schaffer's Vector Evaluated Genetic Algorithm (VEGA), Hajela and Lin's Weighting-based Genetic Algorithm, Fonseca and Fleming's Multi objective Genetic Algorithm (FFGA).Horn and Controlled Elitist Non-dominated Sorting Genetic Algorithms [15]. Additional operators were adopted in these methods to modify GA for implementation multi-objective optimization, to see advantages and disadvantages of these methods [16].

The aim of the current study is to find the optimum values of two objective functions in terms of their design variables. The first objective is the surface roughness of the workpiece and the second objective is the tip tool temperature. The design variables are spindle speed, feed rate and depth of cut. The optimization process has been performed by multi-objective optimizer *(gamultiobj)* in MATLAB R2010a software that optimizes the two objectives simultaneously. The *gamultiobj* uses

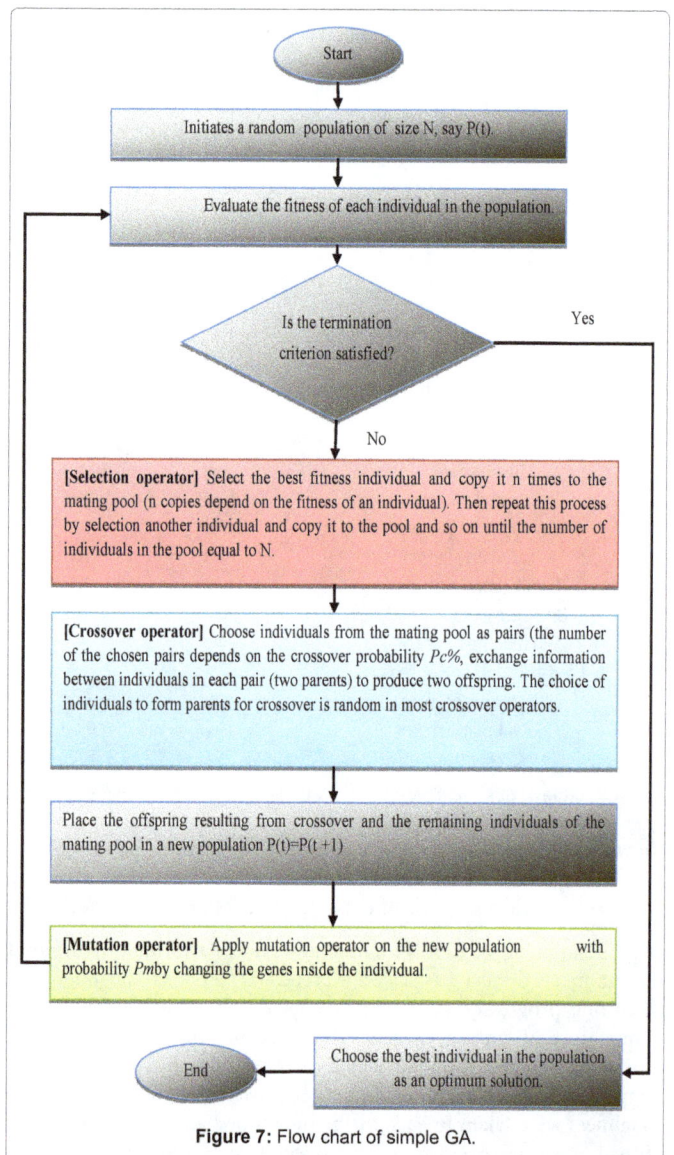

Figure 7: Flow chart of simple GA.

a *Controlled Elitist Genetic Algorithm* (a variant of *NSGA-II*) [15]. The options in multi-objective optimizer *(gamultiobj)* in MATLAB were specified as following:

1. Population Size=100.

2. Population Initial Range=[200 0.5 50 ; 600 1.5 150].

3. Lower Bound=[200; 0.5; 50], Upper Bound=[600; 1.5; 150].

4. Creation Function=Feasible population.

5. Selection Function=Tournament, Tournament size=10.

6. Crossover Function=Two point crossover, Crossover Fraction=0.7.

7. Mutation Function=Adaptive feasible.

8. Pareto Fraction=0.7.

9. Plot Function=Pareto front.

Results and Discussion

Temperature of cutting tool

In all machining processes most of the cutting energy converts to heat between tool, chip and work piece. Therefore, it is important to measure the temperature at the tip of cutting tool to control the process and avoid high temperatures of tool that may cause wearing and reduce tool hardness. By using infrared thermometer device, temperature of cutting tool was recorded at different cutting parameters. The multiple non-linear regression method has been used to predict a mathematical model for cutting tool temperature by using LAB-fit Curve Fitting software. Cutting tool temperature model is given by the equation (1) under effect of three variables spindle speed, feed rate and depth of cut with multiple correlation coefficient (R) equal to 98.83%, constants values of the equation are shown in Table 4. The predicted values of equation (1) and their errors are shown in Table 5. Figure 8 shows a

comparison between the predicted values of the tool temperature empirical model with the experimental values.

$$T = A + Bx_1^2 + Cx_2x_3^2 + D(x_1 - x_3) + x_1^{(E+Fx_2)} + G\frac{x_1}{x_2x_3} + H\sqrt{\frac{x_1}{x_3}} \quad (1)$$

The effect of spindle speed on tool temperature is shown in Figure 9. Increasing in spindle speed causes increasing in tool temperature and this can be explained as following, the total energy in cutting process can be partitioned into shear energy, friction energy, kinetic energy and surface energy. Both the kinetic energy and surface energy represent a small amount and they are usually neglected except kinetic energy that is take into consideration at high speeds (900 to 1200 m/min) [17]. The shear energy is the largest one where more than 75% of total energy is related to shear action and all this energy transforms into heat because of plastic deformation of material in shear zone. The friction energy represent the energy dissipated in sliding the chips on the tool face and as a result, heat will generates due to friction action, this energy is sensitive to the velocity of chips as it flows over the tool. So when spindle speed increases, the cutting speed increases which cause increasing in the shear and chip velocities so the shear and friction energies that will be dissipated in the form of heat will increase and as sequences the tool temperature will increase [17,18].

Figures 9 and 10 show that tool temperature increases with increasing feed rate and depth of cut respectively. The tool temperature will increase due to increase cutting forces where the amount of heat generated during turning increases directly with the cutting forces according to André Stefenon [19] and Harinath Gowd [20]. The cutting forces increase as feed rate and/or depth of cut increase due to increasing in the shear plane area as Awadhesh Pal [18] said.

Constants	Values
A	-5.54E+01
B	3.45E-04
C	8.95E-04
D	-2.11E-01
E	8.32E-01
F	-2.18E-01
G	-8.49E+00
H	6.20E+01

Table 4: Constants values of the equation.

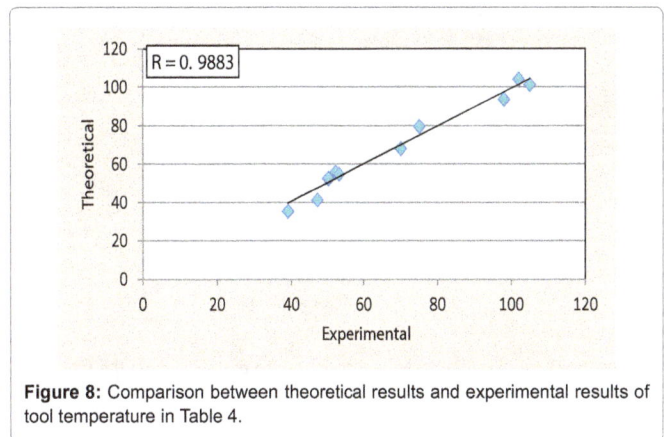

Figure 8: Comparison between theoretical results and experimental results of tool temperature in Table 4.

No.	Speed (rpm)	Depth (mm)	Feed (mm/min)	Tool temperature (°C) (Experimental)	Tool temperature (°C) (Theoretical)	Error %
1	300	0.5	50	39	35.64	9.41
2	500	0.5	50	50	52.44	4.66
3	400	1.5	50	75	79.72	5.92
4	500	1.5	50	105	101.46	3.48
5	300	0.5	100	52	56.03	7.2
6	500	0.5	100	98	93.85	4.41
7	200	1.5	100	47	41.47	13.33
8	300	1.5	100	53	54.9	3.49
9	200	0.5	150	50	52.88	5.44
10	300	1.5	150	70	68.22	2.6
11	500	1.5	150	102	104.34	2.24
-	-	-	-	-	-	Mean Error=5.65%

Table 5: Experimental and theoretical cutting tool temperatures at different cutting parameters.

In addition, an increase in the area of contact between tool and chip yield from increasing feed rate and depth of cut that lead to a higher frictional energy [21]. In addition, some of the generated heat dissipates in conduction to work piece material and tool holder according to their properties in conduction, when feed rate increases the cutting time will decrease which means less amount of heat will conduct to work piece material and cutting tool holder and so the temperature of the tool tip will increase.

From the proceeding discussion, the temperature increases with increasing each of spindle speed, feed rate and depth of cut, and as mentioned in chapter four that chip removes 60% from heat generated in cutting, the heat removed by chip will increase as cutting parameters increase. Because of dry cutting condition during turning process, the chip will exposed to atmosphere and due to high temperatures, chip oxidation will take place producing chip with blue colour surface [22], as shown in Figure 4. This type of chip formation leads chip to accumulate on the cutting tool and welding to the edge of tool. This state was noticed at reading 4 and 11 in Table 5, so in optimization it is desirable to avoid this unfavorable chip formation by choosing temperature below 100°C.

Surface roughness of work piece

Surface roughness is an important requirement in turning process due to its effect on mechanical properties of worpiece as mentioned before; therefore it has been depended as quality indictor in this study. Surface roughness data was collected during experiments by using surface roughness tester in turning AISI 1045 steel work piece at

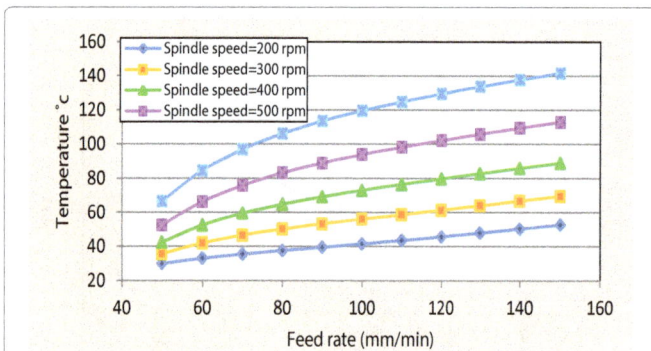

Figure 9: Effect of feed rate on tool temperature at depth of cut 0.5 mm and different cutting speed.

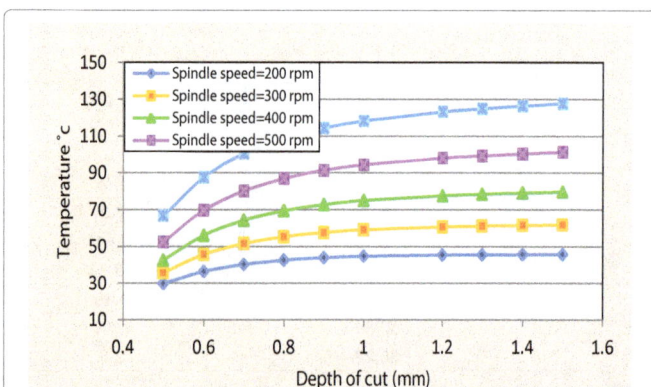

Figure 10: Effect of depth of cut on tool temperature at feed rate 50 mm/min and different cutting speed.

different cutting conditions. Multiple linear regression analysis was used to establish the mathematical model. By using LAB-fit Curve Fitting software, an empirical equation for predication surface roughness of the work piece was obtained. Ra model is given by equation (2) which describes the behaviour of surface roughness under the effect of spindle speed, depth of cut and feed rate with multiple correlation coefficient R equal to 98.22%, constants values of the equation are shown in Table 6. The predicted values of equation (2) and their errors are shown in Table 7. Figure 11 shows the comparison between the experimental values of surface roughness with the theoretical values obtained by empirical model.

$$SR = A + Bx_1 + Cx_2 + Dx_3 + \frac{E}{x_1^4} + F\,exp\left(\frac{1}{x_2}\right) + \frac{G}{x_3^2} + H\ln(x_1) + K\ln\left(\frac{x_1}{x_3}\right) \quad (2)$$

Figure 12 explains the behavior of surface roughness with spindle speed at different feed rates. It is obvious that when spindle speed increases surface roughness decreases. This can be attributed to two reasons:

The first reason is that ductile materials have high coefficient of friction, so when cut starts, some of the material because of high friction coefficient, built up ahead of the cutting edge, some of the material may even weld onto the tool point, and thus known as built up edge BUE. As the cutting proceeds, the chips flow over this edge and up along the face of the tool. Periodically a small amount of this BUE separates and may be embedded in the turned surface or leaves with the chip. Some of the tool material can be torn away with chips leading to wear in tool edge or weld into the edge and alters tool's geometry leading to inaccuracy in the required dimension. Because of these actions, BUE increases surface roughness. The coefficient of friction that related to BUE is velocity dependent. With increasing speeds, temperature will increase and the coefficient of friction decreases yielding lower friction,

Constants	Values
A	1.59E+01
B	2.19E-02
C	4.26E-01
D	-2.27E-02
E	-5.70E+09
F	2.01E-01
G	3.49E+04
H	3.43E-01
K	-1.74E+01

Table 6: Constants values of the equation.

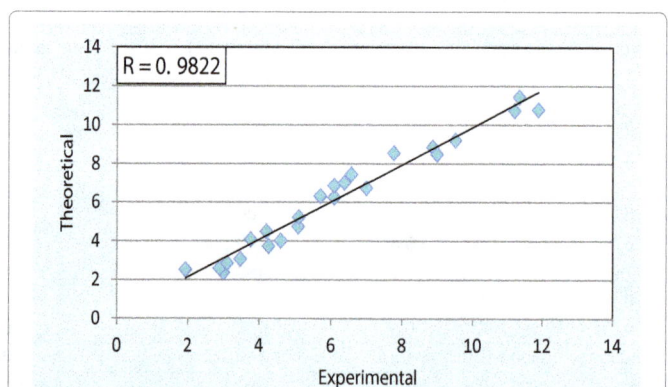

Figure 11: Comparison between theoretical results and experimental results of workpiece surface roughness.

No.	Speed (rpm)	Depth (mm)	Feed (mm/min)	Ra(μm) (Experimental)	Ra(μm) (Theoretical)	Error %
1	200	0.5	50	8.87	8.86	0.11
2	400	0.5	50	5.09	4.76	6.93
3	600	0.5	50	3	2.4	25
4	300	1	50	6.09	6.27	2.87
5	400	1	50	4.6	4.03	14.14
6	500	1	50	1.93	2.55	24.31
7	300	1.5	50	5.7	6.33	9.95
8	400	1.5	50	3.76	4.09	8.06
9	500	1.5	50	2.89	2.61	10.72
10	300	0.5	100	6.57	7.47	12.04
11	400	0.5	100	5.1	5.23	2.48
12	500	0.5	100	4.26	3.75	13.6
13	600	0.5	100	3.1	2.88	7.63
14	300	1	100	6.99	6.74	3.7
15	400	1	100	4.2	4.51	6.87
16	500	1.5	100	3.47	3.08	12.66
17	300	0.5	150	11.34	11.46	1.04
18	400	0.5	150	9.51	9.22	3.14
19	600	0.5	150	6.09	6.86	11.22
20	300	1	150	11.21	10.73	4.47
21	400	1	150	8.99	8.5	5.76
22	500	1	150	6.37	7.01	9.12
23	300	1.5	150	11.89	10.79	10.19
24	400	1.5	150	7.77	8.55	9.12
-	-	-	-	-	-	Mean Error=8.96%

Table 7: Experimental and theoretical workpiece surface roughness (SR) at different cutting parameters.

and that means the size of the BUE decreases as cutting speed increases and the surface finish is improved [4,17,23].

The second reason is that when cutting speed increases the shear angle will increase resulting in shorter shear plane and hence lower shear force, so lower cutting forces are required and this mean less vibration leading to produce smoother surface roughness [24].

Figure 12 shows the effect of feed rate on surface roughness. It can be seen that surface roughness increases with increasing feed rate. This behavior can be attributed to the increasing in cutting forces due to increase the amount of material in contact with the tool. These forces cause vibration leading to a higher surface roughness according to Murat Sarıkaya [1] and M. CemalCakir [25], also more heat is generated due to these forces, this heat contributes in tool wear and thus may produce high surface roughness.

When depth of cut increases the cutting forces increase because larger material has to be removed. As mentioned before that higher cutting forces lead to a higher vibration and thus rough surface is yield, but what has been observed in this study during experiments is that as depth of cut increases from 0.5 mm to 0.7 mm the surface roughness decreases as shown in Figure 13. Suresh [24] said that effective material removal might not have taken place at lower depth of cut, mainly due to predominant rubbing and plugging actionand hence higher surface roughness is obtained. After 0.7 mm the increasing in depth of cut has no significant effect on surface roughness.

Theempirical model of surface roughness was compared to surface roughness models of Young K. and Choon M [26]. In their research, they adopted MQL and wet conditions in turning AISI 1045 carbon steel as workpiece. Figure 14 shows the three models of surface roughness as dry of this work and MQL and wet of Young K. and Choon M. work at depth of cut 1mm and feed rate 0.2 mm/rev. The

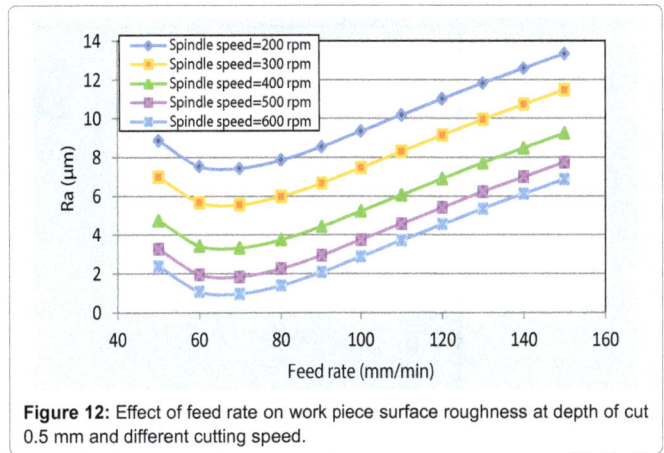

Figure 12: Effect of feed rate on work piece surface roughness at depth of cut 0.5 mm and different cutting speed.

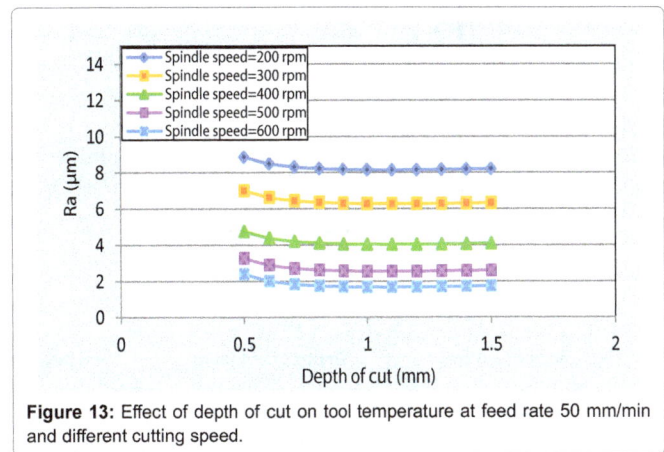

Figure 13: Effect of depth of cut on tool temperature at feed rate 50 mm/min and different cutting speed.

curves of dry, MQL and wet in Figure 14 show the same behaviour, where surface roughness decreases as cutting speed increases.

Optimization using genetic algorithm

Multi- objective optimization was performed by applying multi-objective optimizer *(gamultiobj)* of Genetic Algorithm package in MATLAB R2010a software. Two objectives were chosen, surface roughness of workpiece and tool temperature. The models of the two objectives were in terms of spindle speed, depth of cut and feed rate.

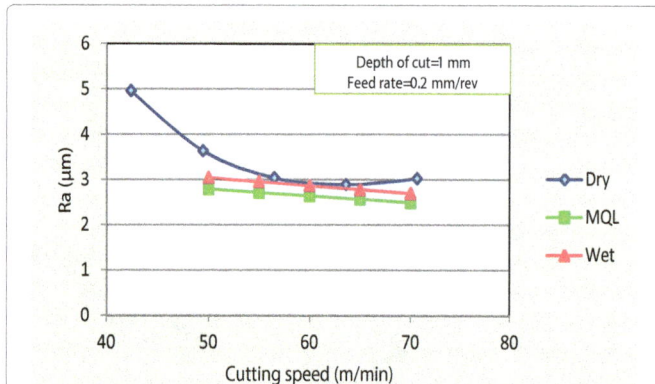

Figure 14: Comparison between the current work (dry condition) and Young K. et al. [26] MQL and Wet conditions for surface roughness models with cutting speed.

Figure 15 shows the result of optimization operation (Pareto optimal set). The optimal cutting parameters shown in Table 8 are chosen as compromise solution to have a low surface roughness at temperature below the values that causing continuous blue chip which is considered undesirable because of its harmful properties for machine operator and cutting tool. The optimum cutting parameters were fulfilled the sustainable manufacture conditions.

Conclusions

From the current study, the more important conclusions can be drawn as following:

1- Sustainable manufacturing can be obtained if dry cutting condition is adopted during machining which minimizes cost, safes operators' health and eliminates environment pollutions.

2- Finding the optimum of cutting parameters contributes in minimizing cost of machining by specifying the desired surface roughness for customer.

3- From the empirical model of cutting tool temperature it has been noticed that tool temperature increases with increasing cutting parameters (spindle speed, feed rate and depth of cut).

4- From the surface roughness empirical model to the workpiece it has been noticed that surface roughness decreases with increasing spindle speed and depth of cut while it increases with increasing feed rate.

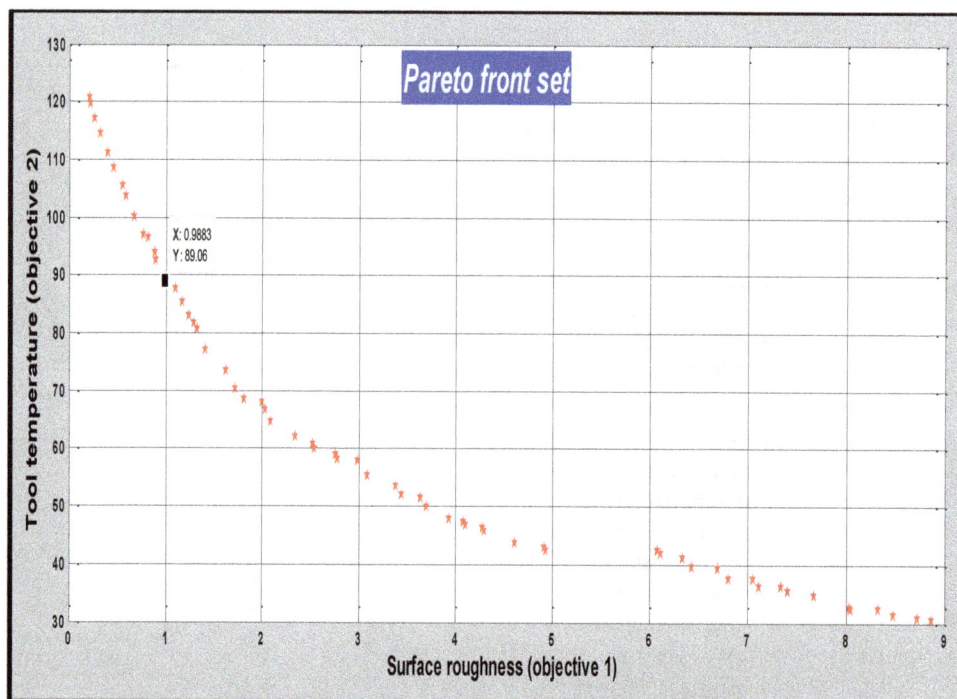

Figure 15: Pareto front set by applying GA in MATLAB software.

Spindle speed (rpm)	Depth of cut (mm)	Feed rate (mm/min)	Surface roughness Ra (µm)	Tool temperature T (°C)
588.96	0.5	64.55	0.98	89.05

Table 8: The optimum cutting parameters to have optimum surface roughness.

5- Genetic Algorithm is an active tool for optimization multi-objectives problems.

6- The best compromise optimum cutting parameters were spindle speed 588.96 rpm, depth of cut 0.50 mm and feed rate 64.55 mm/min that gave surface roughness (Ra) 0.98 μm and tool temperature 89.05°C.

References

1. Sarıkaya M, Güllü A (2014) Taguchi design and response surface methodology based analysis of machining parameters in CNC turning under MQL. J Cleaner Prod 65: 604-616.

2. Makadia AJ, Nanavati JI (2013) Optimisation of machining parameters for turning operations based on response surface methodology. Measurement 46:1521-1529.

3. Kumar NS, Shetty A, Shetty A, Ananth K, Shetty H (2012) Effect of spindle speed and feed rate on surface roughness of Carbon Steels in CNC turning. Procedia Engineering 38: 691-697.

4. Amsted BH, Phillip FO, Myron LB (1987) Manufacturing Processes. 7th ed. John Wiley &Sons, USA.

5. Janez K, Franci P (2009) Concepts of Sustainable Machining Processes. 13th International Research/Expert Conference, Trends in the Development of Machinery and Associated Technology, Tunisia.

6. Rosen MA, Kishawy HA (2012) Sustainable Manufacturing and Design: Concepts, Practices and Needs. Sustainability 4: 154-174.

7. Hemant S, Mahendra S (2014) Influence of Cutting Fluids on Quality and Productivity of Products in Manufacturing Industries. Int J Eng Manag Sci 1: 8-11.

8. Sahin Y, Motorcu AR (2005) Surface roughness model for machining mild steel with coated carbide tool. Mat & Des 26: 321-326.

9. Gökkaya H, Nalbant M (2007) The effects of cutting tool coating on the surface roughness of AISI 1015 steel depending on cutting parameters. Tur J Eng Environ Sci 30: 307-316.

10. Guo Y, Loenders J, Duflou J, Lauwers B (2012) Optimization of energy consumption and surface quality in finish turning. Procedia CIRP 1: 512-517.

11. ASTM International (1998) Specification for Test Methods, Practices, and Terminology F or Chemical Analysis of Steel Products. ASTM SA-751.

12. ASTM International (1998) Specification for Steel Bars, Carbon and Alloy, Hot-Wrought and Cold- Finished. ASTM SA-29/SA-29M.

13. Rao SS (1996) Engineering Optimization. 4th ed., John Wiley & Sons, USA.

14. Long Q (2015) A Genetic Algorithm for Unconstrained Multi-Objective Optimization. Swarm and Evolutionary Computation 22: 1-14.

15. Deb K, Tushar G (2001) Controlled Elitist Non-dominated Sorting Genetic Algorithms for Better Convergence. Springer-Verlag Berlin Heidelberg, USA.

16. Abdullah K, Coitb DW, Smithc AE(2006) Multi-Objective Optimization Using Genetic Algorithms: A Tutorial. Reliability Engineering and System Safety 91: 992-1007.

17. ASM Metals Handbook (1992) Machining processing 16.

18. Awadhesh P, Choudhury SK, Chinchanikar S (2014) Machinability Assessment through Experimental Investigation during Hard and Soft Turning of Hardened Steel. Procedia Materials Science 6:80-91.

19. André S (2015) A Qualitative Analysis of Cutting Parameters Influence in Temperature of Stainless Steel AISI 420C Turning. 23rd ABCM International Congress of Mechanical Engineering, Brazil.

20. Hraninath G, Vali SS, Ajay V, Mahesh GG (2014) Experimental Investigations and Effects of Cutting Variables on MRR and Tool Wear For AISI S2 Tool Steel. Procedia Materials Science 5: 1398-1407.

21. Abhang LB, Hameedullah M (2010) Chip-Tool Interface Temperature Prediction Model for Turning Process. Int J Eng Sci & Tech 2: 382-393.

22. Siow PC, Ghani JA, Tan SCV (2011) The Wear Progression of TIN Coated Carbide Insert in Turning FCD 700 Cast Iron. Regional Tribology Conference, Malaysia.

23. Debnath S, Reddy MM, Yi QS (2016) Influence of cutting fluid conditions and cutting parameters on surface roughness and tool wear in turning process using Taguchi method. Measurement 78: 111-119.

24. Suresh R, Basavarajappa S, Gaitonde VN, Samuel GL (2012) Machinability Investigations on Hardened AISI 4340 Steel Using Coated Carbide Insert. Int J Ref Metals & Hard Mat 33: 75-86.

25. CemalCakir M, Ensarioglu C, Demiraya I (2009) Mathematical Modeling of Surface Roughness For Evaluating The Effects of Cutting Parameters and Coating Material. J Mat Proc Technology 209.

26. Young K. and Choon M. (2010) Surface Roughness and Cutting Force Prediction in MQL and Wet Turning Process of AISI 1045 Using Design of Experiments. J Mech Sci Tech 24: 1669-1677.

25

Synthesis of Alumina Fibre by Annealing Method using Coir Fibre

Khan A[1]*, Ahmad MA[2], Joshi S[1] and Lyashenko V[3]

[1]*Department of Physics, Motilal Vighyan Mahavidhyalaya, Bhopal, India*
[2]*Department of Physics, Faculty of Science, University of Tabuk, Saudi Arabia*
[3]*Laboratory "Transfer of Information Technologies in the Risk Reduction Systems", Kharkov National University of Radioelectronics, Kharkov, Ukraine*

Abstract

The present study report the synthesis of alumina fibre by annealing method using coir fibre. In this study chemical treatment of coir fibre has been done using a nitro compounds at higher temperature. The composite formed after the chemical treatment has been characterized by Xrd, SEM and FTIR technique. Successful synthesis of alumina fibre has been confirmed after the chemical treatment of coir fibre.

Keywords: Coir; Annealing; Nitro compound; Alumina fibre

Introduction

Currently scientist are showing ardent attention in synthesizing composites from recycled waste specially by means of the natural environment-friendly fibres as reinforcing fillers and thermosetting polymers as matrix [1]. Natural fibres play a vital role as fillers in the fabrication of composites. The use of these fibres is beneficial not only as biodegradable substance but also it affects the cost of composite and also boost life of the composites. Natural fibres are being used as a reinforcing agent in place of synthetic fillers such as glass, aramid, talc, and silica; as they are having low density and they do not leave any by-product at the time of manufacture of composites as they are bio-degradable in nature. Also, processing of these fibres can be done at low temperature and thus may play an essential role in energy consumption for the production of composites [2-9]. Natural fibres has been used as reinforcing agent in polymer composites and becomes the point of attention and attraction among several researchers during last many years owing to their easy accessibility and biodegradable nature [3,4].

As natural fibres are having low density so they can be used, easily available, renewable and biodegradable in nature. Natural fibres (NFs) have provided raw materials to meet the human requirements of fibres in their life. The first utilization of natural fibre composite (NFC), made with clay in Egypt, can be dated back to 3000 years ago. With the high-tech developments of man-made fibres, NF lost much of its interest and many of the ancient natural fibres are no longer in use. However, as a result of a growing awareness of the interconnectivity of global environmental factors, the principles of sustainability, industrial ecology, eco-efficiency, and green chemistry and engineering are being integrated into the development of the next generation of materials, products, and processes. Though their use as a reinforcing agent is strongly reduced because of their hydrophobic nature, their tendency to amass at the time of processing and their poor resistance to moisture [10]. Chemical treatments has been done for treating natural fibres as the bonding between the fibre and matrix can be improved by physical and chemical modification of fibre surface. A study has been done to show the effect of chemical treatment on the structure and morphology of coir fibre [5]. The present study shows some important effects of chemical treatment on the structure and morphology of coir fibre. The objective of the present study is to optimise overall properties of coir fibre so as to use coir fibre as a reinforcing agent in thermoplastic and thermosetting polymers. In the present study, coir fibre is treated with ferric nitrate salt. A thermal treatment has been done at temperature of 1000°C by using annealing method. X-ray diffraction of the treated coir fibre reveals the crystalline nature of the fibre. Change in morphology

has been found in coir fibre when subjected to scanning electron microscopy. Finally, the Fourier transform and infrared spectrographs show the presence of traces of iron oxide:fibre in the prepared composite.The present study deals with the chemical treatment of coir fibre at higher temperature by nitro compounds and the synthesis of alumina fibre after the treatment of coir fibre has been done using coir fibre.

Materials and Method

In the present study chemical treatment of coir fibre has been done with aluminum nitrate salt in order to find the change in the structure and morphology of fibre embedded composites.

Processing of coir fibre

The fibre used for the experimental studies was kept dipped in water for 24 hours and then washed so as to remove impurities like dust etc. On drying in direct sunlight chemical treatment as under was given to it [11].

Chemical treatment of coir fibre

Chemical treatment done by aluminum nitrate salt: Aluminium nitrate [Nonahydrate, Extra pure, $Al(NO_3)_3.9H_2O$] and ammonium chloride (NH_4Cl) salts were used for the chemical treatment of the fibre. The coir fibre used was collected from the Temples of Bhopal city.

Process - In 500 ml of distilled water dissolved aluminum nitrate salt and ammonium chloride salt in the proportion of 10:4 respectively. 100 g of fibre was kept in this solution and then 100 drops of ammonia was poured to it and left it for one hour. After one hour the mixture was dried for 48 hours at room temperature and then annealed the mixture in a muffle furnace at 1000°C. The fired mixture was kept for 15 minutes at that temperature [11].

The scanning electron microscopy of the untreated coir fibre and

---

*Corresponding author: Khan A, Department of Physics, Motilal Vighyan Mahavidhyalaya, Bhopal, India, E-mail: khanalveeera@gmail.com

chemically treated coir fibre was carried out by JSM 6390A (JEOL Japan) at different magnifications. Gold coating was done on the samples by gold coating unit before taking it for examination. Figures 1a-1b show the pictorial view of crushed raw coir fibre and chemically treated coir fibre. One can notice from the figures that there is change in the morphology of coir fibre embedded composites by chemical treatment done on it. The Scanning Electron pictographs of crushed to powder form of raw coir fibre shows needle shaped structure at small intervals. It is clear that after chemical treatment of coir fibre done by aluminum nitrate salt the composite material appeared to be in fibrous form and the composite is said to be synthesized alumina fibre as the presence of alumina has been confirmed by Xrd of the composite using coir fibre as template. Similar work has been done by Khan et al., [12] in which synthesis of gamma alumina fibre has been done by bio-replica technique using sisal fibre as template.

X-Ray Diffraction Analysis

The XRD measurements were done with the help of Bruker D8 Advance X-ray diffractometer. The X-rays used had wavelength of 0.154 nm (Cu-Kα), produced in a sealed steel tube.

The diffracted X-rays were detected by a fast counting detector based on silicon strip technology (Bruker Lynx Eye Detector). Measurements of diffraction parameters of the samples were carried out in the range from 5° to 70° (of 2θ) at a scanning rate of 1°/min from Figure 2. it is clear that after the chemical treatment of coir fibre by aluminium nitrate salt, the XRD peaks were obtained at 2θ values of 10.60°, 20.06°, 25.44°, 32.82°, 37.00°, 39.48°, 45.72°, 61.08° and at 67.04°. These indicated the extent of crystallinity of the composite. The XRD peaks resembled with the peaks obtained for alumina. Hence we can expect the presence of Al_2O_3 in the composites. The peaks corrosponds to 20.06°, 37.00°,45.72° and 67.04° resembled the peaks of γ-alumina as Sivadasan et al., [13] reported similar peaks of γ-alumina in their work of synthesis of γ-alumina. They inferred from the studies done by microwave assisted hydrolysis of aluminum metal. While the peaks at 39.48° resembled the peak of α-alumina as is reported by Hong et al., [14] in their work in which they prepared the precursor sol of alumina by sol gel method by taking aluminum nitrate and malic acid as raw materials [14]. Similar peaks of α-alumina and γ-alumina found by Guerro et al., [12] fired at 1000°C in their work in which synthesis of γ-alumina fibre has been done using sisal fibre as template [12]. Similar results have been found by other scholars too [15,16]. As the composite material consists of phases of α-alumina and γ-alumina so the resultant composite may be considered as alumina [12-14].

The XRD peaks of composites with raw coir are shown in Figure 3. It shows a peak at 12.9° while after chemical treatment of coir fibre by aluminium nitrate, there is a shift in the peak of coir fibre which is seen at 10.66° after the treatment. Thus, it is obvious from the above discussion that the cystallinity of composites made from coir fibre are affected on chemical treatment. It may thus be said that presence of coir fibre along with the synthesis of alumina affects the crystallanity

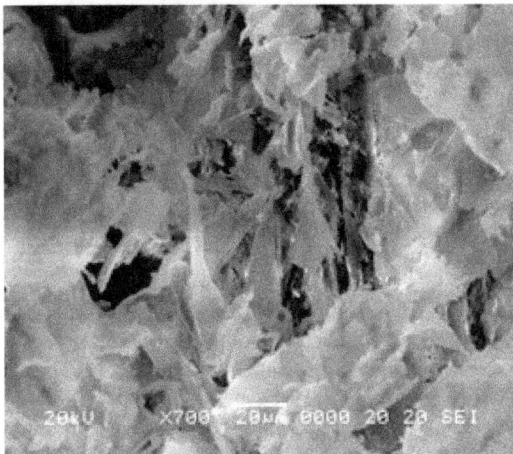

Figure 1a: Crushed Raw Coir Fibre.

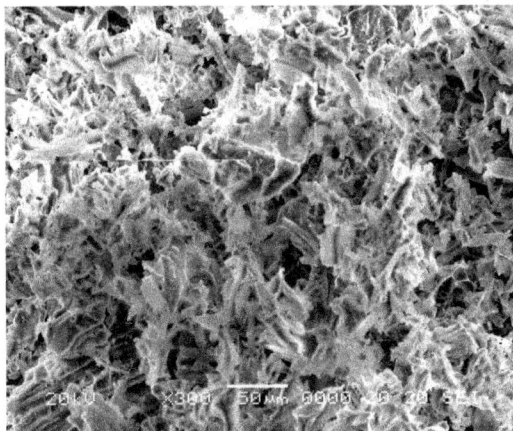

Figure 2: XRD pattern of Alumina (chemical treatment done by aluminum salt).

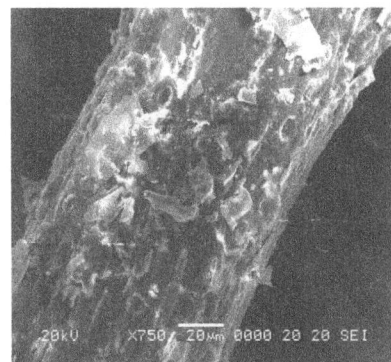

Figure 1b: Synthesized Alumina Fibre.

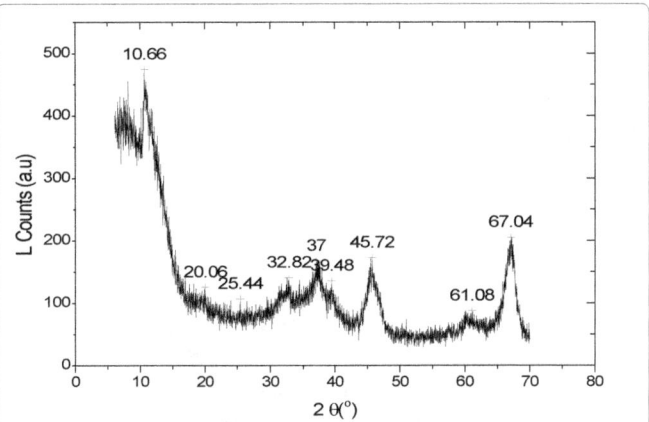

Figure 3: XRD of Raw Coir Fibre.

of the final composites. The change in the *d* value in composites with coir fibre after chemical treatment may be due to the carbonization of coir fibre.

FT-IR analysis was done for our samples. The results are shown in Figure 4. The 3446 cm^{-1} broad intense peak, in raw coir fibre included sample is due to the O-H stretching for hydrogen-bonded hydroxyl group present in polysaccharides. The broad, strong band emanates from cellulose, hemi-celluloses and lignin present in coir fibre. The weak intensity peak occurring at 1386 cm^{-1} in raw coir fibre used sample may be due to the presence of hemi-celluloses. It can be assign to the group of C = O stretching bonds. However, the broad intense peak at 1049 cm^{-1} is the absorption peak of (C-OH) bonds. The peak at 2928 cm^{-1} refers to alkyl C-H group. The peak at 2359 cm^{-1} refers to the presence of water. The peak at 1633 cm^{-1} in composites with raw coir is due to the C = C aromatic skeletal ring-vibration, which is most likely due to the presence of lignin[17-19].

FT-IR response is as shown in Figure 5. The FT-IR study reveals that the peak 3446 cm^{-1} shifts to 3517 cm^{-1} and becomes narrower. It suggests that after the chemical treatment to the fibre, the peak gets narrower due to the reduction of O-H bond, and is shifted due to partial removal of lignin, cellulose and hemi-cellulose. Similar results have been reported by one of the authors of our laboratory [18]. After the chemical treatment the peak at 1633 cm^{-1} in composite formed from raw coir fibre, shifts to 1636 cm^{-1}. Hence chemical treatment on fibre can be assumed to be responsible to this kind shift due to structural change.

The other cause can be removal of lignin on chemical treatment.

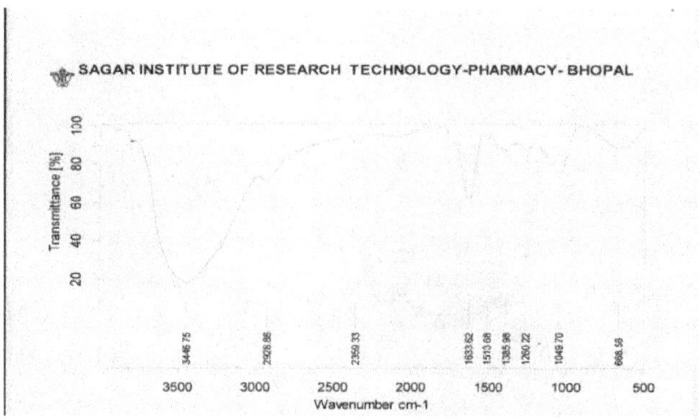

Figure 4: FT-IR of Raw Coir.

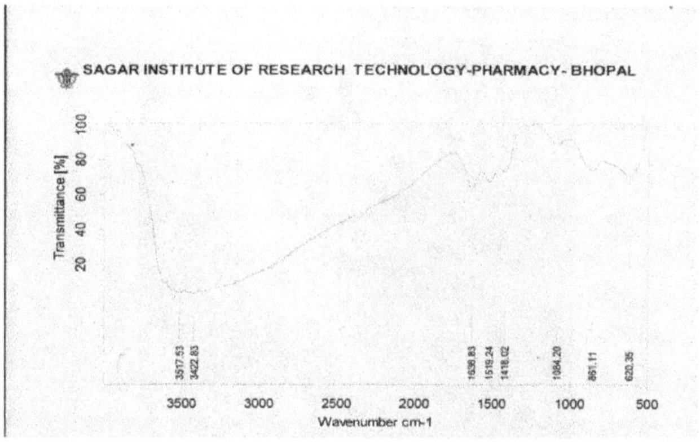

Figure 5: FT-IR of Chemically treated coir fibre.

However, the broad intense peak at 1049 cm^{-1} shifts to 1084 cm^{-1} and becomes narrower due to absorption of (C-OH) bond. However, the peak at 1418 cm^{-1} refers to Al-OH bonding mode [15,20,21], while the peak at 620 cm^{-1} and at 861 cm^{-1}, suggests the presence of stretching mode of Al-O-Al [21,22].

From the above study it is clear that there can be a trace of coir fibre in the sample (may be in oxidized form), which might be carbonized coir fibre, that causes succinct removal of impurities present in the fibre. However the peaks at 1418 cm^{-1}, 620 cm^{-1} and at 861 cm^{-1} suggest probable presence of Al$_2$O$_3$ in the sample. So, it can be inferred that there has been successful synthesis of Alumina, Al$_2$O$_3$ fibre by annealing method.

Results and Discussions

Present study was carried out with a primary objective of undertaking chemical treatment to coir fibre, to use a waste product by nitro-composites in order to synthesize metal-oxides and to convert a non-degradable waste product into useful product for the welfare of the society. It was specifically done in order to improve the structure and morphology of coir fibre composites. It, in future, can be used as a good replacement of synthetic fibre in polymers. Some more details are given in our recent publications [23-28]. The important findings of the present investigation are as follows:

Satisfactory modification in the morphology of the coir fibre was noticed. The newly formed composites results in the form of alumina fibre may be used in polymers as fillers in place of synthetic fibres so as to monitor the properties of polymers to our advantage. Crystalline nature of the newly synthesized composites by chemically treated coir fibre may help in synthesizing of different composite by different permutations in further studies.

Conclusions

Chemical treatment prior to composite synthesis is an effective method which is very likely to change the morphology of coir fibre. The treatment done by nitro compounds improves the surface coarseness of the surface of coir fibre in contrast to untreated coir fibre composites. It is clear that after chemical treatment of coir fibre done by aluminum nitrate salt the composite material appeared to be in fibrous form while powdered form of raw coir fibre shows needle shaped structure at small intervals

The XRD studies reveal the crystallinity of the resulting composites. The composite formed by the treatment of aluminium nitrate salt is crystalline in nature. Similar peaks of α-alumina and γ-gamma alumina was found in the composite material so the resultant composite may be said to be alumina. So, it can be concluded that when coir fibre is treated with aluminum nitrate, alumina fibre is found after the treatment.

FT-IR study done on chemically treated coir fibre (treatment was done by aluminum nitrate) showed traces of coir fibre which may be in the form of carbonized coir fibre. There are various samples in which there was progressive removal of impurities with each successive treatment on the fibre. However, the peaks obtained at 1418 cm^{-1}, 620 cm^{-1} and 861 cm^{-1} suggest the presence of Al$_2$O$_3$ in the sample. As a result it can be inferred that the synthesis of alumina (Al$_2$O$_3$) was successful by annealing.

References

1. Son J, Kim HJ, Lee PW (2001) Role of paper sludge particle size and extrusion temperature on performance of paper sludge–thermoplastic polymer composites. J Appl Polym Sci 82: 2709-2718.

2. Choi NW, Mori I, Ohama Y (2006) Development of rice husks–plastics composites for building materials. Waste Manage 26: 189-194.

3. Zeng Z, Ren W, Xu C, Lu W, Zhang Y, Zhang Y (2010) Maleated natural rubber prepared through mechanochemistry and its coupling effects on natural rubber/cotton fiber composites. J Polym Res-Taiwan 17: 213-219.

4. Thwe MM, Liao K. (2002) Effects of environmental aging on the mechanical properties of bamboo–glass fiber reinforced polymer matrix hybrid composites. Compos Part A-Appl S 33: 43-52.

5. Yang HS, Kim HJ, Son J, Park HJ, Lee BJ, et al. (2004) Rice-husk flour filled polypropylene composites; mechanical and morphological study. Compos Struct 63: 305-312.

6. Rana AK, Mandal A, Bandyopadhyay S (2003) Short jute fiber reinforced polypropylene composites: effect of compatibiliser, impact modifier and fiber loading. Compos Sci Technol 63: 801-806.

7. Karmarkar A, Chauhan SS, Modak JM, Chanda M (2007) Mechanical properties of wood–fiber reinforced polypropylene composites: Effect of a novel compatibilizer with isocyanate functional group. Compos Part A-Appl S 38: 227-233.

8. Haque MM, Rahman R, Islam MN, Huque MM, Hasan M (2009) Mechanical properties of polypropylene composites reinforced with chemically treated coir and abaca fiber. J Reinf Plast Comp

9. Rimdusit S, Wongsongyot S, Jittarom S, Suwanmala P, Tiptipakorn S (2011) Effects of gamma irradiation with and without compatibilizer on the mechanical properties of polypropylene/wood flour composites. J Polym Res 18: 801-809.

10. Choudhury A, Kumar S, Adhikari B (2007) Recycled milk pouch and virgin low-density polyethylene/linear low-density polyethylene based coir fiber composites. J Appl Polym Sci 106: 775-785.

11. Khan A, Joshi S (2014) Mechanical and Morphological Study of Coir Fiber Treated with Different Nitro Compounds. International Journal of Advancement in Electronics and Computer Engineering 2: 276-279.

12. Alveera K, Mohammad A, Shirish J, Lyashenko V (2015) Dielectric and Electrical Characterization Study of Synthesized Alumina Fibre Reinforced Epoxy Composites. Elixir Crystal 87: 35801-35805.

13. Sivadasan AK, Selvam IP, Potty SN (2010) Microwave assisted hydrolysis of aluminium metal and preparation of high surface area γ Al2O3 powder. B Mater Sci 33: 737-740.

14. Hong-bin T, Congsheng G (2011) Malic acid sol-gel method of long alumina fibers. Transactions of Non Ferrous Metals Society of China 21: 1563-1567.

15. Sedaghat A, Taheri-Nassaj E, Naghizadeh R (2006) An alumina mat with a nano microstructure prepared by centrifugal spinning method. J Non-Cryst Solids 352: 2818-2828.

16. Hosseini SA, Niaei A, Salari D (2011) Production of γ-Al_2O_3 from Kaolin. Open Journal of Physical Chemistry 1: 23-27.

17. Choudhury R (2012) Fabrication and characterization of raw and dewaxed coir fiber reinforced polymer composites. Doctoral dissertation, National Institute of Technology Rourkela.

18. Samal N(2012) Fabrication and characterization of acetone treated natural fibre reinforced polymer composites. Department of physics, National Institute of Technology.

19. Zuraida A, Norshahida S, Sopyan I, Zahurin H (2011) Effect of fiber length variation on coir fiber reinforced cement–albumen composites. IIUM Engineering Journal 12: 63-75.

20. Chandradass J, Balasubramanian M (2006) Sol–gel processing of alumina fibres. Journal of Materials Processing Technology 173: 275-280

21. Tan H, Ma X, Fu M (2013) Preparation of continuous alumina gel fibres by aqueous sol–gel process. B Mater Sci 36: 153-156.

22. Padmaja P, Anilkumar GM, Mukundan P, Aruldhas G, Warrier KGK (2001) Characterisation of stoichiometric sol–gel mullite by fourier transform infrared spectroscopy. Int J Inorg Mater 3: 693-698.

23. Alveera Khan M, Ahmad A, Shrish J, Abd. El-Khalek AM (2013) Study of Mechanical and Electrical Behavior of Chemically Treated Coir Fibre Reinforced Epoxy Composites. Int J of Multidisciplinary Research & Advcs in Engg 5: 171-180.

24. Khan A, Ahmad MA, Joshi S, Al Said SA (2014) Abrasive wear behavior of chemically treated coir fibre filled epoxy polymer composites. Amer J of Mech Eng and Auto 1: 1-5.

25. Ahmad MA, Lyashenko VV, Lyubchenko VA, Khan A, Kobylin OA (2016) The Methodology of Image Processing in the Study of the Properties of Fiber as a Reinforcing Agent in Polymer Compositions. Int J of Adv Res in Computer Science 7: 15-18.

26. Khan A, Ahmad MA, Joshi S (2015) A systematic study for electrical properties of chemically treated coir fiber reinforced epoxy composites with ANN model. Int J of Sci and Res 4: 410-414.

27. Alveera KS, Vyacheslav JL, Nicolina P, Ayaz Ahmad M(2015) Artificial Neural Networking (ANN) Treatment on Electrical Properties of Coir Fiber Reinforced Epoxy Composites. Saudi International Meeting on Frontiers of Physics 17-19.

28. Agag T, Takeichi T (2003) Synthesis and characterization of novel benzoxazine monomers containing allyl groups and their high performance thermosets. Macromolecules 36: 6010-6017.

26

Melt Infiltration Casting of Alumina Silicon Carbide and Boron Carbide Reinforced Aluminum Matrix Composites

Ali Kalkanlı[1]*, Tayfun Durmaz[1], Ayşe Kalemtaş[2] and Gursoy Arslan[3]

[1]*Department of Metallurgical and Materials Engineering, Middle East Technical University, Ankara, Turkey*
[2]*Department of Metallurgical and Materials Engineering, Bursa Technical University, Turkey*
[3]*Materials Science and Engineering, Anadolu University, Eskişehir, Turkey*

Abstract

This paper discuss the effect of processing details such as particle size, sintering temperature, preform preparation, aluminum alloy characteristics and melt temperature on the final mechanical properties of ceramic phase reinforced metal matrix composites. Since alloy composition was determined as 7075 and 7085 optimum solutionizing and ageing temperatures were studied to determine maximum hardness values. For only 7085 alloy best solutionizing temperature is 465°C and for 7075 alloy the maximum hardness achived as 178 BHN after heat treatment at 475°C. Alloys were heat treated for recyctallization after hot rolling grain size were measured as 100-120 µm for 7085 alloy matrix.

Various sintering temperatures were used for preform preparation such as 1300-1450°C. In 85% Al_2O_3 reinforced 7085 Alloy based MMCs preforms sintered at 1450°C high hardness values were achieved as 545 BHN. Intermetallic phase was determined in 7075 and 7085 alloys selected as alloy matrix. Al_2Cu intermetallic pecipitate (θ phase) was determined as dominant second phase after T6 heat treatment but highly expected phase in 7000 series alloys $MgZn_2$ (η phase) was not determined by XRD and SEM analysis techniques due to ultrafine precipitate size and homogeneous distribution.

Keywords: Composite materials; Armor materials; Ductility

Introduction

Development of metal ceramic composite materials with possible lowest density and higher energy absorbing capacity is highly important issue for defense industry. So, many criteria's should be considered when selecting of materials that are used in armor system. The impact resistance of shield materials against projectile is strongly required to be determined.

Scientific research on metal matrix composites and mechanical characterization by three point bending, hardness and impact toughness measurements are typical tests for performance characterization apart from ballistic tests.

Boron carbide is known with its extreme hardness of 30 GPa. It is the hardest third material after diamond and cubic-BN which are very expensive and hard to prepare. At temperatures above 1200°C its hardness value even exceeds that of diamond [1]. Combination of high hardness and low density makes boron carbide top candidate of armor materials. However low strength, high price, poor fracture toughness, sinterability and machinability of boron carbide limits its industrial applications [2]. Also, boron carbide doesn't provide efficient protection to stop armor piercing bullet with high velocity due to amorphisation process that occurs in boron carbide in the presence of high pressure. At this pressure, ballistic performance of boron carbide drops because shear strength of boron carbide decreases. Another problem with boron carbide is its brittleness which makes them not suitable for multi-hit protection [3]. The popularity of silicon carbide for armor technologies has increased due to its improved cost/performance ratio relative to other candidate materials like alumina [4]. Silicon carbide is produced in larger scales because it has many fields of application areas compared with boron carbide. The price of silicon carbide is lower than that of boron carbide and ballistic performance of silicon carbide is

very close to that of boron carbide. Thus many researchers have worked for production of armor system with as low boron carbide content as possible [5]. Silicon carbide is often mixed with boron carbide, but monolithic silicon carbide is also used in production of armor materials. Density of silicon carbide is 3.21 g/cm³ which are between that of boron carbide and alumina. Hardness of silicon carbide is very close to that of boron carbide. Thus for higher level ballistic threats silicon carbide is better alternative than boron carbide in spite of its higher density [3].

Although ceramic armors are used currently, low fracture toughness and high expense of them limits their widespread use. They do not provide efficient protection against multi-hit in a short space of time. After the first hit, armor system is expected to heavily damage and then any bullet would penetrate armor system which have already fractured [6]. Combining ceramic materials with a metal may provide better ballistic efficiency. So creating of porous ceramic preform, then infiltrating it with a ductile metal is considered as solution. Aluminum is most widely used infiltrated metal because of its low density, low melting point and outstanding ductility. It is also non-toxic, relatively inexpensive and easy to obtain. Molten aluminum reacts with boron

*Corresponding author: Ali Kalkanli, Department of Metallurgical and Materials Engineering, Middle East Technical University, Ankara, Turkey
E-mail: kalkanli@metu.edu.tr

carbide easily, thus infiltration process may be achieved. Resulted B_4C-SiC-Al_2O_3-Al composites exhibit combination of high hardness and high toughness without defeating the aim of obtaining lightweight structure [7]. It is very hard to obtain 100% dense B_4C-SiC-Al_2O_3 composites due to presence of strong covalent bonds, high resistance to grain boundary sliding and absence of plasticity which limit their diffusion coefficients in sintering process. Combination of high temperatures and high pressures used in sintering process is the most important economic problem besides the high cost of powders [7]. Compared with traditional sintering techniques, melt infiltration is a promising process to produce composite with porous ceramic preforms due to its several advantages. Near or near-net shape composites with high volume fraction ceramics can be obtained. And these ceramic phases can be uniformly distributed in composite structure. Resulted composite exhibits high dimensional stability [8]. Residual stress build up due to different thermal expansion of dissimilar materials can be eliminated, thus residual porosities are prevented. Mechanical properties of composite can be arranged via addition of appropriate compounds or elements [9]. Using high strength aluminum alloy such as 7075 as matrix alloy with addition of %10 SiC displays high flexure strength values 600 MPa reported [9]. Reaction products and reaction rate between ceramic and metal phases can be controlled. The most important advantage of pressureless melt infiltration is that there is no need to use high temperatures and high pressures (when wetting condition is provided there is no need to apply pressure-pressureless melts infiltration) that makes this process very economic. The most important criteria of melt infiltration process are the wetting behavior of the system. Wettability is the ability of a liquid to spread on a solid surface and it demonstrates the extent of close contact between a liquid and a solid [5]. The driving force for wetting is the reduction in free energy of the system. Wetting of the ceramic phase by the metal must be achieved for infiltration process because in the absence of wetting there is no interfacial reaction between ceramic and metal phases [10]. With appropriate temperature and atmosphere conditions, wettability between ceramic and metal phases is achieved and liquid metal is drawn into the porous ceramic preform via capillarity thermodynamic criteria [9] (Table 1).

Lee and Hong worked on production of high volume fraction SiC/Al metal matrix composites by pressure infiltration method. High volume fraction of metal matrix composites such as SiC/Al composites containing nearly 70 vol% SiC particles could be fabricated without forming residual porosity and Al vein layers by pressure infiltration method controlling such process parameters for Al melt temperature 800°C, SiC preform preheat temperature 550°C, infiltration pressure 30-50 MPa and infiltration time 20-70 seconds after pouring Al melt into ceramic preform [10]. The maximum pressure (130 MPa) is maintained until the melt has solidified. With this procedure, the whole infiltration cycle does not take more than 120 s. [11] Pressure infiltration is useful technique for fast and high volume production of metal matrix composite systems having high liquid/solid surface tension and difficult wetting conditions.

Melt infiltration with and without pressure techniques are not always succesful for all combinations, Since aluminum alloy surface tension is a strong function of surface active elements and Mg content of the alloy, wetting and liquid penetration is improved by high pressure application on top of liquid aluminum during squeeze casting [12].

Experimental Procedure

In this work, melt infiltration method was used as major technique to produce high strength aluminum matrix composites. During these experiments, squeeze casting technique was used to achieve melt infiltration method with ceramic preforms. There are some reasons why squeeze casting process was preferred to form metal matrix composites. Wetting between ceramic preform and liquid Al melt can be increased by applying high pressure such as 150-170 MPa. It is a way of rapid process to produce aluminum matrix composite components. One of the most important advantages is that this casting technique is highly suited to mass production in the industry. This means that it is very economical process if it is aimed to produce great number of products. Furthermore, there is no shrinkage in squeeze casting process because of applied pressure during solidification. During squeeze casting process, the mould is heated up to 250°C in the beginning in order to prevent high heat flow from liquid metal to the metallic mould. At the same time, ceramic preforms were also heated up to 1000°C in the muffle furnace. After that, preform is placed into the mould cavity and hot metal is poured immediately. At the end, pressure was applied by squeeze casting technique yielding full penetration of liquid aluminum alloy. So composites were obtained after solidification. During these experiments, thermal paper on the lower punch was used in order to eliminate hot tears associated with differential thermal shrinkage. Al_2O_3 and B_4C ceramic preforms were infiltrated with 7085 and 7075 aluminum alloys respectively (Figure 1).

Results

Aluminum 7075 matrix with Al_2O_3, SiC and B_4C_3 ceramic preforms were infiltrated with and without pressure. In the beginning, melt infiltration process was done by the first group of alumina preforms that were sintered at 1000, 1100, 1200 and 1300°C whereas, only two of them were characterized. Metal matrix composites that were produces by the preforms (Figure 2).

After melt infiltration method, ceramic preform is confined by molten metal. In order to be able to investigate the microstructure and interfaces in between ceramic reinforcement and aluminum metal, these specimens were machines and metal matrix ceramic was revealed as it can be seen in below Figures 3 and 4.

Boron Carbide - 7075 Aluminum composite structure after the production of alumina - 7085 aluminum matrix composites, the surfaces were prepared by metallographic methods. The microstructures of these composites were examined after processing to reveal interface microstructures.

Composites, produced by preform whose sintering temperatures are 1300 and 1450°C. Obtaining high hardness values can be a promising property for the ballistic performance. High hardness values are needed in order to abrade the projectiles. As it was thought that the hardness of the tool steels is about 530 HB, the average composite hardness 545 HB can be considered as sufficient results for ballistic armor plate material. Therefore, Alumina - 7085 aluminum composites can be a good candidate for these applications (Figure 5).

Even if there is incomplete penetration along the interfaces, melt infiltration was achieved successfully in most specimens since aluminum phases were detected in the pores of alumina structure as it can be seen in below figures. In other words, 7085 aluminum alloy was completely sucked by alumina preform with the help of squeeze casting. This shows that squeeze casting is a good method to produce metal matrix composites if ceramic preform is wetted and completely infiltrated (Figure 6).

In the beginning, ceramic plate was preheated at 500°C during 90 minutes. At the same time, mold was also heated up to 250°C. Composite

Materials	Density (g/cm³)	Melting point (°C)	Hardness (Knoop)	Fracture Toughness (MPa.m$^{-1/2}$)
B4C	2.52	2445	2750	2.9-3.7
SiC	3.21	2730	2480	4.3
Al_2O_3	4.00	2070	2100	3.3-5.0
Al	2.70	660	120	29

Table 1: Some important properties of boron carbide, silicon carbide, alumina and aluminium.

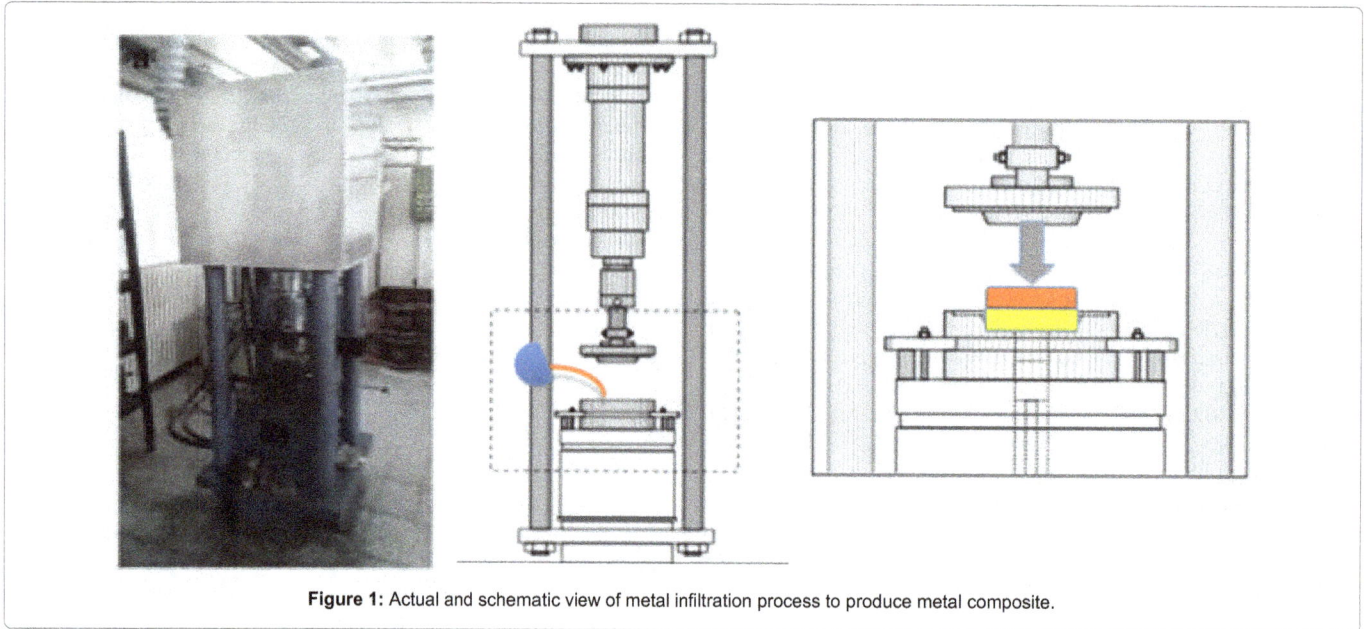

Figure 1: Actual and schematic view of metal infiltration process to produce metal composite.

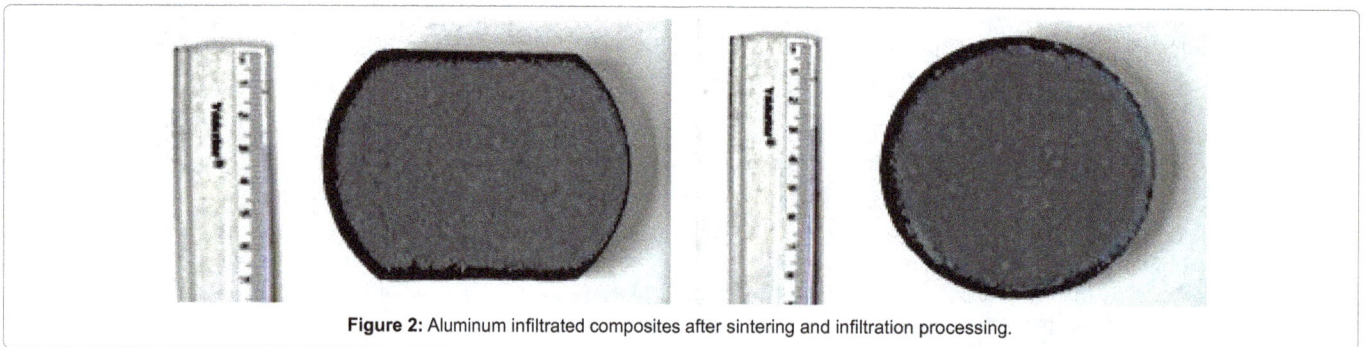

Figure 2: Aluminum infiltrated composites after sintering and infiltration processing.

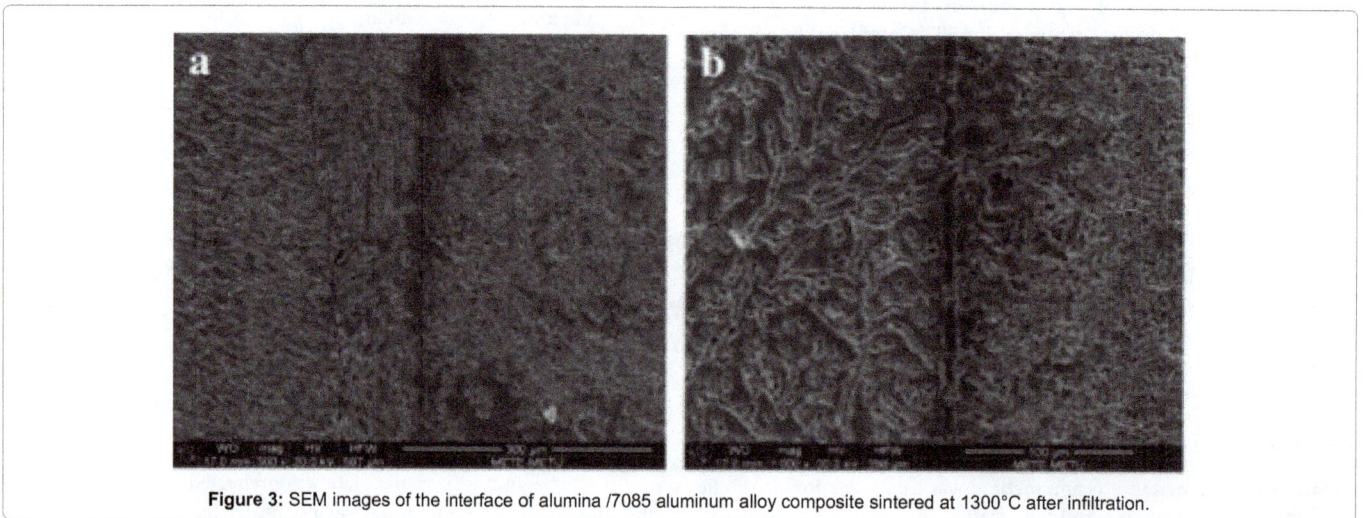

Figure 3: SEM images of the interface of alumina /7085 aluminum alloy composite sintered at 1300°C after infiltration.

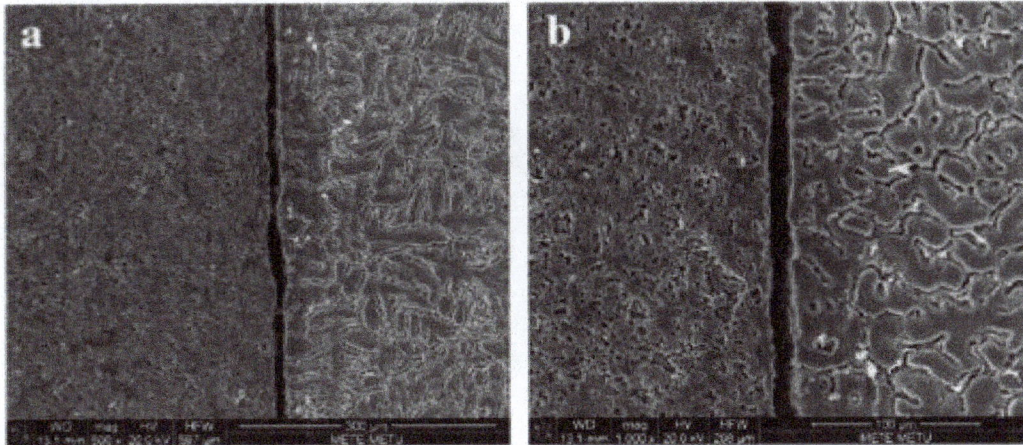

Figure 4: SEM images of the interface of alumina /7085 aluminum alloy composite sintered at 1450°C after infiltration revealed incomplete penetration of liquid aluminum alloy.

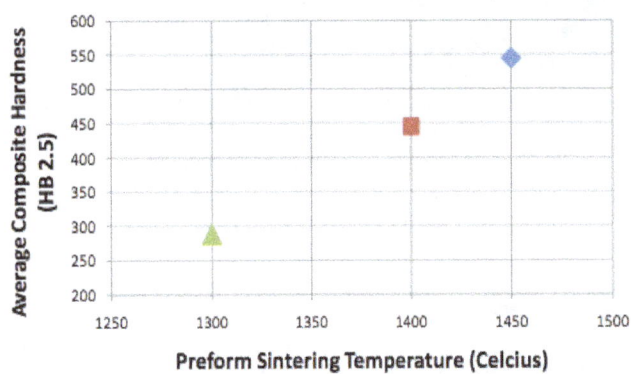

Figure 5: Distribution of composite hardness values according to Al$_2$O$_3$ preform sintering temperature.

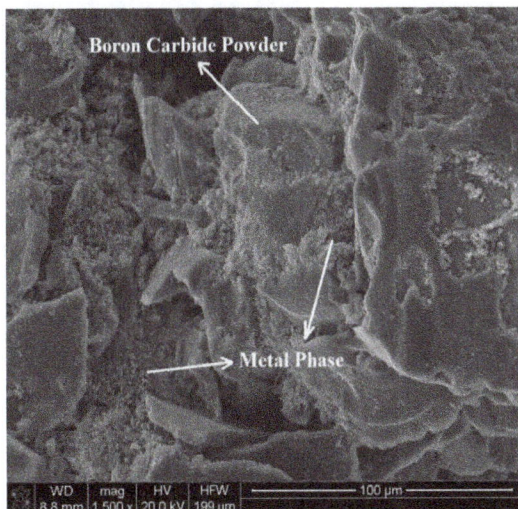

Figure 6: Microstructure of Boron Carbide – 7075 Aluminum composite.

plate was just put inside the mold and molten 7085 aluminum is poured into the cavity and squeeze casting was performed. At the end of this process crack formation was seen at the center of composite. Therefore, it should be noted that confinement operation is not possible by squeeze

casting process for boron carbide/aluminum composite structures because of different thermal expansion coefficient, too fine carbide size and wetting difficulties for these materials (Figure 7).

Since heat treatments to achieve T6 condition is not convenient for infiltrated composites infiltrated ceramic preform and backing 7075 plate should be prepared separately, then two layers should be integrated in the end. Therefore plain 7075 squeeze cast plates without ceramic preforms were studied to determine the optimum solutionizing time and temperature for peak hardness condition of T6 heat treatment. It can be seen that 475°C is the best temperature for solution heat treatment of 7075 aluminum alloy according to the hardness values. The average hardness of the specimens that were solutionized at 475°C was measured as 178 HB and it was noted as the best hardness result after T6 heat treatment of squeeze casted 7075 aluminum alloy (Figure 8).

After the production of the boron carbide-7075 aluminum composites, confinement experiment was done for these composites. In this experiment, square shaped boron carbide-aluminum composite that was produced by melt infiltration without pressure. By confinement of infiltrated composites it was aimed to improve corrosion behavior of these composite structures. The main aim is to cover boron carbide composite with aluminum alloy.

High pressure die casting machine was used as an alternative way to study and develop high strength aluminum alloy as matrix alloy for defence and automotive applications. During the production, the speed of injection piston of high pressure die caster was decreased to 0.7 m/sec and rheocasting was performed. Therefore, specimens were produced by rheocasting with high pressure die casting machine (Figure 9).

During this experiment, a special mold was used that give billets to machine tensile test specimens that were produced by rheocasting technique. Before machining, T6 heat treatments were operated for these specimens with the best conditions. These means that, 475°C was used as solutionizing temperature for 7075 aluminum alloy whereas, 465°C was used as solutionizing temperature for 7085 aluminum alloy. After that specimens were quenched in the water and they were aged at 120°C during 24 hours.

Conclusions

1. Both pressure and pressureless melt infiltration techniques yielded at least 85% ceramic phase containing composites.

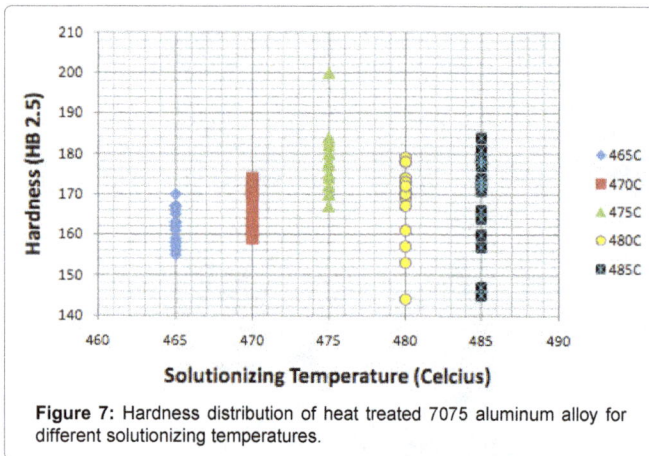

Figure 7: Hardness distribution of heat treated 7075 aluminum alloy for different solutionizing temperatures.

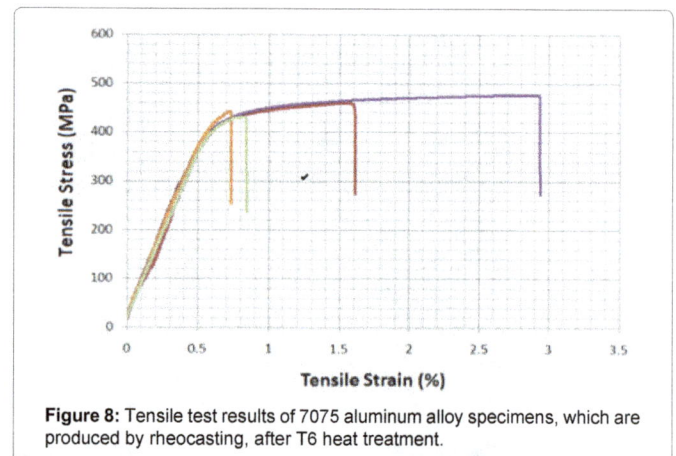

Figure 8: Tensile test results of 7075 aluminum alloy specimens, which are produced by rheocasting, after T6 heat treatment.

Figure 9: (a) Tensile test specimens obtained side by side 7075 aluminum alloy castings produced by rheocasting performed by HPDC, (b) machined tensile test specimens.

2. Pressureless infiltration of B_4C preforms having 1-3 μm carbide powder size produced for with 7075 alloy yielded complete infiltration. On contary pressure infitration of the same combination was failed due to low liquid phase penetration.

3. Pressure infiltration of Al_2O_3 preforms with coarse ceramic powders 20-50 μm was completely infiltrated with 7075 alloy

4. To develop high strength backing alloy plate, rheocasting of 7075 alloy was performed and this process yielded 400-500 MPa tensile strength after solutionizing at 475°C temperature for 90 mins then quenched in the water and aged at 120°C during 24 h.

5. T6 Heat treatment of infiltated 7075 based composites is found to be unsuccesful due to large difference in thermal expansion coefficients of metallic and ceramic phases.

6. For integration of plain alloy backing plate and infiltared composites gluing with adhesive materials can be used

7. Highest composite hardness values (550 HB) were obtained at ceramic (Al_2O_3) preforms sintered at maximum sintering temperature of 1450°C.

References

1. Balakrishnarajan MM, Pancharatna PD, Hoffmann R (2007) Structure and bonding in boron carbide: The invincibility of imperfections. New Journal of Chemistry 31: 473-485.

2. Arslan G, Kara F, Turan S (2003) Quantitative X-ray diffraction analysis of reactive infiltrated boron carbide-aluminium composites. Journal of the European Ceramic Society 23: 1243-1255.

3. Dennis B, Rahbek B, Johnsen B (2015) Dynamic behaviour of ceramic armour systems, Norwegian Defence Research Establishment (FFI).

4. Roberson CJ, Hazell PJ (2003) Resistance of silicon carbide to penetration by a tungsten carbide cored projectile.

5. Arslan G, Kalemtas A (2009) Processing of silicon carbide-boron carbide-aluminium composites. Journal of the European Ceramic Society 29: 473-480.

6. Nanda H, Appleby-Thomas GJ, Wood DC, Hazell PJ (2011) Ballistic behaviour of explosively shattered alumina and silicon carbide targets. Advances in Applied Ceramics 110: 287-292.

7. Mashhadi M, Taheri-Nassaj E, Sglavo VM, Sarpoolaky H, Ehsani N (2009) Effect of Al addition on pressureless sintering of B4C. Ceramics International 35: 831-837.

8. Kalkanlı A, Yılmaz S (2008) Synthesis and characterization of aluminum alloy 7075 reinforced with silicon carbide particulates. Materials & Design 29: 775-780.

9. Bilici İ, Metin GÜRÜ, Tekeli S (2015) Production of Ti-Fe Based MgAl2O4 Composite Material by Pressureless Infiltration Method. Gazi University Journal of Science 28: 295-299.

10. Halverson (1986) Boron-Carbide-Aluminum and Boron Carbide Reactive Cermets, United States Patent, Patent Number: 4, 605,440, Date of Patent: Aug. 12.

11. Lee HS, Hong SH (2003) Pressure infiltration casting process and thermophysical properties of high volume fraction SiCp/Al metal matrix composites. Materials Science and Technology 19: 1057-1064.

12. Chedru M, Vicens J, Chermant JL, Mordike BL (1999) Aluminium-aluminium nitride composites fabricated by melt infiltration under pressure. Journal of microscopy 196: 103-112.

27

Fiber Reinforced Composites

Prashanth S[1], Subbaya KM[2], Nithin K[3]* and Sachhidananda S[4]

[1]Department of Mechanical Engineering, Gargi Memorial Institute of Technology, Bharathinagar, Mandya, India
[2]Department of Industrial and Production Engineering, The National Institute of Engineering, Mysuru, India
[3]Department of Chemistry, The National Institute of Engineering, Mysuru, India
[4]Department of Polymer Science and Technology, Sri Jayachamarajendra College of Engineering, JSS S&T University, Mysuru, India

Abstract

Fiber-reinforced composites are essentially axial particulates embedded in fitting matrices. The primary objective of fiber-reinforced composites it to obtain materials with high strength in conjunction with higher elastic modulus. The strength elevation is however affected with applied load transiting from matrix to fibers, interfacial bonding between fiber-matrix, their relative alignment and nature of fiber scheming the overall material behaviors. The alignment of fibers may however be continuous or random depending on the end applications. The choice of the fiber reinforcement and its fitting matrix also depends on application requirements. In recent years, the advent of composite technology has led to the development of different fiber reinforced composite systems via varying manufacturing methodologies to obtain advanced material behaviors. Herein, we present a comparative account on various kinds of synthetic fibers and their significance as potential reinforcements with special emphasis on carbon fibers.

Keywords: Fibers; E-glass; Carbon; Kevlar; Low elastic modulus (LM)

Introduction

Composite materials are considered as one of the most potential candidates for aerospace applications owing to their high strength-to-weight ratio and excellent fatigue resistance [1,2]. Polymer based composite systems offer multiple functionalities, owing to the synergistic combination of functional fillers with highly process able polymers, which in-turn widens their application window [3,4]. In general, composite materials employed for structural applications are best classified as high performance systems and are made of synthetic materials that offer high strength-to-weight ratios, but often demands controlled manufacturing environments for optimum performance. Fiber reinforced polymers are also known for their potential as high-tech, high-quality materials in electrical as well as military applications. Although, the applications of fiber reinforced plastics are still limited to smaller parts made of reinforced plastics, such as parts of the bridge deck, girders, reinforcement bars, staying cables or handrails [5-11]. In recent years, reinforced plastics are also established to be well drafted for building and construction, in addition to transport and electronic applications. Some of the major application areas of Fiber Reinforced Plastics (FRP) in transportation are automotive, aviation, shipping and other related sectors. In energy/electronic sector, FRP are employed towards the fabrication of high voltage switches, cryostats, dry transformers and many more. Lately, carbon fiber reinforced polymers are also used for advanced technical applications such as in rocket nozzles. The pie chart below gives a detailed picture of market shares of FRPs in various application areas [12-14] (Figure 1).

The reinforcement of fiber upon polymeric matrix is found to bring about significant advancements in mechanical behaviors of polymeric host with added advantages of light weight, high strength to weight ratio, excellent weathering stabilities and enhanced dimensional stabilities [15], in addition to low maintenance cost and tailor made material behaviors. However, to obtain tailor made properties that suit specified application requirements, various types of fibers with varying polymers have been tried, with fibers contributing towards the betterment of mechanical, tribological, thermal and water sorption behaviours of resulting composites. Nevertheless, one could expect herculean results, when the fibers are of near-to-infinite length,

isotropic and are inserted unidirectional. Conversely, the greater anisotropic and shorter their length of fibers, lower would be their overall mechanical performance. The observed large leeway regarding length, direction and type of fiber widens the application window of FRP composites. Further, the composite can be completely tailor made to suit specific mechanical needs for any given project, which in turn aids far better efficiency towards end applications. The most common fibers exercised in recent times are glass fibers, carbon fibers, aramid fibers and natural fibers, next to which are the nylon-and polyester fibers. All these reinforcements are ideally less dense and hence present

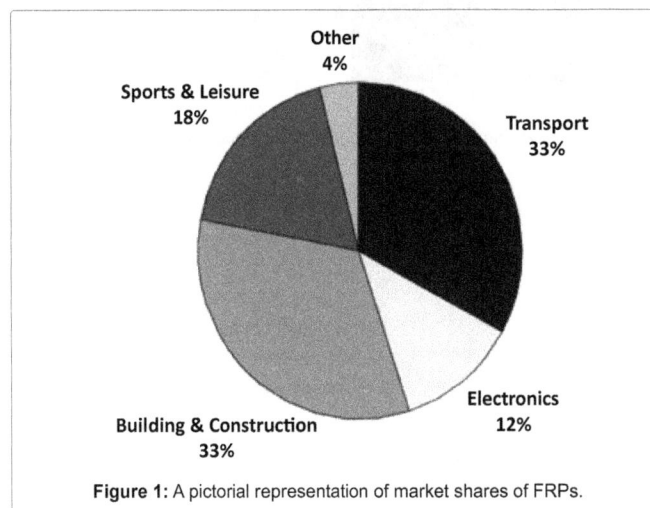

Figure 1: A pictorial representation of market shares of FRPs.

*Corresponding author: Nithin Kundachira, Department of Chemistry, The National Institute of Engineering, Mysuru-570008, India
E-mail: nithukundachira@gmail.com

the matrix with a higher strength and stiffness. The excellent strength/stiffness to weight ratio of composites is however credited to low fiber densities. Nevertheless, all fibers behave quasi-elastically until breakage, with carbon fibers much stiffer and lighter that glass fibers. This in turn, accounts for their increased preference for many high performance applications.

Important Types of Fiber Reinforcements

Glass fibers

The mechanical behaviours of fiber-reinforced composites are primarily dependent on their inherent abilities to enable stress transfer, which in turn depends on the fiber strength, matrix strength and the strength of interfacial adhesion between the fiber/matrix [16]. Glass fibers (GFs) have been employed in various forms such as longitudinal, woven mat, chopped fiber (distinct) and chopped mats to enhance the mechanical and tribological properties of the fiber reinforced composites. The properties of such composites was however dependent on the nature and orientation of the fibers laid during composite preparation [17]. Glass fibers are one of the most widely used polymer reinforcements with nearly 90% of all FRPs made of glass fibers. Of which, the oldest and the most popular form is the E-glass or electrical grade glass. Other types of glass fibers include A-glass or alkali glass, C-glass or chemical resistant glass, and the high strength R-glass and/or S-glass. Under laboratory circumstances glass fibers can resist tensile stresses of about 7000 N/mm², whereas commercial glass fibers reach 2800 to 4800 N/mm² [18] (Figure 2).

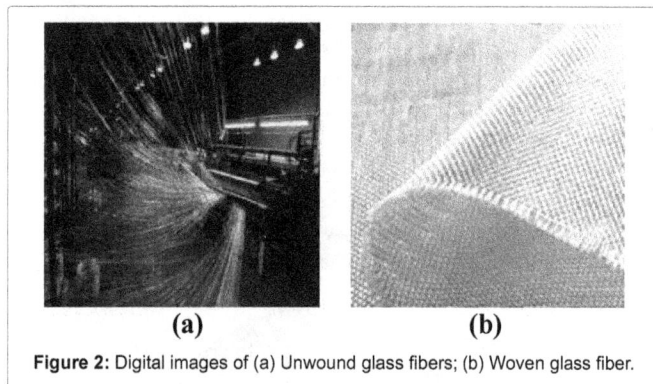

(a)   **(b)**

Figure 2: Digital images of (a) Unwound glass fibers; (b) Woven glass fiber.

Classes of GFs	Physical properties
A glass	High durability, strength and electrical resistivity
C glass	High corrosion resistance
D glass	Low dielectric constant
E glass	Higher strength and electrical resistivity
AR glass	Alkali resistance
R glass	Higher strength and acid corrosion resistance
S glass	Higher tensile strength

Table 1: Physical properties of various classes of glass fibers.

Classification of glass fibers: The major classes of GFs and their inherent physical properties are depicted in Table 1 [19], while their chemical composition is tabulated in Table 2. The physical and mechanical properties of different classes of GFs are shown in Table 3.

Various commercially imperative products have been developed using glass fibers. Each of which defer in its overall material performance. Table 4 summarises the mechanical properties of various glass fibre products.

Carbon fibers

Carbon fibers (GFs) are the new breed of high strength materials made of graphitic and non-crystalline regions. Of all reinforcing fibers, carbon fibers offer the highest specific modulus and strength. Additionally, carbon fibers have the ability to retain its tensile strength even at high temperatures and are independent of moisture. Carbon fibers do not necessarily break under stress in contrast to glass and other organic polymer fibers [20]. Carbon fibres also offer high electrical and thermal conductivities with relatively low coefficient of thermal expansion [21-24]. This innate property of carbon fibers makes them ideal for applications in aerospace, electronics and automobile sectors. The carbon fibres offer a maximum strength of 7Gpa, axial compressive strength is 10-60% of their tensile strength [25,26] and transverse compressive strength is 12-20% of their axial compressive strength [27]. Poly-acrylonitrile (PAN) is one of the most common precursors employed in carbon fiber production, which offers high tensile strength and higher elastic modulus, extensively applied for structural material composites in aerospace and sporting/recreational goods. Depending on the final curing temperature, different classes of carbon fibers namely high tenacity (HT) fibers, intermediate modulus (IM) fibers, high modulus (HM) fibers and ultra-high modulus (UHM) fibers are formed with PAN precursors. Another production technique, takes advantage of petroleum-pitch (PP) as precursors. These types of carbon fiber have a higher E-modulus and lower tensile strength and are extensively adopted in high stiffness components that utilize high thermal and electrical conductivities [28] (Figure 3).

Characteristics of carbon fibers: Carbon fibers are very versatile, because of their extremely high strength to weight and stiffness to weight ratios. Moreover, they are chemically inert, electrically conductive and infusible. The stiffness and modulus of elasticity of carbon fibers can range from glass to three times that of steel. The most widely used types have a modulus of 200.000-400.000 N/mm². Table 5 gives a detailed representation of mechanical behaviours of various carbon fiber products.

Carbon fibers may either be directly processed as finished product or as pre-product. Finished products may appear as woven tubes or pultruded sections. While, pre-products may include short fibers, twisted or non-twisted yarns, continuous filament, tows and so forth. Table 6 gives a detailed picture of important types of commercially important carbon fiber tows, with a comparative view of its physico-mechanical behaviours.

Type of GF	SiO$_2$	Al$_2$O$_3$	Fe$_2$O$_3$	TiO$_2$	B$_2$O$_3$	CaO	MgO	Na$_2$O	K$_2$O
E-glass	55	14	-	0.2	7	22	1	0.5	0.3
C-glass	64.6	4.1	-	-	5	13.4	3.3	9.6	0.5
S-glass	65	25	-	-	-	-	10	-	-
A-glass	67.5	3.5	-	-	1.5	6.5	4.5	13.5	3
R-glass	60	-	-	-	-	9	6	0.5	0.1
EC-Glass	58	12.4		-		23		-	-
AR-glass	61	1	-	-	-	5	1	14	3

Table 2: Chemical compositions (wt %) of glass fiber types.

Type	Density (g/cm³)	Tensile strength GPa	Young's modulus (GPa)	Elongation (%)	Coefficient of thermal expansion (10⁻⁷/°C)	Poison's ratio	Refractive index
E-glass	2.58	3.445	72.3	4.8	54	0.2	1.558
C-glass	2.52	3.310	68.9	4.8	63	-	1.533
S-glass	2.46	4.890	86.9	5.7	16	0.22	1.521
A-glass	2.44	3.310	68.9	4.8	73	-	1.538
R-glass	2.54	4.135	85.5	4.8	33	-	1.546
EC-Glass	2.72	3.445	85.5	4.8	59	-	1.579
AR-glass	2.70	3.241	73.1	4.4	65	-	1.562

Table 3: Physical and mechanical properties of GFs.

Types	Unit	Woven cloth	Chopped strand mat	Continuous roving
Glass content	%	55	30	70
Tensile strength	N/mm²	300	100	800
Compressive strength	N/mm²	250	150	350
Flexural strength	N/mm²	400	150	1000
Flexural modulus	N/mm²	15000	7000	40000
Impact strength	kJ/m²	150	75	250
Coefficient of linear thermal expansion	×10⁻⁶/°C	12	30	10
Thermal conductivity	W/mK	0.28	0.2	0.29

Table 4: Mechanical properties of various glassfiber products.

Types	Ultra high elastic modulus type (UHM)	High elastic modulus type (HM)	Intermediate elastic modulus type (IM)	Standard elastic modulus type (HT)	Low elastic modulus type (LM)
Tensile elastic modulus	600 GPa	350-600 GPa	280-350 GPa	200-280 GPa	200 GPa
Tensile strength	2,500 MPa	2,500 MPa	3,500 MPa	2,500 MPa	3,500 MPa

Table 5: Mechanical properties of different types of carbon fiber tows.

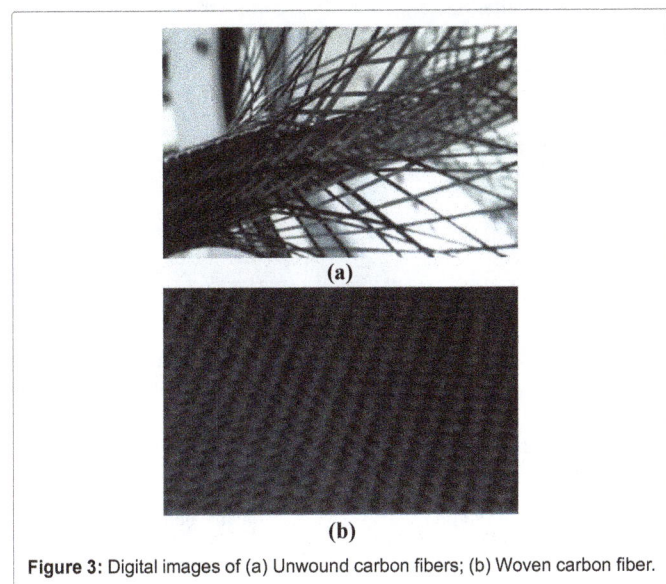

Figure 3: Digital images of (a) Unwound carbon fibers; (b) Woven carbon fiber.

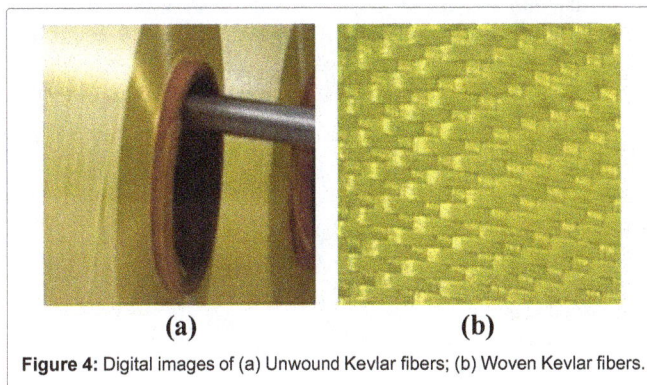

Figure 4: Digital images of (a) Unwound Kevlar fibers; (b) Woven Kevlar fibers.

Kevlar fibers

Kevlar is an aramid fiber of Poly-para-phenylen-terephtalamide (PPTA) with a rigid molecular structure. Aramid like fibers were first developed in 1960s as an alternate for steel reinforcements in rubber tyres. However, once developed the aramid fibers were also found suitable for ballistics and as surrogate for asbestos. Kevlar fibers are often used for high-performance composite applications where light weight, high strength and stiffness, damage resistance, and resistance to fatigue are of utmost importance. Aramid fibers are often employed in those applications that demands high strength and low weight together with a high impact resistance. Some of the most frequent applications of aramid materials include bullet proof vests, cooling vehicles, ship hulls and lately towards structural strengthening of civil structures.

Nevertheless, aramid fibers often present low compression strength. Further, the compressive modulus of aramid is of the same order as its tensile modulus [18] (Figure 4).

Types of aramid fibers: In general, aramid fibers are grouped into two categories namely,

1. Meta-aramid

2. Para-aramid

The termsmeta and para refers to the relative position of chemical bonds in the aramid fiber structure. The chemical bonds of para-aramid fibers are more aligned in the long direction of the fibers. While, the meta-aramid fibers are relatively less and/or not aligned and hence they do not they develop higher tensile strength in contrast to para-aramid bonds.

Meta-aramids: Fibers made from the meta-aramid fibers often present excellent thermal, chemical and radiation resistance. They are usually employed in the fabrication of fire retardant textiles such as outer wear for fire fighters and racing car drivers. Nomex and teijiconex are well established examples of meta-aramids.

Para-aramids: Fibers made of para-aramids offer higher strength. These are more commonly used in fibers reinforcement plastics for engineering structures, Stress skin panels, and other high tensile strength applications. Kevlar and Technora are commercially important para-aramid fibers [28-31]. Table 7 gives a comparative account on different grades (kevlar 29, kevlar 49, and kevlar 149) of commercially important kevlar fibers and their innate mechanical properties.

Although, aramid fibers are compatible with almost any kind of polymeric matrix. Yet, thermoplastic matrices host aramid fibers with nearly 20% more efficiency. Owing to their higher ductility, they are more compatible to aramids [16,18]. Table 8 presents a comparative view of tensile properties of epoxy composites reinforced with different types of aramid fibers.

Comparative Account of Glass, Carbon and Kevlar Fibers

Fibers in general play a crucial role in deciding the end property of the fiber reinforced composite systems. Thus, appropriate selection of fiber material and their relative orientation can lead to composite to composites with tailor made properties to suit specific application requirements. Herein, we present a comparative account of various properties of some of the technologically significant fibers (E-Glass, Low elastic modulus type carbon fiber (LM) and kevlar149), so as to ease the selection of appropriate fibers to satisfy the real time application demands. However, during, comparative studies, one has to take into account of property changes arising from difference in processing techniques, variation in precursor materials and also post-fabrication treatments, since all these factors influence the end properties of resultant fibers. E-Glass is commonly used glass fiber that offers higher strength and high electrical resistivity, LM is the type of carbon fiber with low elastic modulus, while kevlar 149 is one of aramid fibers which show significantly higher tensile strength.

Density

The density values of E-Glass, Low elastic modulus type carbon fiber (LM) and kevlar149 display significant difference. Furthermore, kevlar and carbon fibre are much lighter, while E-Glass is the heaviest (Table 9) [32-35].

Thus, carbon fiber or kevlar fibers are preferred over glass fibers in those applications that involve any structures demanding higher strength within smaller sizes.

Tensile strength

The tensile strength of glass, carbon and kevlar fibers are however similar. If tensile strength is only criteria for material selection, then cost of the material will be absolute (Table 10) [35].

Tensile modulus

All details related to Tensile Modulus mentioned in Table 11 [35].

Electrical conductivity

Carbon fibers are excellent conductors in contrast to kevlar and glass fibers. Nevertheless, kevlar is used for guy lines in transmission towers, owing to its ability to absorb water and the absorbed water conducts electricity. Since carbon fiber conducts electricity, galvanic corrosion is a concern when in contact with other metallic parts. Boaters, who have carbon fiber masts and spars, will have to insulate their aluminium fasteners and connections to avoid corrosion.

Type	Unit	55% CF SMC	Lowest grade CF epoxy fabric	Highest grade CF epoxy fabric	Carbon epoxy
Carbon fiber content	%	55	n.a	n.a	67
Density	g/cm^3	1.45	1.15	1.8	n.a
Tensile strength ultimate	N/mm^2	289	50	2.100	1.362
Modulus of elasticity	10^3 N/mm^2	55.1	6.6	520	140
Flexural yield strength	N/mm^2	613	110	1600	1383
Flexural modulus	10^3 N/mm^2	34.5	6.410	125	122
Compressive yield strength	N/mm^2	275	50	1720	1084
Compressive modulus	10^3 N/mm^2	31.7	8.2	140	136
Shear strength	N/mm^2	65.6	0.8	120	87.4

Table 6: Comparison between three types of carbon fiber/epoxy products.

Grade	Density (g/cm^3)	Tensile modulus (GPa)	Tensile strength (GPa)	Tensile elongation (%)
Kevlar 29	1.44	83	3.6	4
Kevlar 49	1.44	131	3.6-4.1	2.8
Kevlar 149	1.47	186	3.4	2

Table 7: Mechanical properties of different types of aramid fibers.

Type	Units	Aramid fiber-standard modulus-UD	Aramid fibers-high modulus-Woven
Density	g/cm^3	1.31	1.27
Tensile strength	N/mm^2	1400	560
Tensile modulus	10^3 N/mm^2	76	30

Table 8: Comparison between two aramid fiber reinforced epoxy composites.

Type of fiber	Density (g/cm^3)
E glass	2.58
LM-CF	1.45
Kevlar 149	1.47

Table 9: Computed density of commercially important fibers.

Fiber	Tensile strength (GPa)
E glass	3.445
LM-CF	3.5
Kevlar 149	3.4

Table 10: Tensile strength of various types of fibers.

Fiber	Tensile modulus (GPa)
E glass	72.3
LM-CF	200
Kevlar 149	186

Table 11: Tensile modulus of various types of fibers.

Fatigue resistance

If a composite part is made to bend and straighten repeatedly, it is bound to fail due to fatigue. However, when reinforced with carbon fibers is somewhat sensitive to fatigue and tends to fail catastrophically without showing many signs of distress, while kevlar is more resistant to fatigue. Glass is somewhere in between and can be quite fatigue resistant depending on the type of glass and the setup conditions.

Abrasion resistance

Kevlar has a strong abrasion resistance, which makes it difficult to cut. One of the common applications of kevlar is as a protective glove, while using sharp blades. Carbon fiber and glass fiber have less abrasion resistant.

Chemical resistance

Aramids are sensitive to strong acids, bases, and some oxidizers, such as the chlorine bleach. Carbon fiber is however more stable and is not sensitive to chemical degradation.

Conclusion

The rationale of the presented review is to focus on the importance and inherent properties of various types of fibers. Accordingly, three synthetic fibers such as glass fiber, carbon fiber and kevlar fiber are presented, as they are the most regularly employed fiber reinforcements. Each fiber is unique in their manufacturing methods, types and properties. Further, the property of the fiber also varies with their manufacturing methods and types.

Herein, we presented a detailed tabulation of fiber types and their inherent properties, with special emphasis on their mechanical behaviors. For better understanding and easy selection, a comparative account on properties of these fibers is presented. Among various types of fiber reinforcements, E-Glass, Low elastic modulus type carbon fiber (LM) and kevlar149 fiber are compared. According to the literatures reviewed,

- Glass fibers are denser than carbon or kevlar fibers.
- Tensile strength of all the fiber types presented is almost similar.
- Tensile modulus of LM carbon fiber is relatively higher, when compared to glass and kevlar fibers.
- Carbon fibre is very good conductor, while kevlar and glass are non-conductors.
- Kevlar is more fatigue resistant, while carbon and glass are somewhat but not totally sensitive to fatigue.
- Kevlar offers strong abrasion resistance, while carbon and glass fibers are less resistant.
- Carbon fiber is highly stable and is relatively non-sensitive to chemical degradation.

As per the comparative account, carbon fibers have salient properties than glass and kevlar fibers. This provides a reference for selection of carbon fibers as the main reinforcing material in FRPC's.

Further research with nano-metric carbon reinforcements may open up the scientific window towards the development of advanced multifunctional composites for broader technological applications.

References

1. Banakar P, Shivanand HK, Niranjan HB (2012) Mechanical Properties of Angle Ply Laminated Composites-A Review.International Journal of Pure and Applied Sciences and Technology 9: 127-133.

2. Subramani K, Siddaramaiah N (2015) Opto-Electrical Characteristics of Poly (vinyl alcohol)/Cesium Zincate Nanodielectrics. The Journal of Physical Chemistry C 119: 20244-20255.

3. Subramani NK, Kasargod Nagaraj S, Shivanna S, Siddaramaiah H (2016) Highly flexible and visibly transparent poly (vinyl alcohol)/calcium zincate nanocomposite films for UVA shielding applications as assessed by novel ultraviolet photon induced fluorescence quenching. Macromolecules 49: 2791-2801.

4. Shilpa KN, Nithin KS, Sachhidananda S, Madhukar BS, Siddaramaiah (2017) Visibly transparent PVA/sodium doped dysprosia (Na2Dy2O4) nano composite films, with high refractive index: An optical study. Journal of Alloys and Compounds 694: 884-891.

5. Subramanian C, Senthilvelan S (2011) Joint performance of the glass fiber reinforced polypropylene leaf spring. Composite Structures 93: 759-766.

6. Zhang Q, Liang Y, Warner SB (1994) Partial carbonization of aramid fibers. Journal of Polymer Science Part B: Polymer Physics 32: 2207-2220.

7. McGee AC, Dharan CKH, Finnie I (1987) Abrasive wear of graphite fiber-reinforced polymer composite materials. Wear 114: 97-107.

8. Danna Q, Limin B, Takatera M, Kemmochi K (2009) Particle erosion behavior of unidirectional CF and GF hybrid fiber-reinforced plastic composites. Journal of Textile Engineering 55: 39-44.

9. Rosato DV (2003) Plastics engineered product design. Elsevier.

10. Mourit AP, Gibson AG (2006) Fire reaction properties of composites. Fire Properties of Polymer Composite Materials 59-101.

11. Rosato DV, Rosato DV (2004) Reinforced plastics handbook. Elsevier.

12. Brydson JA (1999) Plastics materials. Butterworth-Heinemann.

13. Karger-Kocsis J (1995) Microstructural aspects of fracture in polypropylene and its filled, chopped fiber and fiber mat reinforced composites. In Polypropylene Structure, blends and composites Springer Netherlands, pp: 142-201

14. Mathew MT, Padaki NV, Rocha LA, Gomes JR, Alagirusamy R, et al. (2007) Tribological properties of the directionally oriented warp knit GFRP composites. Wear 263: 930-938.

15. Kendall D (2006) Fiber reinforced polymer composite bridges.

16. Erden S, Sever K, Seki Y, Sarikanat M (2010) Enhancement of the mechanical properties of glass/polyester composites via matrix modification glass/polyester composite siloxane matrix modification. Fibers and Polymers 11: 732-737.

17. Alam S, Habib F, Irfan M, Iqbal W, Khalid K (2010) Effect of orientation of glass fiber on mechanical properties of GRP composites. J. Chem. Society Pakistan 32: 265-269.

18. Rosato DV, Rosato DV (2004) Reinforced plastics handbook. Elsevier.

19. Sathishkumar TP, Satheeshkumar S, Naveen J (2014) Glass fiber-reinforced polymer composites: A review. Journal of Reinforced Plastics and Composites 33: 1258-1275.

20. Hart-Smith LJ (1987) Engineered Materials Handbook (Vol. 1). ASM International, Ohio 479-495.

21. Donnet JB, Bansal RC (1998) Carbon fibers. CRC Press.

22. Allred RE (2005) Carbon-reinforced composite recycling: process and business development. In Global Outlook for Carbon Fibers 2005, Intertech Conferences, San Diego, CA.

23. Hajduk F (2005) Carbon fibres overview; global outlook for carbon fibres 2005. In: Intertech conferences. San Diego: CA.

24. Carolin A (2003) Carbon fibre reinforced polymers for strengthening of structural elements (Doctoral dissertation, Luleå tekniska universitet).

25. Chand S (2000) Review carbon fibers for composites. Journal of Materials Science 35: 1303-1313.

26. Ohsawa T, Miwa M, Kawade M, Tsushima E (1990) Axial compressive strength of carbon fiber. Journal of Applied Polymer Science 39: 1733-1743.

27. Shinohara AH, Sato T, Saito F, Tomioka T, Arai Y (1993) A novel method for measuring direct compressive properties of carbon fibres using a micro-mechanical compression tester. Journal of Materials Science 28: 6611-6616.

28. Hearle JW (2001) High Performance Fibers. Woodhead Publishing. Cambridge, England.

29. Aramide FO, Atanda PO, Olorunniwo OO (2012) Mechanical Properties of a Polyester Fibre Glass Composite. International Journal of Composite Materials 2: 147-151.

30. Wu SR, Sheu GS, Shyu SS (1996) Kevlar fiber-epoxy adhesion and its effect on composite mechanical and fracture properties by plasma and chemical treatment. Journal of Applied Polymer Science 62: 1347-1360.

31. Kodama M, Karino I (1986) Polar-polar interaction between the reinforcement and matrix for kevlar fiber-reinforced composite: Effect of using the blend of polar polymers as matrix. Journal of Applied Polymer Science 32: 5345-5355.

32. Kim EY, An SK, Kim HD (1997) Graft copolymerization of ϵ-Caprolactam onto Kevlar-49 fiber surface and properties of grafted Kevlar fiber reinforced composite. Journal of Applied Polymer Science 65: 99-107.

33. Li XG, Huang MR (1999) Thermal degradation of Kevlar fiber by high-resolution thermogravimetry. Journal of Applied Polymer Science 71: 565-571.

34. Tanner D, Fitzgerald JA, Phillips BR (1989) The kevlar story-an advanced materials case study. Angewandte Chemie International Edition in English 28: 649-654.

35. Hillermeier K (1984) Prospects of Aramid as a Substitute for Asbestos. Textile Research Journal 54: 575-580.

28

Synthesis and Crystallization Behavior of 3 mol% Yttria Partically Stabilized Zirconia (3Y-PSZ) Nanopowders by Microwave Pyrolysis Process

Bingbing Fan[1]*, Fan Zhang[1,2], Jian Li[1], Hao Chen[1] and Rui Zhang[1,3]

[1]*School of Materials Science and Engineering, Zhengzhou University, Zhengzhou, Henan 450001, China*
[2]*Henan Information and Statistics Vocational College, Zhengzhou, Henan, 450002, China*
[3]*ZhengZhou Institute of Aeronautical Industry Management, Zhengzhou, Henan 450015, China*

Abstract

A crystalline Nano powders of 3 mol% yttria-partially stabilized (3Y-PSZ) has been synthesized using $ZrOCl_2$ and $Y(NO_3)_3$ as raw materials by microwave pyrolysis with a TE666 resonant mode at 700-900°C. The frequency of the microwave was 2.45 GHz with the maximum power of 10 KW, and a hybrid heating structure was used with insulation of porous mullite and SiC aided heaters. For comparison, conventional heating was performed in air at 750°C for 20 min. The as-synthesized products were characterized by SEM and TEM images, XRD patterns. It was found that microwave energy promotes the conversion of tetragonal ZrO_2 (t-ZrO_2) to monoclinic ZrO_2 (m-ZrO_2) phase compared with conventional pyrolysis. TEM images showed that highly dispersed 3Y-ZrO_2 powders with ~23 nm in size were obtained by microwave pyrolysis at 750°C for 20 min.

Keywords: 3Y-ZrO_2 powders; Microwave pyrolysis; Highly dispersed nano powders

Introduction

Zirconia (ZrO_2) is an important material possessing many excellent properties, including high melting point, high hardness and strength, high fracture toughness, low thermal conductivity, high chemical stability, ionic conductivity, and excellent corrosion and abrasion resistance [1,2]. It is thus extensively used in many important areas, e.g., functional ceramics, high-temperature and corrosion resisting components, abrasive and insulating material, dielectric element, catalysts, and ion exchanger [3,4].

To date, several techniques have been developed to prepare ZrO_2 nanoparticles, including sol-gel, flame spray, combustion, glycothermal process, hydrothermal processing, precipitation and other techniques [5-10]. Unfortunately, these techniques all suffer from various disadvantages, such as strong agglomerates, difficulty in particle size control, complex drying procedures, requirement of high energy and/or long reaction time, and low production efficiency. To overcome these drawbacks, it is necessary to develop other alternative techniques.

Microwave method has recently attracted an increasing amount of interest [11-15] owing to the advantages, such as cost-effective, energy efficient, rapid and convenient method of heating, and results in higher yields in shorter reaction times. In this work, 3Y-ZrO_2 Nano powders were prepared by microwave pyrolysis combined with a co-precipitation process using $ZrClO_2 \cdot 8H_2O$ as the starting material, and NH_4OH as the mineralizer.

Experimental Procedure

Preparation of 3Y-ZrO_2 powders

Commercially available zirconium oxychloride octahydrate ($ZrClO_2 \cdot 8H_2O$, purity: 99.2%, Zibo Huantuo Chemical Co. Ltd., Shandong, China), yttriumnitrate hexahydrate ($Y(NO_3)_3 \cdot 6H_2O$, A.R., Tianjin Guangfu Fine Chemical Research Institute, Tianjin, China), and ammonia solution (NH_4OH, A.R., Xilong Chemical Co., Ltd., Guangdong, China) were used in the preparation of the precursor. $ZrClO_2 \cdot 8H_2O$ and $Y(NO_3)_3 \cdot 6H_2O$ were used as received and dissolved in DI water. The concentration of zirconium ion was 1.0 mol/L, to which 3 mol% $Y(NO_3)_3 \cdot 6H_2O$ was added. 1 M ammonia solution was added dropwise in given solution with continuous stirring, adjusting its Ph. value at 12-13. After co-precipitation, the precursor solution was filtered and washed with ethanol repeatedly until no Cl^- was detected in the filtrate by an $AgNO_3$ solution. The resulting precursor powder was oven-dried at 80°C for 24 h. Subsequently, the dried powder was kept in a microwave chamber with the resonant mode of TE666 (WXD20S-07, Nanjing Sanle Microwave Technology Development Co., Ltd., Jiangsu, China) at 700-900°C for 20 min. The frequency of the microwave oven was 2.45 GHz with the maximum power of 10 KW. The temperature was monitored by using an infrared radiation thermometer (OI-T6I2-B-1-type, GOIDSUN, USA) with initial display of 700°C. A thermal insulation structure based on a hybrid heating mode was well designed with the wall material of porous mullite and aided heaters of SiC rods. For comparison, 3Y-ZrO_2 was also prepared via conventional pyrolysis at 750°C for 20 min.

Characterization of 3Y-ZrO_2 powders

Phases in the as-prepared product powders were identified by powder X-ray diffraction (XRD) analysis (XD-3, Persee, China) with Cu K_a radiation (λ=1.5406 Å). Morphologies and microstructures were observed by using a field emission electron microscope (SEM) (JSM-7001F, JEOL, and Japan) and a transmission electron microscopy (HRTEM) (Tecnai G2 F20, Philips Co. Holland).

---

*Corresponding author: Bingbing Fan, School of Materials Science and Engineering, Zhengzhou University, Zhengzhou, Henan 450001 China
E-mail: fanbingbing@zzu.edu.cn

Results and Discussion

Microwave pyrolysis behavior

It is known that microwave absorption strongly depends on the dielectric loss factor of the material of interest [16]. At low temperature, ZrO_2 precursor cannot effectively absorb microwaves. However, with the help of hybrid heating by SiC aided heaters, heat is transferred to the precursor in the low temperature region, heating at this stage is analogous to conventional processes. After reaching the critical temperature, ZrO_2 precursor couples with the electromagnetic field and a higher heating rate is obtained due to the increased dielectric loss factor [17]. This change is consistent with the heating curve shown in Figure 1. In Figure 1, the temperature was well monitored by manual control of input powder. Owning to the rapid sintering characteristics of microwave sintering, it took only 30 min to reach the temperature of 750°C. It was observed that an efficient forward power profile requires a high power initial segment, the reflected power increases synchronously with input power. After about 4 minutes, the reflect power was suitably reduced; thermal runaway could be prevented albeit quite fast temperature increase to 750°C. After about 35 minutes, the input power of the system decreased as well as reflected power concomitantly decreased, but the temperature was kept at 750°C, which was due to selective absorbing phenomena, which is the unique heating feature of microwave sintering.

The DTA/TG curves of the 3Y-PSZ precursor amorphous powders at heating rate 10 k/min in air are shown in Figure 2. An endothermic peak at about 94°C is accompanied with a weight loss of 10.5% which is attributed to the evaporation of water. The exothermic peak at 305°C is attributed to the dehydration of precursors. The second exothermic peak at 450°C due to the formation of the tetragonal phase of ZrO_2 in the 3Y-PSZ freeze dried precursor powders.

X-ray diffraction analysis

As shown in Figure 3, the observed diffraction peaks at $2\theta=30.2°$, 35.0°, 50.4°, 58.9°, and 62.9° are associated with -111, -200, -220, -311, and -222 plane of t-ZrO_2 (JCPDS No. 50-1089). The peaks are at around 28.2 and 31.4 correspond to the (-111) and -111 planes of m-ZrO_2 (JCPDS No. 37-1484). According to the XRD analysis, both the 3Y-ZrO_2 powders obtained by microwave pyrolysis and conventional pyrolysis are a mixture of monoclinic and tetragonal phases. As shown in Figure 3a, the metastable tetragonal phase is the main phase in the powders by CS method when the heating temperature is less than

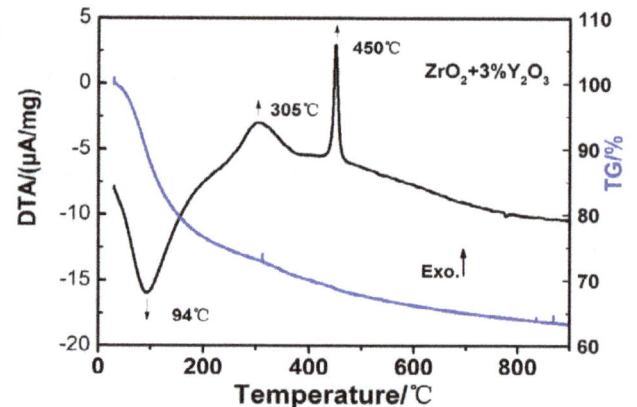

Figure 2: DTA/TG curves for 3Y-PSZ precursor amorphous powders at heating rate 10 k/min in air.

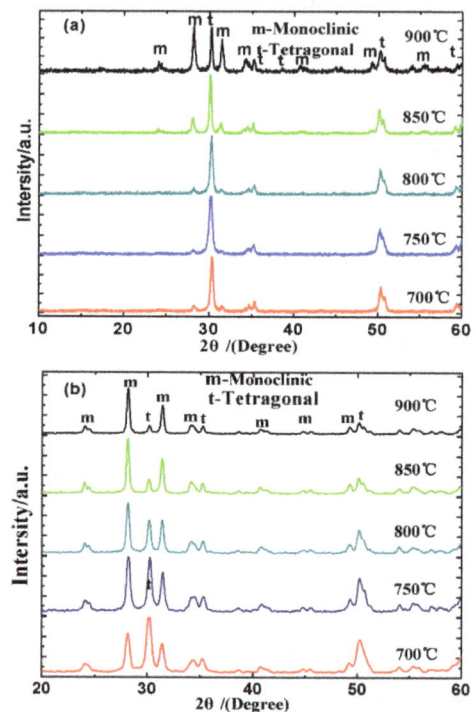

Figure 3: X-ray diffraction patterns of 3Y-ZrO_2 powders synthesized at different temperature by (a) conventional sintering (CS), (b) microwave pyrolysis (MP).

900°C, whereas the monoclinic phase of zirconia appears when the heating temperature is enhanced to 900°C, the prepared powder is a mixture of m-ZrO_2 and the t-ZrO_2, which is the same with the reference [18]. Compared with the CS method, the obtained powders by the MP method are multi-phase, and m-ZrO_2 with t-ZrO_2 are coexisted at every heating temperature, as shown in Figure 3b. It is found that the intensity of the tetragonal phase reflection peaks is greater than the monoclinic phase peaks at 700°C, the relative peaks of tetragonal phase decreased with increasing calcinations temperature, which means the m-ZrO_2 content increased while the t-ZrO_2 content decreased. This results show that the microwave energy can accelerate the formation of m-ZrO_2 phase.

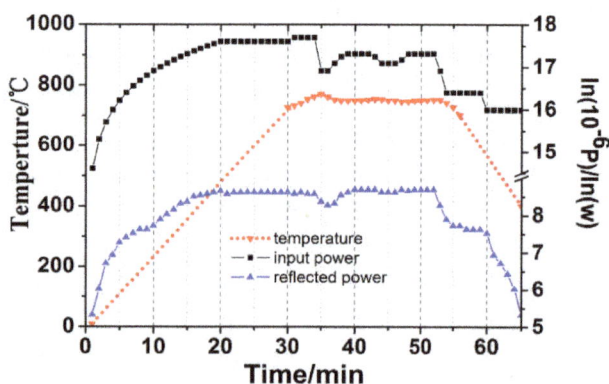

Figure 1: Heating cycle and power variation during microwave pyrolysis of ZrO_2 precursor.

Morphology of ZrO$_2$ powders

The surface morphology and particle sizes of the prepared ZrO$_2$ were examined by SEM images, as shown in Figure 4. Figure 4a shows that ZrO$_2$ powders have been partially crystallized and some residue of precursor coexist with ZrO$_2$ powders. Powders sintered at 750°C show fine crystalline and high dispersity with uniform particle size, as shown in Figure 4b, the average particle size of ZrO$_2$ powders is found to be less than 25 nm. In Figure 4c and 4d, the images of the zirconia powder obtained at 800°C and 850°C. It can be clearly seen that there is an appreciable formations of agglomeration. As the temperature further increases, non-uniform powders with agglomeration and abnormal growth of crystalline grain are observed from Figure 4d and 4e. Therefore, it is clearly understood that in the MP method, the molecular dipoles are induced to oscillate by microwave. This oscillation caused a higher rate of molecular collision which generates enormous amounts of heat. Consequently, the temperature distribution is homogeneous and is transferred to the materials interior, making the ZrO$_2$ particles synchronous growth, and with high dispersity.

The sample sintered by conventional method at 750°C shows bigger particles with serious agglomeration in Figure 3f. Other samples obtained at different temperatures by conventional sintering method present the similar phenomenon (not shown here). There is a temperature gradient between the heat source and the mass to be heated. Thus, during the heating process, the temperature distribution is not homogeneous and cannot be transferred to the materials interior, but spreads more to the particle surface. Consequently, an increase in the size of the crystals takes place owing to solid-state diffusion.

To provide further evidence for the formation mechanism, TEM analysis was carried out only for MP samples. A TEM image of ZrO$_2$ obtained at 750°C is presented in Figure 5a and 5b, indicating that the nearly spherical nano-crystals are uniformly formed. Moreover, the average size of Nano crystallites obtained from the SEM is in a relative agreement with the TEM studies which show the size in the range of 15-30 nm, the average size was 23 nm. The high-resolution TEM image of nanoparticles in Figure 5b shows that the nanoparticles are highly crystalline. The marked lattice fringes correspond to (-111) plane in m-ZrO$_2$ with a d-spacing of 3.1 Å.

Conclusion

The 3Y-ZrO$_2$ Nano sized powder prepared by microwave pyrolysis was thoroughly investigated. Optimized microwave pyrolysis condition is around 750°C for 20 min. The powders were characterized by a narrow particle size distribution, high dispersive and the average

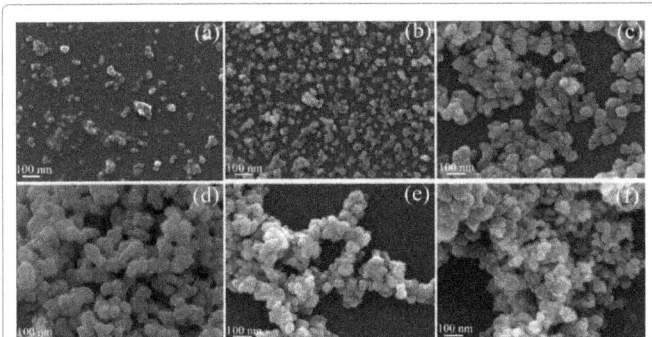

Figure 5: TEM image (a) and HRTEM image (b) of ZrO$_2$ powders.

size was 23 nm. The microwave pyrolysis YSZ powders consisted of tetragonal and monoclinic phase. In the microwave field, the stability of tetragonal phase was weak and debased during the formation of ZrO$_2$.

The results of the XRD, Raman spectra, and SAED show the tetragonal ZrO$_2$ formation when the 3Y-TZP freeze-dried precursor powders calcined at 773-1273 K for 5 min. Moreover, the RAMAN spectrum shows that the tetragonal ZrO2 had already formed in the 3Y-TZP freeze-dried precursor powders. The crystallization activation energy of the tetragonal phase from the 3Y-TZP freeze-dried precursor powders when using a non-isothermal method was 169.2 ± 21.9 kJmol^{-1}. The crystallite growth morphology parameter (n) and crystallization mechanism index (m) were approximated as 2.0. This result means that the tetragonal ZrO$_2$ crystallites have a growth mechanism with a plate-like morphology.

Acknowledgements

This work was sponsored by the National Natural Science Foundation of China (NSFC) (51172113, 51602287 and 51672254).

References

1. Maheswari AU, Kumar SS, Sivakumar M (2013) Influence of alkaline mineralizer on structural and optical properties of ZrO2 nanoparticles. Journal of Nanoscience and Nanotechnology 13: 4409-4414.

2. Singh AK, Nakate UT (2014) Microwave synthesis, characterization, and photoluminescence properties of nanocrystalline zirconia. The Scientific World Journal 2014: 7.

3. Gole JL, Prokes SM, Stout JD, Glembocki OJ, Yang R (2006) Unique properties of selectively formed zirconia nanostructures. Advanced Materials 18: 664-667.

4. Dutta G, Hembram KPSS, Rao GM, Waghmare UV (2006) Effects of O vacancies and C doping on dielectric properties of ZrO2: A first-principles study. Applied physics letters 89: 202904.

5. Gajović A, Furić K, Štefanić G, Musić S (2005) In situ high temperature study of ZrO2 ball-milled to nanometer sizes. Journal of molecular structure 744: 127-133.

6. Taguchi M, Takami S, Adschiri T, Nakane T, Sato K, et al. (2012) Simple and rapid synthesis of ZrO2 nanoparticles from Zr (OEt)4 and Zr(OH)4 using a hydrothermal method. Cryst Eng Comm 14: 2117-2123.

7. Zevert WFMG, Winnubst AJA, Theunissen G, Burggraaf AJ (1990) Powder preparation and compaction behaviour of fine-grained Y-TZP. J Mater Sci 25: 3449-3455.

8. Manivasakan P, Rajendran V, Ranjan Rauta P, Bandhu Sahu B, Krushna Panda B (2011) Synthesis of monoclinic and cubic ZrO2 nanoparticles from zircon. Journal of the American Ceramic Society 94: 1410-1420.

9. Song SH, Gu HZ, Tang XH, Yuan ZX, Wang HZ, et al. (2005) Production of nano-sized yttria-stabilised zirconia powder by means of sol-gel supercritical fluid drying. Journal of materials science 40: 1547-1548.

10. Hsu YW, Yang KH, Chang KM, Yeh SW, Wang MC (2011) Synthesis and crystallization behavior of 3mol% yttria stabilized tetragonal zirconia polycrystals (3Y-TZP) nanosized powders prepared using a simple co-precipitation process. Journal of Alloys and Compounds 509: 6864-6870.

11. Santos T, Valente MA, Monteiro J, Sousa J, Costa LC (2011) Electromagnetic

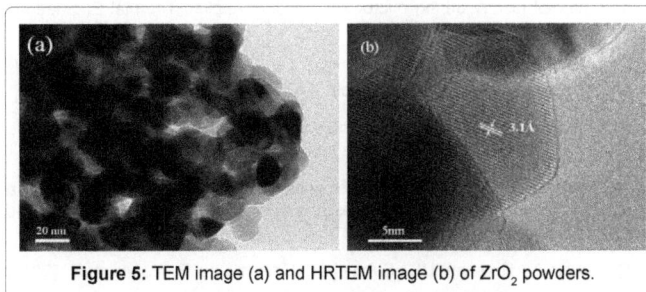

Figure 4: The SEM micrographs of 3Y-PSZ powders by microwave pyrolysis at different temperature: 700°C (a); 750°C (b); 800°C (c); 850°C (d); 900°C (e) and conventional sintered at 750°C (f).

and thermal history during microwave heating. Applied Thermal Engineering 31: 3255-3261.

12. Rybakov KI, Olevsky EA, Krikun EV (2013) Microwave sintering: fundamentals and modeling. Journal of the American Ceramic Society 96: 1003-1020.

13. Upadhyaya DD, Ghosh A, Gurumurthy KR, Prasad R (2001) Microwave sintering of cubic zirconia. Ceramics International 27: 415-418.

14. Oghbaei M, Mirzaee O (2010) Microwave versus conventional sintering: a review of fundamentals, advantages and applications. Journal of Alloys and Compounds 494: 175-189.

15. Rajeswari K, Hareesh US, Subasri R, Chakravarty D, Johnson R (2010) Comparative evaluation of spark plasma (SPS), microwave (MWS), two stage sintering (TSS) and conventional sintering (CRH) on the densification and micro structural evolution of fully stabilized zirconia ceramics. Science of Sintering 42: 259-267.

16. Zhao C, Vleugels J, Groffils C, Luypaert PJ, Van der Biest O (2000) Hybrid sintering with a tubular susceptor in a cylindrical single-mode microwave furnace. Acta materialia 48: 3795-3801.

17. Jinsong Z, Yongjin Y, Lihua C, Shengqi C, Xiaoping S, et al. (1994) Microwave Sintering of Nanocrystalline ZrO_2 Powders. In MRS Proceedings 347: 591-593.

18. Patil RN, Subbarao EC (1970) Monoclinic–tetragonal phase transition in zirconia: mechanism, pretransformation and coexistence. Acta Cryst 26: 535-542.

29

Microhardness and Adhesion Strength of PMC's Coatings by NiCr Alloy

Kareem AA* and Raheem Z

Department of Physics, College of Science, University of Baghdad, Iraq

Abstract

The use of polymer matrix composites (PMCs) in the gas flow path of advanced turbine engines offers significant benefits for aircraft engine performance but their useful lifetime is limited by their poor environmental resistance. Flame sprayed NiCr graded coatings are being investigated as a method to address this technology gap by providing high temperature and environmental protection to polymer matrix composites. In this research coating was spread with two configuration, coating with bound coat and coating without bound coat.

In general the coating with bound coat and coating without bound coat showed increase in micro hardness and adhesion with increase curing temperature; this is due to the microstructural changes the physical splat structure of the coating also changes with heat treatment. All coating failed at the interface between the composites and the coating, failure occurs along the weakest plane within the system, some of the coating systems that have presented fracture at the bond coat/top coat interface. The surface topography of NiCr films was further examined by using AFM atomic force microscopy as a function of curing temperature at 100,200 and 300°C for 1 h each, it can be clearly seen that the island structure was observed and the R_{max} increase, the surface became rougher with increasing curing temperature. The surface morphology and microstructure of the coating were examined using SEM.

Keywords: Protective polymer fiber composites; Polymer matrix composites in aerospace applications; High temperature flame spray coating; Hard coating

Introduction

Coating and surface modification technologies allow the engineer to improve the performance, extend component. Surface engineering is defined as the design of a composites system of a surface and a substrate together to give a performance which cannot be achieved breather the surface or the substrate alone [1]. The primary benefit in replacing metals with lower density, higher specific strength PMC's is the weight savings. Additional advantages are the lower processing and fabrication costs [2]. Polymer matrix composites can be successfully deposited by with Thermal spray coating. Successful deposition of a wide array of materials shows that thermal spray coating is available technology for the polymer composites surface protection [3]. A graded coating composition or structure improves the load coatings is astright forweard process and not as defecult as metallographic prepareion. The system can consist of a coating with or without an interface [4]. Since polyimides are thermally stable at high temperature they are a popular choice for structural parts in aerospace applications, where metal replacement is required with lightweight materials. Polyimide adhesives are used for joining metals and high temperature composites because their coefficient of thermal expansion is comparable to that of metals [5].

Applications of these coatings are widespread and can be found in aerospace, petrochemical [6]. The material selection for turbine engines is a balance between the cost and efficiency, high-strength NiCr alloys are often used in the aero-engine applications for weight reduction [7]. The micro-indentation indentation technique has been used to characterize the material properties and of coating materials because it is simple and can be performed on small specimens [8].

Experimental

A woven Carbon fiber epoxy composite was selected as substrate; the hand lay-up technique was used to prepare these composites with volume fraction 30%. The composites specimen was cleaned with acetone to remove moisture, dirt oil and other foreign particle. The coating that improves the adherence of the subsequent deposited is called bond coat. Polyimide are used as bond coat, In this work, pyromellitic dianhydride (PMDA) and p-phenylene diamine (PDA), which are commercially available from Sigma-Aldrich are used to prepare polyimide by thermal evaporation technique. These two monomers, 2 g each, were evaporated from two separated boats to form a poly (amic acid) (PAA) thin film on substrate. The deposition process began at vacuum of 2×10^{-5} mbar. The resultant polyamic acid PAA film was then soft baked to remove nH_2O from the substrate followed by a thermal treatment at 250°C for 1 h each in an air circulating oven, and deposited polyimide film into the composites substrate. The final thickness of films is 5 ± 0.1 μm.

On the other hand NiCr is used as atop coat. The elemental composition of NiCr alloy samples used in this work was made by using X-ray fluorescence (XRF) analysis technique as shown in Table 1. Spray Gun (rototec 80), it's used for thermal spraying by flame which was made in Germany by (Castolin+Eutectic) Company. In this process oxygen-acetylene mixture is passed through a nozzle and ignited to form a combustion flame. Ni-Cr Coating powder with particle sizes ranging from 50 to 90 μm were used is fed into the flame, accelerated and projected onto the substrate to form a top coating with thickness about 70 ± 2 μm calculated by magnetic induction measurement methods. The flame temperature is limited to around 1400°C, particle velocities are relatively slow.

Operating parameters during coating deposition process are listed

*Corresponding author: Kareem AA, Department of Physics, College of Science, University of Baghdad, Iraq, E-mail: aseelalobaedy@yahoo.com

in Table 2. Before coating the samples are cured at (100, 200 and 300°C). Hardness type Vickers was conducted for all samples by using (Hensddt Wetzlar). Vickers hardness values were calculated according to the following equation:

$$HV = 1.8544 \frac{F}{d^2} \left(kgf/mm^2 \right) \tag{1}$$

Where F is applied load (kgf) and d is the main diagonal of indentation (mm). The controlled electronic universal testing machine used for pull off adhesion tests, and it is type is (WDW- 200E). The bond strength is found from the simple relation between the composites and the NiCr top coating [9,10].

$$UTS = L/A \tag{2}$$

Where: UTS=cohesive or adhesive strength - force per unit of surface area; L=load to failure (force); A=cross sectional area of specimen.

Results and Discussions

Hardness is described as resistance to surface indentation of the material. In Figure 1 the response of the uncoated composites to heat treatment induced softening of the microstructure and account for the reduction in hardness. Heat treatment in air generated higher average hardness values in coating systems, the coating with bound coat and coating without bound coat showed increase in micro hardness with increase temperature; this is due to the microstructural changes the physical splat structure of the coating also changes with heat treatment [11]. It is found that the degree of fusion of the particles and the presence of an oxide phase have effect on the microhardness of the coatings [7,12]. It can clearly see in Figure 2. At room temperature PMCs with NiCr coatings had enhanced high hardness, this is due to the hardness of NiCr. The increase in the hardness in the composites coating with polyimide bound coat is the indication of good polyimide bonding.

The results of pull off tests are shown in Figure 3. Adhesion strength for the PMCs coating with polyimide bound coat is higher than PMCs coating without bound coat. The adhesive strength between the polyimide and metal was affected by the chemical state of bonding on the surface in polyimide films; the hydrophilic bonding such as C-O bonding is believed to be suitable for enhanced adhesion between polyimide thin films and NiCr [13]. During the spray process, there is some partial formation of intermetallic phases. Subsequent fusing of the coating causes a complete transformation of the materials [14].

We can see from Figure 4. When curing temperature increase the interlocking (and then adhesion) increase because of diffusion into the substrate also occurs, improving bonding. Porosity is nearly eliminated, with no interconnecting porosity and the formation of hard oxide phases leads to increases the roughness of substrate surface [10,15].

Figure 5 shows that all coating failed at the interface between the composites and the coating failure occurs along the weakest plane within the system, some of the coating systems that have presented fracture at the bond coat/top coat interface. In most cases there is a cohesive failure occur of the substrate [15]. Figure 6 gives 3D topography of films. For film surface, R_{max} is explained as maximum

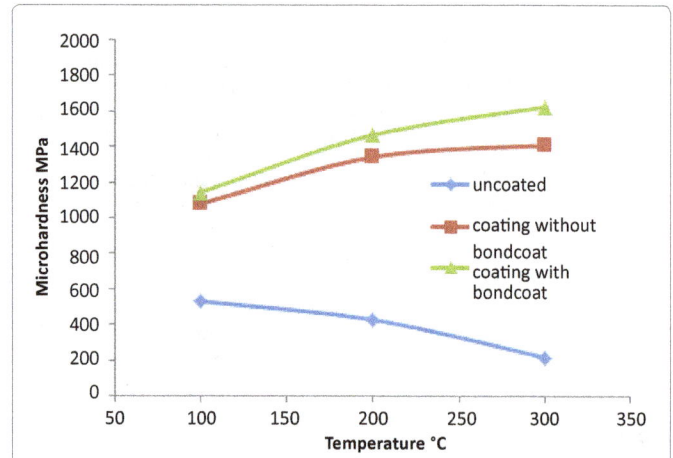

Figure 1: Microhardness of coated and uncoated PMCs as a function of temperature.

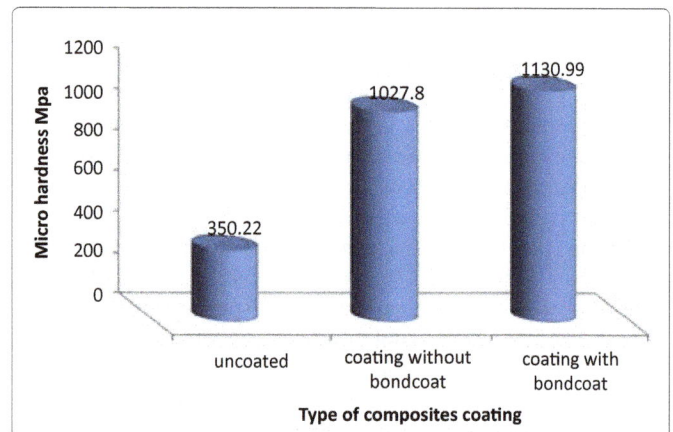

Figure 2: Type of Composite Coating.

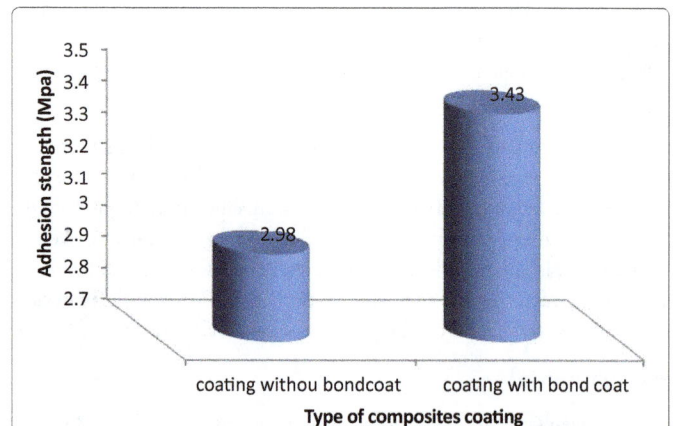

Figure 3: Adhesion strength of coated and uncoated PMCs at room temperature.

Table 1: Elemental composition of the powder used for deposition of coatings.

Powder	Elemental Composition (%)					
	Ni	Cr	Si	C	Fe	other
NiCr	43.4	52.6	0.13	0.62	0.17	0.08

Table 2: Operating parameters during coating deposition process.

Operating Parameters	Values
Oxygen pressure	4 bar
Acetylene pressure	0.7 bar
Standoff distance	200 mm

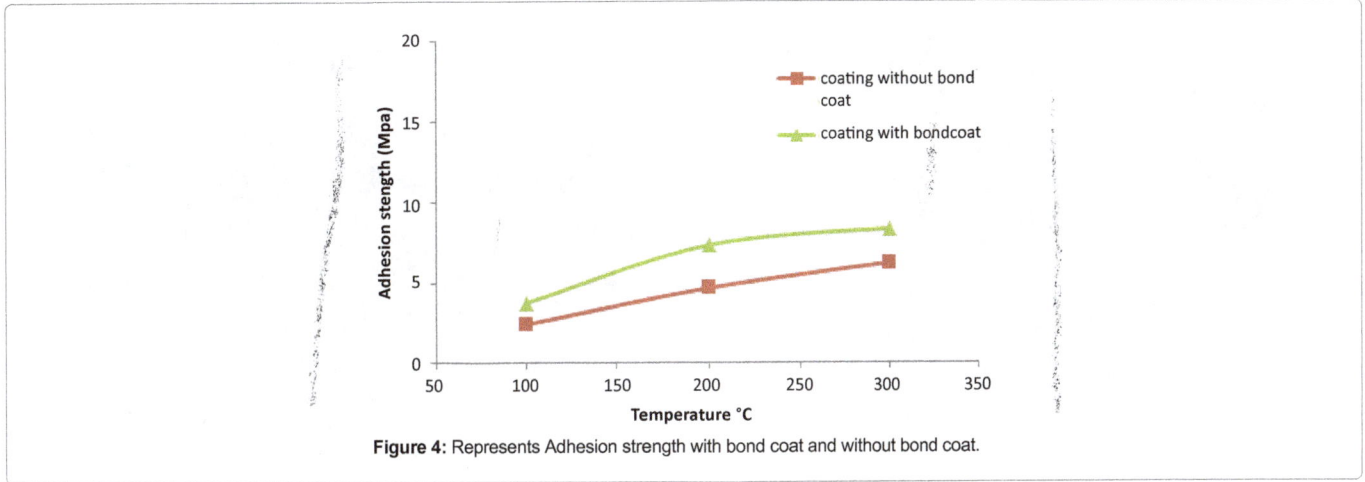

Figure 4: Represents Adhesion strength with bond coat and without bond coat.

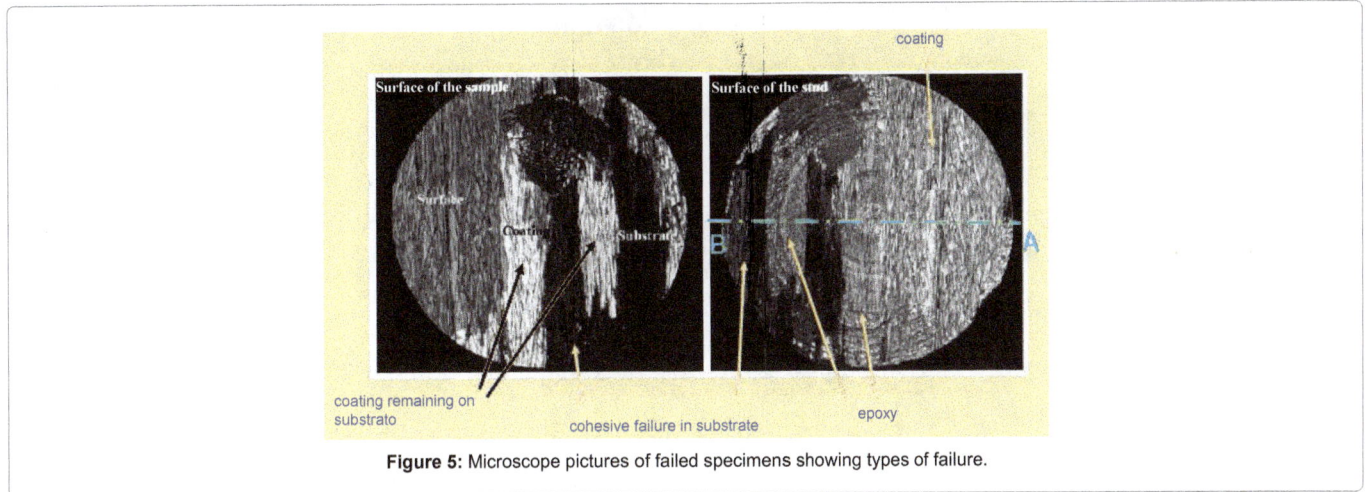

Figure 5: Microscope pictures of failed specimens showing types of failure.

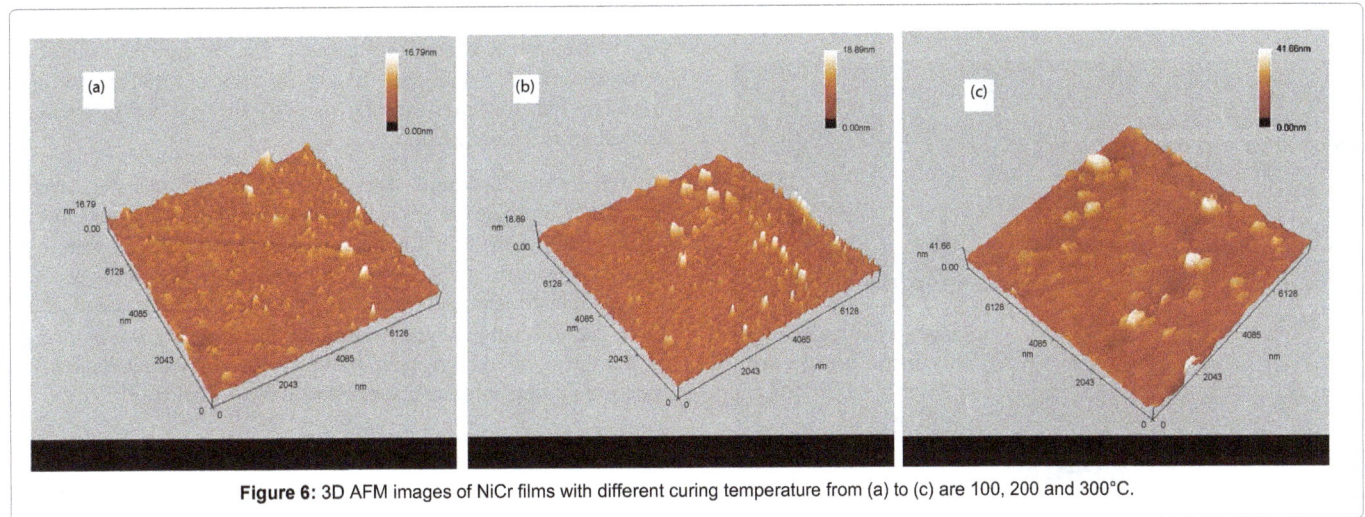

Figure 6: 3D AFM images of NiCr films with different curing temperature from (a) to (c) are 100, 200 and 300°C.

height of peak to valley for the depicted surface. σ is the root mean square roughness. With curing temperature ranging from 100°C to 300°C, it can see the R_{max} is equal (0.8, 0.9 and 2.28) nm and σ is equal (1.16, 1.47 and 3.54) nm. When the film is cured at 100°C, islands with small size are observed. However, when the film is cured at 300°C, the islands have agglomerated or coalesced to form bigger structure. The phenomenon can be explained by film growth process: during deposition process, particles are deposited and form nucleus first and then islands on substrate. This is mostly caused by atomic shadowing effects, which makes R_{max} reach 2.28 nm and σ 3.54 nm, and the film surface turns rough correspondingly as shown in Figure 6c.

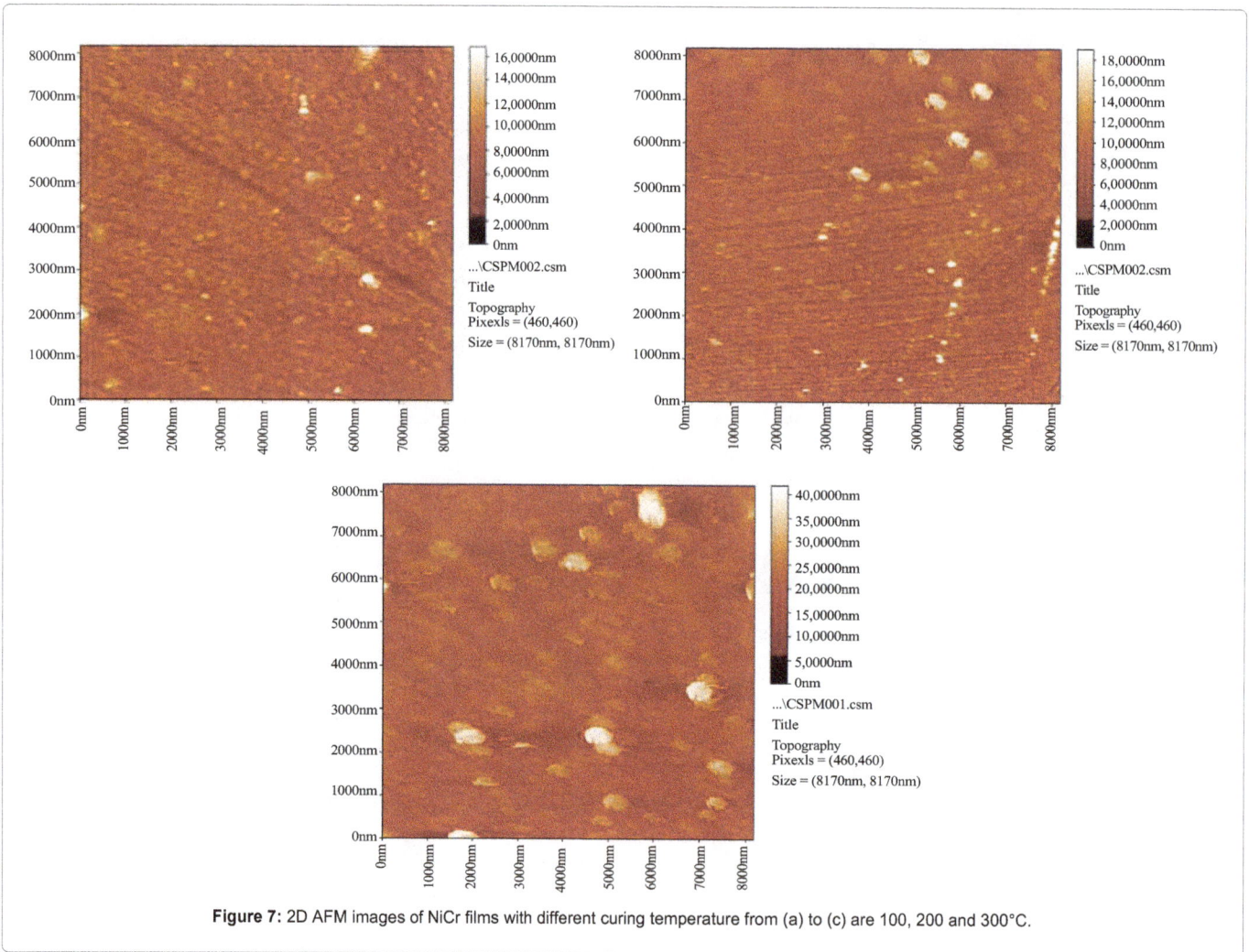

Figure 7: 2D AFM images of NiCr films with different curing temperature from (a) to (c) are 100, 200 and 300°C.

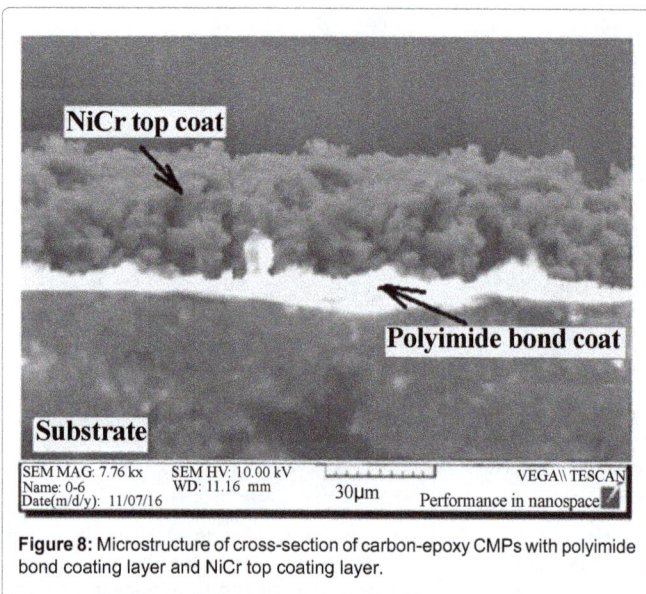

Figure 8: Microstructure of cross-section of carbon-epoxy CMPs with polyimide bond coating layer and NiCr top coating layer.

However, when surface diffusion dominates the growing process, the coalescence of neighboring islands makes the valleys become higher and the peak become lower, consequently the surface becomes flat and

R_{max} is decrease as known in Figure 6a. The film growth is finished by coalescence of neighboring islands. Surface morphology not only relates to film thickness but also to substrate type, works pressure, annealing and so on [14,16].

When the temperature reaches 100°C, the substrate was covered completely by spherical grains with similar radius. With an increase of temperature, the lateral grain size tends to increase. As seen from Figure 7, the lateral grain size changes from about 14.8 nm to 36.7 nm when temperature ranges from 100°C to 300°C. The increase of lateral grain size with temperature is common for films [16]. In Figure 8 an abrupt transition from the bond coating to the top coating that leads to their top coating in intimate contact with bond coating [2].

Conclusions

This paper presents an experimental process to protect polymer matrix composite (PMCs) by metallic flame spray coating. The results of the investigations provide useful information for applying the NiCr coating for the improvement of the hardness of PMCs. According to the results of this study, In general the coating with bound coat and coating without bound coat showed increase in micro hardness and adhesion strength with increase temperature. The adhesion strength for the PMCs coating with polyimide bound coat is higher than PMCs coating without bound coat.

The AFM analysis also provides information on the changes in the surface morphology and roughness introduced by the heat treatment. When a temperature change from 100°C to 300°C, the island structure was observed and the R_{max} increase from (0.8 to 2.28) nm, and σ increase (1.16 to 3.54) nm.

References

1. Muktinutalapati NR (2011) Materials for Gas Turbines–An Overview. Intech, Croatia.

2. Miyoshia K, Suttera J, Horanb R, Naikb S, Cupp R (2004) Assessment of erosion resistance of coated polymer matrix composites for propulsion applications. Tribol Lett 17: 377-387.

3. Amado J, Montero J, Tobar M, Yáñez A (2012) Ni-based metal matrix composite functionally graded coatings. Phys Proced 39: 362-367.

4. Hetmańczyk M, Swadźba L, Mendala B (2007) Advanced materials and application, protective coatings in aero-engines. J Achiev Mater Manufact Eng 24: 372-381.

5. Ivosevic M, Knight R, Kalidindi S, Palmese G, Sutter J (2005) Adhesive / cohesive properties of thermally sprayed functionally graded coating for polymer matrix composites. J Therm Spray Technol 14: 45-51.

6. Picas J, Forna A, Matth¨aus G (2006) HVOF coatings as an alternative to hard chrome for pistons and valves. Wear 261: 477-484.

7. Hadad M, Marot G, De´mare´caux P, Chicot D, Lesage J, et al. (2007) Adhesion tests for thermal spray coatings:correlation of bond strength and interfacial Toughness. Surf Eng 23: 279-283.

8. Sidhu H, Sidhu B, Prakash S (2006) Comparative Characteristic and Erosion Behavior of NiCr Coatings Deposited by various High-Velocity Oxyfuel Spray Processes. J Mater Eng Perform 15: 699-704.

9. Brossard S, Munroe P, Tran A, Hyland M (2010) Study of the Splat-Substrate Interface for a NiCr Coating Plasma Sprayed onto Polished Aluminum and Stainless Steel Substrates. J Therm Spray Technol 19: 24-30.

10. Richert M, Leszczyńska-Madejj B (2011) Effect of the annealing on the microstructure of HVOF deposited coatings. Achiev Mater Manufact Eng 46: 95-102.

11. Sidhu B, Prakash S (2006) Nickel-Chromium Plasma Spray Coatings:A Way to Enhance Degradation Resistance of Boiler Tube Steels in Boiler Environment. J Therm Spray Technol 15: 131-140.

12. Harsha S, Dwivedi D, Grawal A (2007) Influence of WC addition in Co–Cr–W–Ni–C flame sprayed coatings on microstructure, microhardness and wear behavior. Surf Coat Technol 201: 5766-5775.

13. Nakamura Y, Suzuki Y, Watanabc Y (1996) Effect of oxygen plasma etching on adhesion between polyimide films and metal. Thin Solid Films 290-291: 367-369.

14. Jicheng Z, Li T, Jianwu Y (2008) Surface and Electrical Properties of NiCr Thin Films Prepared by DC Magnetron Sputtering. J Wuhan Univer Technol Mater Sci Ed 23:159-162.

15. Lesagea J, Staiab M, Chicota D, Godoyc C, De Miranda P (2000) Effect of thermal treatments on adhesive properties of a NiCr thermal sprayed coating. Thin Solid Films 377-378: 681-686.

16. Patil A, Patil V, Choi J, Kim H, Cho B, et al.(2009) Structural and electrochemical properties of Nichrome anode thin films for lithium battery. J Electroceram 23: 230-235.

30

Improvement in Tensile Strength and Microstructural Properties of SAW Welded Low Alloy Steels by Addition of Titanium and Manganese in Agglomerated Flux

Chandra RK*, Majid M, Arya HK and Sonkar A

Department of Mechanical Engineering, SLIET, India

Abstract

The present work is an effort to study the effect of titanium and manganese powder addition in agglomerated flux, on the mechanical properties, of MS 1025 steel welds made by submerged arc welding. The effect of titanium and manganese powder addition on the agglomerated fluxes by varying the welding parameters like welding voltage and welding speed has been evaluated. Taguchi technique has been used for the design of experiments. The effects of flux, voltage and travel speed have been evaluated on the tensile strength and on the microstructure refining. The effect of all the input parameters on the output responses have been analyzed using the analysis of variance (ANOVA).

Keywords: SAW; Tensile strength; Micro structure and various properties using Taguchi technique

Introduction

Submerged arc welding is an arc welding process that uses an arc between a bare metal electrodes and the weld pool. The arc is maintained in a cavity of molten flux or slag which refines the weld metal and also protects it from atmospheric contamination [1-3]. Flux is the main ingredient on which the stability of the arc depends. The Unused flux can be extracted from left behind the welding head and subsequently recycled. The filler material is an uncoated, continuous wire electrode, applied to the joint together with a flow of granular flux, which is supplied from a flux hopper through a tube. The electrical resistance of the electrode should be as low as possible to facilitate welding at a high current, and so the welding current is supplied to the electrode through contacts very close to the arc and immediately above it. The current can be direct current with electrode positive (reverse polarity), with negative (straight polarity), or alternating current. The arc burns in a cavity which, apart from the arc itself, is filled with gas and metal vapors [4-6]. The size of the cavity in front of the arc is delineated by the scrap basic material and behind it by the molten weld.

Welding of 1.25Cr-0.5Mo steel with SiO_2 and TiO_2 based flux increases the heat input, lowers the yield strength and decreases the % elongation [7]. Manganese fluxes give lower residuals of sulphurwhere as calcium silicate fluxes were responsible for removal of phosphorus [4]. Because of small grain size and high angle grain boundaries acicular ferrite is desired microstructure in weld for high impact strength and ductility [8-10]. Low level of oxygen is desired in the weld metal for good impact strength [11]. The effect of wire/flux combination on chemical composition, tensile strength and impact strength of weld metal were investigated and interpreted in terms of element transfer between the slag and weld metal [12-14]. Chemical composition of some flux were studied for microstructure and tensile strength in submerged arc welded metal and compared with those off commercial available flux [9]. Non active flux enhance the formation of pearlite and ferrite in weld having the highest toughness and ductility where as active flux with Cr and Mo promoted the formation of acicular ferrite and fine carbide in the weld showing higher tensile strength and hardness [12]. After literature study we find that very few efforts have been made to understand mechanical properties using Taguchi Technique and Very little work is made to improve the weld joint strength in single pass [15-17]. In this experiment we added titanium and manganese metal powder in 9%, 14% and 18% concentration of titanium and 1.4%, 1.7% and 2% manganese in AUTOMELT B31 flux

and investigate its effect on the mechanical properties. The optimization of result has been done by using Taguchi's Philosophy [18,19].

Methodology

Material selection

1025 mild steel is selected over other materials because of its distinct properties, cheaper cost and its Availability in the market.1025 plain carbon steel is used in the manufacturing of ships hulls Table 1.

Flux preparation

The agglomerated flux is prepared by the addition of the TiO_2 and MnO_2 powder in auto melt B- 31. Both the ingredients and Auto melt B-31 were crushed and wet mix in sodium silicate to form chemical bonding. Mixture is then passed through a 10 mesh screen to form small pallets (up to 2 mm diameter). The pellets of the flux were dried in air for 24 hours and then baked in the muffle furnace between 750°C (Figure 1 and Table 2).

Taguchi method analysis

In Taguchi method first optimal parameters were determined by using L9 orthogonal array. L9 means that it will investigate for 3 levels and 3 factors on qualitative index for each factor. Table 3 gives the levels and factors which are employed for welding the samples. Table 4 gives the experimental data that is taken for analysis [20].

Experimentation

In this study L 9 was chosen as the preferred array, 18 plates were cut to size of dimension 150 × 120 × 14 mm for experiment because four

*Corresponding author: Chandra RK, Department of Mechanical Engineering, SLIET, India, E-mail: Rinku007chandra@gmail.com

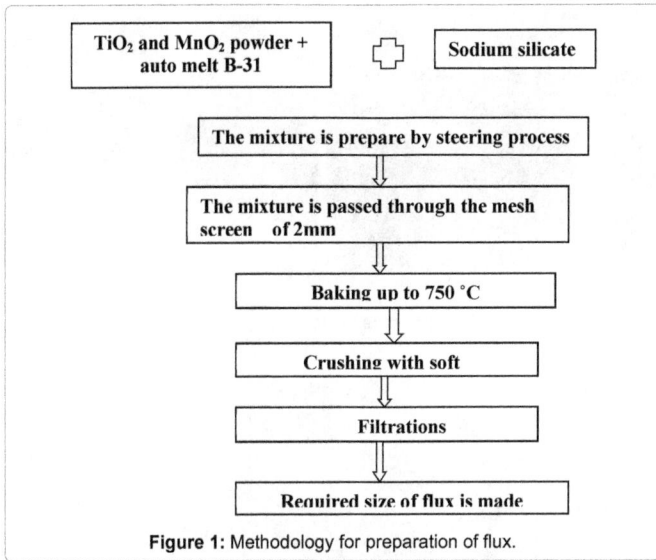

Figure 1: Methodology for preparation of flux.

Elements	C	Si	Mn	Ti	S	Fe
Base plate	0.250	0.1728	0.5050	0.0022	0.0426	Rest
Electrode	0.068	0.17	0.4	-	-	Rest

Table 1: Chemical composition of base plate.

SiO_2+ TiO_2	Al_2O_3+MnO	CaO + MgO	CaF_2
15%	30%	20%	35%

Table 2: Chemical composition of auto melt B31.

Parameter	% age composition of TiO_2	% age composition of MnO_2
Level 1	9	1.4
Level 2	13.5	1.8
Level 3	18	2.0

Table 3: Levels and factors used in design.

Sample no.	% age composition of TiO_2	% age composition of MnO_2
1	9	1.4
2	9	1.7
3	9	2.0
4	13.5	1.4
5	13.5	1.7
6	13.5	2.0
7	18	1.4
8	18	1.7
9	18	2.0

Table 4: Design of taguchi orthogonal array.

plates were required for welding in each trial. Edge preparation was completed on each of the 18 plates as per the requirement of 9 trial condition given by Taguchi orthogonal array. The edge preparation of 45 degree was made on 120 mm side Edge preparation was done on shaper. Groove between two plates. After edge preparation, the plates were tacked through manual arc welding to avoid misalignment. Following tests are carried out in the present study to study mechanical properties.

a) Tensile strength test

b) Micro structural study

c) Spectroscopy

Results and Discussion

Table 5 shows percentage composition of TiO_2 and MnO_2.

Analysis of tensile strength

Tensile strength analysis done here is dependable variable on two factors:

1. Percentage composition of TiO_2

2. Percentage composition of MnO_2

Analysis of variance for S/N ratios for tensile strength calculated from software is given in Table 6. Since the p -value in the Table 6 is less than 0.05; there is a statistically significant relationship between the variables at the 95.0% confidence level. From the Table 6 it can be easily figure out that percentage composition of MnO_2 are most significant.

The main effect of MnO_2 is to increase ductility of the weld metal and slight loss in the tensile strength which is overcome by TiO_2 so nominal is better option is selected from Taguchi design and accordingly response for signal to noise ratio is generated. It was found that level-3 of Percentage composition of TiO_2 and level-3 of e composition of MnO_2 gives the optimum values [21].

In response Table 7 Delta value indicates the variability of factors. Table 7 shows that percentage composition of MnO_2 has highest variability. Also rank of parameters in response Table 7 indicates the order at which the parameters influence the quality characteristics. So MnO_2 has greatest influence on the Tensile strength of the weld.

Mathematical model

Tensile strength = 20 + 3.63 % composition of TiO_2 + 197.8% composition of MnO_2. Table 8 shows the model summary.

Main effects for S/N ratios

Figure 2 shows main effect plots for S/N ratio of tensile strength.

Percentage composition of TiO_2	Percentage Composition of MnO_2	Tensile Strength(MPa)	S /N ratio of tensile strength
9	1.4	320	49.6575
9	1.7	445	52.9672
9	2.0	465	52.1919
13.5	1.4	359	51.1019
13.5	1.7	490	50.9061
13.5	2.0	529	54.4691
18	1.4	330	51.7990
18	1.7	393	51.8879
18	2.0	472	53.4788

Table 5: Shows percentage composition of TiO_2 and MnO_2.

Parameter	Degree of freedom(DF)	Sum of square(SS)	Mean square(MS)	F– value	P –value
Percentage composition of TiO_2	2	6291	3145.4	7.37	0.046
% age composition of MnO_2	2	36692	18314	42.91	0.002
Residual error	4	1707	426.8		
Total	8	44626			

Table 6: Analysis of variance for S/N ratios for tensile strength.

Interaction effect for tensile strength

Microstructures analysis of various specimens was done with an optical microscope of magnification up to 100 X to compare microstructure of the welded samples. Images were taken that are shown in Figure 4-7. All the welded samples show a microstructure composed of acicular ferrite. It is conforming that Ti containing white inclusions in the sub arc weld metals plays a very important role The term interaction expressed by inserting "x" mark between the two interacting factors is used to explain a condition in which the influence of one factor upon the result, is dependent on the condition of the other [22-24]. The two factors A & B are said to interact written as (A×B) when the effects of changes in the levels of A, determines the influence of B and vice versa. If there is absolutely no interaction, these lines would be parallel. Figure 3 shows

Level	Percentage composition of TiO_2	Percentage composition of MnO_2
1	51.61	50.85
2	52.16	51.92
3	52.39	53.38
Delta	0.78	2.53
Rank	2	1

Table 7: Response for signal to noise ratios: nominal is better.

S	R-sq	R-sq(adj)	R-sq(predicted)
20.6586	96.17%	92.35%	80.63%

Table 8: Model summary.

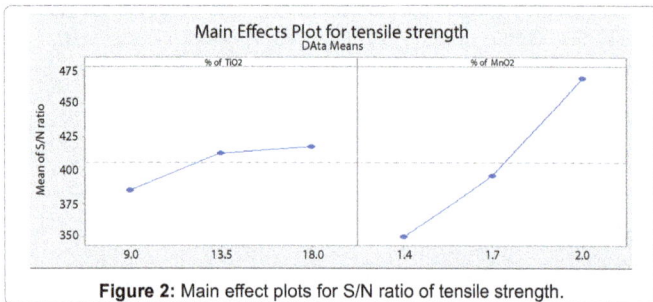

Figure 2: Main effect plots for S/N ratio of tensile strength.

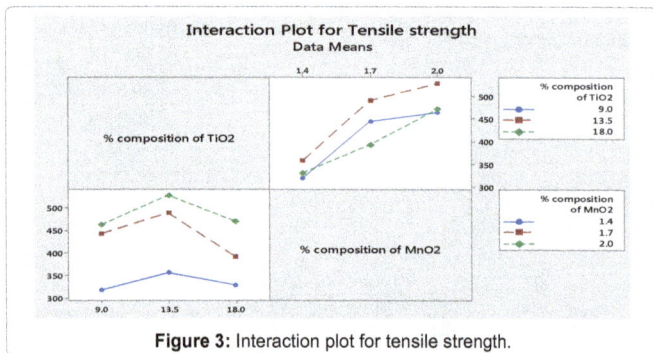

Figure 3: Interaction plot for tensile strength.

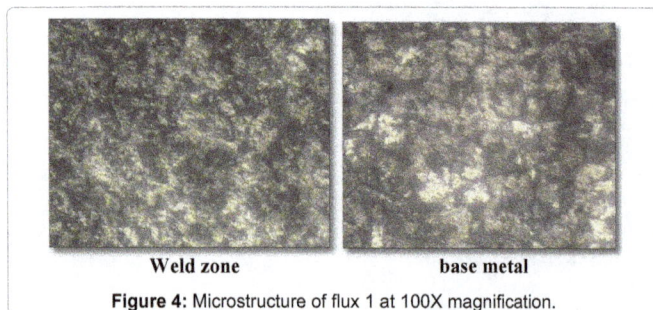

Figure 4: Microstructure of flux 1 at 100X magnification.

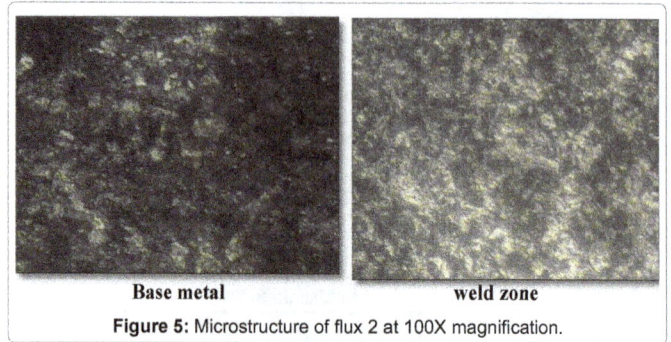

Figure 5: Microstructure of flux 2 at 100X magnification.

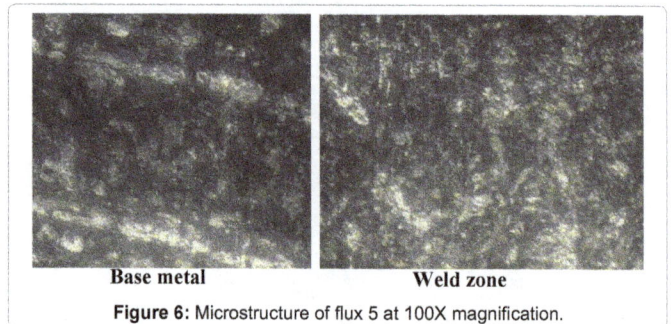

Figure 6: Microstructure of flux 5 at 100X magnification.

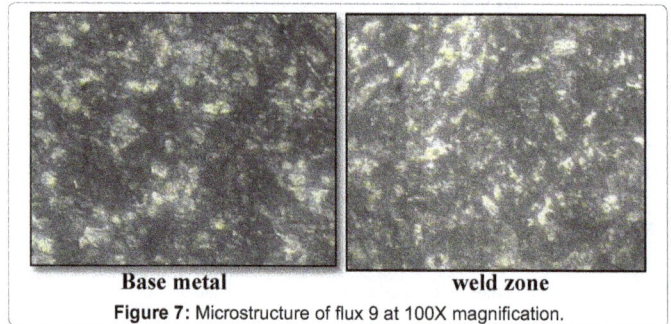

Figure 7: Microstructure of flux 9 at 100X magnification.

the interaction plots for S/N ratio. The interaction plot for the tensile strength shows that the tensile strength is improved at the less amount of percentage of TiO_2 and then it becomes reduced due to the quasi cleavage fracture occurring in the microstructure to overcome this issue we add the manganese in the weld metal to improve the tensile strength it becomes highest at the moderate value of manganese [25].

Micro-structural analyses

Micro-structural analyses for heterogeneous nucleation of acicular ferrite. Figure 7 shows the microscopic magnification graphs of the welded steels corresponding to the fluxes with different percentage composition of TiO_2. The microstructure of the weld metal for each flux consisted mainly of equiaxed ferrite and acicular ferrite. Some small pearlite colonies were also observed in the welds. The micro constituent's volume percentage, as well as mean radius of equiaxial ferrite and mean length of acicular determined by means of quantitative metallographic. The lowest and highest volume percentages of acicular ferrite corresponded to the weld with 9% and 18 % of TiO_2 fluxes respectively.

Chemical composition analysis of welds

The chemical composition of the weld metal is obtained from the spectroscopy of the weld joints and the specimen of the spectroscopy (Figure 8 and Table 9).

% age composition	C%	Si%	Mn%	Cr%	Mo%	Ni%	Cu%	Al%	S%	Ti%
base plate	0.25	0.1728	0.505	0.1406	0.0055	0.0485	0.1232	0.0098	0.0426	0.0022
F1 (9, 1.4%)	0.151	0.2361	1.1115	0.0971	0.0052	0.0416	0.0963	0.0278	0.0217	0.0143
F2 (9, 1.7%)	0.141	0.2441	1.1358	0.0987	0.006	0.039	0.089	0.0333	0.0214	0.0195
F3 (9, 2%)	0.125	0.2048	1.0745	0.0889	0.0034	0.0327	0.0957	0.0121	0.0254	0.0092
F4 (13.5, 1.4%)	0.143	0.1691	0.9941	0.0987	0.0026	0.0279	0.1127	0.0121	0.0281	0.0096
F5 (13.5, 1.7%)	0.125	0.1829	1.0682	0.0923	0.0026	0.0244	0.1109	0.0169	0.0272	0.0132
F6 (13.5, 2%)	0.14	0.1961	1.0312	0.0773	0.0023	0.0247	0.1126	0.0148	0.0277	0.0165
F7 (18, 1.4%)	0.125	0.1611	0.9461	0.088	0.0034	0.0327	0.0957	0.0121	0.0254	0.0092
F8 (18, 1.7%)	0.133	0.1675	1.0074	0.0932	0.0025	0.0215	0.1135	0.0131	0.0327	0.0125
F9 (18, 2%)	0.146	0.2351	1.0754	0.0945	0.005	0.0412	0.0878	0.0199	0.0212	0.0178

Table 9: Chemical composition of the base metal and weld metals.

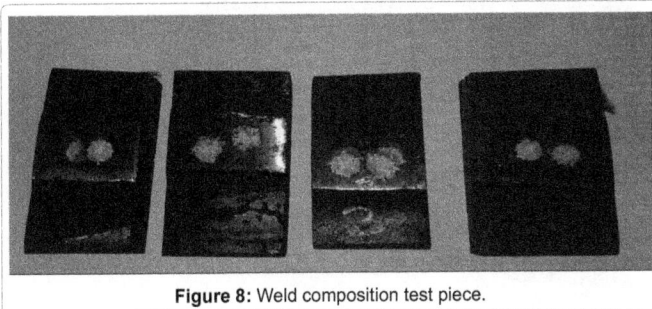

Figure 8: Weld composition test piece.

Conclusions

The present study was carried out to study the effect of titanium and manganese addition in submerged arc welding flux on the mechanical properties of AISI 1025 welds by keeping all other variables constant like current, speed and arc voltage. The tensile strength value is maximum (529Mpa) at flux composition having titanium 13.5% and 2% manganese at 400 A and 27 V. As a result of increasing the titanium content in the flux, the amount of acicular ferrite was increased in the weld metal. With the further addition of titanium, the microstructure the weld has changed from a mixture of ferrite, grain-boundary ferrite to a mixture of acicular ferrite, bainite and ferrite with M/A micro-constituents. The higher percentage of titanium in weld metal encouraged formation of hard phases in the weld microstructure which increases the micro hardness of the weld. The increase in the titanium content in the fluxes improved the ductility and toughness of the weld due to the formation of white inclusions in the sub-arc weld metal and heterogeneous nucleation of acicular ferrite but tensile strength is slightly loss which was overcome by the addition of the moderate % age of manganese in the flux, it refines the microstructure and micro constituent in the flux. The micro structural refining improved the impact strength properties of the weld but the hardening effect produced by the manganese reduced the toughness which is overcoming by % age composition of titanium improves the toughness of the weld. From the spectroscopy of each weld it is found that the carbon percentage in weld is reduced due to use of low carbon electrode for welding the plates. Also the percentage composition of titanium and manganese is increased in the weld when compared with the base metal of each weld.

From the results it concluded that the titanium addition and manganese addition in the flux helped to refine the micro structure of the weld moreover it improves the mechanical properties like tensile strength which is necessary objective of the research work.

Acknowledgement

I would like to express a deep sense of gratitude and thanks profusely to my esteemed guide Mr. Mohd. Majid, Assistant professor and Mr. H.K. Arya, Assistant Professor Department of Mechanical Engineering, SLIET Longowal, for his exceptional patience , for all kind of help and for being there to encourage me all the times. His enthusiasm and optimism made this experience both rewarding and enjoyable. Thank you for all what you have done for me and for providing me with valuable feedback in my research which made it stronger and more valuable. I humbly express my thanks to Dr. Kulwant Singh (HOD), Department of Mechanical Engineering and Technology, SLIET, for his concern at every step of my project and providing me necessary facilities.

References

1. Pandey ND, Bharti A (1994) Effect of submerged arc welding parameters and fluxes on element transfer behavior and weld-metal chemistry. Journal of Materials Processing Technology 40: 195-211.

2. Dallas CB (1995) Studied the effect of flux composition on microstructure and toughness of HSLA weldment by submerged arc welding. Journal of Materials Processing Technology 62: 875-878.

3. Tuseka J (2003) Studied the multiple-wire submerged-arc welding and cladding with metal-powder addition. Journal of Materials Processing Technology 35: 135-141.

4. Singh K, Pandey S, Arul MR (2005) Effect of recycled slag on bead geometry in submerged arc welding. Proceedings of International Conference on Mechanical Engineering in Knowledge Age.

5. Mercado A, Lopez-Hirata VM, Saucedo Munoz ML (2005) Influence of the chemical composition of flux on the microstructure and tensile properties of submerged-arc welds. Journal of Materials Processing Technology 169: 346-351.

6. Kanjilal P, Pal TK, Majumdar SK (2005) Combined effect of flux and welding parameters on chemical composition and mechanial properties of submerged arc weld metal. Journal of Materials Processing Technology 171: 223-231.

7. Prasad K, Dwivedi DK (2006) Some investigations on microstructure and mechanical properties of submerged arc welded HSLA steel joints. Journal of Advanced Manufacturing Technology 36: 475-483.

8. Bhole SD, Nemade JB, Collins L, Liu C (2006) Effect of nickel and molybdenum additions on weld metal toughness in submerged arc welded HSLA line-pipe steel. Journal of Materials Processing Technology 173: 92-100.

9. Shao- Huo S (2006) Studied the effects of submerged arc welding flux components on softening temperature. Transactions of the Faraday Society 134: 22-78.

10. Marcado AM, Lopez VM, Dorantes-Rosales HJ, Valdez ED (2008) Effect of titanium containing fluxes on the mechanical properties and micro structures in submerged arc weld steels. Materials Characterization 60: 36-39.

11. Bang K, Park C, Jung H, Lee J (2009) Effects of flux composition on the element transfer and mechanical properties of weld metal in submerged arc welding. Metals and Materials International 15: 471-477.

12. Beidokhti (2009) The effect of titanium addition on the SAW weld metal microstructure of API 5L-X70 pipeline steel. Journal of Materials Processing Technology 33: 450-459.

13. Mohan N, Khamba JS (2009) Microstructure/Mechanical property relationships of submerged arc welds in HSLA 80 steel. Proceedings of world congress on Engineering, London, UK.

14. Datta S, Bandyopadhyay A, Pal PK (2009) Application of Taguchi philosophy for parametric optimization of bead geometry and HAZ width in submerged arc welding using a mixture of fresh flux and fused flux. Int J Adv Manufacturing Technology 36: 689-698.

15. Kumar V, Mohan N, Khamba JS (2009) Development of cost effective agglomerated fluxes from waste flux dust for submerged arc welding. Preceding the World Congress of Engineering.

16. Aghakhani M, Ghaderi MR, Karami A, Derakhshan AA (2013) Combined effects of TiO_2 nano particles and input welding parameters on the weld bead penetration in submerged arc welding process using fuzzy logic. The International Journal of Advanced Manufacturing Technology 70: 63-72.

17. Kumar A, Bhardwaj D (2014) Enrichment of flux by nickel to improve impact strength in submerged arc welding and its effect on the bead height and bead width. International Journal for Research Publication & Seminar 5: 1-6

18. Zhanga T, Li Z, Kou S, Jing H, Li G, et al. (2015) Effect of inclusions on the microstructure and toughness of the deposited metals of self-shielded flux cored wires. Materials Science and Engineering: A 628: 332-339.

19. Ward RG (1962) An introduction to the physical chemistry of iron and steel making. Publisher English Book Society, London.

20. Tarlinsku (1980) The effect of viscosity of slag on the welding and technological properties of electrodes. Svar Proiz 9: 21-22.

21. Tandon S, Kanshal GC, Gupta SR (1988) Effect of flux characteristics on HAZ during submerged Arc welding. Proceedings of the International Conference on Welding Technology.

22. Schwemmer DD, Olson DL, Williamson DL (1979) Relationship of weld penetration to welding fluxes. Welding Journal 58: 153s-160s.

23. Potapov, Lazerev BI (1981) The metallurgical properties of silicon fluxes containing halides salts and alkaline metal oxides. Automatic Welding 6: 38-40.

24. Patchett BM (1974) Some influences of slag composition on heat transfer and arc stability. Welding Journal 53: 203s-210s.

25. Liu S, Olson DL (1986) The role of inclusions in controlling HSLA steel weld microstructures. Weld J Suppl Res 65: 139s-141s.

31

Pt-Shell Nanowires for Fuel Cell Electrodes

James CM Li*

Materials Science Program, Department of Mechanical Engineering, University of Rochester, Rochester, New York, USA

Abstract

Gibbs idea of surface excess is reexamined to discover that the surface excess is not the same as surface enrichment. This finding could help us design Pt shell nanowires for fuel cell electrodes which should have the best performance than any other kinds of Pt catalysts. All the reasons behind this possibility are collected and discussed. It is hoped that this analysis will convince you to make such wires as fuel cell electrodes.

Keywords: Pt-shell nanowires; Fuel cell electrodes; Gibbs surface excess; Surface enrichment; Fuel cell cars; Self driving cars; Pt price

Introduction

As analyzed by Ugurlu and Oztuna [1], fuel cells definitely can be used in automobiles. In fact the Toyota Mirai and Hyundai ix35 FCEV are already fuel cell cars. They can be driven 250 miles between fueling and the fueling takes only about 5 minutes. Currently $1 worth of hydrogen can drive 146 miles while $1 worth of gasoline can only drive 10 miles. The source of hydrogen is unlimited but gasoline will one day be gone. With hydrogen the exhaust is only water, no pollution of air as with gasoline. Hence fuel cells are the future for ground transportation.

In a fuel cell, hydrogen flows at one electrode and air flows at the other. With a catalyst the hydrogen ionizes into H^+ and an electron which flows through the external circuit to the other side. The H^+ diffuses through a PEM (Proton Exchange Membrane) to the other side and reacts with oxygen and an electron to form water. The second catalytic reaction is 6 orders slower than the first catalytic reaction.

The catalyst is Pt and the price of Pt can increase as much as 4% a day. If we produce more fuel cells, the price of Pt will increase rapidly. Zhu et al. [2] found a correlation between oil prices and Pt prices. They did not know why. But this is probably the reason. Luckily we need Pt only on the surface. The question is how to make a catalyst with Pt only on the surface. To increase the surface/volume ratio, we have been using Pt nanoparticles which need a support structure. The support is usually carbon. But the adhesion between Pt and C is not so good and the Pt particles can agglomerate and grow. The carbon can be oxidized also. We are looking into the self-supporting nanowires. The question is whether we can make Ni (or other metal) nanowires with only one atomic layer of Pt on the surface. The radius of a Pt atom is 135 pm and the atomic weight is 195 so a close packed layer of Pt atoms is 5 mg per square meter area, the lowest possible Pt loading achievable. This paper is to try to shed some light on this question. We are going to try to make it. You should try it too.

Gibbs Surface Excess

Gibbs started with two homogeneous phases 1 and 2 and then let them touch each other to form an interface. Consider phase 1 of energy U_1, entropy S_1, volume V_1, chemical components n_{i1} (i=1 to c, c being the number of components):

$$dU_1 = TdS_1 - PdV_1 + \sum_{i=1}^{c}\mu_i dn_{i1} \tag{1}$$

This equation just says for phase 1 at constant temperature T, pressure P and all the chemical potentials μ_i (i=1 to c) the reversible heat absorbed is TdS_1, the reversible mechanical work done is PdV_1

and the reversible chemical work is $\mu_i dn_{i1}$ (i=1 to c) by diffusing in the component i. Similarly for phase 2:

$$dU_2 = TdS_2 - PdV_2 + \sum_{i=1}^{c}\mu_{i2} dn_{i2} \tag{2}$$

Now at the same temperature, pressure and all the chemical potentials, combine the two phases to form an interface. All the energy, entropy, volume, and the chemical components may change. Gibbs defined them as excess quantities due to the interface: Let U, S, V and n_i be the total energy, entropy, volume and the number of moles of component i of the combined system:

Excess energy $U_{xs=}U-U_1-U_2$ (3)

Excess entropy $S_{xs=}S-S_1-S_2$ (4)

Excess volume $V_{xs}=V-V_1-V_2$ (5)

Excess ith component $n_{ixs}=n_{i1}-n_{i2}$ (6)

These excess quantities may be negative. The system now also has an interface of area A and energy γ per unit area.

So for the combined system

$$dU = TdS - PdV + \gamma dA + \sum_{i=1}^{c}\mu_i dn_i \tag{7}$$

Hence $dU_{xs} = TdS_{xs} - PdV_{xs} + \gamma dA + \sum_{i=1}^{c}\mu_i dn_{ixs}$ (8)

As you may remember from your thermodynamics course, Gibbs integrated this equation by adding small quantities together into the big system at the same temperature, pressure, and all the chemical potentials. So,

$$U_{xs} = TS_{xs} - PV_{xs} + \gamma A + \sum_{i=1}^{c}\mu_i n_{ixs} \tag{9}$$

He then differentiated this equation and compared with eqn. (8) to obtain:

$$S_{xs}dT - V_{xs}dP + Ad\gamma + \sum_{i=1}^{c}n_i d\mu_i = 0 \tag{10}$$

---

*Corresponding author: James CM Li, Professor Emeritus, Department of Mechanical Engineering, Materials Science Program, University of Rochester, Rochester, New York, USA, E-mail: li@me.rochester.edu

At constant temperature, pressure and the chemical potential of all other components, the surface excess of component i is:

$$\frac{n_{ixs}}{A} = -\left(\frac{\partial \gamma}{\partial \mu_i}\right)_{T,P,\mu_j} \qquad (11)$$

Which is the famous Gibbs adsorption equation. It is applicable to interfaces, grain boundaries and free surfaces.

Surface Excess and Surface Enrichment

While thermodynamics is never wrong, the prediction may not be exactly what you expected. Let us look at some of the PtNi alloys. Gauthier et al. [3,4] studied the {111} surface of 50-50 NiPt alloy. They found the surface layer contains 88 ± 2% Pt, the second layer 9 ± 5% Pt, the 3rd layer 65 ± 10% Pt and the 4th layer and inside 50% Pt. Since they started with 50% Pt and the bulk is 50% Pt, the three surface layers averaged 54 ± 7% Pt so the surface excess was 12 ± 7% Pt if concentrated on the surface. Yet there was actually strong surface enrichment of Pt (88%). For the {111} surface of 50-50 NiPt alloy Van et al. [5] also found Pt enrichment on the surface. The Gibbs eqn. (11) predicts only the surface excess, not the surface enrichment. But the catalytic properties depend on the surface enrichment, not the surface excess.

For the {111} surface of an 22-78 NiPt alloy, Gauthier et al. [3,4] found the first layer 1 ± 1% Ni, the second layer 70 ± 5% Ni, the 3rd layer 13 ± 10% Ni and the 4th layer and inside 22% Ni. So the surface excess is 18 ± 6% Ni but the surface enrichment is 99 ± 1% Pt. Based on the surface excess, this alloy may not be a good catalyst. But based on the surface enrichment, it is an excellent catalyst.

Similar results were found by computer simulation as reported by Wang et al. [6]. At 600 K, a nano particle (2 to 5 nm) of 25-75 Ni-Pt alloys has Pt strongly enriched on the surface layer and the third layer and the Ni enriched on the second layer. For a nano particle of 25-75 Re-Pt alloys, a nearly pure Pt shell surrounds a more uniform Pt-Re core. For a nano particle of 20-80 Mo-Pt alloys, the facets are fully occupied by Pt atoms, the Mo atoms are at the edges and vertices. The amount of Pt atoms on the surface increases with the size of the particle. So when it is a flat surface it will be pure Pt.

Experimentally by using LEIS (Low Energy Ion Scattering) in the same system $Pt_{75}Ni_{25}$, Stamenkovic et al. [7] found pure Pt on all 3 surfaces, (100), (110) and {111}. Their CTR (Crystal Truncation Rods) analysis for the {111} surface showed the second layer to consist of 45% Pt, the third layer 82% Pt and the fourth layer and inside were normal (75% Pt). Since the average of the first 3 layers was about 75% Pt, there was no surface excess yet there was strong surface enrichment of Pt (100%).

Gallego et al. [8] deposited Mn on Pt {111} by evaporation and did prolonged annealing at 950 K. LEED analysis showed first layer 100 ± 1% Pt, 2nd layer 75 ± 4% Pt, 3rd layer 100 ± 10% Pt, 4th layer 75 ± 10% Pt and 5th layer and inside 100% Pt. So the surface excess is 50 ± 8% Mn but the surface enrichment is 100 ± 1% Pt. The surface Pt atoms have the ability to push Mn atoms inside even though Mn atoms were coated on the Pt surface.

Some Properties of Pt Shells

Shui et al. [9] made $PtNi_5$ nanowires by electro spinning and then treated in 0.001 M sulfuric acid for 8 minutes and heated to 60°C for 4 minutes to remove the Ni atoms and Ni oxides on the surface. Then the nanowires were cleaned several times in deionized water and ethanol before they were heat treated in a hydrogen (5 vol%) argon (95 vol%) mixture at 300°C for 15 hours to allow Pt atoms to reach the surface to replace Ni atoms. Sure enough the Pt atoms can protect Ni in hot (60°C) 1 M sulfuric acid for hours.

Even for an alloy with 90% Ni, Tammermann et al. [10] and Gauthier et al. [11] still reported Pt enrichment on the surface. Hebenstreit et al. [12] suspected that Pt atoms on the (100) surface tend to be reconstructed into a close-packed surface layer.

For Ni-core Pt-shell nanoparticles of about 7.5 nm size, Godinez-Salomon et al. [13] made Ni core by colloidal reduction of $NiCl_2$ with $NaBH_4$ and coated with Pt by subsequent reduction of H_2PtCl_6. Cyclic voltammetry on thin film rotating disk electrode revealed that these particles had a more than twice enhanced catalytic activity than Pt nanoparticles synthesized by the same way.

Pt-skin surfaces were fabricated on Pt-Ni nanoparticles (2-3 nm) by Jung et al. [14] by using chemical deposition. They found that the chemically tuned Pt skin had a higher Pt coordination number and surface crystallinity which resulted in better oxygen reduction reaction activity and durability compared to the Pt-skin formed by heat annealing.

Pt-shell Ni-core nanoparticles supported on C were prepared by Kang et al. [15] with either an amorphous or crystalline Ni core while the thickness and structure of the Pt shell were similar. They compared the methanol oxidation activities by using cyclic voltammetry and chrono-amperometry and found that the amorphous core had better performance. Ni nanowires can be made by electro-spinning [9,16] and these wires can be made amorphous by straining [17].

Serra et al. [18] made mesoporous CoNi and Pt Nano rods by electrochemical synthesis and then coated the mesoporous CoNi Nano rods with a layer of Pt shell by interfacial replacement reaction. They found the later was better than the mesoporous pure Pt Nano rods. In fact they found the best ones were obtained from the water-in-ionic liquid micro emulsion which gave a mass activity of 1.3 A/mgPt.

From these considerations we suspect that Pt can protect a pure Ni wire by depositing on the surface.

Stability of the Surface Layer

Recently, Tao et al. [19] observed compositional changes near the surface of a nanoparticle for a free surface (in vacuum) and for surfaces in different environments. So the surface layer is the most important part of the surface which is affected by the environment and can be treated as a thermodynamic phase in equilibrium with the bulk and with the environment. The Gibbs equation of state and the associated Helmholtz and Gibbs free energies are:

$$dU_S = TdS_S - PdV_S + \gamma_S dA_S + \sum_i \mu_{iS} dn_{iS} \qquad (7.1)$$

$$dF_S = d(U_S - TS_S) = -S_S dT - PdV_S + \gamma_S dA_S + \sum_i \mu_{iS} dn_{iS} \qquad (8.1)$$

$$dG_S = d(U_S - TS_S + PV_S) = -S_S dT + V_S dP + +\gamma_S dA_S + \sum_i \mu_{iS} dn_{iS} \qquad (9.1)$$

Unlike the Gibbs excess quantities which belong to the geometric surface, the surface quantities here are for the surface layer atoms including the surface energy γ_s which is not the usual surface energy but the energy for the surface layer atoms. At constant temperature, pressure, surface energy and all the chemical potentials, eqn. (7.1) can be integrated by putting small systems together and give:

$$U_S = TS_S - PV_S + \gamma_S A_S + \sum_i \mu_{iS} n_{iS} \qquad (10.1)$$

Which can be differentiated and compared with eqn. (7.1) to give:

$$A_S d\gamma_S = -S_S dT + V_S dP - \sum_i n_{iS} d\mu_{iS} \qquad (11.1)$$

At constant temperature and pressure, the surface composition still can be found from the following equation:

$$\frac{n_{iS}}{A_S} = -\left(\frac{\partial \gamma_S}{\partial \mu_{iS}}\right)_{T,P,\mu_{jS}} \qquad (12)$$

but this is no longer the Gibbs adsorption equation since γ_s is not the usual surface energy and n_{iS} is no longer the surface excess quantity. However now n_{iS} is the real surface composition. So the surface layer is like another phase (a homogeneous part) in equilibrium with the rest of the system. The equilibrium with the interior could be an exchange equilibrium in a substitutional binary alloy of A and B such as Pt-Ni:

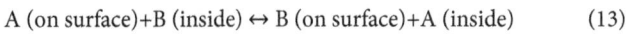

A (on surface)+B (inside) ↔ B (on surface)+A (inside)   (13)

This exchange equilibrium can be described by:

$$\mu_{AS}+\mu_B=\mu_{BS}+\mu_A \text{ or } \mu_{AS}-\mu_{BS}=\mu_A-\mu_B \qquad (14)$$

So the difference between chemical potentials of any two components is a constant throughout the system.

At constant temperature, pressure and the area of the surface layer, eqn. (9) can be integrated by adding up small systems and then differentiated to compare with eqn. (9) to obtain the Gibbs-Duhem type equation:

$$\sum_i n_{iS} d\mu_{iS} = 0 \qquad (15)$$

For any changes of surface composition, for a binary alloy A and B, this equation can be re-arranged into:

$$x_{BS} = \frac{n_{BS}}{n_{BS}+n_{AS}} = \left[\frac{\partial \mu_{AS}}{\partial(\mu_{AS}-\mu_{BS})}\right]_{T,P,A} = \left[\frac{\partial \mu_{AS}}{\partial(\mu_A-\mu_B)}\right]_{T,P,A} \qquad (16)$$

Where μ_A and μ_B are the chemical potentials for A and B in the bulk, respectively. So by changing the bulk alloy composition and examining the effect on the surface composition, it is possible to calculate the change of chemical potentials on the surface.

However, it is possible that A prefers to stay on the surface and B has zero solubility in the surface layer so that eqn. (13) is an inequality:

B (on surface)+A (next layer) → A (on surface)+B (next layer)   (17)

This inequality can be described by:

$$\mu_{BS}+\mu_{A1}>\mu_{AS}+\mu_{B1} \text{ or } \mu_{AS}-\mu_{BS}<\mu_{A1}-\mu_{B1} \qquad (18)$$

Then the surface layer may not respond to the change of bulk composition. It is possible that for Pt-Ni alloy, the Pt surface layer is a stable layer obeying the inequalities represented by eqn. (18).

However, the surface layer may be in equilibrium with the surroundings such as the fluid in contact with the surface. If C is such a component, then:

$$\mu_{CS}=\mu_{CF} \qquad (19)$$

Where μ_{CF} is the chemical potential of C in the contacting fluid. For a two-component system, A and C, the Gibbs-Duhem eqn. (15) shows:

$$n_{AS}d\mu_{AS}+n_{CS}d\mu_{CS}=n_{AS}d\mu_{AS}+n_{CS}d\mu_{CF}=0 \qquad (20)$$

So by changing the composition of the fluid phase in contact, the surface composition or the chemical potentials could be affected.

The Stable Surface Layer

Which atoms should be enriched on the surface and form a stable layer? The usual finding is that the metal with lower surface energy or the larger atoms are enriched on the surface. Ma and Balbuena [20] collected data for some Pt_3M systems and compared with their theoretical predictions with two notable exceptions, Pt_3Mn {111} and Pt_3Ti {111}. While their theory predicts Mn and Ti segregation, experiments showed Pt segregation [8,21]. So the surface segregation is governed, not by the atomic size mismatch and the surface energy differences but by the subsurface atomic arrangement, namely the shell structure. In other words, the shell free energy determines the surface composition.

Li et al. [22] discovered in the oxidative steam reforming of methane, a Pt coated Ni catalyst was better than a Pt-Ni alloy catalyst. Comparing to the activity in the steam reforming of methane without oxygen, the presence of oxygen led to decreased reforming activity in both pure Ni (2.6 wt% in catalyst) and Pt (0.1)+Ni (2.6) alloy due mainly to the oxidation of nickel, whereas Pt (0.1) coated over Ni (2.6) exhibited a high resistance to oxidation and maintained the activity even in the presence of oxygen. The mole ratio of Pt (0.1) to Ni (2.6) is only 0.01 and yet it was sufficient to protect Ni from oxidation. Considering the fact that the catalyst was calcined in air at 573 K for 3 hours and the bed temperature went as high as 1123 K in the oxidative steam reforming of methane we suspect that Pt segregates over the Ni surface and may not diffuse in as an alloy element.

Similarly Mukainakano et al. [23] found in the same steam reforming of methane, hysteresis with respect to the addition and removal of the methane oxygen mixture was clearly observed on a Pt (0.1)+Ni (2.6) alloy catalyst and pure Ni (2.6) catalyst but no hysteresis was observed for a Pt (0.1) coated Ni (2.6) catalyst. They also believed that the Pt coated Ni catalyst had Pt segregated on the surface which enhanced the reducibility of Ni drastically and eliminated the hysteresis behavior.

For steam reforming of ethanol, Soyal-Baltacioglu et al. [24] found the best catalyst was 0.3 wt% Pt-15 wt% Ni over δ-Al_2O_3. Here the Pt/Ni atomic ratio was only 0.006. It shows the importance of understanding the surface enrichment rather than the surface excess.

Pt-Shell Co-Core Nanoparticles

For Co core, Pt shell nanoparticles prepared by electroless deposition, Beard et al. [25] found that the lower Pt:Co ratio (monolayer Pt on Co) Pt-Co/C catalysts outperformed a commercial Pt/C catalyst.

Wang et al. [26] made ordered Pt_3Co intermetallic particles coated with 2-3 atomic layers of Pt. These nanocatalysts exhibited over 200% increase in mass activity when compared with the disordered Pt_3Co alloy nanoparticles.

So the monolayer Pt shell over Co nanowire is a distinct possibility. The size effect is shown by Frankenburg et al. [27]. Larger wires have better surfaces.

Pt-Shell Pt-Cu Nanoparticles

Strasser et al. [28] made $Pt_{25}Cu_{75}$, $Pt_{50}Cu_{50}$ and $Pt_{75}Cu_{25}$ nano particles and annealed them at 800°C and 950°C and then dealloyed electrochemically to remove Cu from the surface. They found a Pt-enriched surface layer (about 0.6 nm thick) in all these 4 nm particles. The fcc unit cell parameter for Pt on the surface (0.388 nm for $Pt_{75}Cu_{25}$, 0.382 nm for $Pt_{50}Cu_{50}$ and 0.375 nm for $Pt_{25}Cu_{75}$) was smaller than the bulk Pt (0.392 nm) and they attributed the superb catalytic activity of these particles for the oxygen reduction reaction in fuel cell electrodes to the reduction in atomic spacing. For a thin layer of Pt on Cu the Pt spacing would be the smallest and the catalytic activity would be the best.

Yeh et al. [29] found that the oxygen reaction reactivity of the Pt-Cu nanorods was 2.2 times higher than that of the Pt nanoparticles after 1000 potential cycles. They attributed this to the 1-D morphology and the low Pt unfilled d-states by alloying.

Pt Shell Ag Core Nanoparticles

Wojtysiak et al. [30] tried to cover Ag particles with Pt shell by the galvanic replacement reaction between Ag and $PtCl_4^{-2}$. However, the coverage was not good and the shell had a lot of holes. To improve the integrity of the shell, seeded growth of Pt on the surface of Ag by the reduction of $PtCl_4^{-2}$ with ascorbic acid was used at room temperature. To cover the surface of an 11 nm Ag particle, the number of atoms of Pt in the final Ag@Pt core-shell particle must be at least the same as that of Ag.

Abkhalimov and Ershov [31] coated 6.3 nm Ag nanoparticles with Pt by treating with aqueous K_2PtCl_4 with hydrogen. The Ag/Pt atomic ratio ranged from 1/9 to 9/1. They used these particles to catalyze the reduction of methyl viologen with hydrogen in an alkaline solution. They found a critical thickness of about 1 nm for Pt below which no catalysis will take place. This critical thickness can be reduced when the nanoparticles are replaced by nanowires of large diameters.

Pryadchenko et al. [32] made C supported PtAg nanoparticles by the chemical reduction of H_2PtCl_6 and $AgNO_3$ in a mixture of water: ethylene glycol=5:1. A 0.5 M $NaBH_4$ solution was used as reducing agent. If both Ag and Pt salts were used together, the particles were solid solutions. But if the Ag salt was reduced first to form Ag nanoparticles and then Pt salt was added to be reduced, the nanoparticles had a core-shell structure. The shell had a thickness of at least 3 atomic layers of Pt.

Pt Shell Ag Core Nanotubes

Kim et al. [33] made Ag nanowires by heating 20 mL ethylene glycol (EG) to 151.5 C and adding 4 mM $CuCl_2 \cdot 2H_2O$ in 0.16 mL EG, 147 mM polyvinyl pyrrolidone (PVP, MW 55,000) in 6 mL EG, and 94 mM $AgNO_3$ in 6 mL EG and maintaining at this temperature for one hr. resulting in a gray solution of Ag nanowires. See Korte et al. [34] for more details. To coat the Ag nano wires with Pt, the Ag nanowire solution (32.16 mL) was cooled to 100 C, added slowly 20 mM K_2PtCl_4 in 6 mL EG and heated for 1 hr. The Pt shell Ag core nanotubes were obtained by centrifuge and washing several times with water and ethanol and NH_4OH to remove AgCl precipitates. To make pure Pt nanotubes, the Pt coated Ag nanotubes were treated with 6 M HNO_3 solution to remove Ag. In alkaline media, the Pt coated Ag nanotubes showed 50% better activity than pure Pt nanotubes and Pt/C.

Pt Shell Au Core Nanoparticles

Min et al. [35] overgrew Pt on the surface of Au nanocrystals of cubic, octahedral and spherical shapes. Different modes of overgrowth were observed depending on the shape of the gold core. It occurred on the planar surfaces of Au cubes, at the vertices of the Au octahedral and over the entire surface of the Au spheres.

Banerjee et al. [36] made 5 nm gold particles by the reduction of chloroauric acid with tannic acid and coated these particles by Pt using different amounts of chloroplatinic acid and hydrazine. For Pt/Au

ratios of 0.19, 0.39, 0.58 and 0.88, the Pt thickness was half a monolayer, 1, 1.5 and 2 monolayers, respectively. The electrochemical activity was the best for the 2 monolayers with a mass activity 3 times better than the pure Pt nanoparticles.

Roy et al. [37] used 40 nm Au nanocrystals with extensive {111} facets and deposited 5 atomic layers of Pt on the surface by reducing hexachloroplatanic acid with ascorbic acid. Electrochemical evaluations revealed a compact Pt shell with a mass activity 4 times better than Pt black and comparable to that of Pt bulk metal.

Hartl et al. [38] used commercial 30 nm Au particles, highly crystalline and stably dispersed. To the dispersion they added hexachloroplatinic acid (mass of Pt=20% of Au), heated to 70-80°C, added 0.01 M ascorbic acid to reduce Pt to coat on Au particles. The Pt shell was about 3 atomic layers thick. The electrocatalytic activity for the oxygen reduction reaction using these core-shell particles equals that of bulk Pt.

Xiao et al. [39] fabricated nanoporous film of 100 nm thick by dealloying $Au_{50}Ag_{50}$ leaf at room temperature for 8 hours, rinsed and treated with 0.31 mM H_2PtCl_6 and 0.13 mM HCOOH in the dark to deposit Pt on Au. The catalytic activity of the nanoporous AuPt film towards electrochemidal oxidation of methanol increases with the loading level of Pt, resulting in the highest electochemical area of 70.4 m^2/g Pt, about 3 monolayers. Compared to the Pt nanoparticles supported in C, this self-supporting film uses much less Pt.

Kulp et al. [40] made Au nanoparticles by adding 1 mL of 1 wt% solution of $HAuCl_4 \cdot 3H_2O$ to 250 mL water with vigorous stirring. Then 1 mL of 1 wt% $Na_3C_6H_5O_7 \cdot 2H_2O$ in water was added and after 1 min, 1 mL of 1 wt% $NaBH_4$ and 1 wt% of $Na_3C_6H_5O_7 \cdot 2H_2O$ was added. The color change of the solution from light yellow to orange red indicated the formation of Au nanoparticles. The solution was stirred for another 10 min followed by adding 90 mg of Vulcan XC72 with continued stirring for 45 more min. The solution was filtered yielding a clear colorless filtrate which was dried for 3 hours at 90°C. TEM images showed homogeneously distributed colorless Au particles of about 5.5 nm adsorbed on C. To coat Pt shells on these particles, they used glassy carbon electrode in a solution of Pt $(NO_3)_2$ and $NaNO_3$ by pulsed electro deposition. See paper for details.

Zhang et al. [41] coated 6 nm Au particles with Pt resulting in particles of 9.0 ± 2.4 nm, 10.4 ± 2.8 nm and 13.0 ± 3.2 nm sizes. These particles stabilized by polyaryl ether trisacetic acid ammonium chloride dendrons and had higher catalytic activity than monometallic Pt nanoparticles. They attributed this to the fact that Au core attracts electrons from Pt.

Li et al. [42] made 55 nm Au particles by reducing $AuCl_4^-$ with sodium citrate. Then different amounts of 1 mM of H_2PtCl_6 were added and the mixture was heated to 80°C. Then while stirring, a solution of 10 mM of ascorbic acid was slowly dropped into the mixture until half of the volume of H_2PtCl_6 was added. The mixture was stirred for another 30 min. and should change from red brown to dark brown indicating the coating was complete. The final diameter of the core-shell particle can be estimated from the volume ratios of Au and Pt. The mixture was then centrifuged 3 times before coated on a smooth and clean glass carbon surface for testing.

Gao et al. [43] coated only 2 monolayers of Pt on 16 nm Au particles by reducing H_2PtCl_6 with ascorbic acid. The uncoated Au nanoparticles exhibited a strong localized surface Plasmon resonance (LSPR) peak at 520 nm. After coating, the LSPR peak shifted to 508 nm.

Zhang et al. [44] made 55 nm Au particles by reducing $HAuCl_4$ with sodium citrate. Then 30 mL of sol containing the 55 nm Au seeds were mixed with 0.76 mL of 1 mM H_2PtCl and heated to 80°C for a few minutes. Ascorbic acid (0.4 mL, 10 mM) was slowly dropped into the mixture with vigorous stirring. Th coated Pt shell over the Au surface was about 0.7 nm thick. The coated Au particles were almost spherical. The sol was centrifuged 3 times to remove excess reactants.

From these observations, it seems likely that gold wires can be coated with only one mono layer of Pt to increase its catalytic activity.

Pt and Pt Alloy Nanowires

Higgins et al. [45] made Pt-Co alloy nanowires by mixing 0.001 m Pt acetylacetonate and 0.001 m Co carbonyl, dissolved in ethylenediamine under nitrogen protection and transferred to an autoclave reactor at 160°C for 1 hour under 600 W powers. Pt-Co alloy nanowires of about 35-60 nm diameters were formed and collected by filtration and washing followed by annealing at 600°C in Ar for 1 hr. These nanowires were found much more stable catalytically than Pt nanoparticles supported on C.

Yaipimai and Pornprasertsuk [46] made Pt, Pt-Cu and Pt-Sn alloy nanowires by electro spinning using the salts and PVP, Poly (vinyl pyrrolidone) with molecular weights 1.3×10^6 and 4×10^4 g/mole.

Pt Shell Pd Core Nanoparticles

Cao et al. [47] made the Pd core Pt shell nanoparticles by using an area selective atomic layer deposition method. The Pt shell thickness could be monitored by an in-situ quartz crystal microbalance. The catalyst with one monolayer Pt shell showed the best mass activity and selectivity and the lowest barrier for CO oxidation.

Choi et al. [48] did electro less deposition of Cu on Pd Nanoparticles and galvanic displacement of Cu by Pt. The catalyst is active toward electro-oxidation of methanol and is more stable against CO poisoning than a commercial Pt/C catalyst.

D'Souza and Sampath [49] made highly uniform, stable Nano bimetallic dispersions using organically modified silicates as the matrix and the stabilizer. They found that the structure of the particles consists of a Pt shell and a Pd core. No aggregation or segregation of the particles was observed after prolonged storage of several months.

Shao et al. [50] found that the specific oxygen reduction reaction activity of Pt shells over Pd octahedra enriched with {111} facets was 28 times higher than that of Pd octahedral without Pt. It was only 3 times better than Pt coated Pd cubes enriched with {100} facets. The Pt coated Pd octahedra also showed excellent durability during potential cycling suggesting their great potential for application in fuel cells.

Wongkaew et al. [51] made electro less deposition of Pt on Pd surfaces of 30 wt% Pd/C. Pt loadings of 6.0, 11.7, 17.2 and 22.7 wt% corresponded to Pt shells of 0.9, 1.7, 2.7 and 3.4 monolayers on Pd. These core/shell catalysts were very active, especially the sample of 0.9 monolayer coverage which had a mass activity of 329 A/gPt as compared to 183 A/gPt for a conventional 50.5 wt% Pt/C sample.

Li et al. [52] synthesized C supported Pd_3Au@Pt core-shell electrocatalyst by chemical reduction of K_2PtCl_4, K_2PdCl_4 and $NaAuCl_4$ with ascorbic acid. The resultant particles (3.4 nm diameter core) had a thin layer (less than 1 nm) of Pt shell. They had a mass activity of 939 A/gPt for oxygen reduction reaction, 4.6 times that of commercial Pt/C (203 A/gPt). But the durability was about the same. They proposed that

the tension in the Pt shell and the electron transfer from the core to the shell contributed to the improved electrocatalytic activity.

Pt Shell Ru Core Nanoparticles

Chen et al. [53] coated Ru nanoparticles (3.2 nm diameter) with 1.5-3.6 atomic layers of Pt. The sample with 1.5 atomic layers showed a 3.2 fold improvement in CO tolerance and 2.4 fold current enhancements during methanol oxidation as compared to the commercial Pt/C.

Yang et al. [54] found that the Pt shell grew on <111> radial facets and <200> face facets of the Ru core if the incubation time was short. For 2 hours incubation, severe chemical etching occurred prior to shell growth. So the dynamic rearrangement at the core-shell interface is important for the final structure.

Wang et al. [55] made Pt shell (0.42 nm thick or about 1.5 atomic layers) and Ru core (3.18 nm diameter) nanoparticles (4.02 nm diameter total). Compared to the pure Pt nanoparticles of 4.38 nm diameter, these Pt-shell Ru-core nanoparticles showed 4.5 fold more power density for the direct methanol fuel cell. The open circuit voltage was improved by 0.18 V (from 0.49 to 0.67 V). They attribute the improvement to the lattice compression in the Pt shell due to the core.

Huang et al. [56] made 15 wt% $Pt_{50}Ru_{50}$/C nano particles (2 nm) by the method of incipient wetness impregnation and activated by hydrogen reduction at 620 K. The reduced catalyst with Pt rich in the shell and Ru rich in the core was subsequently modified by oxidation in air. This oxidation enhanced significantly the electrochemical activity of Pt-Ru/C for electro-oxidation of methanol. Such enhancement was attributed to the segregation of Ru and the formation of RuO_2.

Pt Shell Pd Core Nanowires

Guo et al. [57] started with Te nanowires (11 nm diameter) produced by a hydrothermal route and used them as both reducing agent and sacrificial template to make Pd nanowires in aqueous solution at room temperature in less than 5 min. The Pd nanowires were used as seeds to direct dendritic growth of Pt upon the reduction of K_2PtCl_4 with ascorbic acid in aqueous solution.

Liao and Hou [58] made Pt-on-$Pd_{0.85}Bi_{0.15}$ nanowires by a facile, one pot, wet-chemical and templateless method in the presence of oleylamine and NH_4Br. These nanowires had 8.3 ± 1.1 nm diameters and 387 ± 105 nm length. Small Pt nanobranches (5 nm) grew on the Pd nanowires at the end of which Pt nanoflowers grew. They could see also about 2 nm thick of amorphous C on the nanowires. Depending on the composition and the ratio of Pd/Bi/Pt they could grow nanowires, nanoflowers, nanoparticles or nanoplates. They all demonstrated high electro-chemical activity and durability for the oxygen reduction reaction.

Xia et al. [59] reported a facile solvothermal synthesis of nanowire assemblies composed of ultra-thin (3 nm) and ultra-long (10 μm) Pt, Pt-Au and Pt-Pd nanowires without involving any template. These nanowires can be easily cast into a free-standing membrane which exhibits excellent electro catalytic activity and very high stability for formic acid and methanol oxidation and the oxidation reduction reaction.

Pt Shell Cu Core Networks

Feng et al. [60] made nanoporous Cu by electrodepositing Zn on a 0.1 mm Cu plate, making Cu-Zn alloy by heating and then removing Zn by HCl. They followed the method described by Leaman [61].

Then they deposited Pt onto the Cu surfaces by electroless plating. This NPCu-Pt catalyst can reduce CO_2 in the ionic liquid BMIMBF$_4$ (1-butyl-3-methyl-imidazolium-tetra-fluoborate) with more stable current, higher current density and efficiency compared to the pure Pt catalyst.

Pt Shell Cu Core Nanowires

Alia et al. [62] and Wittkopf et al. [63] already made Pt shell (14 monolayers) over Cu core nano wires with mass activities of 0.1-1.0 A/mgPt which may be improved by reducing the number of monolayers of Pt on the surface.

Conclusions

It is seen that there are many possibilities to make a single layer Pt Shell over large nanowires of a cheaper metal so a commercial catalyst for a fuel cell can be made to make self- driving automobiles widely used soon. It is anticipated that we will have cheaper and safer ground transportation available in the very near future.

References

1. Ugurlu A, Oztuna S (2015) A comparative analysis study of alternative energy sources for automobiles. International Journal of Hydrogen Energy 40: 11178-11188.

2. Zhu XH, Chen JY, Zhong MR (2015) Dynamic interacting relationships among international oil prices, macroeconomic variables and precious metal prices. Transactions of Nonferrous Metals Society of China 25: 669-676.

3. Gauthier Y, Joly Y, Baudoing R, Rundgren J (1985) Surface-sandwich segregation on nondilute bimetallic alloys: Pt 50 Ni 50 and Pt 78 Ni 22 probed by low-energy electron diffraction. Physical Review B 31: 6216.

4. Gauthier Y, Baudoing-Savois R, Rundgren J, Hammar M, Gothelid M (1995) Reconstruction of the Pt50Ni50 (100) surface: A LEED and STM study. Surface Science 327: 100-120.

5. Van de Riet EGJP, Deckers S, Habraken FHPM, Niehaus A (1991) The atomic surface structure of Pt0. 5Ni0. 5 (111). Surface Science 243: 49-57.

6. Wang G, Van Hove MA, Ross PN, Baskes MI (2005) Quantitative prediction of surface segregation in bimetallic Pt-M alloy nanoparticles (M=Ni, Re, Mo). Progress in Surface Science 79: 28-45.

7. Stamenkovic VR, Fowler B, Mun BS, Wang G, Ross PN, et al. (2007) Improved oxygen reduction activity on Pt3Ni (111) via increased surface site availability. Science 315: 493-497.

8. Gallego S, Ocal C, Munoz MC, Soria F (1997) Surface-layered ordered alloy (P t/P t 3 Mn) on Pt (111). Physical Review B 56: 12139.

9. Shui JL, Zhang JW, Li JC (2011) Making Pt-shell Pt 30 Ni 70 nanowires by mild dealloying and heat treatments with little Ni loss. Journal of Materials Chemistry 21: 6225-6229.

10. De Temmerman L, Creemers C, Van Hove H, Neyens A, Bertolini JC, et al. (1986) Experimental determination of equilibrium surface segregation in Pt-Ni single crystal alloys. Surface Science 178: 888-896.

11. Gauthier Y, Hoffmann W, Wuttig M (1990) Structure and composition of Pt10Ni90 (100): A low energy electron diffraction study. Surface Science 233: 239-247.

12. Hebenstreit W, Ritz G, Schmid M, Biedermann A, Varga P (1997) Segregation and reconstructions of PtxNi1-x (100). Surface Science 388: 150-161.

13. Godínez-Salomón F, Hallen-López M, Solorza-Feria O (2012) Enhanced electroactivity for the oxygen reduction on Ni@ Pt core-shell nanocatalysts. International Journal of Hydrogen Energy 37: 14902-14910.

14. Jung N, Chung YH, Chung DY, Choi KH, Park HY, et al. (2013) Chemical tuning of electrochemical properties of Pt-skin surfaces for highly active oxygen reduction reactions. Physical Chemistry Chemical Physics 15: 17079-17083.

15. Kang J, Wang R, Wang H, Liao S, Key J, et al. (2013) Effect of Ni core structure on the electrocatalytic activity of Pt-Ni/C in methanol oxidation. Materials 6: 2689-2700.

16. Shui J, Li JC (2009) Platinum nanowires produced by electrospinning. Nano Letters 9: 1307-1314.

17. Ikeda H, Qi Y, Cagin T, Samwer K, Johnson WL, et al. (1999) Strain rate induced amorphization in metallic nanowires. Physical Review Letters 82: 2900.

18. Serrà A, Gómez E, Vallés E (2015) Novel electrodeposition media to synthesize CoNi-Pt Core@ Shell stable mesoporous nanorods with very high active surface for methanol electro-oxidation. Electrochimica Acta 174: 630-639.

19. Tao F, Grass ME, Zhang Y, Butcher DR, Renzas JR, et al. (2008) Reaction-driven restructuring of Rh-Pd and Pt-Pd core-shell nanoparticles. Science 322: 932-934.

20. Ma Y, Balbuena PB (2008) Pt surface segregation in bimetallic Pt 3 M alloys: a density functional theory study. Surface Science 602: 107-113.

21. Chen W, Severin L, Göthelid M, Hammar M, Cameron S, et al. (1994) Electronic and geometric structure of clean Pt 3 Ti (111). Physical Review B 50: 5620.

22. Li B, Kado S, Mukainakano Y, Miyazawa T, Miyao T, et al. (2007) Surface modification of Ni catalysts with trace Pt for oxidative steam reforming of methane. Journal of Catalysis 245: 144-155.

23. Mukainakano Y, Yoshida K, Kado S, Okumura K, Kunimori K, et al. (2008) Catalytic performance and characterization of Pt-Ni bimetallic catalysts for oxidative steam reforming of methane. Chemical Engineering Science 63: 4891-4901.

24. Soyal-Baltacıoğlu F, Aksoylu AE, Önsan ZI (2008) Steam reforming of ethanol over Pt-Ni Catalysts. Catalysis Today 138: 183-186.

25. Beard KD, Borrelli D, Cramer AM, Blom D, Van Zee JW, et al. (2009) Preparation and structural analysis of carbon-supported Co core/Pt shell electrocatalysts using electroless deposition methods. ACS Nano 3: 2841-2853.

26. Wang D, Xin HL, Hovden R, Wang H, Yu Y, et al. (2013) Structurally ordered intermetallic platinum–cobalt core-shell nanoparticles with enhanced activity and stability as oxygen reduction electrocatalysts. Nature Materials 12: 81-87.

27. Frankenburg WG, Komarewsky VI, Rideal EK (1952) Advances in catalysis (Vol. 4). Academic Press.

28. Strasser P, Koh S, Anniyev T, Greeley J, More K, et al. (2010) Lattice-strain control of the activity in dealloyed core–shell fuel cell catalysts. Nature Chemistry 2: 454-460.

29. Yeh TH, Liu CW, Chen HS, Wang KW (2013) Preparation of carbon-supported PtM (M=Au, Pd, or Cu) nanorods and their application in oxygen reduction reaction. Electrochemistry Communications 31: 125-128.

30. Wojtysiak S, Solla-Gullón J, Dłużewski P, Kudelski A (2014) Synthesis of core-shell silver-platinum nanoparticles, improving shell integrity. Colloids and Surfaces A: Physicochemical and Engineering Aspects 441: 178-183.

31. Abkhalimov EV, Ershov BG (2014) The size effect in the catalytic activity of AgcorePtshell nanoparticles. Colloid Journal 76: 381-386.

32. Pryadchenko VV, Srabionyan VV, Mikheykina EB, Avakyan LA, Murzin VY, et al. (2015) Atomic Structure of Bimetallic Nanoparticles in PtAg/C Catalysts: Determination of Components Distribution in the Range from Disordered Alloys to "Core-Shell" Structures. The Journal of Physical Chemistry C 119: 3217-3227.

33. Kim Y, Kim H, Kim WB (2014) PtAg nanotubes for electrooxidation of ethylene glycol and glycerol in alkaline media. Electrochemistry Communications 46: 36-39.

34. Korte KE, Skrabalak SE, Xia Y (2008) Rapid synthesis of silver nanowires through a CuCl-or CuCl 2-mediated polyol process. Journal of Materials Chemistry 18: 437-441.

35. Min M, Kim C, Yang YI, Yi J, Lee H (2009) Surface-specific overgrowth of platinum on shaped gold nanocrystals. Physical Chemistry Chemical Physics 11: 9759-9765.

36. Banerjee I, Kumaran V, Santhanam V (2015) Synthesis and characterization of Au@ Pt nanoparticles with ultrathin platinum overlayers. The Journal of Physical Chemistry C 119: 5982-5987.

37. Roy RK, Njagi JI, Farrell B, Halaciuga I, Lopez M, et al. (2012) Deposition of continuous platinum shells on gold nanoparticles by chemical precipitation. Journal of Colloid and Interface Science 369: 91-95.

38. Hartl K, Mayrhofer KJ, Lopez M, Goia D, Arenz M (2010) AuPt core-shell

nanocatalysts with bulk Pt activity. Electrochemistry Communications 12: 1487-1489.

39. Xiao S, Xiao F, Hu Y, Yuan S, Wang S, et al. (2014) Hierarchical nanoporous gold-platinum with heterogeneous interfaces for methanol electrooxidation. Scientific Reports, p: 4.

40. Kulp C, Chen X, Puschhof A, Schwamborn S, Somsen C, et al. (2010) Electrochemical synthesis of core-shell catalysts for electrocatalytic applications. ChemPhysChem 11: 2854-2861.

41. Zhang W, Li L, Du Y, Wang X, Yang P (2009) Gold/platinum bimetallic core/shell nanoparticles stabilized by a fréchet-type dendrimer: Preparation and catalytic hydrogenations of phenylaldehydes and nitrobenzenes. Catalysis Letters 127: 429-436.

42. Li JF, Yang ZL, Ren B, Liu GK, Fang PP, et al. (2006) Surface-enhanced Raman spectroscopy using gold-core platinum-shell nanoparticle film electrodes: toward a versatile vibrational strategy for electrochemical interfaces. Langmuir 22: 10372-10379.

43. Gao Z, Tang D, Tang D, Niessner R, Knopp D (2015) Target-induced nanocatalyst deactivation facilitated by core@ shell nanostructures for signal-amplified headspace-colorimetric assay of dissolved hydrogen sulfide. Analytical Chemistry 87: 10153-10160.

44. Zhang P, Cai J, Chen YX, Tang ZQ, Chen D, et al. (2009) Potential-dependent chemisorption of carbon monoxide at a gold core-platinum shell nanoparticle electrode: A combined study by electrochemical in situ surface-enhanced Raman spectroscopy and density functional theory. The Journal of Physical Chemistry C 114: 403-411.

45. Higgins DC, Ye S, Knights S, Chen Z (2012) Highly durable platinum-cobalt nanowires by microwave irradiation as oxygen reduction catalyst for PEM fuel cell. Electrochemical and Solid-State Letters 15: B83-B85.

46. Yaipimai W, Pornprasertsuk R (2013) Fabrication of Pt, Pt-Cu, and Pt-Sn nanofibers for direct ethanol protonic ceramic fuel cell application. Journal of Materials Science 48: 4059-4072.

47. Cao K, Liu X, Zhu Q, Shan B, Chen R (2015) Atomically Controllable Pd@ Pt Core–Shell Nanoparticles towards Preferential Oxidation of CO in Hydrogen Reactions Modulated by Platinum Shell Thickness. ChemCatChem.

48. Choi I, Ahn SH, Kim MH, Kwon OJ, Kim JJ (2014) Synthesis of an active and stable Pt shell–Pd core/C catalyst for the electro-oxidation of methanol. International Journal of Hydrogen Energy 39: 3681-3689.

49. D'Souz L, Sampath S (2000) Preparation and Characterization of Silane-Stabilized, Highly Uniform, Nanobimetallic Pt-Pd Particles in Solid and Liquid Matrixes. Langmuir 16: 8510-8517.

50. Shao M, He G, Peles A, Odell JH, Zeng J, et al. (2013) Manipulating the oxygen reduction activity of platinum shells with shape-controlled palladium nanocrystal cores. Chemical Communications 49: 9030-9032.

51. Wongkaew A, Zhang Y, Tengco JMM, Blom DA, Sivasubramanian P, et al. (2016) Characterization and evaluation of Pt-Pd electrocatalysts prepared by electroless deposition. Applied Catalysis B: Environmental 188: 367-375.

52. Li H, Yao R, Wang D, He J, Li M, Song Y (2015) Facile Synthesis of Carbon Supported Pd3Au@ Super-Thin Pt Core/Shell Electrocatalyst with a Remarkable Activity for Oxygen Reduction. The Journal of Physical Chemistry C 119: 4052-4061.

53. Chen TY, Lin TL, Luo TJM, Choi Y, Lee JF (2010) Effects of Pt Shell Thicknesses on the Atomic Structure of Ru-Pt Core-Shell Nanoparticles for Methanol Electrooxidation Applications. ChemPhysChem 11: 2383-2392.

54. Yang PW, Liu YT, Hsu SP, Wang KW, Jeng US, et al. (2015) Core-shell nanocrystallite growth via heterogeneous interface manipulation. CrystEngComm 17: 8623-8631.

55. Wang JJ, Liu YT, Chen IL, Yang YW, Yeh TK, et al. (2014) Near-Monolayer Platinum Shell on Core–Shell Nanocatalysts for High-Performance Direct Methanol Fuel Cell. The Journal of Physical Chemistry C 118: 2253-2262.

56. Huang SY, Chang SM, Lin CL, Chen CH, Yeh CT (2006) Promotion of the electrochemical activity of a bimetallic platinum-ruthenium catalyst by oxidation-induced segregation. The Journal of Physical Chemistry B 110: 23300-23305.

57. Guo S, Dong S, Wang E (2010) Ultralong Pt-on-Pd bimetallic nanowires with nanoporous surface: nanodendritic structure for enhanced electrocatalytic activity. Chemical Communications 46: 1869-1871.

58. Liao H, Hou Y (2013) Liquid-phase templateless synthesis of Pt-on-Pd0. 85Bi0. 15 nanowires and PtPdBi porous nanoparticles with superior electrocatalytic activity. Chemistry of Materials 25: 457-465.

59. Xia BY, Wu HB, Yan Y, Lou XW, Wang X (2013) Ultrathin and ultralong single-crystal platinum nanowire assemblies with highly stable electrocatalytic activity. Journal of the American Chemical Society 135: 9480-9485.

60. Feng Q, Liu S, Wang X, Jin G (2012) Nanoporous copper incorporated platinum composites for electrocatalytic reduction of CO 2 in ionic liquid BMIMBF 4. Applied Surface Science 258: 5005-5009.

61. Leaman FH (1972) US Patent 3698939.

62. Alia SM, Jensen K, Contreras C, Garzon F, Pivovar B, et al. (2013) Platinum coated copper nanowires and platinum nanotubes as oxygen reduction electrocatalysts. Acs Catalysis 3: 358-362.

63. Wittkopf JA, Zheng J, Yan Y (2014) High-performance dealloyed PtCu/CuNW oxygen reduction reaction catalyst for proton exchange membrane fuel cells. Acs Catalysis 4: 3145-3151.

γ – Irradiated Jute Reinforced Polypropylene Composites: Effect of Mercerization and SEM Analysis

Islam Bossunia MT[1], Poddar P[1]*, Hasan MM[3], Hossain MT[4], Gulenoor F[1], Khan RA[2] and Sarwaruddin Chowdhury AM[1]

[1]Department of Applied Chemistry and Chemical Engineering, Faculty of Engineering, University of Dhaka, Bangladesh
[2]Institute of Nuclear Science and Technology, Bangladesh Atomic Energy Commission, Bangladesh
[3]Institute of Leather Engineering and Technology, University of Dhaka, Bangladesh
[4]Bangladesh Jute Research Institute, Bangladesh

Abstract

Jute fibres (*Corchorus olitorious* L.), an environmentally and ecologically friendly product, were chemically modified and treated as surface treatment which was carried out by mercerizing jute fabrics with aqueous solutions of NaOH (5,10 and 20%) at different soaking times (30, 60 and 90 minutes) and temperatures (0, 25 and 50°C). These mercerized jute fabric reinforced polypropylene composites were fabricated by composition molding technique and equated with virgin jute fabric reinforced polypropylene composites, fabricated by same technique. The above composite samples ware compared by evaluating the mechanical parameters such as tensile strength, tensile modulus, bending strength, bending modulus. The effect of mercerization on weight and dimension of jute fabrics was studied. Mechanical properties of mercerized jute-PP composites ware measured and found highest at 20% NaOH at 0°C for 60 min soaking time. Alkali treatment helped in the development of hydrophobicity and reduction in volume fraction of the porosity. This may be due to the better fibre matrix interface adhesion caused due to the fibre surface treatment by alkali. The optimized formulation was irradiated by γ radiation at different dosage (100, 150, 200, 250 & 500 Krad). Among them 250 Krad showed highest mechanical properties.

Keywords: Jute fiber; Polyethylene; Composite; Tensile strength; Bending strength; Bending modulus

Introduction

Environmental awareness and depletion of the petroleum resources are among vital factors that motivate a number of researchers to explore the potential of reusing natural fiber [1-4] as an alternative composite material in industries such as automotive component [5,6], packaging materials [7],insulation [8,9], acoustic absorption panel [10,11] and building materials [12,13]. Natural fibers are available in abundance, low cost, lightweight polymer composite and most importance its biodegradability features [14-17], which often called "ecofriendly" materials. Handling of the natural fibers causes little concerns in terms of health, safety and energy can be recovered from it in an environmentally friendly way. Natural fibers like jute, kenaf, flax, hemp, coir and sisal have already earned a testimony of success as reinforcing material in engineering markets such as in automotives, construction as well as in packaging industries.

Among all the natural fibers, jute appears to be the most useful, economical and exhibits moderately higher mechanical properties. High cellulose content and low microfibril angle of jute fiber provide good reinforcement in polymer matrix. Moreover it is fully bio degradable and eco-friendly. However, jute fiber due to hydrophilic nature is inherently incompatible with hydrophobic thermoplastics, such as polyolefin. So, jute fibers suffer poor interfacial adhesion with non-polar hydrophobic matrix. To overcome this limitation, development of strategies for surface modification of natural fibers or polymer matrices is needed in order to create a strong adhesion at the interfaces of two different phases. Many attempt such as physical and chemical treatments have been adopted to improve the interfacial adhesion. Mercerization is one of the most used chemical treatments of natural fibers when they are used as reinforcement in polymer matrices [18-21]. Treatment of jute and sisal fibers with 5% aqueous NaOH solution showed significant change on the crystallinity of fibers. Alkaline treatment also significantly improved the mechanical and dynamic mechanical behaviours of natural fibre reinforced composites.

The major fibers constituents (α-cellulose,hemicellulose and lignin) of the untreated and alkaline-treated jute fibre samples were determined by Exequiel S. Rodríguez et al. [22]. The results are listed as in Table 1.

Gassan et al. [23] observed that shrinkage of the fibers during mercerization had the most significant effect on the fiber structure and as a result, on the fiber mechanical properties such as tensile strength, modulus and toughness. Rajulu et al. [24] showed that the maximum tensile strength, modulus and density of the mercerized lignocelluloses fabric ware attained at 4 hour of alkali treatment when fabric ware treated with 5% aqueous NaOH solution for 0, 2,4,6 and 8 hours. The alkali treatment also increased the surface roughness of the fibers. Khan et al. [1] prepared a series of composites jute, mercerized jute and manmade cellulose tyre cord yarn Codenka in polypropylene/ethylene block copolymer (PP) with maleic acid anhydride grafted PP (MAPP) as a coupling agent. They observed that the partial substitution of jute instead of Codenka leads to enhance stiffness properties of the composites. On the other hand, impact strength decreases with the increasing jute fraction. Liu et al. [25] showed that combination of NaOH and MAPP emulsion is a good adhesion promoter for jute fiber mat/PP composite. The present research work deals with the extensive study on the properties of mercerized jute fabrics in terms of weight and dimension at different soaking time and temperature. The consequence of the mercerization on the mechanical and thermal properties of the mercerized jute composites was explored.

*Corresponding author: Poddar P, Department of Applied Chemistry and Chemical Engineering, Faculty of Engineering, University of Dhaka, Dhaka-1000, Bangladesh, E-mail: p.pinku@yahoo.com

Component (%)	Treated	Untreated
α-Cellulose	80.3	71.1
Hemi-cellulose	9.6	15.9
Lignin	7.6	11.8
Other components (ash, water, pectins)	2.5	1.2

Table 1: Major fibers constituents of the treated and untreated alkaline treated juice fibre sample.

Mercerization Treatment and γ Irradiation

Mercerization is an alkali treatment process. It is widely used in textile industry [26]. The standard definition of mercerization as proposed by ASTM D1965 is: the process of subjecting a vegetable fiber to an interaction with a fairly concentrated aqueous solution of strong base, to produce great swelling with resultant changes in the fine structure, dimension, morphology and mechanical properties [2]. Therefore, mercerization is a chemical modification process that changed the chemical constituent behavior in natural fiber. The effect of alkali on cellulose fiber is a swelling reaction, during which the natural crystalline structure of the cellulose relaxes. Native cellulose (i.e., cellulose as it occurs in nature) shows a monoclinic crystalline lattice of cellulose-I, which can be changed into different polymorphic forms through chemical or thermal treatments.

The type of alkali and its concentration will influence the degree of swelling, and hence the degree of lattice transformation into cellulose-II. It has been reported that Na+ has got a favorable diameter, able to widen the smallest pores in between the lattice planes and penetrate into them. Consequently, sodium hydroxide (NaOH) treatment results in a higher amount of swelling. This leads to the formation of new Na-cellulose-I lattice, a lattice with relatively large distances between the cellulose molecules, and these spaces are filled with H_2O molecules. In this structure, the OH groups of the cellulose are converted into O-Na-groups, expanding the dimensions of molecules. Subsequent rinsing with water will remove the linked Na-ions and convert the cellulose to a new crystalline structure, i.e., cellulose-II, which is thermodynamically more stable than cellulose-I. NaOH can cause a complete lattice transformation from cellulose-I to cellulose-II. Addition of aqueous sodium hydroxide (NaOH) to natural fiber promotes the ionization of the hydroxyl group to the alkoxide. The following reaction takes place as a result of alkali treatment [27-29]:

$$Fiber - OH + NaOH \longrightarrow Fiber–O–Na + H_2O \quad (1)$$

As reported in much literature, natural fiber chemical constituent consists of cellulose and other non cellulose constituent like hemicellulose, lignin, pectin and impurities such as wax, ash and natural oil [30,31]. This non cellulose material could be removed by appropriate alkali treatments, which affect the tensile characteristic of the fiber [29,32].

Mercerization was found to change fiber surface topography, and the fiber diameter was reported to be decreased with increased concentration of sodium hydroxide concentration [33]. Mercerization treatment also results in surface modifications leading to increase wet ability of coir fiber polyester resin as reported by Prasad et al. [34]. It is reported in that alkaline treatment has two effects on the henequen fiber: (1) it increases surface roughness, resulting in better mechanical interlocking; and (2) it increases the amount of cellulose exposed on the fiber surface, thus increasing the number of possible reaction sites [35]. Consequently, alkaline treatment has a lasting effect on the mechanical behavior of natural fibers, especially on their strength and stiffness. In

order to find out the effect of γ radiation, optimized formulation was irradiated at different dosage (100, 150, 200, 250 & 500 Krad).

Materials and Experiment

Materials

Jute fibers (Hessian Cloth) were supplied by Jute Corporation, Dhaka, Bangladesh. Area density of jute fabrics is 0.0224 g/cm². Sodium hydroxide and Polypropylene pellets (Trade name: Cosmoplene) were procured from Merck, Germany and Polyolifen Company, Singapore. Pte Ltd., respectively. The melt flow index (g/ 10 mins) and density (g/ cm²) of PP are 3.0 and 0.9 respectively.

Surface modification of jute fabrics

Jute fabrics were cut into the size of 40×20 cm and treated with different concentrations of NaOH (5, 10, and 20%) in aqueous medium for soaking time of 30, 60 and 90 mins at room temperature. Jute fabrics were also soaked with NaOH solution at 0°C, 25°C and 50°C for a specific concentration and soaking time. After soaking jute fabrics were washed with running tape water until there was no trace amount of NaOH present in the fabrics. The detection of NaOH was carried out by litmus test. The jute fabrics were then dried in an oven at 105°C until the fabrics reached a constant weight. After drying jute fabrics were kept in a desiccator for composite fabrication and γ irradiation.

γ Irradiation of mercerized jute fiber

The optimized formulation was irradiated by γ radiation (Co-60) from 100 Krad to 500 Krad doses at a dose rate of 300 Krad / h.

Fabrication of composites

PP sheets of desired size (15 cm × 13 cm) were prepared by compressing PP granules in the heat press (Carver.Inc., USA, Model 3856) at 190°C for 5 mins under 5 ton pressure. Jute fabrics were cut into small pieces of 15×13 cm size. For composite fabrication, sandwich was made by alternatively placing four layers of pre-weight jute fabrics inside five layers of PP sheets. The sandwich was then kept between two steel plates with a composite fabrication mold (thickness: 2 mm). The sandwich with this arrangement was compressed in the heat press at the same conditions applied for PP sheet preparation. After being heat pressed, PP sheets and composites were cooled to another press (Carver, Model: 4128) operated in a cooling mode. The average thickness of the composites is 2 mm.

Mechanical testing

Tensile test of composite sample was carried out by a universal testing machine using a gauge length of 20 mm and cross head speed of 10 mm/min. Three point bending strength of the composites sample was measured in the same machine with a span length of 58 mm and cross head speed of 2 mm/min. The test was carried out according to DIN (Deutsches Institut fur Normung meaning German Institute for Standardisation) 53455 and DIN 53452 standard methods, respectively. Impact strength (Charpy) of the composites and PP was carried out in an impact tester (MT-3016, Pendulum type, Germany) following ASTM (American Society for Testing and Materials) D 6110-97. All the results were taken as the average value of five samples.

Scanning electron microscopic (SEM) investigation

The surfaces of the composites of both treated and untreated composite samples were examined using a Hitachi S-4000 field emission scanning electron microscope, operated at 5 kV.

Results and Discussion

Effect of mercerization on physical properties of hessian cloth

Weight of jute cloth: Weight of the jute fiber is largely depends on the concentration of NaOH and the time period of treatment. Weight of the jute fiber decreases with the increasing of concentration of NaOH. At the same concentration of NaOH the weight depends on the time periods. The weight of jute fiber follows a decreasing trend from 5% to 20% of NaOH and at the same concentration it decreases from 30 min to 90 min. the maximum weight loss is 20.37% which is for 20% NaOH (90 min) (Figure 1).

Dimension of jute cloth: The dimension of the jute fiber are also depends on the concentration of NaOH and the time period of treatment. Dimension of the jute fiber decreases with the increase of concentration of NaOH. At the same concentration of NaOH the dimension depends on the time periods. The dimension of jute fiber follows a decreasing trend from 5% to 20% of NaOH and at the same concentration it decreases from 30 min to 90 min. the highest value obtains is 44.01% which is for 20% NaOH (60 min) (Figure 2).

Effect of mercerization on mechanical properties of jute composites

Tensile strength of jute/PP composite: Mechanical properties of composites were strongly influenced by adhesion between the matrix and fibers. The Figures 3 and 4 showed that the treatment with NaOH had the highest effect on tensile strength (TS) producing composites with the best tensile properties. The tensile strength also depends on the concentration of NaOH, time period and finally on temperature. It is clearly indicated in Figure 3 and 4, tensile strength is highest in 60 min for mercerized condition of 20% NaOH and for 20% conc. NaOH (60 min) it is higher at 0°C (59.87 MPa) than 25°C and 50°C. This may be due to the interfacial adhesion between jute fiber and PP was increased by treatment of the jute fibers with alkali and it is optimum for 20% conc. NaOH (60 min) at 0°C mercerized condition.

Bending strength of jute/PP composite: The dependency of bending strength on mercerization of jute fiber, of jute composite is investigated. Mercerized jute composites have higher bending strength than the untreated jute composites for max cases. The bending strength also depends on the concentration of NaOH, time period and finally on temperature. At 5% and 10% concentration of NaOH, it decreases from 30 min to 90 min. while at 20% conc. NaOH, it is highest for 60 min. On the other hand, for 20% conc. NaOH (60 min) it is higher at 0°C than 25°C and 50°C. This value is 72.41 MPa. Observation proves that there is a poor interaction between fiber and matrix in the untreated composites. In the chemically treated composites, interaction between fibers and matrix is strong and fibers are uniformly spread in the composites thus prevented to form a bundle (Figures 5 and 6).

Bending–E-modulus of jute/PP composite: Bending- E-modulus of both untreated and chemically treated composites was studied (Figure 7). Mercerized jute composites have higher bending -E-modulus than the untreated jute composites showed in Figure 8. The variation of Bending –E-modulus with the concentration of NaOH, time period and finally on temperature is given in Figures 7 and 8 respectively. The mercerized jute composite with different mercerized condition of concentration of NaOH did not show the same pattern with increase of time period. The highest value obtained is 1.96 GPa at 0°C (for

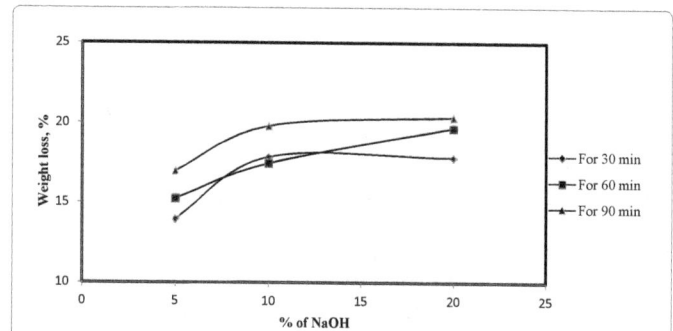

Figure 1: Weight loss of jute cloth at different mercerized condition.

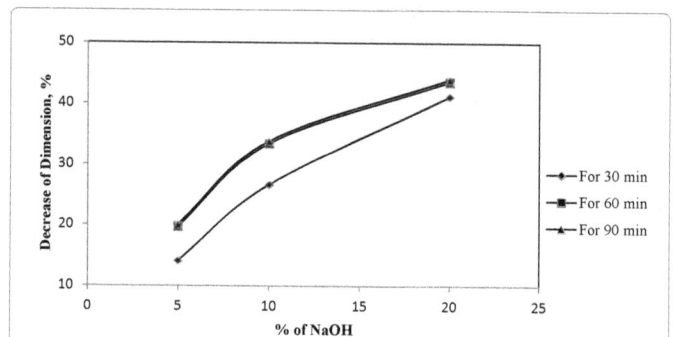

Figure 2: Change in dimension of jute cloth at different mercerized condition.

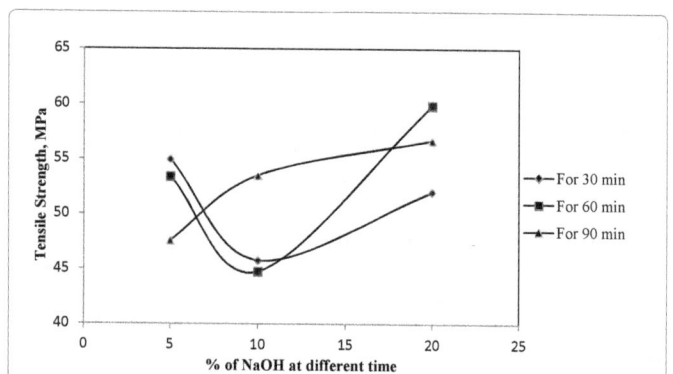

Figure 3: Tensile strength of Mercerized Jute-PP composite at different mercerized condition.

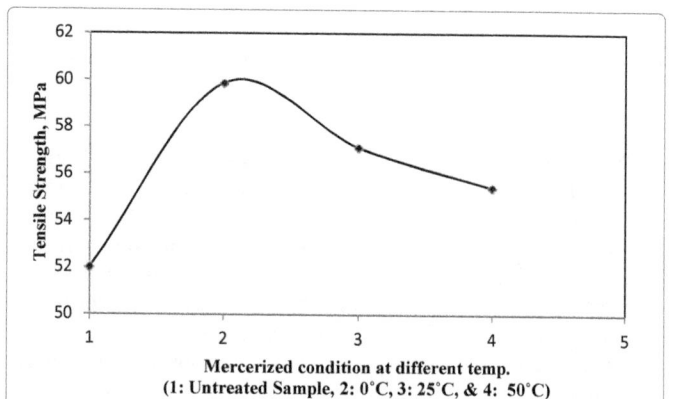

Figure 4: Comparison of tensile strength between untreated Jute-PP composite and Mercerized Jute-PP composite (at different temperature, °C).

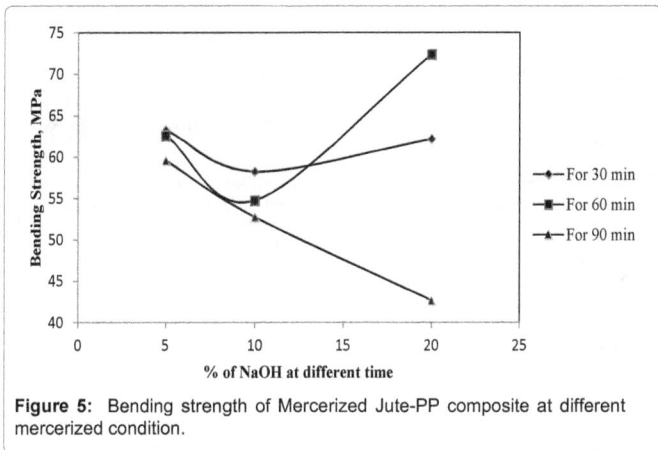

Figure 5: Bending strength of Mercerized Jute-PP composite at different mercerized condition.

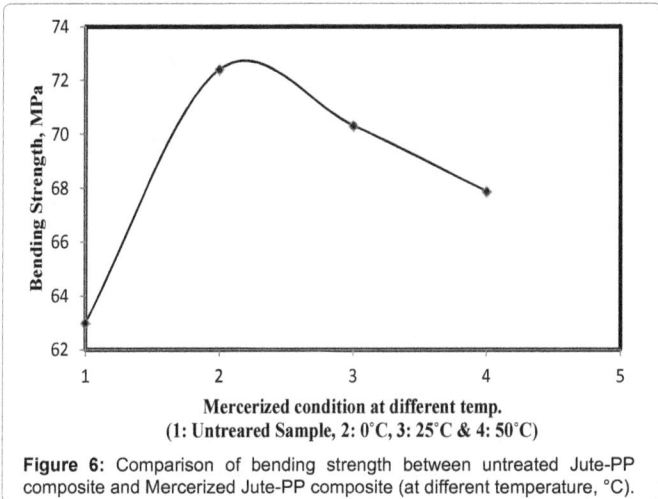

Figure 6: Comparison of bending strength between untreated Jute-PP composite and Mercerized Jute-PP composite (at different temperature, °C).

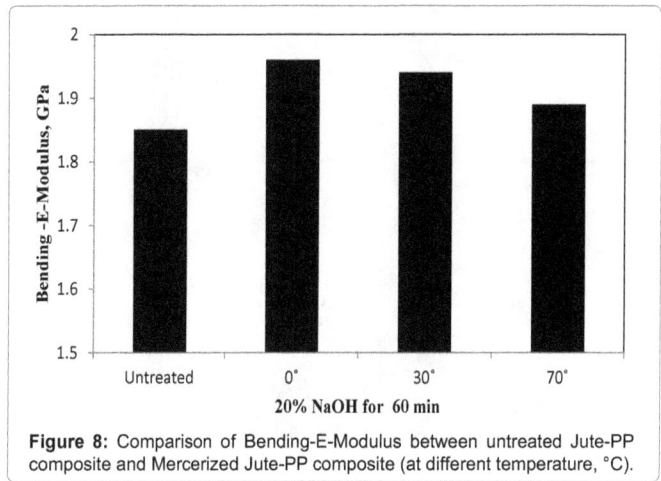

Figure 7: Bending -E- modulus of Mercerized Jute-PP composite at different mercerized condition.

20% NaOH treated for 60 min.) This may be due to elasticity of jute fibers which depends simultaneously both on amounts of fiber and chemical treatment process. The significant information of this part of research is the improvement of BE for jute/PP composites. Hence, it can be concluded that using chemical treatment have the potentially to improve BE to the required range.

Tensile strength and bending strength of γ irradiated composites: Effect of γ radiation on mechanical properties of optimized (20% NaOH

treated at 0°C for 60 min.) mercerized jute composites at different doses was studied. The highest value of TS and BS was 58.36 and 70.47 MPa respectively for 250 Krad. Whereas the TS and BS for untreated sample was 52 and 63 MPa respectively. γ radiation dose may be due to the intercross-linking between the fiber and matrix which enhance the mechanical properties of the composites (Figure 9).

SEM analysis

Figures 10 and 11 indicates some gaps between jute fiber and matrix which is responsible for the low mechanical properties. From the treated SEM image it is found that gaps between fiber and matrix are not observed. Due to NaOH treatment, organic constituents of jute such as lignin, pectin, traces of metals and impurities, etc. might be dissolved and the surface of the cellulose in jute become sleek which enhanced fiber matrix adhesion. Again, γ radiation may generate some active sites and reduce more moisture which might be responsible for better fiber matrix bond.

Conclusion

Mercerization of jute fiber carried out in different concentration (5, 10 and 20%) of NaOH and for different time periods (30, 60, 90 min) and also at different temperatures (0°, 25° and 50°C). Polypropylene (PP) based composites are fabricated by using untreated and treated jute fibers by compression molding. Tensile strength (TS), bending strength (BS) and Bending E Modulus (BE) of the composites studied.

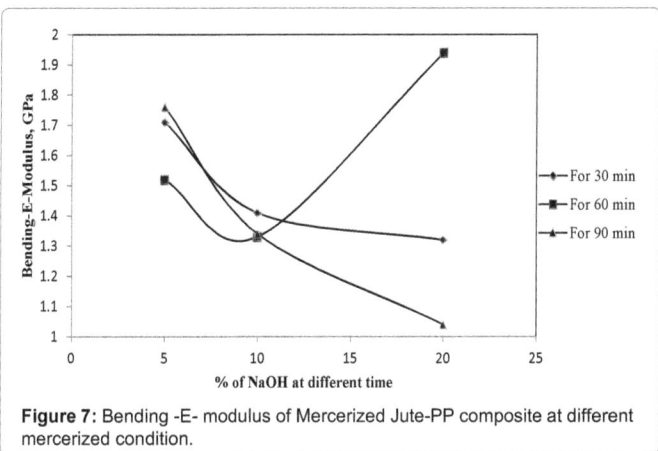

Figure 8: Comparison of Bending-E-Modulus between untreated Jute-PP composite and Mercerized Jute-PP composite (at different temperature, °C).

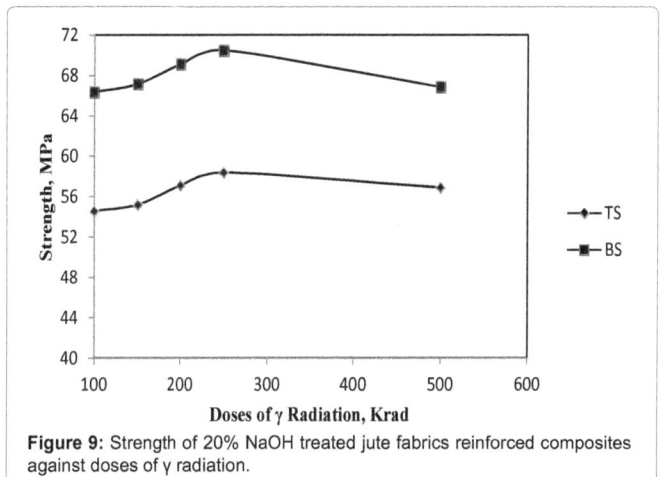

Figure 9: Strength of 20% NaOH treated jute fabrics reinforced composites against doses of γ radiation.

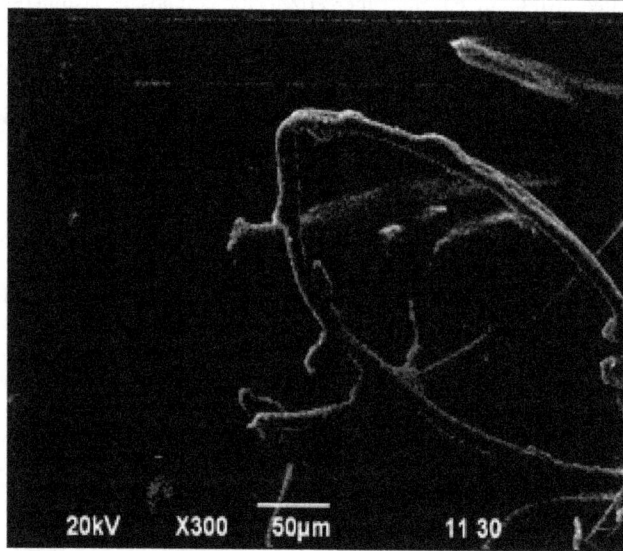

Figure 10: SEM image of untreated jute-PP composite.

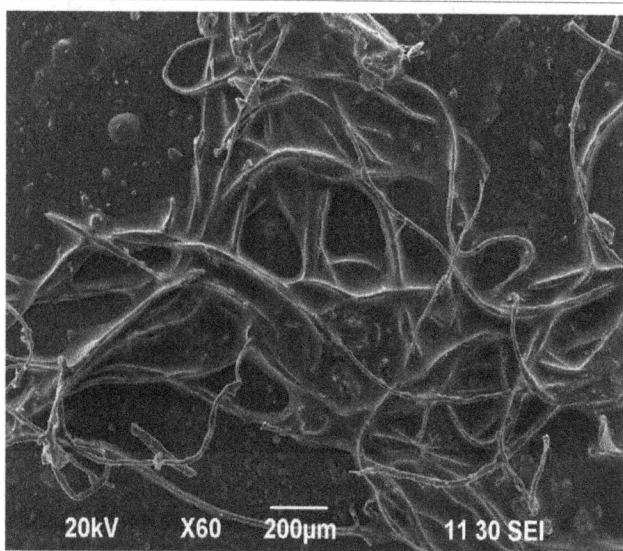

Figure 11: SEM image of γ treated (250 Krad) on mercerized (20% NaOH at 0°C for 60 min) jute-PP composite.

At 0°C temperature the tensile strength, bending strength and Bending E Modulus are the highest for 20% NaOH treated jute composite with a time period of 60 min. It decreases when treated at 25 and 50°C. Again, optimized (20% NaOH treated for 60 min.) mercerized jute fibers were irradiated under γ radiation and mechanical properties were measured. But the mechanical properties of the composites were found lower than that of 20% NaOH treated for 60 min jute PP composites. SEM images supported the idea that the improvement of interfacial fiber matrix adhesion was most likely responsible for those improvements.

References

1. Khan JA, Khan MA, Islam R (2012) Effect of mercerization on mechanical, thermal and degradation characteristics of jute fabric-reinforced polypropylene composites. Fibers and Polymers 13: 1300-1309.

2. Bledzki AK, Gassan J (1999) Composites reinforced with cellulose based fibres. Progress in Polymer Science 24: 221-274.

3. Dweib MA, Donnell AO (2004) All natural composite sandwich beams for structural applications. Composite Structures 63: 147-157.

4. Graupner N, Herrmann AS, Müssig J (2009) Natural and man-made cellulose fibre-reinforced poly (lactic acid) (PLA) composites: An overview about mechanical characteristics and application areas. Composites Part A: Applied Science and Manufacturing 40: 810-821.

5. Holbery J, Houston D (2006) Natural-fiber-reinforced polymer composites in automotive applications. Journal of the Minerals, Metals and Materials Society 58: 80-86.

6. Davoodi MM, Sapuan SM, Ahmad D, Jonoobi M (2010) Mechanical properties of hybrid kenaf/glass reinforced epoxy composite for passenger car bumper beam. Materials and Design 31: 4927-4932.

7. Chaudhary SN, Borkar SP, Mantha SS (2010) Sunnhemp fiber-reinforced waste polyethylene bag composites. Journal of Reinforced Plastics and Composites 29: 2241-2252.

8. Kymäläinen HR, Sjöberg AM (2008) Flax and hemp fibres as raw materials for thermal insulations. Building and Environment 43: 1261-1269.

9. Zhou XY (2010) An environment-friendly thermal insulation material from cotton stalk fibers. Energy and Buildings 42: 1070-1074.

10. Koenig C, Müller DH, Thoben KD (2008) Acoustical parameters of automotive interiors using hybrid fleeces basing on natural fibres. Journal of the Acoustical Society of America 123: 3675.

11. Hosseini Fouladi M, Mohd Nor MJ, Ayub M, Lema ZA (2010) Utilization of coir fiber in multilayer acoustic absorption panel. Applied Acoustics 71: 241-249.

12. Elsaid A, Dawood M, Seracino R, Bobko CP (2011) Mechanical properties of kenaf fiber reinforced concrete. Construction and Building Materials 25: 1991-2001.

13. Rodríguez NJ, Yáñez-Limón M, Gutiérrez-Miceli F, Vazquez JA (2011) Assessment of coconut fibre insulation characteristics and its use to modulate temperatures in concrete slabs with the aid of a finite element methodology. Energy and Buildings 43: 1264-1272.

14. Joshi S, Drzal LT, Mohanty AK (2004) Are natural fiber composites environmentally superior to glass fiber reinforced composites? Composites Part A: Applied Science and Manufacturing 35: 371-376.

15. Mwaikambo L (2006) Review of the history, properties and application of plant fibres. African Journal of Science and Technology 7: 121.

16. John M, Anandjiwala R (2008) Recent developments in chemical modification and characterization of natural fiber reinforced composites. Polymer composites 29: 187-207.

17. Satyanarayana KG, Arizaga GGC, Wypych F (2009) Biodegradable composites based on lignocellulosic fibers--An overview. Progress in Polymer Science 34: 982-1021.

18. Bachtiar D, Sapuan SM, Hamdan MM (2008) The effect of alkaline treatment on tensile properties of sugar palm fibre reinforced epoxy composites. Materials and Design 29: 1285-1290.

19. Islam MS, Pickering KL, Foreman NJ (2010) Influence of alkali treatment on the interfacial and physico-mechanical properties of industrial hemp fibre reinforced polylactic acid composites. Composites Part A: Applied Science and Manufacturing 41: 596-603.

20. Akil HM, Omar MF, Mazuki AAM, Safiee S, Ishak ZAM, et al. (2011) Kenaf fiber reinforced composites: a review. Materials and Design 32: 4107-4121.

21. Cho D, Myung Kim J, Song IS, Hong I (2011) Effect of alkali pre-treatment of jute on the formation of jute-based carbon fibers. Materials Letters 65: 1492-1494.

22. Rodríguez ES, Vázquez A (2006) Alkali treatment of jute fabrics: influence on the processing conditions and the mechanical properties of their Composites. The 8th International Conference on Flow Processes in Composite Materials (FPCM8) Douai, France.

23. Gassan J, Bledzki AK (1999) Posibilities for improving the mechanical properties of jute/epoxy composites by alcali tratment of fibres. Comp Sci Technol 59: 1303-1309.

24. Varada Rajulu A, Meng YZ, Li XH, Rao B, Ganga Devi L, et al. (2003) Effect of alkali treatment on properties of the lignocellulose fabric Hildegardia. Appl Polym Sci 90: 1604-1608.

25. Liu XY, Dai GC (2007) Surface modification and micromechanical properties of jute fiber mat reinforced polypropylene composites. eXPRESS Polymer Letters 1: 299-307.

26. Wang HM, Postle R, Kessler RW, Kessler W (2003) Removing pectin and lignin during chemical processing of hemp for textile applications. Textile Research Journal 73: 664-669.

27. Mwaikambo LY, Ansell MP (2002) Chemical modification of hemp, sisal, jute, and kapok fibers by alkalization. Journal of Applied Polymer Science 84: 2222-2234.

28. Li X, Tabil LG, Panigrahi S (2008) Chemical treatments of natural fiber for use in natural fiber-reinforced composites: a review. Journal of Polymers and the Environment 15: 25-33.

29. Sreenivasan S, Iyer PB, Krishna Iyer KR (1996) Influence of delignification and alkali treatment on the fine structure of coir fibres. Journal of Materials Science 31: 721-726.

30. Khalil A, Alwani MS, Mohd Omar AK (2006) Chemical composition, anatomy, lignin distribution, and cell wall structure of Malaysian plant waste fibers. BioResources 1.

31. Abdul Khalil HPS, Yusra AFI, Bhat AH, Jawaid M (2010) Cell wall ultrastructure, anatomy, lignin distribution, and chemical composition of Malaysian cultivated kenaf fiber. Industrial Crops and Products 31: 113-121.

32. Gassan J, Bledzki AK (1999) Architecture and surfactant behavior of amphiphilic prototype copolymer brushes. Composites Science and Technology 59: 1303-1309.

33. Mwaikambo LY, Ansell M (2006) Mechanical properties of alkali treated plant fibres and their potential as reinforcement materials. I. hemp fibres. Journal of Materials Science 41: 2483-2496.

34. Prasad SV, Pavithran C, Rohatgi PK (1983) Alkali treatment of coir fibres for coir-polyester composites. Journal of Materials Science 18: 1443-1454.

35. Valadez-Gonzalez A, Cervantes-uc JM, Olayo R, Herrera-franco PJ (1999) Effect of fiber surface treatment on the fiber-matrix bond strength of natural fiber reinforced composites. Composites Part B: Engineering 30: 309-320.

33

Synthesis of MgO Nanoparticles from Different Organic Precursors; Catalytic Decontamination of Organic Pollutants and Antitumor Activity

Islam MI Moustafa[1]*, Ihab A Saleh[1] and Mohamed R Abdelhamid[2]

[1]*Chemistry Department, Faculty of Science, Benha University, 13518, Egypt*
[2]*Department of Medicine and Hepatology, Faculty of Medicine, Minia University, Egypt*

Abstract

The present work deals with the synthesis of nanostructured MgO from different organic precursors, by a facile precipitation method as catalysts for the decontamination of Malathion (stimulant of chemical warfare agents) (VX) and orange G as organic pollutants. The as-prepared nanoparticles were obtained by thermal decomposition of the oxalate, tartarate, citrate, succinate, malate, malonate and glycinate precursors at ≈650°C and were characterized by thermal analysis, FTIR, X-ray diffraction, high resolution transmission electron microscope (HRTEM) and absorption spectra. The morphology and crystal sizes were found to be highly affected by the starting organic precursors. The results revealed that the prepared inexpensive magnesium oxides have high potential as catalysts for photo degradation of both Malathion and orange G from water samples. Inhibitory activity against Breast Carcinoma MCF-7 cell line was detected using some selected nanosized MgO and compared to that of Vinblastine as a standard drug.

Keywords: Nano sized MgO; Catalytic photo degradation; VX stimulant (Malathion); Orange G; Antitumor activity

Introduction

Researches on the synthesis and application of nanomaterials had experienced tremendous growth in recent years, owing to their unique properties making them suitable for applications in almost every field of science [1-11]. Photocatalytic degradation of dyes and toxic organic materials from aqueous solution is considered one of the most effective methods for water treatment especially using nanosized metal oxides as catalysts [12-18]. Nanocrystalline MgO is an interesting functional material due to its low heat capacity, chemical inertness, optical transparency and high thermal stability. Due to its high surface area, it is used as an efficient adsorbent for numerous toxic chemicals and acid gases. Recently, MgO nanoparticles have shown promise for application in tumor treatment and also have considerable potential as an antibacterial agent [19-24]. In the present work and in continuity to our previous work [25], we study the effect of starting organic precursors on the crystal size and morphology of nanosized MgO. The prepared nanooxides were tested successfully as adsorbents for the decontamination of Malathion as stimulant of chemical warfare agents (CWA) (VX) and orange G as organic pollutant. Inhibitory activity against Breast Carcinoma MCF-7 cell line was also tested using some selected nanosized MgO.

Experimental Section

Synthesis of magnesium oxide nanoparticles

AR grade chemicals obtained from Merck and Aldrich were used for the preparation of the nano particles of magnesium oxide (M1-M7). The individual oxides were prepared via the precipitation method by titration of 50 ml of 0.1 M magnesium acetate drop by drop to 50 ml of 0.2 M of different organic acids viz. oxalic acid (M1), tartaric acid (M2), citric acid (M3), succinic acid (M4), malic acid (M5), malonic acid (M6) and glycine (M7), respectively. The mixtures were stirred for 1 hr. and the precipitates formed were filtered, washed thoroughly using bidistilled water, air dried, thoroughly grounded in agate mortar and finally dried at 120°C for two hours in the form of fine powder. The powder so obtained was annealed at 650°C for six hours in muffle furnace to obtain the corresponding nanosized magnesium oxides.

Physical measurements

FT-IR spectra of both of the precipitated and ignited samples were recorded on a Nicolet iSio FT-IR spectrophotometer in the 4000-400 cm⁻¹ region using KBr disk technique (Chemistry department, Faculty of science, Benha University, Egypt). Electronic absorption spectra of the prepared nanooxides were recorded on a Jasco (V-530) UV-Vis spectrophotometer (Chemistry department, Faculty of Science, Benha University, Egypt). Thermogravimetric analysis (TG-DT) for the organic precursors was recorded on Shimadzu TA-60 WS thermal analysis (Micro analytical unit, Menofia University, Shebin El-Kom, Egypt). Elemental analysis for C and H of the nanooxides were carried out using Elementer Vario EL III Carlo Erba 1108 instrument (The Regional Center for Mycology and Biotechnology, Al-Azhar University, Cairo, Egypt). X-ray powder diffraction (XRD) was recorded on a 18 kW diffractometer (Bruker; model D8 Advance) with monochromated Cu K$_a$ radiation (λ) 1.54178 Å (Central metallurgical research institute, Helwan, Egypt). The HR-TEM images of some selected nanooxides were taken on a transmission electron microscope (JEOL; model 1200 EX) at an accelerator voltage of 220 kV (Egyptian Petroleum Research Institute, Cairo, Egypt).

Photocatalytic degradation of Orange G dye

For a typical photocatalytic experiment, 100 mg of the nanosized photocatalyst was added to 25 ml of 20 ppm aqueous dye solution which was kept in dark for 6 hrs to allow the system to reach an adsorption desorption equilibrium then 2 ml of 0.5 M hydrogen peroxide solution was added. The degradation process was investigated in a Pyrex beaker under the UV illumination using a 250 W xenon arc lamp (Thoshiba, SHLS-002) (λ=365 nm). After recovering the catalyst by centrifugation,

*Corresponding author: Islam MI Moustafa, Chemistry Department, Faculty of Science, Benha University, 13518, Egypt, E-mail: islamshahin84@outlook.com

the absorption spectra of the clear solution was measured at 485 nm (λ_{max} for Orange G dye) at different time intervals using a UV-Vis spectrophotometer.

Photocatalytic degradation of Malathion

The photocatalytic degradation of 0.06 ml/l of Malathion solution was performed using the smallest crystal sized magnesium oxide (M7) sample (prepared from glycine). For a typical experiment, 100 mg of the nanooxide sample was added to 25 ml of 0.06 ml/l Malathion solution which was kept in dark for 6 hrs. The degradation process was investigated as previously mentioned for Orange G dye and the concentration of the remaining pollutant was followed up by measuring absorption spectra at 238, 265 nm (λ_{max} for Malathion) at different time intervals.

Antitumor activity

Antitumor activity against Breast Carcinoma MCF-7 cell line was measured for some selected samples at The Regional Center for Mycology and Biotechnology, Al-Azhar University, Cairo, Egypt and compared to that of Vinblastine as a standard drug. The number of viable cells and the percentage of viability were calculated as $(1-\frac{ODt}{ODc})\times100$; where ODt is the mean optical density of the wells treated with the tested samples and ODc is the mean optical density of the untreated cells. The relation between surviving cells and drug concentration is plotted to get the survival curve of each tumor cell line after treatment with the nanooxides. The 50% inhibitory concentration (IC_{50}), the concentration required to cause toxic effect in 50% of intact cells, was estimated from graphic plots of the dose response curve for each concentration.

Results and Discussion

Characterization

Thermogravimetric analysis: The thermogravimetric-differential thermal analysis was performed on the organic precursors to follow their thermal decomposition to the final oxide forms. Inspection of the thermograms (Figure 1) showed that the organic precursors were thermally degraded through three main steps. The first within the temperature range 66.75-184.78°C due to the dehydration of humidity and crystallinity water (this step is sometimes a composite of two steps). The beginning of the thermal degradation of the unhydrous compounds took place within the second step within the temperature range 114.72-344.33°C by the evolution of gases such as N_2 and CO_2. Complete decomposition of the organic precursors occurred in the third step within the range 232.99-565.99°C which led to the formation of the nanosized MgO as final product. Example of thermal decomposition process of tartaric acid precursor is represented as:

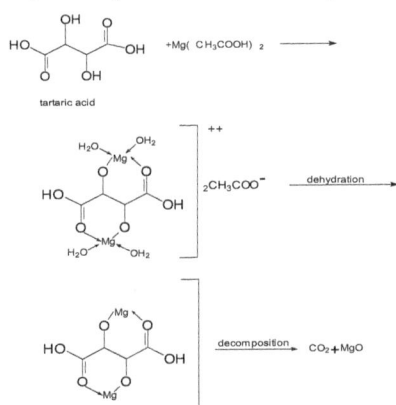

Fourier transforms infrared spectra (FTIR): The FTIR spectra of the magnesium-organic precursors were studied and compared to those of the corresponding nanooxides. The most important band frequencies (cm⁻¹) are listed in Table 1. The spectra of the organic precursors show weak absorption bands within the wavenumber ranges 3373-3430 cm⁻¹ and 1033-1171 cm⁻¹ due to the stretching and bending vibrations of the trace water molecule, respectively. These two bands appeared within the ranges 2955-3430 cm⁻¹ and 1032-1176 cm⁻¹ as very weak bands in the spectra of the ignited samples. The strong bands within the range 1570-1677cm⁻¹are due to the stretching vibration of C=O group ($V_{C=O}$). These bands, more or less, disappeared in case of calcinated samples. It is worthy to mention that there is a shift in the IR active mode, which is due to nano size grain. For a nano size grain, the atomic arrangement on the boundaries differ greatly from that of the bulk crystals, both in coordination number and bond lengths, showing some extent of disorder [26]. Crystal symmetry is thus, degraded in nano size grains. The degradation in crystal symmetry results in the shifting of the IR active mode [27].

X-ray diffraction analysis (XRD): The X-ray diffraction patterns of the nano sized magnesium oxides are shown in Figure 2. The XRD sharp lines reveal that the oxide nanoparticles are crystalline. The relative crystalline sizes are determined from the XRD lines broadening using the Scherrer equation [28]. From calculation, the average crystalline size was found to be 44.8, 33.94, 44.60, 46.32, 47.16, 60.80 and 53.64 nm for magnesium oxides (M1-M7), respectively (Table 2). The phase purity of all the samples was established by comparison of the X-ray diffraction patterns with JCPDS international data value.

High resolution transmission electron microscopy (HRTEM): The micro structural analysis of the synthesized samples calcined at 650°C for 6 hours was investigated using HRTEM. The HRTEM images of some selected samples are shown in Figure 3. It can be seen from the graphs that magnesium nanoparticles have narrow size distribution and are rectangular rode shapes with weak agglomeration. The average particle sizes ranged from 33.93-60.80 nm.

Optical analysis: A fundamental property of nanosized metal oxides is the band gap energy. The band gap energy is the energy separation between the filled valence band and the empty conduction band. Optical excitation of electrons across the band gap is strongly allowed, producing an abrupt increase in the absorption at the wavelength corresponding to the band gap energy. The UV-Visible spectra allow direct determination of band gap using the relation

Compound	IR frequency (cm⁻¹)				
	v_{OH}	δ_{OH}	$V_{C=O}$	v_{C-O}	Mg-O
Mg-oxalate	3389 br	1128.96	1637	829.92	501
NP (M1)	2955 v.w	1166.68	---	---	541
Mg-tartarate	3408 br	1090.45	1617	832.88	486
NP (M2)	2979 v.w	1135.71	---	---	542
Mg-citrate	3425.52	115.64	1624	885	544
NP (M3)	3430 v.w	1085	---	---	553
Mg-succinate	3430.71	1176.47	1691	893.54	522
NP (M4)	3110 v.w	1152.45	---	---	544
Mg-malate	3404.33 br	1084.14	1570	870.23	579
NP (M5)	3052 v.w	1171.25	---	---	570
Mg-malonate	3428	1032.77	1677	913.15	586
NP (M6)	3362 br	1136.70	---	---	583
Mg-glycinate	3373 br	1033.78	1617	924.83	512
NP (M7)	3381	1095.12	---	---	522

Table 1: IR frequencies (cm⁻¹) of some important groups in the organic precursors and the corresponding nanooxides.

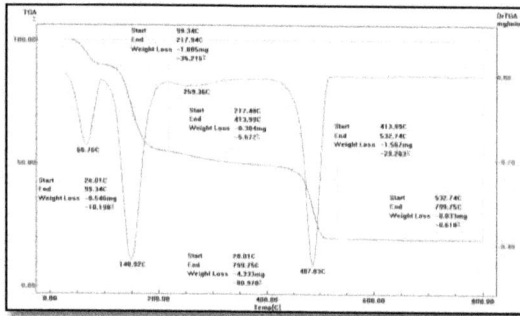

M1

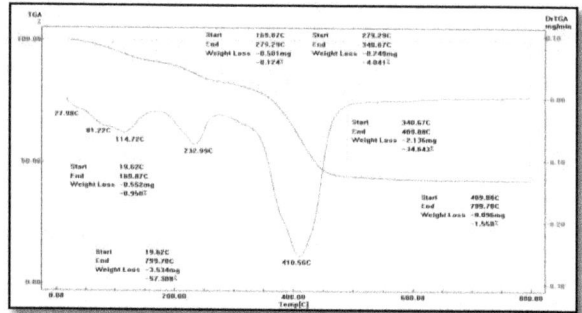

M2

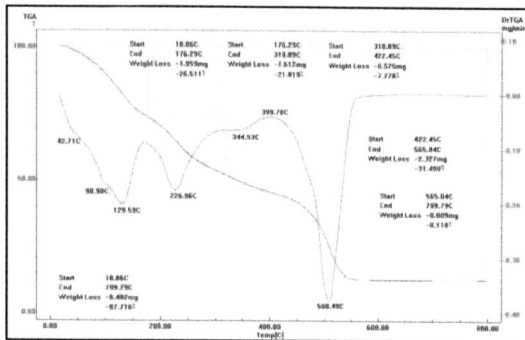

M3

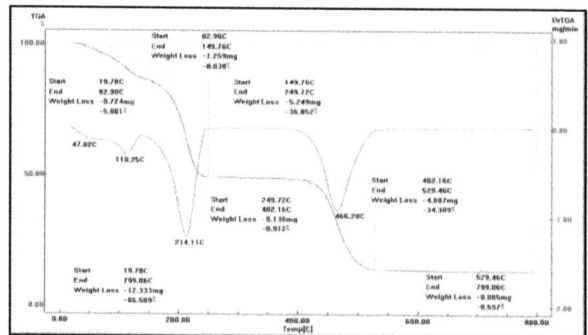

M4

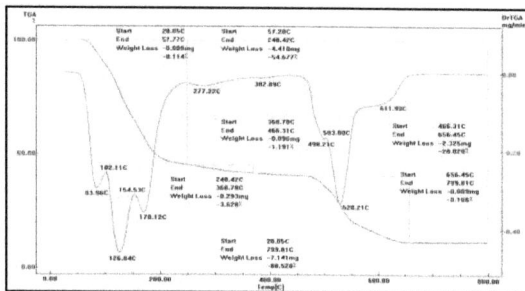

M5

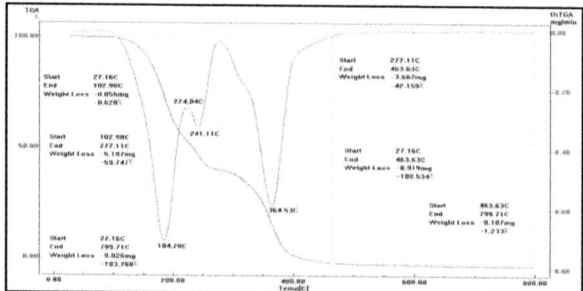

M6

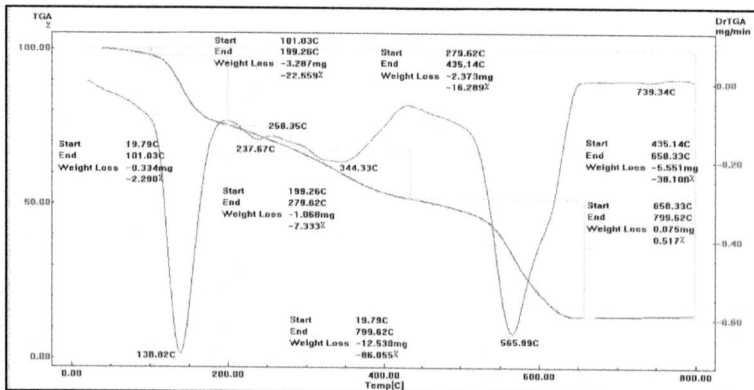

M7

Figure 1: TGA-DTA curves for precursors M1- M7.

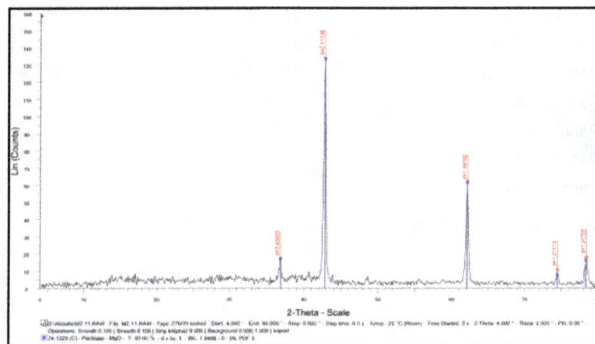

M1

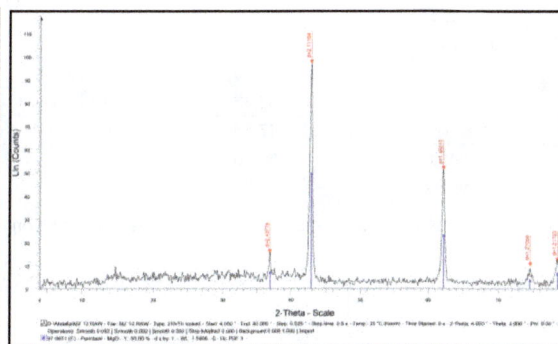

M2

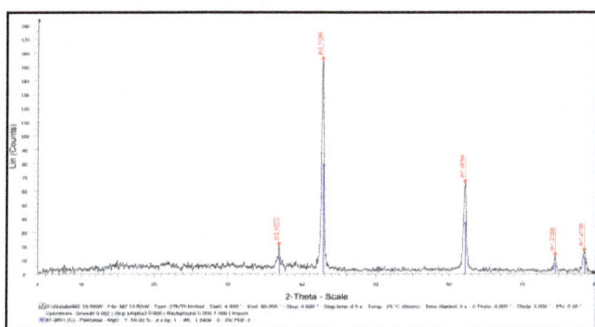

M3

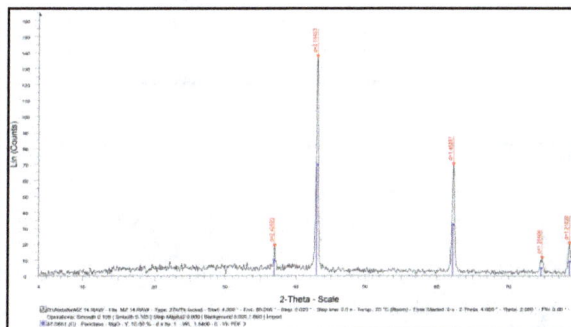

M4

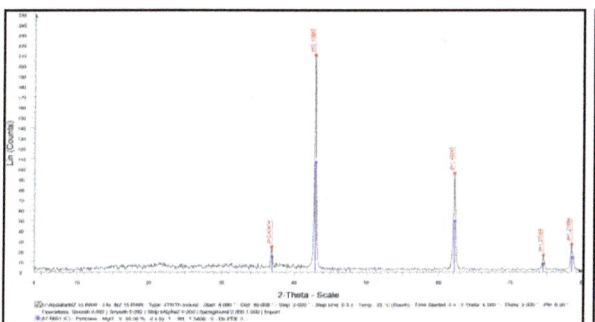

M5

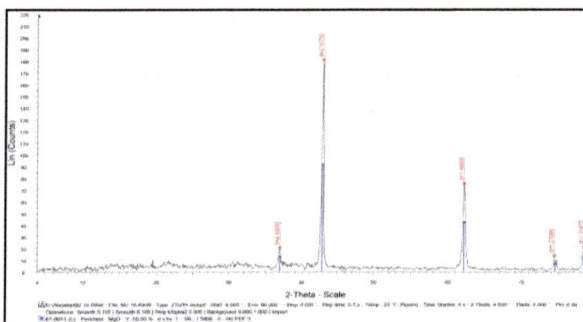

M6

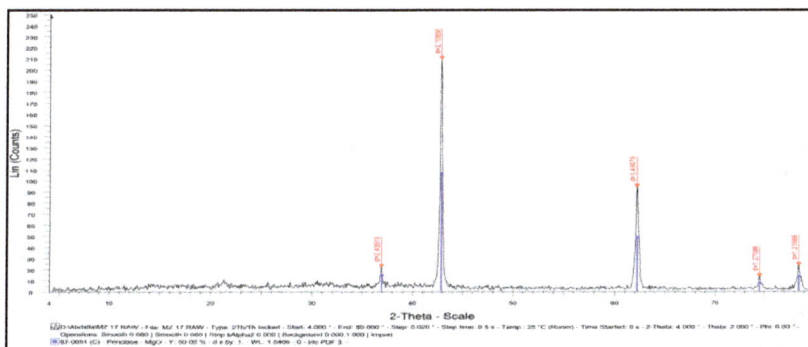

M7

Figure 2: XRD of the samples M1- M7 prepared from organic precursors.

Sample	Precursor	crystal size (nm)	Phase produced by XRD	Card No.	Crystal system
M1	Oxalate	44.80	Periclase-MgO	01-074-1225	Cubic
M2	Tartarate	33.94	Periclase-MgO	01-087-0651	Cubic
M3	Citrate	44.60	Periclase-MgO	01-087-0651	Cubic
M4	Succinate	46.32	Periclase-MgO	01-087-0651	Cubic
M5	Malate	47.16	Periclase-MgO	01-087-0651	Cubic
M6	Malonate	60.80	Periclase-MgO	01-087-0651	Cubic
M7	Glycinate	53.64	Periclase-MgO	01-087-0651	Cubic

Table 2: Effect of organic acids and glycine precursors on the crystal size and morphology of the prepared nanooxides.

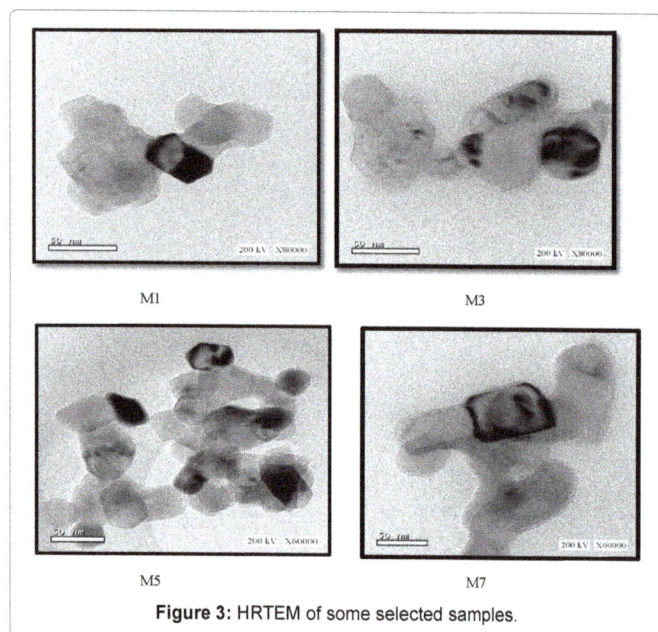

Figure 3: HRTEM of some selected samples.

Oxide	IC$_{50}$ (µg/ml)
Vinblastine	4.6
M2	213
M3	398
M7	391

Table 3: Lethal concentration (IC50) of the nanosized oxides M2, Z3 and MZ7.

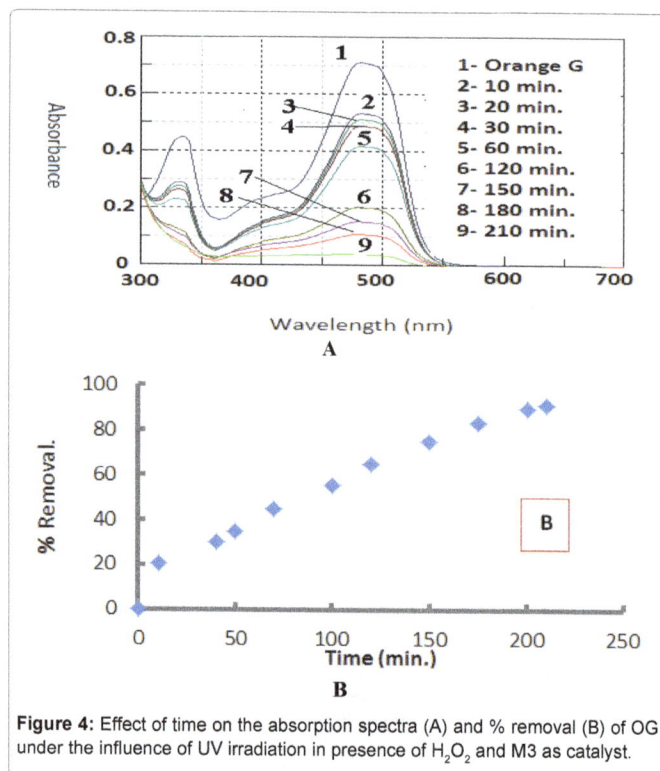

Figure 4: Effect of time on the absorption spectra (A) and % removal (B) of OG under the influence of UV irradiation in presence of H_2O_2 and M3 as catalyst.

between the absorption coefficient (α) and the incident photon energy (hv) represented by Tauc equation [29]:

$$(\alpha hv)=A(hv-Eg)^n$$

Where, A is a constant, E_g is the band gap of the material and exponent n depends on the type of transition, n is either 2 for an indirect transition or ½ for a direct transition. Here the transition is direct so, n is taken to be ½. The value of optical band gap is calculated by plotting the relation between $(\alpha hv)^2$ vs. hv and extrapolating the straight line portion to the hv axis. The extrapolation of linear portion to the hv axis gives values of energy gap for magnesium oxide nano particles. The obtained E_g values (within the range 2.91-3.12 eV) show the semiconductor nature of the samples and are in an excellent agreement with the reported data [30].

Application

Photocatalytic degradation of organic G dye: The photodegradation efficiency of selected zirconium oxide nanoparticle (sample M3; prepared from citric acid as organic precursor) was tested using Orange G (OG) as model. The experiments were done at different conditions, namely; (UV+H$_2$O$_2$+M3), (UV+H$_2$O$_2$), (UV only), (UV+M3) where best results were obtained in the case of (UV+H$_2$O$_2$+M3). At periodic intervals of time, aliquots of the sample were withdrawn and the absorption spectra were recorded. Clearly, the absorbance decreases and the photodegradation efficiency increases (reaching a plateau) as a function of time (Figure 4). The results showed that the maximum percent of degradation of OG dye was 93% after 210 min indicating the very high efficiency of the nanoparticle used.

Decomposition of chemical warfare agents on metal oxides: Metal oxides demonstrate superior ability to adsorb and decompose CWA compared to pure metal surfaces. This is often attributed to reactive sites on the metal oxide surface through which organophosphonate species (nerve agents) can adsorb and subsequently undergo a hydrolysis reaction. The use of MgO nanoparticle, M7, (prepared from glycine as precursor) as catalyst for the decontamination of VX stimulant Malathion was carefully investigated. The effect of different factors affecting the removal efficiency, such as: time, pH, temperature and initial dose of Malathion was studied. The results obtained (Figures 5 and 6) showed that such nanooxides have high ability for the removal of this pollutant from water samples.

Antitumor activity

Inhibitory activity against Breast Carcinoma MCF-7 cell line was detected using some selected nanosized metal oxides viz; M2

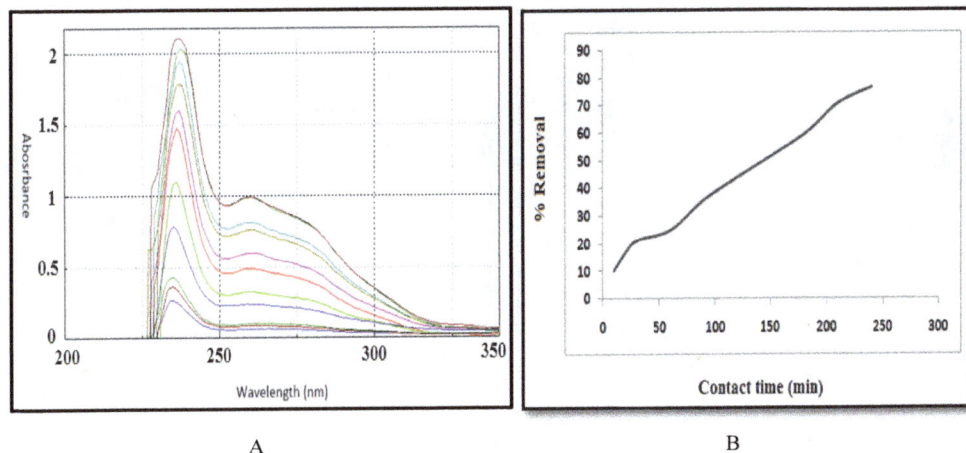

Figure 5: Effect of contact time on the absorption spectra of Malathion (A) and its % removal (B).

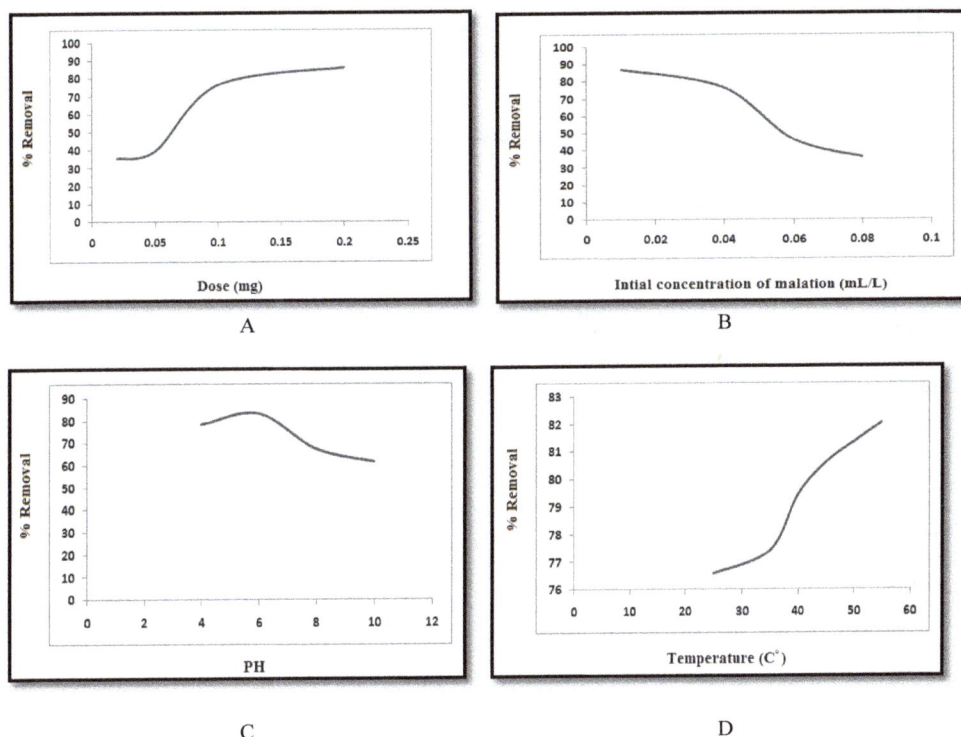

Figure 6: Effect of different factors on the % removal of Malathion; (A); adsorbent dose, (B); initial concentration, (C); pH and (D); temperature.

and M3 and M7 and compared to that of Vinblastine as a standard drug. The relation between surviving cells and drug concentration is plotted to get the survival curve of each tumor cell line after treatment with the nanooxides. The 50% inhibitory concentration (IC$_{50}$), the concentration required to cause toxic effect in 50% of intact cells, was estimated from graphic plots of the dose response curve for each concentration. The results are represented graphically in Figure 7 and the lethal concentrations (IC$_{50}$) values are listed in Table 3.

Inspection of the cytotoxic data, it is found that magnesium oxide M2 (prepared from tartaric acid) is, in general, more effective than those of M3 (prepared from citric acid) and M7 (prepared from

glycine). Shier [31] suggested that compounds having IC$_{50}$ values 10-25 µg/ml are considered to have weak cytotoxic activities, while those having intermediate values (ranging from 5-10 µg/ml) are classified as moderately active. On the other hand, compounds with IC$_{50}$ values less than 5 µg/ml are considered to be very active. Consequently, the nanosized metal oxides under study are considered to have weak activity with IC$_{50}$ values higher than 213 µg/ml.

Conclusion

MgO nanoparticles (M1-M7) were prepared by precipitation of Mg as oxalate, tartarate, citrate, succinate, malate, malonate and glycinate then ignition at 650°C. The magnesium-organic precursors and the

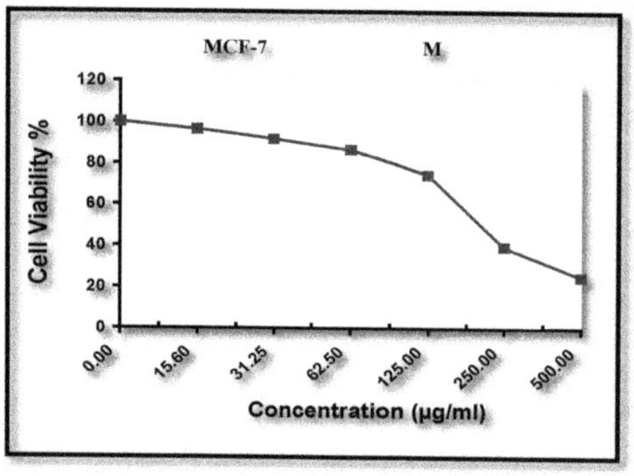

Figure 7: Inhibitory activity of nanooxideM2 (A), (IC50=213 µg/ml), M3 (B), (IC50=398 µg/ml) and M7 (C), (IC50=391 µg/ml) against Breast carcinoma cells MCF-7.

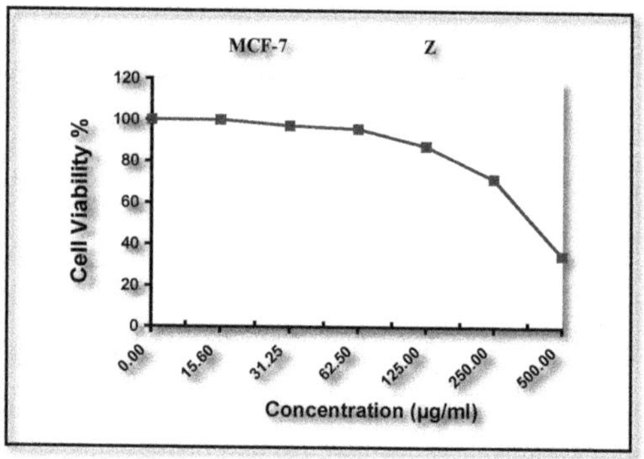

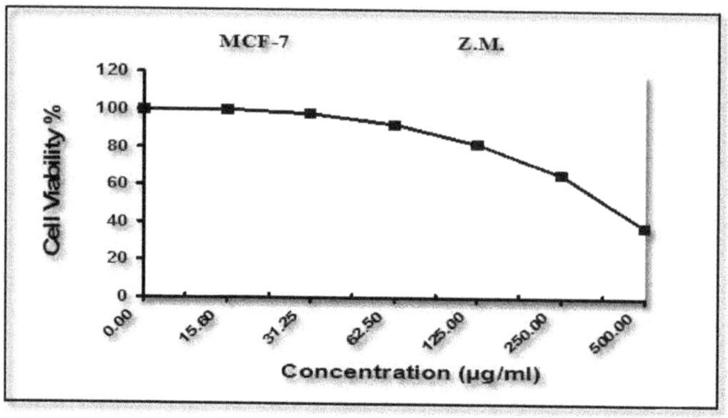

corresponding nanooxides were characterized by thermal analysis and different spectroscopic techniques. The morphology and crystal sizes were found to be highly affected by the starting organic precursors. The optical energy gaps (E_g) calculated from electronic absorption spectra ranged from 2.91 to 3.12 eV suggesting the semiconductor nature of the nanooxides. The results revealed that the prepared inexpensive magnesium oxides have high potential as catalysts for photo degradation of both Malathion and orange G from water samples. Inhibitory activity against Breast Carcinoma MCF-7 cell line was detected using some selected nanosized MgO and compared to that of Vinblastine as a standard drug.

Acknowledgement

We would like to acknowledge the financial support from the MSP (Management of Scientific Projects), Benha University, Benha, Egypt (project presented by I. M. Ibrahim).

References

1. Neppolian B, Wang Q, Yamashita H, Choi H (2007) Synthesis and characterization of ZrO2-TiO2 binary oxide semiconductor nanoparticles: application and interparticle electron transfer process. Applied Catalysis A: General 333: 264-271.

2. Lara-García HA, Romero-Ibarra IC, Pfeiffer H (2014) Hierarchical Na-doped cubic ZrO2 synthesis by a simple hydrothermal route and its application in biodiesel production. Journal of Solid State Chemistry 218: 213-220.

3. Słońska A, Kaszewski J, Wolska-Kornio E, Witkowski B, Mijowska E, et al. (2016) Luminescent properties of ZrO2: Tb nanoparticles for applications in neuroscience. Optical Materials 59: 96-102.

4. Xiong C, Wang W, Tan F, Luo F, Chen J, et al. (2015) Investigation on the efficiency and mechanism of Cd(II) and Pb(II) removal from aqueous solutions using MgO nanoparticles. Journal of hazardous materials 299: 664-674.

5. Li S, Jiao Y, Wang Z, Wang J, Zhu Q, et al. (2015) Performance of RP-3 kerosene cracking over Pt/WO3–ZrO2 catalyst. Journal of Analytical and Applied Pyrolysis 113: 736-742.

6. Renuka L, Anantharaju KS, Sharma SC, Nagaswarupa HP, Prashantha SC, et al. (2016) Hollow microspheres Mg-doped ZrO2 nanoparticles: Green assisted synthesis and applications in photocatalysis and photoluminescence. Journal of Alloys and Compounds 672: 609-622.

7. Peng W, Li J, Chen B, Wang N, Luo G, et al. (2016) Mesoporous MgO synthesized by a homogeneous-hydrothermal method and its catalytic performance on gas-phase acetone condensation at low temperatures. Catalysis Communications 74: 39-42.

8. Varshney D, Dwivedi S (2015) On the synthesis, structural, optical and magnetic properties of nano-size Zn–MgO. Superlattices and Microstructures 85: 886-893.

9. Gh AB, Sabbaghan M, Mirgani Z (2015) A comparative study on properties of synthesized MgO with different templates. Spectrochimica Acta Part A: Molecular and Biomolecular Spectroscopy 137: 1286-1291.

10. Klubnuan S, Amornpitoksuk P, Suwanboon S (2015) Structural, optical and photocatalytic properties of MgO/ZnO nanocomposites prepared by a hydrothermal method. Materials Science in Semiconductor Processing 39: 515-520.

11. Li H, Li M, Qiu G, Li C, Qu C, et al. (2015) Synthesis and characterization of MgO nanocrystals for biosensing applications. Journal of Alloys and Compounds 632: 639-644.

12. Jin Z, Jia Y, Luo T, Kong LT, Sun B, et al. (2015) Efficient removal of fluoride by hierarchical MgO microspheres: performance and mechanism study. Applied Surface Science 357: 1080-1088.

13. Ding YD, Song G, Zhu X, Chen R, Liao Q (2015) Synthesizing MgO with a high specific surface for carbon dioxide adsorption. RSC Advances 5: 30929-30935.

14. Verma R, Naik KK, Gangwar J, Srivastava AK (2014) Morphology, mechanism and optical properties of nanometer-sized MgO synthesized via facile wet chemical method. Materials Chemistry and Physics 148: 1064-1070.

15. Yadav LR, Lingaraju K, Manjunath K, Raghu GK, Kumar KS, et al. (2017) Synergistic effect of MgO nanoparticles for electrochemical sensing, photocatalytic-dye degradation and antibacterial activity. Materials Research Express 4: 025028.

16. Devaraja PB, Avadhani DN, Prashantha SC, Nagabhushana H, Sharma SC, et al. (2014) Synthesis, structural and luminescence studies of magnesium oxide nanopowder. Spectrochimica Acta Part A: Molecular and Biomolecular Spectroscopy 118: 847-851.

17. Moussavi G, Mahmoudi M (2009) Removal of azo and anthraquinone reactive dyes from industrial wastewaters using MgO nanoparticles. Journal of Hazardous Materials 168(2): 806-812.

18. Moussavi G, Mahmoudi M (2009) Degradation and biodegradability improvement of the reactive red 198 azo dye using catalytic ozonation with MgO nanocrystals. Chemical Engineering Journal 152: 1-7.

19. Bindhu MR, Umadevi M, Micheal MK, Arasu MV, Al-Dhabi NA (2016) Structural, morphological and optical properties of MgO nanoparticles for antibacterial applications. Materials Letters 166: 19-22.

20. Hamdy MS, Awwad NS, Alshahrani AM (2016) Mesoporous magnesia: Synthesis, characterization, adsorption behavior and cytotoxic activity. Materials & Design 110: 503-509.

21. Rao PV, Nallappan D, Madhavi K, Rahman S, Jun Wei L, et al. (2016) Phytochemicals and biogenic metallic nanoparticles as anticancer agents. Oxidative medicine and cellular longevity.

22. Krishnamoorthy K, Manivannan G, Kim SJ, Jeyasubramanian K, Premanathan M (2012) Antibacterial activity of MgO nanoparticles based on lipid peroxidation by oxygen vacancy. Journal of Nanoparticle Research 14: 1063.

23. Tang ZX, Lv BF (2014) MgO nanoparticles as antibacterial agent: preparation and activity. Brazilian Journal of Chemical Engineering 31: 591-601.

24. Yadav LR, Lingaraju K, Manjunath K, Raghu GK, Kumar KS, et al. (2017) Synergistic effect of MgO nanoparticles for electrochemical sensing, photocatalytic-dye degradation and antibacterial activity. Materials Research Express 4: 025028.

25. Ibrahim IM, Moustafa ME, Abdelhamid MR (2016) Effect of organic acids precursors on the morphology and size of ZrO2 nanoparticles for photocatalytic degradation of Orange G dye from aqueous solutions. Journal of Molecular Liquids 223: 741-748.

26. Yua J, Kimb D (2013) Powder Tech 235: 1030-1037.

27. Motlagh MK, Youzbashi AA, Sabaghzadeh L (2011) Synthesis and characterization of Nickel hydroxide/oxide nanoparticles by the complexation-precipitation method. International Journal of Physical Sciences 6: 1471-1476.

28. Arora K, Devi S (2012) current trends in biotech and chem. Research 2.

29. Chang-Sam K, Dond-Hun Y, Sung-Woon J, Hyok-Bo K, Sang-Hwan P (2010) J Korean Cryst, Growth Cryst Tech 20: 283-288.

30. Dhere SL (2015) Silica–zirconia alkali-resistant coatings by sol–gel route. Curr Sci 108: 1647-1652.

31. Shier WT (1991) Mammalian cell culture on $5 a day: a laboratory manual of low cost methods. Los Banos, University of the Philippines 64: 9-16.

34

Thermodynamic Evaluation Using the Law of Mass Action under Consideration of the Activity Coefficients in the System NdCl$_3$-HCl (or NaOH)-H$_2$O-DEHPA-Kerosene

Scharf C[1]* and Ditze A[2]

[1]*Helmholtz Centre Dresden Rossendorf, Helmholtz Institute Freiberg for Resource Technology, Chemnitzer Strasse 40, 09599 Freiberg, Germany*
[2]*MetuRec, Metallurgie und Recycling, Ingenieurbüro, An den Eschenbacher Teichen 16, 38678 Clausthal-Zellerfeld, Germany*

Abstract

Rare earth elements, including neodymium, are in widespread use today. They serve, for example, as alloying elements in magnesium (WE 43, WE 54, AE44, AE42), in permanent magnets (neodymium-iron-boron) or as luminescent materials. In addition, they are amongst the most important commodities in Europe. The demand for their use is growing and the recycling of these elements is indispensable. Their recycling potential can be increased through detailed scientific studies and to this end this article presents a thermodynamic evaluation of equilibrium data in the system neodymium-chloride-hydrochloric acid (or sodium hydroxide)-water-di-(2-ethylhexyl) phosphoric acid (DEHPA)-kerosene. Considering the relationship between the activity coefficients $\gamma_{Nd^{3+}_{aq}}$, $\gamma_{Nd,org}$ and $\gamma_{(DEHPA)_2}$, which arise from the law of mass action, the deviations of the experimental results from the ideal behaviour can be explained. From the calculation of $\gamma_{Nd^{3+}_{aq}}$ with the expanded Debye-Hückel equation and from literature data for $\gamma_{(DEHPA)_2}$, indications arise for the development of the functions of the activity coefficients of DEHPA and neodymium in the organic phase.

Keywords: Law of mass action; Thermodynamic evaluation

Symbols

c_0 [mol/l]: Initial Concentration of DEHPA;

$a_{(DEHPA)2}$ [mol/l]: Activitiy of Dimeric DEHPA;

$c_{(DEHPA)2}$ [mol/l]: Concentration of Dimeric DEHPA;

$m_{(DEHPA)2}$ [mol/kg]: Molality of Dimeric DEHPA;

$\gamma_{(DEHPA)2}$: Activity Coefficient of Dimeric DEHPA;

ρ [g/ml, g/cm^{-3}]: Mass Density of Solutions;

pH=-log a_H^+: Decimal logarithm of the Reciprocal of the Hydrogen Ion Activity;

V_O/V_A: Volume Ratio of the Organic and the Aqueous Phase;

(DEHPA)$_2$: Dimeric DEHPA;

(DEHP·DEHPA): Anion of Dimeric DEHPA;

K: Equilibrium Constant;

$D_e=c_{Nd,\,org}/c_{Nd^{3+}_{aq}}$: Distribution Coefficient of Neodymium in the Organic and the Aqueous Phase;

$a_{Nd,\,org}$ [mol/l]: Activity of Neodymium in the Organic Phase;

$a_{Nd^{3+}_{aq}}$ [mol/l]: Activity of Neodymium in the Aqueous Phase;

$c_{Nd^{3+}_{aq}}$, $c_{SE^{3+}_{aq}}$ [mol/l]: Concentration of Neodymium Cations in the Aqueous Phase;

$\gamma_{Nd,\,org}$: Activity Coefficient of Neodymium in the Organic Phase;

$\gamma_{Nd^{3+}_{aq}}$: Activity Coefficient of Neodymium in the Aqueous Phase;

c_H^+ [mol/l]: Concentration of Hydrogen Ions in the Solution;

R [J/(mol K)]: Ideal Gas Constant;

$a_\pm, a_+, a_-$: Mean Activity, Activity of the Individual Ions;

$\gamma_\pm, \gamma_+, \gamma_-$: Mean Activity Coefficient, Activity Coefficients of the Individual Ions (which cannot be directly measured);

$\nu_\pm, \nu_+, \nu_-$: Stoichiometric Coefficients;

$m_\pm, m_+, m_-$: Mean Molality, Molality of the Individual Ions;

I_c, I_m [mol/l] bzw. [mol/kg]: Ionic Strength;

z_i: Valency of Ion i;

N_A [mol^{-1}]: Avogadro Constant;

e_0 [g$^{1/2}$ cm$^{3/2}$ s^{-1}]: Elementary Electric Charge;

$\varepsilon_{r,H2O}$: Dielectric Constant of the Water;

T [K(°C)]: Temperature.

Introduction

A thermodynamic evaluation of the system NdCl$_3$-HCl (or NaOH)-H$_2$O-DEHPA-kerosene is required in order to explain the deviations in the evaluation with the simplified mass action law. From the reaction equation for the extraction:

$$Nd^{3+} + 3\,(DEHPA)_2 \rightleftarrows Nd(DEHP\cdot DEHPA)_3 + 3\,H^+ \qquad (1)$$

The equation $\log D_e - 3pH = 3\log\left(\dfrac{c_0}{2} - 3c_{Nd,org}\right) + \log K'$ can be

---

*Corresponding author: Christiane Scharf, Professor, Helmholtz Centre Dresden Rossendorf, Helmholtz Institute Freiberg for Resource Technology, Germany
E-mail: c.scharf@hzdr.de

derived, initially without considering the activity coefficients $\gamma_{Nd_{aq}^{3+}}$, $\gamma_{Nd,org}$ and $\gamma_{(DEHPA)_2}$. The activity coefficient, $\gamma_{Nd_{aq}^{3+}}$ calculated with the expanded Debye-Hückel equation $\log \gamma_+ = -\dfrac{A z_+^2 I_c^{1/2}}{1 + \mathring{a} B I_c^{1/2}}$ and $\gamma_{(DEHPA)_2}$ from literature data [1], $\log \gamma_{(DEHPA)_2} = -0.5227 m_{(DEHPA)_2}^{1/3} + 0.42 m_{(DEHPA)_2}$, provide indications for the development of the functions of the activity coefficients of DEHPA and neodymium in the organic phase. For the quantitative description of equilibriums, the activity coefficients of the components of the aqueous as well as the organic phase must be known. Theories concerning the activity coefficients in aqueous solutions can be found in various studies [2-9]. The influence of hydrochloric acid (HCl) and its salts on the extraction behaviour in studies [5,6] has been sufficiently demonstrated through comparison of the results of the Pitzer and Pitzer-Li equations with three further theories. Moreover, studies [7-9] take a detailed look at the activity coefficients in undiluted aqueous solutions. From this, an overview arises of the validity of the different equations such as the Debye-Hückel equation, the expanded Debye-Hückel equation, as well as the Davies equation. Similarly, for organic solutions there are calculation options for the activity coefficients, which are however specific to each system [10]. For example, the activity coefficient for DEHPA in n-octane for the extraction of zinc, europium, thulium and calcium in trace level is calculated [1,11] and is determined by the isopiestic method.

Experimental Procedure

The organic phase consisted of DEHPA (Merck-Schuchardt, $\rho_{293\,K\,(20°C)}$=0.974 g/mL) and kerosene (Fluka Chemika, bp 473-523 K (200-250°C), $\rho_{293\,K\,(20°C)}$=0.76 g/mL). The aqueous chloride solutions were prepared in different concentrations. The organic and aqueous phases were mixed during the test series at various proportions (V_O/V_A). Through the addition of concentrated hydrochloric acid (Riedel-de Haën) or of 10m sodium hydroxide (Merck, sodium hydroxide for analysis) the pH value was adjusted. After reaching equilibrium the phases were allowed to separate in a separating funnel. Then a sample was taken from the aqueous as well as the organic phase for chemical analysis. The following Tables 1 and 2 show the test conditions.

Thermodynamic Evaluation

Presentation of the measurement results with the terms for the activity coefficients

The deviations from the simplified mass action law can have the following explanations. First, the organic phase is not ideal, which means if reaction (1) is taken as a basis, the activity coefficients of $(DEHPA)_2$ and $Nd(DEHP·DEHPA)_3$ are not constant. Second, the concentrations of Nd^{3+}-, Na^+-, H^+-, OH^-- and Cl^--ions in the aqueous

solution coat such a large area that the activity coefficient Nd_{aq}^{3+} can no longer be used constantly unity. Finally, the extraction can also take place via other reactions as reaction (1), at least in part. That means, in the organic phase in addition to $Nd(DEHP·DEHPA)_3$ there occur other Nd-containing species in notable concentrations [12]. The results are not considering activity coefficients.

Under the condition that the reaction eqn. (1) describes the extraction sufficiently well, for Nd_{aq}^{3+}, $Nd(DEHP·DEHPA)_3$ and DEHPA activity coefficients or activities can be introduced. The mass action law is valid in the form of:

$$K = \frac{a_{Nd(DEHP.\,DEHPA)_3}\, a_{H^+}^3}{a_{Nd_{aq}^{3+}}\, a_{(DEHPA)_2}^3} \tag{2}$$
or,
$$K = \frac{c_{Nd(DEHP.DEHPA)_3}\, \gamma_{Nd(DEHP.DEHPA)_3}\, a_{H^+}^3}{c_{Nd_{aq}^{3+}}\, \gamma_{Nd_{aq}^{3+}}\, c_{(DEHPA)_2}^3\, \gamma_{(DEHPA)_2}^3} \tag{3}$$

It is assumed that the contents of other DEHPA-containing species with respect to $(DEHPA)_2$ und $Nd(DEHP·DEHPA)_3$ are very small, which means $c_{Nd(DEHP·DEHPA)3} \approx c_{Nd,\,org}$. It follows that for the mass balance of DEHPA:

$$c_{(DEHPA)_2} = \frac{c_0}{2} - 3c_{Nd,org} \tag{4}$$

With the definition for the distribution relationship $D_e(p,T) = \dfrac{c_{Nd,org}}{c_{Nd_{aq}^{3+}}}$ eqn. (3) results in

$$K = \frac{c_{Nd,org}\, \gamma_{Nd,org}\, a_{H^+}^3}{c_{Nd_{aq}^{3+}}\, \gamma_{Nd_{aq}^{3+}}\, c_{(DEHPA)_2}^3\, \gamma_{(DEHPA)_2}^3} \tag{5}$$

and in logarithmic form $pH = -\log a_{H^+}$ is valid with $c_{(DEHPA)2}$ from eqn. (4)

$$\log D_e - 3pH = 3\left[\log\left(\frac{c_0}{2} - 3c_{Nd,org}\right) - \frac{1}{3}\log\left(\frac{\gamma_{Nd,org}}{\gamma_{Nd_{aq}^{3+}}\, \gamma_{(DEHPA)_2}^3}\right)\right] + \log K \tag{6}$$

The expression $\dfrac{1}{3}\log\left(\dfrac{\gamma_{Nd,org}}{\gamma_{Nd_{aq}^{3+}}\, \gamma_{(DEHPA)_2}^3}\right) = X$ in eqn. (6) summarises the deviations from the ideal behaviour of neodymium in the aqueous and organic phases and DEHPA. The calculation of X-values took place using the measurement results, which means:

$$X = \log\left(\frac{c_0}{2} - 3c_{Nd,org}\right) - \frac{(\log D_e - 3pH - \log K)}{3}, \tag{7}$$

Where log K=0 was set.

Initially it was examined whether within the concentration

Test designation	Extraction agent DEHPA with c_0 [Vol.-%]	pH	V_O/V_A	T [K (°C)]
Determination of the distribution equilibrium of neodymium depending on c_0	Variable 1, 3, 5, 10, 15, 20, 25, 30, 35 und 40 (0.03 to 1.21 mol/L)	Const. 0.80 ± 0.03	1 (per 100 mL)	291 ± 1 (18 ± 1)
Determination of the extraction isotherms (V_O/V_A-dependence)	Const. 20 (0.604 mol/L)	Const. 0.80 ± 0.02	Variable 0.1-15	
Determination of pH dependence	Const. 20 (0.604 mol/L)	Variable -0.03 to 0.98	1 (per 100 mL)	

Table 1: Test parameters for the extraction of neodymium with DEHPA.

Test designation	Nd [g/L]
Determination of the distribution equilibrium of neodymium depending on c_0	2.00 ± 0.10
Determination of the extraction isotherms (V_O/V_A-dependence)	2.00 ± 0.01; 19.94
Determination of pH dependence	2.00 ± 0.01

Table 2: Mixtures of the pure aqueous starting solutions to determine dependencies.

dependencies of X resulting from the three test series there were already clues for how far the activity coefficients $\gamma_{Nd^{3+}_{aq}}$, $\gamma_{Nd,org}$ and $\gamma_{(DEHPA)_2}$ cause the deviations. In the test series "pH-dependence" the main change is in the aqueous phase. In the organic phase, the DEHPA content ($c_0 = 0.604$ mol/L) is constant and the content of neodymium in the organic phase ($c_{Nd,org}$) is small. Therefore, the free DEHPA content $\left(\dfrac{c_0}{2} - 3c_{Nd,org}\right)$ is relatively constant. It can thus be assumed that also $\gamma_{(DEHPA)2}$ and $\gamma_{(Nd,org)}$ a are relatively constant. That means that a variable X is being largely caused by $\gamma_{Nd^{3+}_{aq}}$. Accordingly, in Figure 1, X was shown against $c_{Nd^{3+}_{aq}}$.

A relatively strong increase in X and $c_{Nd^{3+}_{aq}}$ arose, which indicates that $\gamma_{Nd^{3+}_{aq}}$ declines with increasing $c_{Nd^{3+}_{aq}}$ in this test series. In the test series "c_0-dependence", the organic phase changes strongly. If one assumes that $\gamma_{Nd^{3+}_{aq}}$ is relatively constant in the aqueous phase, a variable X would be caused by $\gamma_{(DEHPA)2}$ and $\gamma_{(Nd,org)}$. As Figure 2 shows,

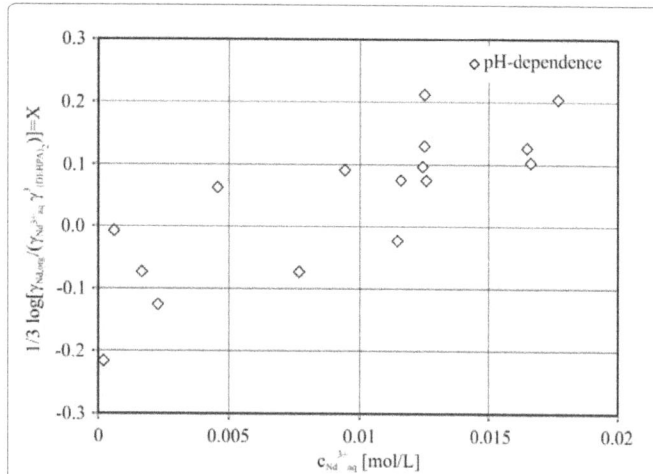

Figure 1: From eqn. (7), X values were calculated (relationship of activity coefficients $\gamma_{Nd^{3+}_{aq}}$, $\gamma_{Nd,org}$ and $\gamma_{(DEHPA)_2}$) depending on $c_{Nd^{3+}_{aq}}$. The calculations were undertaken with the measurements for the pH-dependence.

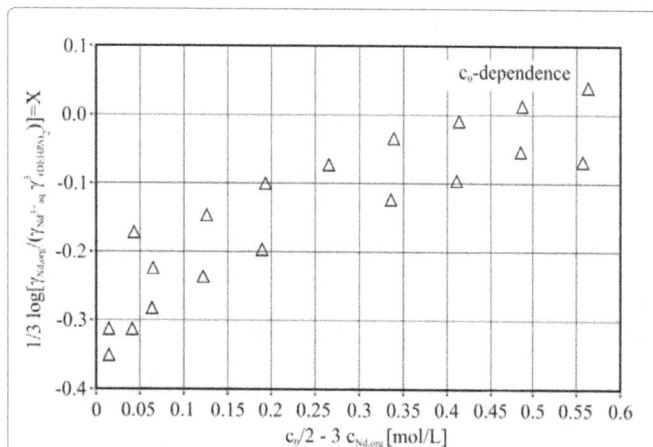

Figure 2: From eqn. (7) X-values are calculated (relationship between activity coefficients $\gamma_{Nd^{3+}_{aq}}$ $\gamma_{Nd,org}$, and $\gamma_{(DEHPA)_2}$ depending on "free" DEHPA content $\left(\dfrac{c_0}{2} - 3c_{Nd,org}\right)$. The calculations were undertaken with the measurements for the c_0-dependence.

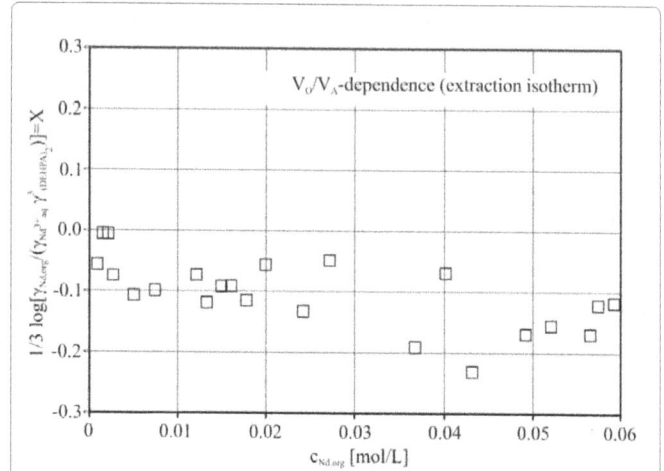

Figure 3: From eqn. (7) X-values are calculated (relationship between activity coefficients $\gamma_{Nd^{3+}_{aq}}$, $\gamma_{Nd,org}$ and $\gamma_{(DEHPA)_2}$) depending on $c_{Nd,org}$Z. The calculations were undertaken with the measurements for the V_O/V_A-dependence.

when X increases with the free DEHPA content $\left(\dfrac{c_0}{2} - 3c_{Nd,org}\right)$ the quotient $\dfrac{\gamma_{Nd,org}}{\gamma^3_{(DEHPA)_2}}$ increases.

Finally in the series, V_O/V_A-dependence "the neodymium contents of both phases increase together at constant c_0. As seen in Figure 3, X remains virtually constant. This indicates that $\dfrac{\gamma_{Nd,org}}{\gamma_{Nd^{3+}_{aq}}}$ is relatively constant, which means since $\gamma_{Nd^{3+}_{aq}}$ decreases with increasing $c_{Nd^{3+}_{aq}}$, $\gamma_{Nd,org}$ will also decrease with increasing $c_{Nd,org}$ (at $\dfrac{c_0}{2} - 3c_{Nd,org}$ = const.) In the following section, these qualitative findings will be quantified.

Activity coefficient of neodymium cation $\gamma_{Nd^{3+}_{aq}}$ in the aqueous phase

First, by using the expanded Debye-Hückel equation, $\gamma_{Nd^{3+}_{aq}}$ should be calculated, in order to extract $\gamma_{Nd^{3+}_{aq}}$ from the total value of X. For this calculation of $\gamma_{Nd^{3+}_{aq}}$ the basis of the theory of aqueous solutions can be used.

The activities of individual ions cannot be measured due to the conditions of electroneutrality. However, using the equations as functions of the so-called ionic strength, they can be determined. According to Lewis and Randall, the ionic strength is

$$I_c = 0.5\sum\left(c_i\, z_i^2\right) \tag{8}$$

c_i=concentration of ion type i

z_i=valency of ion type i

or

$$I_m = 0.5\sum\left(m_i\, z_i^2\right) \tag{8a}$$

m_i=molality of ion type i

z_i=valency of ion type i.

In eqns. (8) or (8a) all ion types present in the solution must be considered [13]. Generally, this means that solutions with<0.1 mol/L are described as "diluted" (e.g. river water with I_c=0.0017 mol/L), with $0.1<I_c<1$ mol/L as "intermediate" (e.g. seawater with I_c=0.67 mol/L) and with $I_c > 1$ mol/L as "concentrated" (e.g. Great Salt Lake/Utah with $I_c = 4.6$ mol/L).

In the area of diluted solutions the activity coefficients are given through the limiting laws of Debye-Hückel, as follows:

$$\ln \gamma_i = -\frac{N_A^2\, z_i^2\, e_0^3}{(\varepsilon\, R\, T)^{3/2}} \left(\frac{2\pi}{1000}\right)^{1/2} I_c^{1/2} \qquad (9)$$

with

$N_A = 6.0221367\ 10^{23}$ mol^{-1} (Avogadro constant),

z_i = valency of ion i,

$e_0 = 4.806\ 10^{-10}$ g$^{1/2}$ cm$^{3/2}$ s^{-1} (elementary electric charge),

$\varepsilon_{r,H2O} = 78.54$ at 298 K (25°C) (dielectric constant of the water),

R = 8.314510 J mol^{-1} K^{-1} and,

$I_c = 0.5\ \Sigma\ (c_i\, z_i^2)$ (ionic strength).

Through summarising the constants and using the decadic logarithm, one calculates

$$\log \gamma_i = -A\, z_i^2\, I_c^{1/2} \quad \text{with A=0.509 for 298 K (25°C).} \qquad (10)$$

The limiting law of Debye-Hückel is thus derived theoretically. It only applies to heavily diluted solutions in the range of ionic strength $I_c < 0.005$ mol/L.

The expanded Debye-Hückel law is half empirical and applies to an ionic strength of ≤ 0.1 mol/L [14] with the equation for the individual cation activity coefficient [14].

$$\log \gamma_+ = -\frac{A\, z_+^2 I_c^{1/2}}{1 + \mathring{a}\, B\, I_c^{1/2}} \qquad (11)$$

The temperature dependent values A and B are tabulated for water as the solvent [15]. For 298 K (25°C), A=0.509 and B=0.3281 10^{-8}. The "effective diameters" å of individual ions in the solution have been determined through a large number of experiments [15,16], which means that they are purely empirical measurements. For the triple positive charged neodymium ion the value is provided in [15] as $\mathring{a}_{Nd_{aq}^{3+}} = 9\,10^8\ \mathring{A}$, which means that for the present case the following expression results:

$$\log \gamma_{Nd_{aq}^{3+}} = -\frac{4.581 I_c^{1/2}}{1 + 2.953 I_c^{1/2}} \qquad (12)$$

If applied to the system NdCl$_3$–HCl (or NaOH)–H$_2$O–DEHPA–kerosene, the following results accordingly. To calculate $\gamma_{Nd_{aq}^{3+}}$ firstly the ionic strengths must be determined, which in this experiment reach up to I_c 1.5 mol/L. To adjust the pH values, for the c_0 und pH measurement series only concentrated hydrochloric acid was used, and for the series of extraction isotherms, also 10m sodium hydroxide was used. The greatest range of ionic strength is covered by the tests on pH dependency (-0.03<pH<1.00).

In the three test series the measurements were in the following ranges:

a) $0.14 < I_c < 0.29$ mol/L for c_0-dependence,

b) $0.04 < I_c < 1.36$ mol/L for pH-dependence and

c) $0.11 < I_c < 1.46$ mol/L for V_O/V_A-dependence (extraction isotherm),

which means that this concerns "diluted" and "intermediate" solutions. Figure 4 shows the calculated values for $\log \gamma_{Nd_{aq}^{3+}}$ against the ionic I_c. When I_c tends to 0, $\log \gamma_{Nd_{aq}^{3+}}$ also tends to 0, which means that

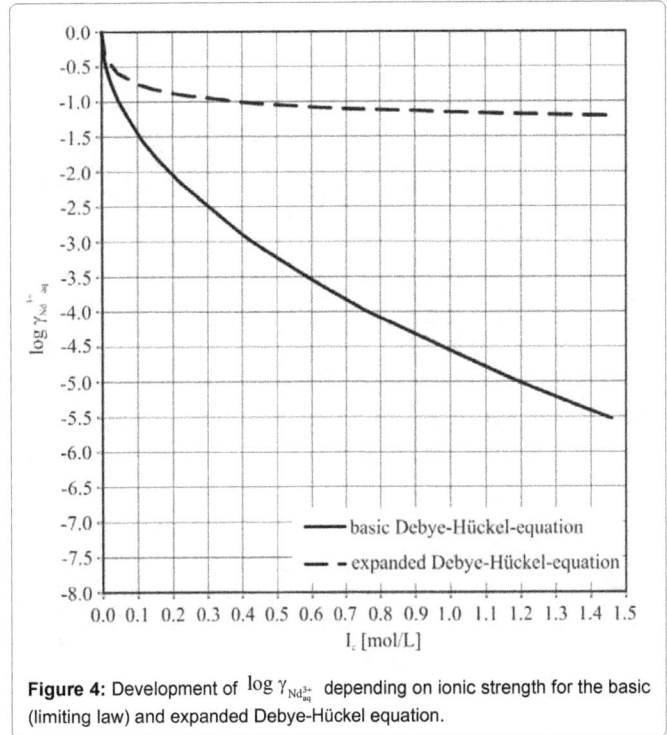

Figure 4: Development of $\log \gamma_{Nd_{aq}^{3+}}$ depending on ionic strength for the basic (limiting law) and expanded Debye-Hückel equation.

$\log \gamma_{Nd_{aq}^{3+}}$ has a limit value of 1.

Returning to the evaluation of the three measurement series for the c_0-, pH-, and V_O/V_A-dependence (extraction isotherm), the activity coefficient $\gamma_{Nd_{aq}^{3+}}$ is calculated under consideration of all the ions present in the solution (Nd^{3+}, Cl$^-$, H$^+$, Na$^+$, OH$^-$) with the expanded Debye-Hückel equation. Following from this, $X + - \log \gamma_{Nd_{aq}}$ for the pH test series is shown against ionic strength I_c (considering all ions present in the aqueous phase) in Figure 5.

If $\gamma_{Nd_{aq}^{3+}}$ is responsible for all deviations from the ideal behaviour, then this should be immediately apparent from this kind of representation, where ΔX becomes smaller. As one can see, the amended X hardly demonstrates any concentration dependency. Thus, the assumption was correct, that in the test series for pH-dependence, the non-constant K' is caused at least in part by $\gamma_{Nd_{aq}^{3+}}$.

Activity coefficient of DEHPA $\gamma_{(DEHPA)2}$ in the organic phase

Next, the influence of $\gamma_{(DEHPA)2}$ on the measurements for c_0-dependence was investigated. The literature contains difference functions to calculate $\gamma_{(DEHPA)2}$ [1,11]. Figure 6 shows $\log \gamma_{(DEHPA)2}$ in dependence with c_0 as per the literature data. The authors [11] fulfil the requirement that one work with the smallest metallic concentrations of europium, thulium and calcium, meaning that the tests occur on "infinitely" diluted aqueous chloride and nitride solutions. However, one assumes that the solution used, n-dodecane, is "inert" in the organic phase. The observed system corresponds to a ternary system and not a binary system. This means that the activity coefficient $\gamma_{i,org}$ of the metallic ions may not be set at a constant, as the authors have done. In comparison [1], an approach of a methodically flawless isopiestic determination of the function for $\gamma_{(DEHPA)2}$ in the system DEHPA–n-octane is used. Here, the DEHPA–n-octane sample to be measured is brought into equilibrium with a phase of known octane vapour pressure (triphenylmethane–n-octane) and the sample composition is subsequently determined.

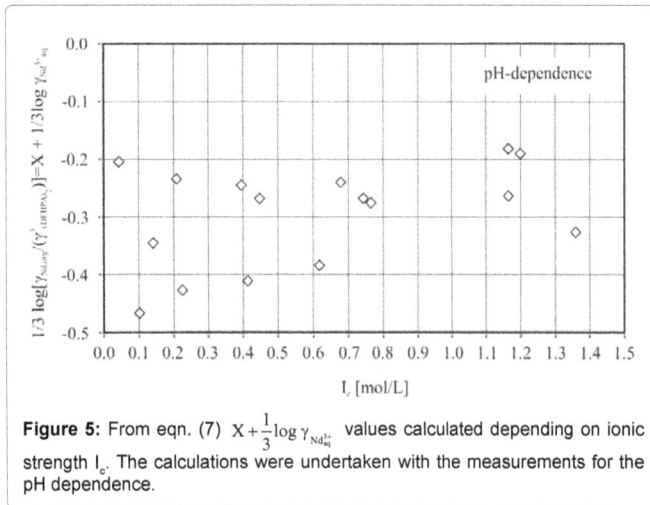

Figure 5: From eqn. (7) $X + \frac{1}{3}\log\gamma_{Nd^{3+}_{aq}}$ values calculated depending on ionic strength I_c. The calculations were undertaken with the measurements for the pH dependence.

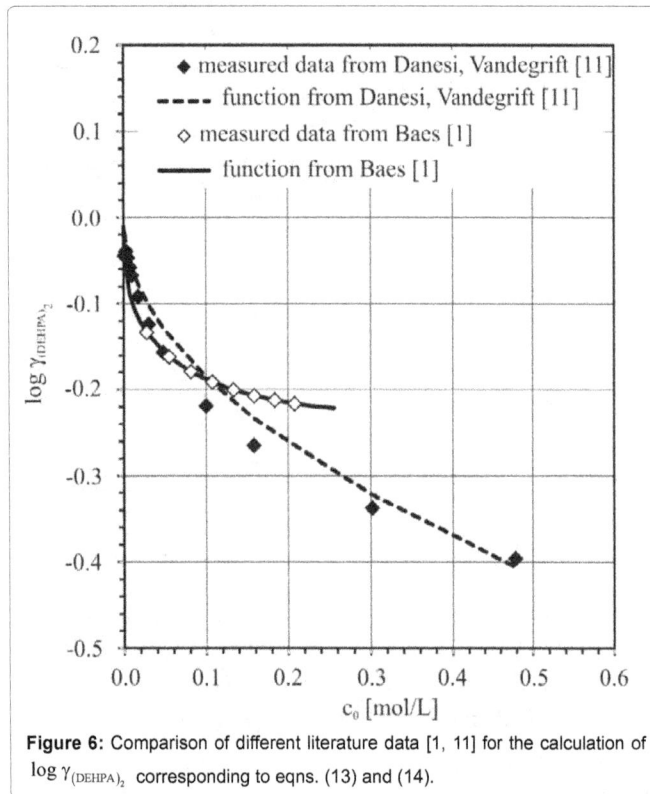

Figure 6: Comparison of different literature data [1, 11] for the calculation of $\log\gamma_{(DEHPA)_2}$ corresponding to eqns. (13) and (14).

The functions for DEHPA are:

a)   from ref. [11] DEHPA in *n*-dodecane

$$\log\gamma_{(DEHPA)_2} = -0.83c^{1/2}_{(DEHPA)_2} \qquad (13)$$

b)   from ref. [1] DEHPA in Tri-*n*-octylphosphine oxide (TPO)

$$\log\gamma_{(DEHPA)_2} = -0.5227m^{1/3}_{(DEHPA)_2} + 0.42m_{(DEHPA)_2} \qquad (14)$$

Where the function $c_{(DEHPA)_2} = 0.6986m_{(DEHPA)_2} - 0.3207m^2_{(DEHPA)_2} + 0.15m^3_{(DEHPA)_2}$ is presented for the conversion of molality $m_{(DEHPA)2}$ in molarity $c_{(DEHPA)2}$ [1]. It results from density measurements.

Activity coefficient of neodymium $\gamma_{(Nd,org)}$ in the organic phase

For the calculation of $\log\gamma_{(DEHPA)2}$ eqn. (14) is used. Unfortunately, it is necessary to extrapolate over a wide area, as the function is only experimentally proven up to $c_{(DEHPA)_2}$ $0.1\,mol/L$. For $c_{(DEHPA)2}$ the free DEHPA content $\left(\frac{c_0}{2} - 3c_{Nd,org}\right)$ is inserted. From the size of X, the activity coefficient $\log\gamma_{(DEHPA)2}$ is calculated. The amended value $X + \log\gamma_{(DEHPA)_2}$ is shown in Figure 7. As a next step, the activity coefficient $\gamma_{Nd^{3+}_{aq}}$ of the aqueous phase is calculated from X, which means that $X + \log\gamma_{(DEHPA)_2} + \frac{1}{3}\log\gamma_{Nd^{3+}_{aq}}$ is determined here, through which $\frac{1}{3}\log\gamma_{Nd,org}$ can be revealed. This function is shown in Figure 8. $\log\gamma_{Nd,org}$ increases considerably with the DEHPA content. In the V_O/V_A test series the content of neodymium in the organic phase

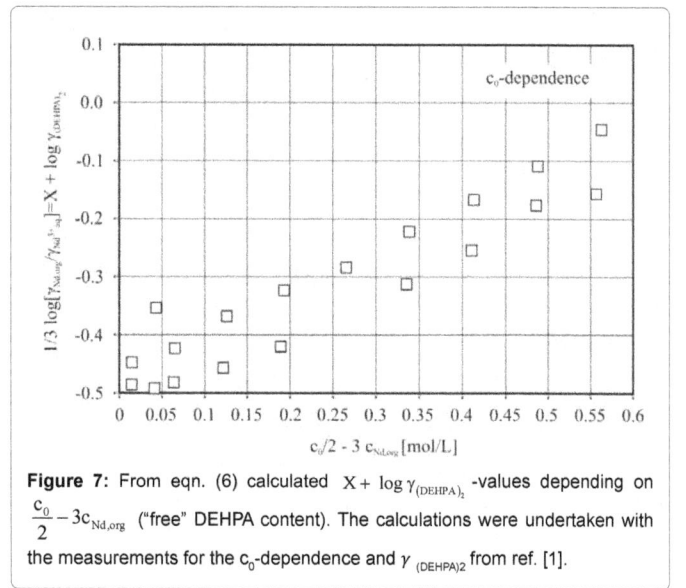

Figure 7: From eqn. (6) calculated $X + \log\gamma_{(DEHPA)_2}$ -values depending on $\frac{c_0}{2} - 3c_{Nd,org}$ ("free" DEHPA content). The calculations were undertaken with the measurements for the c_0-dependence and $\gamma_{(DEHPA)2}$ from ref. [1].

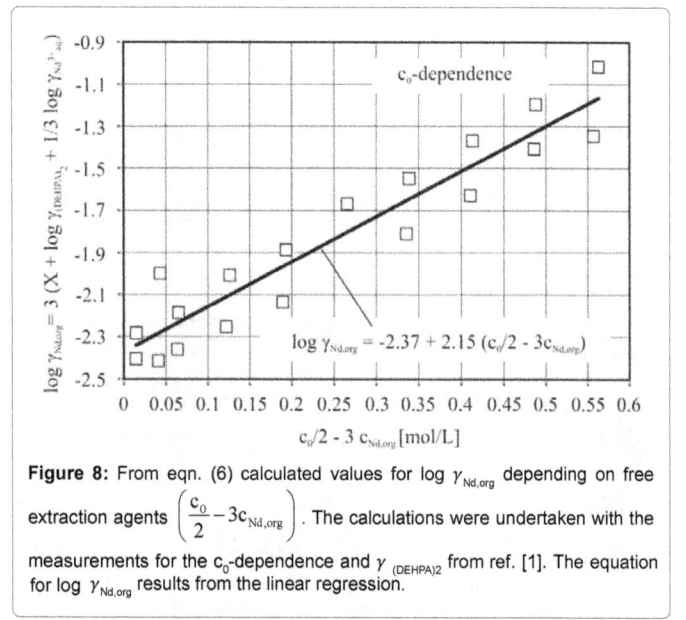

Figure 8: From eqn. (6) calculated values for $\log\gamma_{Nd,org}$ depending on free extraction agents $\left(\frac{c_0}{2} - 3c_{Nd,org}\right)$. The calculations were undertaken with the measurements for the c_0-dependence and $\gamma_{(DEHPA)2}$ from ref. [1]. The equation for $\log\gamma_{Nd,org}$ results from the linear regression.

($c_{Nd,org}$) extends to higher values, which means that the influence of the neodymium content can be determined to $\gamma_{Nd,org}$.

The evaluation corresponds to that for the c_0 test series. Figure 9 shows $X + \frac{1}{3}\log \gamma_{Nd_{aq}^{3+}}$ and Figures 10a and 10b $3\left(X + \log \gamma_{(DEHPA)_2} + \frac{1}{3}\log \gamma_{Nd_{aq}^{3+}}\right) = \log \gamma_{Nd,org}$. As already presumed in the application of X against $c_{Nd,org}$, Figure 3, $\log \gamma_{Nd,org}$ decreases with increasing $c_{Nd,org}$.

From the c_0-dependence, Figure 8, the following relationship results

$$\log \gamma_{Nd,org} = -2.37 + 2.15\left(\frac{c_0}{2} - 3c_{Nd,org}\right) \qquad (15)$$

If one assumes that the same equation can be applied to the V_O/V_A-dependence, then for $c_0 = 0.604$ mol/L it follows that $\log \gamma_{Nd,org} = -1.72 - 6.45\ c_{Nd,org}$. The corresponding line is shown

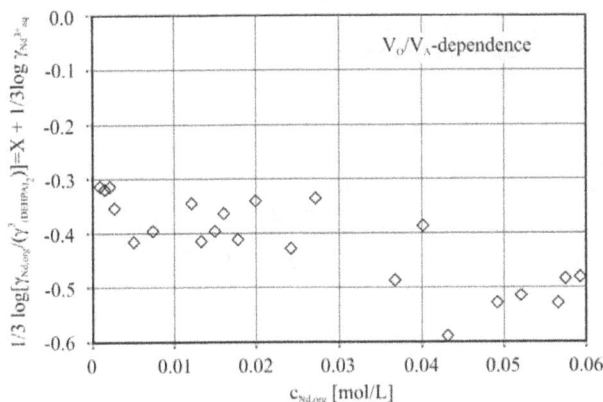

Figure 10b: Application for the V_O/V_A-dependence as in Figure 10a. The equation for $\log \gamma_{Nd,org}$ results from eqn. (15) with $c_0 = 0.604$ mol/L and an additional square term.

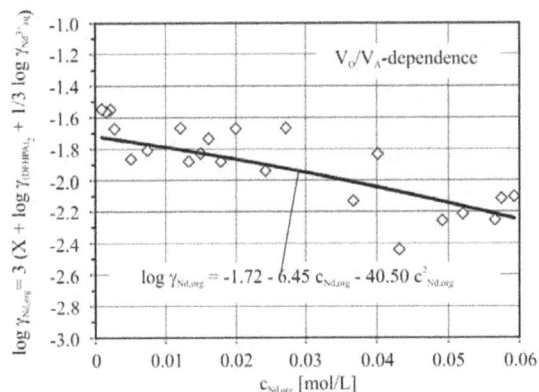

Figure 9: From eqn. (6) the calculated values of $X + \frac{1}{3}\log \gamma_{Nd_{aq}^{3+}}$ depending on the concentration of neodymium in the organic phase $c_{Nd,org}$. The calculations were undertaken with the measurements for the V_O/V_A-dependence and $\gamma_{Nd_{aq}^{3+}}$ using the expanded Debye-Hückel equation.

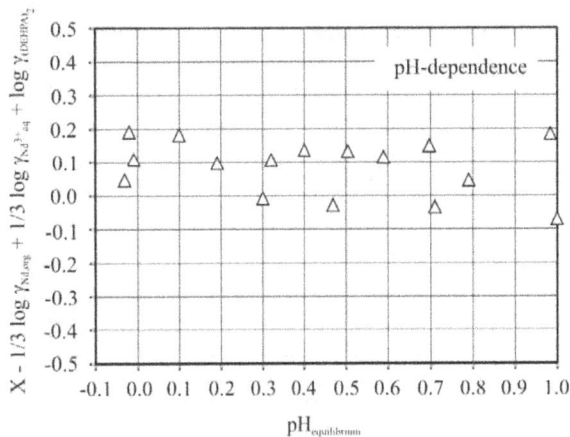

Figure 11: $X - \frac{1}{3}\log \gamma_{Nd,org} + \frac{1}{3}\log \gamma_{Nd_{aq}^{3+}} + \log \gamma_{(DEHPA)_2}$ applied depending on the equilibrium pH value to check the consistency with the other test series (c_0 and V_O/V_A dependencies). The calculations were undertaken with the measurements for the pH-dependence from eqn. (6) with K=1, $\gamma_{Nd,org}$ from eqn. (16), $\gamma_{Nd_{aq}^{3+}}$ from the expanded Debye-Hückel equation and $\gamma_{(DEHPA)_2}$ from ref. [1].

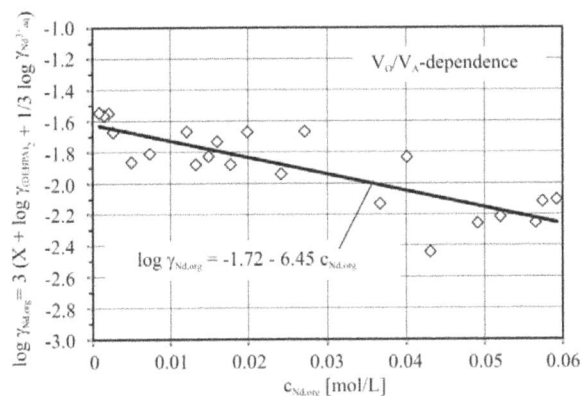

Figure 10a: From eqn. (6) the calculated values of $\log \gamma_{Nd,org}$ depending on the concentration of neodymium in the organic phase $c_{Nd,org}$. The calculations were undertaken with the measurements for the V_O/V_A-dependence, $\gamma_{(DEHPA)_2}$ from [1] and $\gamma_{Nd_{aq}^{3+}}$ using the expanded Debye-Hückel equation. The linear equation for $\log \gamma_{Nd,org}$ results from eqn. (15) with $c_0 = 0.604$ mol/L.

in Figure 10a. It describes the experimental points quite well. One can also introduce a square term (Figure 10b) which results in $\log \gamma_{Nd,org} = -1.72 - 6.45\ c_{Nd,org} - 40.50\ c_{Nd,org}^2$ or rather as a general expression that also applies to the c_0-dependence:

$$\log \gamma_{Nd,org} = -2.37 + 2.15\left(\frac{c_0}{2} - 3c_{Nd,org}\right) - 40.50\ c_{Nd,org}^2 . \qquad (16)$$

It is now necessary to check the consistency for the test series for pH-dependence. To this end, $X - \frac{1}{3}\log \gamma_{Nd,org} + \frac{1}{3}\log \gamma_{Nd_{aq}^{3+}} + \log \gamma_{(DEHPA)_2}$ was calculated, whereby $\log \gamma_{Nd,org}$ was inserted as in eqn. (16). The value must be 0. The evaluation is shown in Figure 11, and indeed, the data is scattered around 0. Finally, $X - \frac{1}{3}\log \gamma_{Nd,org} + \frac{1}{3}\log \gamma_{Nd_{aq}^{3+}} + \log \gamma_{(DEHPA)_2}$ for the c_0- and V_O/V_A-dependence was calculated as well. As Figures 12 and 13 shows, the value within the scattering is 0 [17,18].

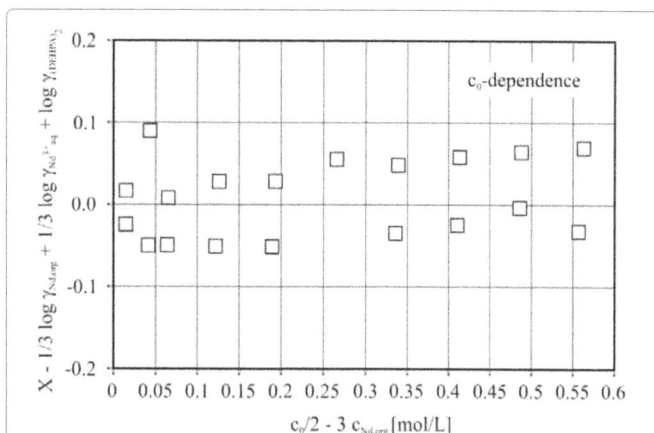

Figure 12: $X - \frac{1}{3}\log\gamma_{Nd,org} + \frac{1}{3}\log\gamma_{Nd_{aq}^{3+}} + \log\gamma_{(DEHPA)_2}$ applied depending on $\left(\frac{c_0}{2} - 3c_{Nd,org}\right)$. The calculations were undertaken with the measurements for the c_0-dependence from eqn. (6) with K=1, $\gamma_{Nd,org}$ from eqn. (16), $\gamma_{Nd_{aq}^{3+}}$ from the expanded Debye-Hückel equation and $\gamma_{(DEHPA)_2}$ from ref. [1].

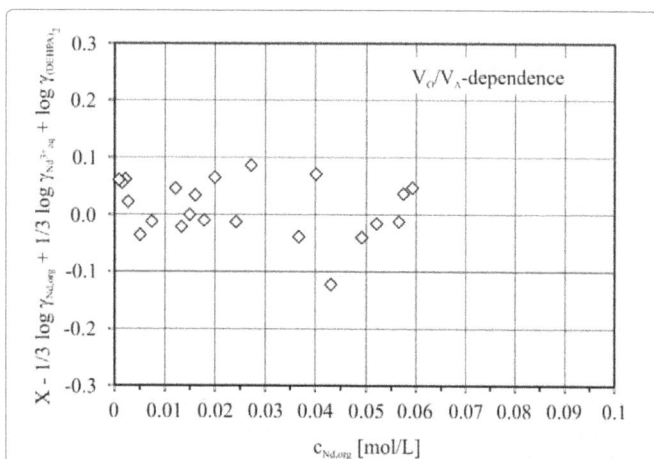

Figure 13: $X - \frac{1}{3}\log\gamma_{Nd,org} + \frac{1}{3}\log\gamma_{Nd_{aq}^{3+}} + \log\gamma_{(DEHPA)_2}$ applied depending on $c_{Nd,org}$. The calculations were undertaken with the measurements for the V_o/V_A-dependence from eqn. (6) with K=1, $\gamma_{Nd,org}$ from eqn. (16), $\gamma_{Nd_{aq}^{3+}}$ from the expanded Debye-Hückel equation and $\gamma_{(DEHPA)_2}$ from ref. [1].

Summary

In sum, it is shown that the measurements for the three series can be made consistent with one another on the basis of the expanded Debye-Hückel theory for the aqueous solution and the eqns. (14) and (16) for the activity coefficients (DEHPA)2 and γNd,org in the organic phase, at least in the concentration range examined here. However, eqn. (16) for log γNd,org is based on the K value of 1. One can remove this restriction if one uses a reference condition (γ=1) for γNd,org. If one selects the infinitely diluted solution of neodymium in pure DEHPA (c0=3.02 mol/L), from eqn. (15) or (16) it follows that and with that K=0.13. As an equation for log γNd,org (undiluted solution) the following applies (instead of eqn. (16)).

Of course one can select other reference conditions e.g. the infinite dilution of neodymium in pure kerosene. However, independent of the choice of the reference condition, the activity coefficient of neodymium

in the organic phase increases with increasing concentration of DEHPA and decreases with the increasing concentration of neodymium in the organic phase, at least in the concentration range examined here and when reaction (1) is used as the only basis reaction. To what extent this evaluation explains the reality well or less well must be subject to further investigation.

Acknowledgments

The authors are thankful to Deutsche Forschungsgemeinschaft for financial support.

References

1. Baes Jr CF (1962) An Isopiestic Investigation of di-(2-ethylhexyl)-phosphoric acid (DPA) and TRI-n-Octylphosphine Oxide (TPO) IN n-OCTANE1. The Journal of Physical Chemistry 66: 1629-1634.

2. Forrest C, Hughes MA (1975) The modelling of equilibrium data for the liquid-liquid extraction of metals Part I. A survey of existing models. Hydrometallurgy 1: 25-37.

3. Pytkowicz RM (1979) Activity coefficients in electrolyte solutions (Vol. 2). CRC.

4. Teng T, Yi-Gui L (1983) Studies on ionic activity coefficients in mixed electrolyte solution by the extraction-distribution method. Hydrometallurgy 10: 69-77.

5. Lu JF, Li YG, Ding HB, Teng T (1989) Salting effect of di (2-ethylhexyl) phosphoric acid in chloride solutions. Fluid phase equilibria 51: 119-131.

6. Lu JF, Niu SF, Li YG (1992) Study on the salting effect of di (2-ethylhexyl) phosphoric acid in MCl2 (MSO4) solutions by the Pitzer-Li and Pitzer equations. Fluid phase equilibria 71: 211-224.

7. C. Spötl: Kapitel Aktivität, Auszug aus der Vorlesung "Grundprinzipien der Sedimentären Geochemie", Institut für Geologie und Paläontologie der Universität Innsbruck, WS 00/01, S. 12-18.

8. Meissner HP, Tester JW (1972) Activity coefficients of strong electrolytes in aqueous solutions. Industrial & Engineering Chemistry Process Design and Development 11: 128-133.

9. Pitzer KS (1973) Thermodynamics of electrolytes. I. Theoretical basis and general equations. The Journal of Physical Chemistry 77: 268-277.

10. Hildebrand JH, Scott RL (1964) The solubility of nonelectrolytes, 3rd edition, Reinhold Publishing Corporation, New York, USA, pp: 419-439.

11. Danesi PR, Vandegrift GF (1981) Activity coefficients of bis (2-ethylhexyl) phosphoric acid in n-dodecane. Inorganic and Nuclear Chemistry Letters 17: 109-115.

12. Scharf C, Ditze A, Schwerdtfeger K, Kaufmann DE, Namyslo JC, et al. (2005) Investigation of the structure of neodymium-di-(2-ethylhexyl) phosphoric acid combinations using electrospray ionization and matrix-assisted laser desorption ionization mass spectrometry and nuclear magnetic resonance spectroscopy. Metallurgical and Materials Transactions B 36: 429-436.

13. Atkins PW (1990) Physikalische Chemie, VCH Verlagsgesellschaft mbH S. 244-253, 770-786.

14. Holleman AF, Wiberg E, Wiberg N. Lehrbuch der Anorganischen Chemie (1995) 101. Auflage. Berlin, Walter de Gruyter, S 187: 199.

15. Garrels RM, Christ CL (1965) Solutions, Minerals, and Equilibria, Harper&Row and John Weatherhill. Inc., p: 450.

16. Klotz IM (1950) Chemical Thermodynamics. Basic Theory and Methods, Prentice-Hall Inc., pp: 300-330.

17. Klotz IM, Rosenberg RM (2000) Chemical Thermodynamics, Basic Theory and Methods, John Wiley & Sons, Inc., (6thedn), pp: 438-471.

18. Scharf C, Ditze A (2017) Metallurgical Research & Technology 114: 404.

35

The ISO Standard 14577 for Mechanics Violates the First Energy Law and Denies Physical Dimensions

Kaupp G*

University of Oldenburg, Germany

Abstract

The basis of the quantitative conical/pyramidal (nano) indentation, without fittings, iterations, or simulations, is the physically founded $F_N = k\, h^{3/2}$ relation. The constant k (penetration resistance, mN/µm$^{3/2}$) from linear plot with excellent regression discards initial surface effects, identifies important phase transformation onsets, conversion and activation energies, and reveals errors. The failing Sneddon theory of ISO with unphysical exponent 2 on h lacks these possibilities, disregards shear-force work, and violates the first energy law since 50 years. The denied but strictly quantified loss of energy (20% for physical $h^{3/2}$; 33.33% at believed h^2) violates the first energy law and disregards the force remaining for penetration. Straightforward correction is performed for the dimensions, by replacing unphysical exponent 2. The correction factors $h_{max}^{1/2}$ and 0.8 are applied via joint maximal force to the universal, FE-simulated, (approximately) ISO hardness, and ISO modulus that unduly rely on h^2, to give the physically founded values with their correct dimensions. Previous corrected k-values obtain H_{phys} directly from the loading curve regression. Previous incomplete corrections are rectified. The new dimensions and daily risk liabilities from ISO versus physics dilemma are discussed, considering the influence on all mechanical parameters from hardness and modulus, regarding technique, biology, medicine, daily life.

Keywords: Correction of ISO hardness and modulus; Energy law violations; Failure risks; False materials parameters; False ISO-standards; Indentation exponent; New hardness and modulus definitions; Penetration resistance; Physical consequences; Physical hardness from loading curve

Introduction

The most basic natural and technical law that can never be dismissed, but must be strictly obeyed, is the energy conservation law. All worlds work on it and must trust in its validity, and that must not be dismissed by any organization. However, ISO and its subsidiary NIST in USA still violate against with ISO standard 14577, claiming the exhaustively complicated mathematical deductions of Sneddon and Love [1,2]. However, these authors obviously missed taking into account the shear-force work, when a rigid indenter is forced to penetrate vertically into a solid. It must be very clear that the pressure (and or plastic deformation) from the rigid indenter against its displaced solid material requires work. Nevertheless, the whole applied force and thus the whole applied energy is still falsely considered to be only acting in vertical direction of the impact. Unfortunately, there was no protest from physics. Rather the work of Oliver and Pharr [3] on the indentation of cones or pyramids was highly acclaimed and adapted by ISO/NIST for ISO 14577. It thus became undisclosed that their assumed relation between force and depth is incorrect and that the hardness and elastic modulus determinations violate the first principle of energy conservation. Such disregard has still been retained till now, even though the unphysical exponent 2 on the depth h had been experimentally demonstrated to be replaced by 3/2 from the present author since 2000 with convincing evidence.

In 1939 and 1965 two mathematicians solved the long standing Boussinesq problem using very complicated mathematics and came (with different constant) to the same exponent 2 on the depth h in relation to the normally applied force in conical indentation when the indenter remains stiff (Figure 1). The Sneddon/Love exponent [1,2] has also been used for partly plastic response (it is a consequence of pressure!) by Oliver and Pharr in 1992 [3], the ISO standard 14577, and finite element (FE) simulations (e.g. ANSYS or ABACUS software),

even though the shear force of the conical (similarly effective cone of pyramids) indenter to the environment did apparently not find any concern in physics. Rather numerous fitting procedures were put forward over the years for the excuse, that the exponent 2 on h could not be found experimentally but only with FE-simulations converging to such exponent. Thus, these mathematical deductions (Figure 1) cannot be correct. It did apparently not help that the energetics of the (pseudo) conical indentation was for the first time quantitatively clarified in a publication from 2013 [4] because the experimental exponent on h was consistently found as 3/2 instead of 2 [5]. The thoughtful convincing

Conical indentation, recent theories

Boussinesq-Sneddon 1965: $F_N = 2\, h^2 E / \pi(1-\nu^2)\tan\alpha$

Boussinesq-Love 1939: $\quad F_N = \pi\, h^2 E \tan\alpha / 2(1-\nu^2)$

F_N: normal force; h: penetration depth; E: elastic modulus

Same exponent on depth h but different constant with the same theory

The exponent 2 on h cannot be found experimentally!
The shear force component is missing!
Both deductions are incorrect!

Figure 1: Previous highest grade mathematic deductions of load-displacement indentation curves with cones and pyramids.

***Corresponding author:** Kaupp G, University of Oldenburg, Diekweg 15 D-26188 Edewecht, Germany, E-mail: gerd.kaupp@uni-oldenburg.de

physical foundation of exponent 3/2 in Equation (1) that followed pre-published since 2015 [6] requires only first grade mathematics.

Materials and Methods

The author's nanoindentations used a fully calibrated Hysitron Inc. TriboScope$^{(R)}$ Nanomechanical Test Instrument with a two-dimensional transducer and leveling device in force control mode after due calibration, including instrument compliance. The samples were glued to magnetically hold plates and leveled at slopes of ±1° in x and y directions under AFM control with disabled plain-fit, and loading times were 10-30 s for 400-500 or 3000 data pairs [5,7]. The radii of the cube corner (55 nm) and Berkovich (110 nm) diamond indenters were directly measured by AFM in tapping mode. Three-dimensional microscopic inspection of the indenter tips secured smooth side faces of the diamonds for at least 2 μm from the (not resolved) apex. The whole data set of the loading curve was used for analysis, using Excel$^{(R)}$. Most analyses were however with published loading curves from the literature, as rapid sketches with pencil, paper, and calculator (10-20 data pairs), but for linear regressions always by digitization to give 50-70 almost uniformly arranged data pairs using the Plot Digitizer 2.5.1 program (www.Softpedia.com), unless complete original data sets could be obtained from the scientists with 400-500 or 3000 data points. They were handled with Excel$^{(R)}$. The distinction of experimental and simulated loading curves succeeded by performing the "Kaupp-plot" (1) revealing $F_N \propto h^{3/2}$ (experimental), surface effects and most important phase changes' onset [8]. The necessary force correction to comply with the energy law is made with the physical k-value (0.8 times the slope). Only FE-simulated or iterated curves gave linear unphysical $F_N \propto h^2$ plots. The linear regressions were calculated with Excel$^{(R)}$. In the case of phase changes the kink positions were precisely calculated by equating the regression lines before and after the kink. Initial surface effects were, of course, exempt from the linear regressions. Previous penetration resistance values k was corrected for complying with the energy/force/depth loss in Figure 2. A 10-figures pocket-calculator was used for the physical calculations, but the final results are reasonably rounded. It was tried to cover all different materials types, all different indentation modes, equipments, response mechanisms, depth ranges, penetration resistance sizes, from numerous authors from all around the globe, in order to show their universal obeying to basic mathematics.

Results

The mathematical clarification of the energetics upon (pseudo)conical indentation

We proceed analogous to the deduction in Kaupp [4]. In force controlled indentations the total force F_N is linearly applied. This can provisionally be imaged together with an assumed normal parabola (with exponent 2) as is used by ISO etc. in a force versus depth diagram, as obtained by a FE-calculation from the literature (Figure 3). Such normal parabola has the Formula (2). The work of the simulated indentation (W_{indent}) gives (3) by integration. The applied work ($W_{applied}$) is the area of the triangle under the applied work (0-F_{Nmax}) in (4). Substitution of F_{Nmax} from (2) into (4) gives (5). The ratio ($W_{applied}$/W_{indent})$_{simul}$=0.5/1/3=3:2. That means: only 2/3 (66.67%) of the applied work (and thus also force) would be left for the indent and 1/3 (33.33%) are for the sum of the reversible pressure and the mostly or completely irreversible plastic deformation energies to the environment. Clearly the disregard of 1/3 from F_N when using the false h^2 for the calculation of e.g. ISO hardness H_{ISO} and ISO modulus $E_{r\text{-}ISO}$, or universal hardness ($H_{univ}=F_N/A_{proj}$, where A_{proj} is projected indenter area, also called Martens hardness), is an obvious and severe violation of the basic energy conservation law. The long-known long-range effects and the elastic deformation would require here 1/3 of the applied energy that would be lost for the penetration depth with ISO, FE-simulation, and universal hardness. But a correction for the false exponent is also required.

$$F_{N\text{-phys}}=k\,h^{3/2} \tag{1}$$

$$F_{N\text{-simul}}=\text{const}\,h^2 \tag{2}$$

$$W_{simul\text{-}indent}=1/3\,\text{const}\,h^3 \tag{3}$$

$$W_{applied}=0.5\,F_{N\text{-}max}\,h_{max} \tag{4}$$

$$W_{simul\text{-}applied}=0.5\,\text{const}^!\,h_{max}^{\ 3} \tag{5}$$

In order to clarify the unlikely objection that the applied force would be parabolic, we plot here in Figure 4 both applied force and depth side by side against the time as these develop. It is, of course, seen that these develop simultaneously with total F_N linearly but depth

Material	Indenter	h_{max}	k $\mu N\,nm^{-3/2}$	r	Reference for original data
PMMA (122 s)	Berkovich	1.6 μm	0.0739	0.9999	Lu 2010
PMMA (33 s)	Berkovich	4.7 μm	0.2626	0.9991	Cook and Oyen 2007
PMMA (28 min)	Berkovich	6.5 μm	0.1554	0.9998	Cook and Oyen 2007
PMMA (33 min)	Berkovich	6.9 μm	0.1418	0.9999	Cook and Oyen 2007
it-PP (10 s)	Cube Corner	433 nm	0.0102	0.9997	Naimi-Jamal and Kaupp 2008
it-PP (10 s)	Berkovich	300 nm	0.0274	0.9997	Tranchida 2010
Polystyrene (1 s)	stiff Si-lever	553 nm	0.9738	0.9997	Kaupp and Naimi-Jamal 2010
Human Bone	Berkovich	500 nm	0.3394	0.9999	Weber et al. 2005
Lobster shell	Cube Corner	270 nm	0.0259	0.9993	Kaupp and Naimi-Jamal 2010
Pistachio	Cube Corner	420 nm	0.0360	0.9998	(Kaupp and
Macadamia	Cube Corner	300 nm	0.0389	0.9995	Naimi-Jamal 2011)
Cherry Stone	Berkovich	400 nm	0.0936	0.9995	Kaupp and Naimi-Jamal 2010
Copper (001)	Berkovich	200 nm	0.3043	0.9999	Shibutani and Tsuru 2007
γ-TiAl	Cube Corner	360 nm	0.2818	0.9996	Zambaldi et al. (2011)
InAs	Berkovich	200 nm	1.0666	0.9999	Le Bourhis and Patriarche 2005
Aluminum (exp.)	Berkovich	250 nm	0.1887	0.9994	Soare et al. 2005
Aluminum	Berkovich	1.2 μm	0.2233	0.9999	Naimi-Jamal and Kaupp 2004
ZnO	Berkovich	200 nm	1.3100	0.9997	Fang and Kang 2008
MgO	Berkovich	530 nm	4.1625	0.9999	Tromas et al. 1999
PDMS	Sphere(192μm)	3.5μm	0.001262	0.9999	Ebenstein et al. 2006
PDMS(1:10)	Cone (90°)	556μm	0.000878	0.9981	Lim, Chaudhri 2004

Figure 2: Slightly supplemented table from Reference [5] with corrected penetration resistance k (1) (factor 0.8), with unchanged correlation coefficients of various materials, indenters, methods, and authors for the whole length, all without phase transition up to h_{max}.

h parabolic. We can thus safely calculate the total applied work from the triangle as in Figure 3 (or Figure 5). Different ways of normal force applications (force controlled, displacement controlled, continuous stiffness, squared progression of the load increments) cannot decrease this applied work. Furthermore, the analysis of strongly creeping loadings (e.g. PMMA data in Figure 2) also gives the unfitted $h^{3/2}$ parabolas (1) with excellent correlation [5] excluding chances to improve the ISO- or FE-indentation efficiency. The formerly forgotten and not considered decreased energy for the indentation and thus also for the actual indentation load part is a striking violation of the first energy law. Only the fraction of the full applied work depends on the exponent on h.

But unfortunately we have to respond against continuing strange attacks on the quantitative treatment of conical or pyramidal indentations without any approximations simulations or fittings, despite the publications [4,5]. The probably last denial of the well-established experimental evidence of the exponent 3/2 on h [9] repeats the offence of Troyon (advocating depth dependent broken exponents such as 1.64533 or 1.75285 on h without discussing the incredibly changing dimensions) [10], which is combined with the violation of the first energy law (not considering [4]). Furthermore, Merle [9] tries to invoke the undisputed self-similarity of cones and pyramids as a theoretical argument. But Merle [9] incorrectly claims that this should be in favor of exponent 2. Self-similarity can by no means decide between the exponents in question. The exponent 3/2 is physically founded [6], and all data relying on the false exponent 2 require correction with the dimensional factor $h^{1/2}$. Furthermore, these unduly opposing authors tried to discredit the successful Kaupp-plot (F_N versus $h^{3/2}$) by calling it "Kaupp's double P-$h^{3/2}$ fit" [9] (P means force, the same as F_N here), even though the "Kaupp-plot" does not fit at all. They pretend that the kink (phase transformation) in the fused quartz example would have been claimed by intersecting an initial surface effect extrapolation line with the second linear branch, instead of equating the first and second linear branches (more of it in the Discussion). Kaupp has always been identifying surface effects and removing them from the regression.

Experimental and physical basis of pyramidal and conical instrumental indentation

The violation of the basic energy law is connected with the use of unphysical exponent 2 on h with implied assumption that the one third loss of the applied energy $\propto$ force (Figure 3) would not count for the peak load in the hardness H and modulus E_r calculations that use F_{Nmax} for the start of the unloading curve. The connection is quite simple and direct with the definition of universal hardness for indentations $H_{universal}=F_{Nmax}/A_{proj}$ (where A_{proj} is the projected area of the indenter). This has been worked out in Kaupp [7] with the formula sequence (6) leading to a disproved ISO $F_N \propto h^2$ relation:

$$F_{Nmax}=\pi R^2 H_{universal} \text{ and } R/h=\tan\alpha \text{ gives } F_{Nmax}=\pi h_{max}^2 \tan\alpha\, H_{universal} \quad (6)$$

The ISO $F_N \propto h_c^2$ relation is also obtained for the ISO-hardness $H_{ISO}=F_{Nmax}/A_{hc}$, where the so called contact height h_c must be adjusted to a standard material in a complicated procedure, including two multiparameter iteration steps [7]. Clearly there are three undisputable flaws against physics with these hardness determinations: 1. the violation of the basic energy law, 2. the use of unphysical exponent and 3. the non-considering of the often occurring phase transformations under load before the chosen peak load is reached, which can only be detected with the Kaupp-plot of (1). The energy law correction will be discussed in the next Section after presenting further support. The dimensional correction will be exemplified in the Sections dealing with the correction of hardness and modulus into physical values.

The convincing physical foundation of exponent 3/2 in the force depth relation (1) [6] (pre-published in 2015) leaves no doubt whatsoever with respect to the present author's analysis of his own and published loading curves from others who wrongly trusted and used the Sneddon/ISO/Oliver-Pharr exponent 2. All details of the loading curves can only be detected when the correct exponent 3/2

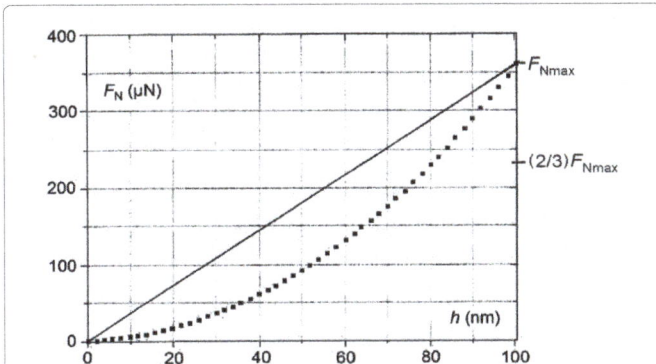

Figure 3: FE-simulated force displacement curve for 500 nm thick gold assuming h^2 with an ideal Berkovich and our comparison with the linearly applied total work (straight line from zero to F_{Nmax} [4]); the force-corrected parabola would end at the (2/3) F_{Nmax} point; evidently a large part of the applied work would be lost for the indentation; the dotted simulated force curve would precisely follow h^2; the simulated data points were taken from Reference [8] (their Figure 3b).

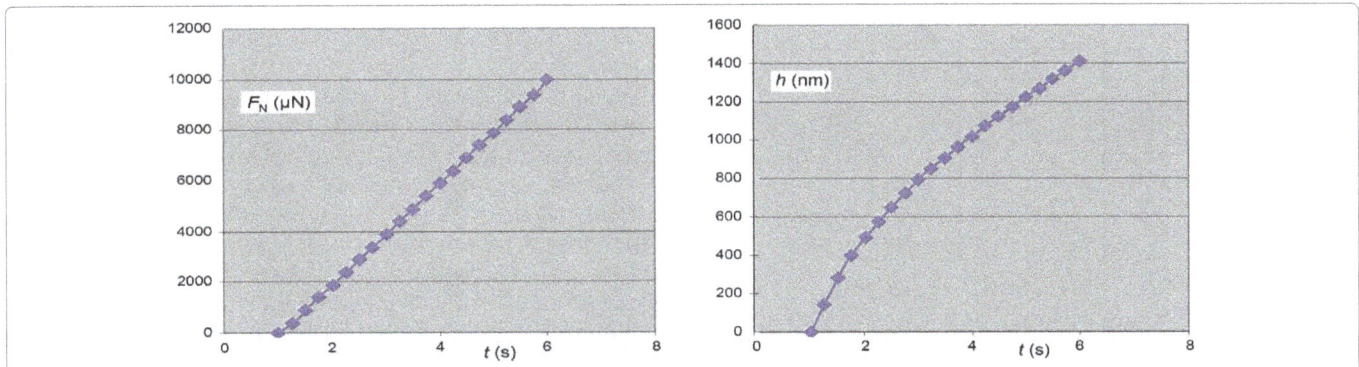

Figure 4: Plots of applied normal force F_N and depth h against time in a typical load-controlled NaCl indentation showing linearity of the applied force and non-linearity of depth.

on h is used for the analysis. The details are lost with unphysical plots and more so with data fitting, iterations, or present FE-simulations. Conversely, the physically founded linear F_N versus $h^{3/2}$ Kaupp-plots, as first introduced in lectures since 2000, correct for initial surface effects, reveal phase transformation if they occur within the chosen force range. Furthermore, they detect alternating layers, gradients, pores, defective tips, tilted impressions, and edge interface or too close-by impressions. For example, fused quartz Berkovich indents exhibit the well-known amorphous to amorphous phase transformation [11,12] at about 2.50 or 2.25 mN applied work and 113 or 107 nm depth (analyzed loading curve of Triboscope or CSIRO-UMIS manual, respectively) [11]. This is indicated by a sharp kink in the Kaupp-plot, as it occurs in the chosen loading range [5,11,13].

The force F_N is linearly applied in force controlled experimental indentations. This can again be imaged together with the exponent 3/2 parabola, which is physically founded [6] and experimentally found (Figure 2 [5,11]) and (1). Similar to Equations (2)-(5) deducing $W_{applied}/W_{indent}$ for the wrongly assumed ISO exponent 2 on h, the energetic deduction for the physical exponent 3/2 on h is given by the formulas (7)-(9). The physical ratio is thus $W_{applied}/W_{indent}$=5:4. The difference 5−4=1 is for the shear force component exerting pressure and plasticization on the adjacent material. That means: precisely 80% of the applied work and (as $W \propto F$) also applied force F_N is left for the penetration. Thus, 20% is for exerting the sum of pressure and plastic deformation energies to the solid environment. This is considerably less loss for the indentation than if the assumed unphysical exponent 2 would apply (33.33%, see above). The new knowledge is expressively supported with Figure 5 that shows the difference in relation to the Figure 3 for the false exponent.

$$F_N = k\, h^{3/2}$$

$$W_{indent} = 0.4\ \text{const}\ h^{5/2} \tag{7}$$

$$W_{applied} = 0.5\ F_{Nmax}\ h_{max} \tag{8}$$

$$W_{applied} = 0.5\ \text{const}\ h_{max}^{5/2} \tag{9}$$

We have now $W_{indent}=0.8\ W_{applied}$. The basic energy law is thus no longer violated when the applied force F_N (and thus also k) is corrected with the factor 0.8. Furthermore the definition of all physical parameters that are related to the indentation force must also not violate the first

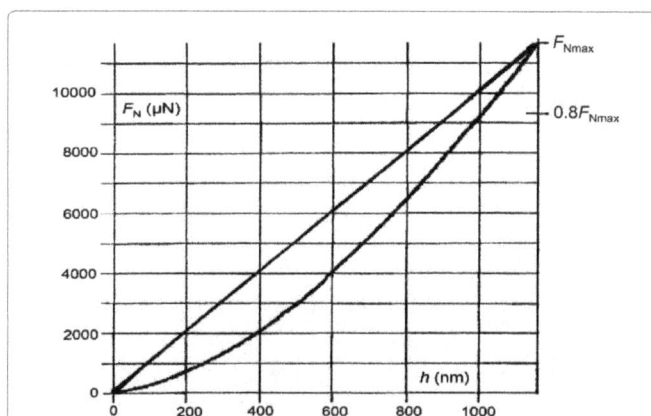

Figure 5: Experimental force displacement curve of aluminum (following the physical exponent 3/2 on the depth h [6]) and the comparison with the linearly applied force line, showing the loss of force (and energy) for the indentation depth; the measurement was with a Hysitron Nanoindenter (R); the force-corrected parabola would end at the 0.8 F_{Nmax} point.

energy law and require the factor 0.8, provided the exponent correction (2 giving 3/2) has also been performed. Importantly, the now deduced universal 5/4/1 ratio (applied/indent/long-range work) for pyramids and cones is valid for all uniform materials, be they elastic, plastic, migrating, viscous, sinking in, piling up, and flowing. Particular cases are surface effects, gradients, tilted or too tight or edge indentations, pores, micro-voids, cracks, defective tips' effects, and most important kink indicating phase transformation onset. It is valid for all differently angled smooth pyramids or cones with mathematical precision. Any deviations are experimental errors. Surface effects include water layers, gradients, oxides, hydroxides, surface compaction, tip rounding (sometimes compensating other surface effects), and the like. They do not belong to the bulk material and must therefore be eliminated from regressions.

Implementation of the first energy law in instrumented indentation

The energetics of the instrumented depth sensing indentation with pyramids or cones has first been published in 2013 [4] for the $F_N=k\, h^{3/2}$ relation. 20% of the applied work is lost for the indentation with mathematical precision due to the shear-force elastic and plastic work, including sink-in or pile-up. This is universal for all different shapes and materials.

As deduced above, the applied force F_N with the directly proportional otherwise physically correct published parameters (including H_{phys} in [7]) must be corrected with the factor 0.8 (5/4 ratio, 80%) (similarly for E_{r-phys}, see below). Thus, Figure 2 (all with correct exponent 3/2) corrects now the data from the originals in Kaupp [5,7]. Considering the advanced knowledge, this includes all the penetration resistance values k and phase-transformation conversion energies W_{conv} (both correction with the factor 0.8) that were published up to 2016. Not affected are the activation energies and the phase-transformation onsets at characteristic depth, because of cancellation. Also most of the other mechanical parameters from indentations in the literature including ISO-hardness and ISO-modulus are affected. The new knowledge that requires a further specification also for the hardness and modulus definitions requires separate treatment in the next Section below, because these require also the above mentioned dimensional correction.

The tip influence on the k-values (Figure 2) and their conversion between different tips has been demonstrated and can be normalized [13]. Creep depends on force and temperature. It is a materials property but does not change the exponent on h of the loading curve, only the penetration resistance k. Loading times should thus not exceed 30 s to avoid such influence. Independent creep measurements and corrections must only be performed for most precise rankings of materials. But it is usually much less severe than with the viscoelastic PMMA (strongly diverging from different authors) and certainly for the PDMS values of Figure 2. Indentation times are in fact generally very fast (10-30 s) and creep is mostly slow even at high temperatures, so that a rating along the k-values is a good choice already without creep corrections. Creep is mostly not corrected for or published, while thermal drift can be easily corrected for. Creep has however great importance for long-term pressure under heat and for the properties of viscoelastic materials with time dependent behavior. Importantly, the exponent on h remains 3/2 also at indentations of organic crystals with lattice guided anisotropic migrations [13,14].

Basic energy law and dimensional corrections of indentation

hardness

A quantitative foundation of conical or pyramidal nanoindentation results as for hardness (and modulus) has to obey the first energy law. All world suffers from such violation that requires correction. The $F_N=k\,h^{3/2}$ relation (1) corrects the fact that only 80% of F_N is used for the indentation with the adjusted k-value in accordance with the energy law. The correction of $H_{phys}=k/\pi\,\tan\alpha^2$ (mN/µm$^{3/2}$), as taken from the correct loading curve, where the factor 0.8 is included in the k-values, is exhaustive and complies with the first energy law. The physical indentation hardness has unavoidably the dimension (µN/nm$^{3/2}$) or (mN/µm$^{3/2}$) (11). The loading curve provides the easiest, most precise, most rapid and cheap way to obtain the correct physical hardness H_{phys}. The deduction of (11) starts with the definition of the universal hardness (F_N/A_{proj}) relying on unphysical h^2 and violating the energy law. This has to be corrected with the dimensional factor $h^{1/2}$ [that is also required to make Equation (6) concur with physics] for exponent correction to concur with the correct exponent 3/2 on h (1) [7] and the factor 0.8 to concur with the first energy law that is already contained in the penetration resistance k. This leads via (10) to (11) after expression of the projected area and insertion of (1) with cancellation of $h^{3/2}$. Importantly, the physical hardness H_{phys} is thus independent of projected area, depth, F_{Nmax}, and standard material. It avoids all iterations or fittings or approximations but is experimentally obtained by linear regression and it becomes a genuine physical quantity for the first time. It is also not falsified by undetected phase transformations, because these would show-up in the linear regression. A sharp kink before F_{Nmax} must be absent! The applications of H_{phys} should be very welcome. It is nothing else than a normalized penetration resistance. For example the physical hardness values can be directly obtained from the examples in Figure 2 by using the α-values of the corresponding indenters (Berkovich is ISO-standard).

$$H_{phys}=0.8\,F_{Nmax}\,h_{max}^{1/2}/\pi\,h_{max}^{2}\tan\alpha^2=k\,h_{max}^{3/2}\,h^{1/2}/\pi h^2\tan\alpha^2 \qquad (10)$$

$$H_{phys}=k/\pi\tan\alpha^2 \text{ (mN/µm}^{3/2}) \qquad (11)$$

The odd appearing dimension mN/µm$^{3/2}$ (also GPa µm$^{1/2}$) of the physical indentation hardness, which does only resemble to a pressure is unavoidable, due to the mathematically fixed shear force component of indentations that cannot be avoided. Nevertheless, indentation remains a very useful particularly precise technique.

Universal hardness, ISO hardness, and FE-simulated hardness would require a factor 2/3 for correction of F_N to give the force for the indentation in order to accept the energy law (Figure 3). But after the necessary multiplication with $h_{max}^{1/2}$ for dimensional correction the force correction becomes 0.8 (Figure 5). However, such corrections of the ISO hardness can only be approximate, because the h_c and thus A_{hc} iterations with respect to a standard material cannot be reverted. Force induced phase-transformations must always be excluded with a Kaupp-plot that at the same time obtains the physical hardness more

safely and directly (11).

The equations (12) and (13) show how easy it is to calculate H_{phys} from published H_{univ} or H$_{simul}$ values, provided the h_{max} values for F_{Nmax} are available, and when phase changes are excluded before F_{Nmax} is reached. The corrections are multiplications with $h_{max}^{1/2}$ for the correct exponent 3/2 and factor 0.8 for the force loss.

$$H_{univ}=F_{Nmax}/\pi\,\tan\alpha^2\,h_{max}^{2} \qquad (12)$$

$$H_{univ\text{-}corr}=H_{phys}=0.8\,h_{max}^{1/2}\,F_{Nmax}/\pi\,\tan\alpha^2\,h_{max}^{2} \qquad (13)$$

This is exemplified in Table 1 with a numerical example from a published indentation onto aluminum, where H_{ISO} and both the FE-simulated H$_{simul}$ (ANSYS software) with exponent 2 on h and the experimental Berkovich loading curves are published (falsely claimed exponent 2 but according to the Kaupp-plot determined with exponent 3/2 on h) [15]. Any universal hardness (H_{univ}) treatment would be the same as the one for H$_{simul}$. The published loading curve was also provisionally analyzed as F$_N$-h^2 plot but only used for numerical achievement of the conversions.

Entry 1 in Table 1 gives the H_{phys} from the analyzed loading curve (11), which is certainly the most reliable value. It does not rely on F_{Nmax}, h_{max}, any h_c or A_{hc} and it secures the absence of a phase change up to the maximal force. And it compares with H_{ISO} and the hardness values that derive from H$_{simul}$ with various stages of correction.

Entry 2 shows that H_{ISO} exhibits a far too high value and an unphysical dimension. The energy correction for leaving exponent 2, removing only the energy law violation, decreases the value insufficiently, still with the unphysical dimension mN/µm^2. A value for h_{max} is not available for a final correction. When exceptionally a guess were tried that it might be in a 0.25 µm region one would guess a further decrease that would look like 0.239 with the changed dimension mN/µm$^{3/2}$. This would be in the region of H_{phys} although with all reservation, because it is only a free guess only indicating the direction. This show the difficulties for the conversion when h_{max} for the used F_{Nmax} is not reported. It is thus much easier to apply the Kaupp-plot to the loading curve (1). We renounce of including the uncorrected simulated value (0.6016 mN/µm^2).

Entry 3 gives only the exponent correction of FE H$_{simul}$ (ANSYS-software) that was probably obtained by using Young's modulus E (either known or iterated) input, with converging criterion to exponent 2 on h.

Entry 4 gives only the energy correction with a rather high value. Table 1 show that neither the exponent correction for exponent 2 alone nor the energy correction (Figure 3) alone (removing energy law violation) is sufficient.

Entry 5, finally with both exponent correction and then smaller energy correction factor for h$^{3/2}$ (Figure 5) provides H$_{simul\text{-}phys}$, with

Number	Technique	h_{max}^{n}	k or h_{max} [a]	Hardness calculations and corrections
1	Experimental linear regression	$h_{max}^{3/2}$	$k=5.9540$(mN/µm$^{3/2}$) (energy corrected)[b]	$H_{phys}=k/\pi\tan\alpha^2=0.24295$(mN/µm$^{3/2}$) independent on F_N and h_{max} (no phase trans.)
2	Experimental with 2/3 factor	h_{max}^{2}	--	$H_{ISO}=0.716$ (GPa) x (2/3) ≈ 0.477 (mN/µm^2) (unphysical dimension) h_{max} not known
3	FE-simul. $h_{max}^{1/2}$ no energy corr.	h_{max}^{2}	$h_{max}=251.984$ nm	$H_{simul\text{-}corr1}$ (as H_{univ})$=F_{Nmax}/\pi\tan\alpha^2 h_{max}^{3/2}=0.2977$ (mN/µm$^{3/2}$) (energy law violation!)
4	FE-simul. 2/3; no exponent corr.	h_{max}^{2}	$h_{max}=251.984$ nm	$H_{simul\text{-}corr2}=2F_{Nmax}/3\pi\tan\alpha^2 h_{max}^{3/2}=0.4011$ (mN/µm^2) (wrong exponent)
5	FE-simul, 0.8, and $h_{max}^{1/2}$	h_{max}^{2}	$h_{max}=251.984$ nm	$H_{simul\text{-}phys}=0.8\,F_{Nmax}/\pi\tan\alpha^2 h_{max}^{3/2}=0.2382$ (mN/µm$^{3/2}$)

Note: [a]Simulated parameters are not italicized; [b]correction factor 0.8.

Table 1: Comparison and correction of unloading H_{ISO} and FE-simulated H$_{simul}$ loading curves of Al on Si [15] with the physical H_{phys}, which is in accordance with the energy law.

surprisingly good match (2%) with H_{phys}. The surprisingly close coincidence of H_{phys} and $H_{simul-phys}$ supports the numerical correctness of the non-fitting (!) straightforward deduction and it also reminds the unbeatable precision of the Kaupp-plot's linear regression (Figure 2). The close correspondence with $H_{simul-phys}$ in this case should however be tested for generality, because this single example could be fortuitous when considering the parameterizations and iteration procedures at FE simulations.

Importantly, the striking dilemma of ISO with physics persists with the false dimension of too large H_{ISO} and unphysical dimension. All of the values and dimensions of the mechanical parameters that depend on it are severely wrong, also those that depend on wrong ISO elastic modulus E_r (see next Section). Clearly, Table 1 and Equations (12) with (13) show an easy way for straightforward corrections of $H_{univers}$, probably H_{simul}, and with reservation H_{ISO}, provided the h_{max} values are known. However, despite the straightforward corrections none of them can handle the very often occurring and so important phase transformations under load (here they were experimentally excluded with Kaupp-plot).

Basic energy law and corrections of indentation elastic modulus

Also the correction of unphysical E_{r-ISO} into a physical value is essential, because elasticity is a technically important materials property. Young's moduli E are required for the deduction of numerous mechanical qualities and for example increasingly as input parameter for numerical FE-simulations, often including FE-iterations with E-Y pairs as free parameters (where Y is yield strength). The determination of the elastic modulus requires the unloading stiffness $S=dF_{Nmax}/dh_{max}$ from the pressure to the displaced material that must be separated from the plastic response. This is achieved for F_{Nmax}, which is a joint quantity of loading and unloading curves. Thus, there must again be corrections for dimension adjustment with $h^{1/2}$ [for not violating (1)] and for shear force loss during the loading (for not violating the energy law). These are not applied in ISO 14577 that applies the Oliver-Pharr iterations [3]. Thus, the slope correction for E_{r-ISO} (14) requires again the exponent correction with $h^{1/2}$ and then the force correction factor 0.8 (Figure 5) to comply with the energy law for obtaining E_{r-phys}. This gives via (14) E_{r-phys} (15) in complete correspondence with the necessary treatment of H_{ISO} (this replaces the incomplete formula 11 in [7]). Again one must be certain that the unloading was performed at F_{max} before any onset of a force derived phase transformation had occurred. By comparing (14) and (15) the correction factors are found to be 0.8 and $h_{max}^{1/2}$. The corrections for obtaining physical modulus values with changed dimension is simply by multiplication with 0.8 $h_{max}^{1/2}$. The unloading stiffness S and h_{max} (before creep) must be known. This is another dilemma between ISO and physics.

$$E_{r-ISO}=S\,\pi^{1/2}/2\,A_{proj}^{1/2}=(dF_N/dh)_{max}/2\,h_{max}\,\tan\alpha \quad (14)$$

$$E_{r-phys}=0.8\,\pi^{1/2}\,h_{max}^{1/2}\,S/2\,\pi^{1/2}\,h_{max}\,\tan\alpha=0.8\,S/2\,h_{max}^{1/2}\tan\alpha\;(mN/\mu m^{3/2}) \quad (15)$$

Discussion

The extremely complicated mathematical deductions of Sneddon/Love ([1,2]; Figure 1) for the conical or pyramidal indentations did not consider the energetics of the process, as illustrated with the Figures 3 and 5. And there was no protest from physicists. Almost all involved people followed Sneddon [1], Oliver Pharr [3], and ISO 14577 all with violating the first energy law for more than half a century. The

general acceptance for half a century of the implied claim that pressure formation and plasticization could be workless achieved is hard to understand. It is apparently the result of hype upon the publication [1] that unfortunately was believed by ISO/NIST. The simple equations as derived starting in 2000 ([14] and before in lectures and in refused manuscripts) and the point by point unraveling of the field until now against strong impediments did not help. The newcomers had to obey ISO 14577 and many very complex rules, and they used the software of the instrument suppliers that had to trust in the ISO/NIST-standards. By doing so they forgot to think on the physical foundations. Thus, the basic formulas (3)-(5) and (7)-(9) that essentially rely on the experimentally (since 2000) (Figure 2) and physically founded (since 2015) Equation (1) [6] found much refusal, various excuses for not experimentally finding exponent 2 with fittings, multi-parameter iterations, and simulations. The actions against the elementary algebraic treatment without any fitting/iterating/simulating were undue repetitive offenses. Rather acknowledgement had to be expected because everything became much easier and quantitative on a sound physical basis with simple closed mathematical formulas, proving the universal validity.

Apparently, nobody else (not even textbook or tutorial writers) asked themselves why all of the applied normal force with cones or pyramids is claimed to be used for the indentation depth, even though the loading curve proceeds not linearly but parabolic. The obvious answer is that well-known long range effects and pressure formation to the environmental solid material require energy that is lost for the indentation depth. When this energy/force/loss was quantified and finally (after difficulties with anonymous Reviewers) published in 2013 [4] with the universal loss of 1/5 for the physical (1) and 1/3 for unphysical (2) equations, there was discussion about the validity for comparing applied work and indentation work. But these proceed at the same time to the same endpoint F_{max}. Surprised about the ease of the mathematical deduction and the strict and universal result, requiring difficult necessary changes, there were objections and much open discussion in plenary lectures from the audience with the guess that all of the linearly applied force might instead go along the parabolic curve during the experiment. This prevented the opponents from recognizing that the first energy law was evidently violated. The linearity of the applied force is however also evident, simply from the additional applied force F_N versus time plot in Figure 4.

The undue opposition against straight forward physics and algebra is surprising even after it was very clear with Kaupp [4] that the ISO-system violates the first energy law (the present author could not dare to verbally express the energy law violation at that time). The offenses have been continuing. For example, the opposing manuscript [9] was received at Scanning on May 27, 2014, whereas the clarifying manuscript [4] was received at Scanning on October 4, 2012 and published on February 25, 2013. The content of Kaupp [4] had thus to be taken up again in Kaupp [7] with more details, because the authors, reviewers, and editors of Merle [9] continued violating the basic energy law. And the Merle [9] continued arguing against the most precise Kaupp-plot that actually was the basis for the quantification of the violation. The opponents tried with iterated own loading curves of fused quartz. But when doing it correctly, even the invoked curve in Troyon [10] would roughly reproduce the well-known transformation onset, despite its using a blunt tip that gave an unusually long initial effect. And Merle [9] tries again with a false intersection at its microindentation "Kaupp-plot" (up to 300 mN and 1600 nm) where the region with all of the nanoindentation details is almost totally obscured in a short unstructured part of it. The false intersection with a remote line far away from the plot is useless. But it is used for falsely criticizing the Kaupp-plots that never used or use such faulty tricks. When properly looking at this linear plot in Reference [9] with a ruler, one recognizes an intersection of two straight lines at about 175 mN and 1225 nm, which the authors do neither trace nor recognize. Four possibilities exist for this kink very close to the plot: either a new high-load phase transition of fused quartz occurred, or a smoothness defect of the tip was present at this depth, or a remote crack at such deep impression was formed, or the impression was too close to an edge/interface/impression. Furthermore, these authors claim and draw a straight single line for their unphysical so called "P-h^2 fit with 0.999 fit quality". However, despite their claimed "three-nines fit", their depicted unphysical "P-h^2 fit" gives two roughly linear branches, intersecting in the region of 60-70 mN force (that is far away from surface effects). This deviation from the claim is easily "overlooked" without a ruler in a wide pencil stroke representation at totally false depth-square scaling (better seen when more precisely drawn, the first part steeper and cutting at small angle). This shall only be a necessary contradiction to the false claim of linearity for a "P-h^2 fit" trying to discredit our simple algebraic treatment on a sound physical foundation. Fitted or FE curves, converging with h^2, must not be used for denying thoughtful and repeatable physically founded [6] and experimental Equation (1). Only untreated experimental loading curves are able to detect surface effects, the important phase changes, conversion energies, etc., when using the physically founded exponent on h.

A problem might arise when fitted, iterated or FE-simulated curves and experimental loading curves might be mixed up in publications. However, when experimental force data are plotted with or fitted to the non-physical h^2, the deviations from a straight line might appear minor for example as in Merle [9]. Also a minor endothermic phase change slightly levels the unphysical F_N-h^2 trial-plot with respect to the stronger curved appearance without phase change [5,11]. Such leveling behavior of the test material fused quartz might have strengthened the belief in h^2, but it reflects the inability to find phase transformation with the physically wrong exponent 2 on h. All of the important details of nanoindentation are lost with h^2. But the kink at $F_N \approx 2.4$ mN (Berkovich) and initial surface effects of the fused quartz standard are easily seen by sharp kinks with the precise Kaupp-plot (1) in nanoindentations, notwithstanding the cases of later or further phase changes in microindentations (e.g. NaCl in [5,11]). But there is no excuse for using the unphysical exponent and thus denial of the phase transitions if these occur, combined with the violation of the first energy law.

The readers of Kaupp [4] and the attendants of the present author's lectures on numerous worldwide conferences were repeatedly urged to think about the unexpected and surprisingly easy deduced energetic facts (2)-(6) and (7)-(9) but the expected response of the scientific establishment is still missing. It appeared unlikely that all of the scientific Celebrities and their successors including textbook authors, ISO/NIST, and numerous anonymous referees have, consciously or not, been violating the first energy law for more than 50 years. Hesitation to use only the normal force left for the indentation depth was thus advisable, before any non-apparent compensation effect for saving the energy law was excluded in the desperate situation. Publications of the truth should stay as close as possible with the current indentation theory unless all objections are removed. Clearly, the believers in exponent 2 on h could for themselves have easily performed the deductions as in Equations (1)-(5) and could have tried to change their minds because of this inexcusable energy law violation. But they did not try to take

into account the always occurring energy loss. Based on their believed exponent 2 on h it would have amounted to 1/3 (33.33%) of F_N due to the work and force proportionality, as shown above with the trial Equations (2)-(5). And they would have found that the violation is also programmed and used in FE-simulations. They refused till now to accept the undeniable wealth of the Kaupp-plot and the physical deduction for the correct exponent 3/2 on h [6] that finally proves energy/force loss of 1/5 (20%) according to Equations (7)-(9), as only the physical exponent is correct. Since ISO/NIST have been reluctant to change their minds, or to announce reconsideration with an alert, there was the urgent preliminary publication in Kaupp [7] for expressively naming the incredible claim of workless pressure formation and plasticization as "violation of the basic first energy law". This is now completed with valid transformation formulas for obtaining the physical values and the necessary conditions for that from unphysical publications. Furthermore, the most easy and precise H_{phys} determination by linear regression of the loading curve (1) (hitherto strongly refuted Kaupp-plot) with energy-based correction is now again strongly advocated for.

Conclusions

The still not settled dilemma between ISO and physics with respect to ISO 14577 (not even an alert has been filed yet) is unbearable due to its enormous risks for science and daily life's safety. It appears unbelievable and even desperate that the first energy law was drastically violated for more than 50 years and none of the physicists protested against such habit. Everything is easily deduced with first grade algebra, avoiding fittings, iterations, simulations, and approximations, making everything much more easy. Hardness is now obtainable by linear regression, no longer by iterations, fittings, approximations, and simulations that are not ready for a controlling assessment. The physical indentation hardness H_{phys} (mN/µm$^{3/2}$) is now for the first time a genuine physical quantity, obeying Equation (1) and the first energy law. The same is true for the indentation modulus $E_{r\text{-phys}}$ (mN/µm$^{3/2}$). The complete, more precise deduction than in Kaupp [7] reveals also the simple conversion from $E_{r\text{-ISO}}$. Only the quantitative indentation on the physical basis reveals numerous otherwise impossible applications. Examples are phase change [4,5,16], conversion energies [4,16] and activation energies [16] of materials, all on the basis of the so-called Kaupp-plot (1) that also checks for correctly performed indentations and provides extrapolation facility up to recognized phase change qualities under pressure. Furthermore, it reveals a large number of special materials' properties and indentation errors that are named above. But it is still being heavily suppressed by the ignoring establishment, including ISO and some anonymous Reviewers with incredible unqualified wording instead of acknowledging this wealth.

The liability with unphysical calculated materials' properties is totally unclear at the present dilemma between ISO and physics, because all safety engineers are falsely trained. That means, the issue counts for every days safety unless ISO files at least an urgent alert. Everybody knows how many materials fail shortly after the warranty period, certainly not purposeful but often with falsely calculated materials. Even worse, falsely calculated components like poorly adjusted medicinal implants or larger scale composites can produce disasters. There is good reason why passenger traffic airplanes require frequent safety checks and complete replacement of all parts within 2 years. For example h goes with $F_N^{2/3}$ not with $F_N^{1/2}$ [5] with all implications for fatigue, and wear, to name a few.

Despite the highly comprehensive results of this paper and numerous worldwide lectures on conferences the ISO versus physics dilemma still remains. The physical indentation H_{phys} and $E_{r\text{-phys}}$ dimensions that only resemble pressures is perhaps difficult to understand at first glance. But it is real and the reasons have been discussed. Importantly this does not detract from indentation as a very precise and reproducible technique, when properly executed, checked, and algebraic evaluated, that means without fittings, iterations, and simulations.. Rather the unavoidable dimensional changes have an enormous bearing for science and practice. The not fitted and not iterated physical quantities must be used to redefine the numerous further mechanical parameters that were deduced from unphysical H_{ISO} or E_{r-ISO}. Further studies are necessary and further important insights are to be expected when the violation of the first and most basic energy law will be removed also for the deduced parameters. This should help for a better understanding and open new horizons. Also textbooks must be rewritten for the sake of physics, compatible materials sciences, and new insights. Since there was the violation of the first energy law, the new results will prove to be more compatible with all related techniques that do not violate physical laws, which is very desirable. The quantitative indentation at the now physical basis has the indispensable advantages of being precise, and in accord with basic principles. This is promising and cannot be denied. The further advancement on the physical basis is a very urgent task that must be pursued, hopefully soon also with ISO/NIST against all of the incredible resistance, because violating the first energy law is an inexcusable fault.

References

1. Sneddon IN (1965) The relation between load and penetration in the axisymmetric Boussinesq problem for a punch of arbitrary profile. Int J Engn Sci 3: 47-57.

2. Love AEH (1939) Boussinesq's problem for a rigid cone. Q J Math (Oxford) 10: 161-175.

3. Oliver WC, Pharr GM (1992) An improved technique for determining hardness and elastic modulus using load and displacement sensing indentation experiments. J Mater Res 7: 1564-1583.

4. Kaupp G (2013) Penetration resistance: a new approach to the energetics of indentations. Scanning 35: 392-401.

5. Kaupp G (2013) Penetration resistance and penetrability in pyramidal (nano) indentations. Scanning 35: 88-111.

6. Kaupp G (2016) The physical foundation of $F_N=k\ h^{3/2}$ for conical/pyramidal indentation loading curves. Scanning 38: 177-179.

7. Kaupp G (2016) Important consequences of the exponent 3/2 for pyramidal/conical indentations-new definitions of physical hardness and modulus. J Mater Sci & Engineering 5: 285-291.

8. Wang TH, Fang TH, Lin YC (2008) Finite-element analysis of the mechanical behavior of Au/Cu and Cu/Au multilayers on silicon substrate under nanoindentation. Phys Appl A 90: 457-463.

9. Merle E, Maier V, Durst K (2014) Experimental and theoretical confirmation of the scaling exponent 2 in pyramidal load displacement data for depth sensing indentation. Scanning 36: 526-529.

10. Troyon M, Abbes F, Garcia Guzman JA (2012) Is the exponent 3/2 justified in analysis of loading curve of pyramidal nanoindentations? Scanning 34: 410-417.

11. Kaupp G, Naimi-Jamal MR (2010) The exponent 3/2 at pyramidal nanoindentations. Scanning 32: 265-281.

12. Trachenko K, Dove M (2003) Intermediate state in pressurized silica glass: reversibility window analogue. Phys Rev B 67: 212203/1-212203/3.

13. Kaupp G (2006) Atomic force microscopy, scanning nearfield optical microscopy and nanoscratching - application to rough and natural surfaces. Springer, Berlin.

14. Kaupp G, Schmeyers J, Hangen UD (2002) Anisotropic molecular movements in organic crystals by mechanical stress. J Phys Org Chem 15: 307-313.

15. Soare S, Bull SJ, Oila A, O'Neill AG, Wright NG, et al. (2005) Obtaining mechanical parameters for metallization stress sensor design using nanoindentation. Int J Mater Res 96: 1262-1266.

16. Kaupp G (2014) Activation energy of the low-load NaCl transition from nanoindentation loading curves. Scanning 36: 582-589.

Permissions

List of Contributors

Hasan MF
Department of Industrial Engineering and Management, Khulna University of Engineering and Technology, Khulna-9203, Bangladesh

Pruthviraj RD
R&D Centre, Department of Chemistry, Raja Rajeshwari College of Engineering, Bangalore, India

Rashmi M
Department of Chemistry, Sri Krishna Institute of Technology, Bangalore, India

Najafov BA and Abasov FP
Institute of Radiation Problems of Azerbaijan Nationale Academy of Science, Baku, Azerbaijan Republic, Russia

Udhayakumar T and Mani E
Corporate Technology Centre, Tube Investments of India Ltd, Chennai-600 054, India

Kouider N, Bengourram J and Mabrouki M
Laboratoire Génie Industriel, Faculté des Sciences et Techniques de Béni Mellal, Morocco

Gamouh A and Chtaini A
Equipe d.Electrochimie Moléculaire et Matériaux Inorganiques, Faculté des Sciences et Techniques de Béni Mellal, Morocco

Forsal I
Laboratoire de Chimie Organique et Analytique, Faculté des Sciences et Techniques de Béni Mellal, Morocco Cruz do Sul, Brazil

Ahmad AF
Materials Processing and Technology Laboratory, Institute for Advance Material, Universiti Putra Malaysia, Serdang, Selangor Darul Ehsan, Malaysia

Abbas Z and Abdalhadi DM
Department of Physics, Faculty of Science, Universiti Putra Malaysia, Serdang, Selangor Darul Ehsan, Malaysia

Obaiys SJ
School of Mathematical and Computer Sciences, Heriot-Watt University Malaysia, Putrajaya, Malaysia

Chandan BR and Ramesha CM
Department of Mechanical Engineering, MS Ramaiah Institute of Technology, Bangalore, Karnataka, India

Baykara T and Bedir HF
Mechanical Engineering Department, Doğuş University, Istanbul, Turkey

Deepak Kumar G
Department of Chemistry, Indian Institute of Space Science and Technology (IIST), Trivandrum, Kerala, India

Singh Arora SP
Scientist/Engineer-SD, Indian Space Research Organisation (ISRO), Govt. of India, Tamil Nadu, India

Sahuban Bathusha MS, Chandramohan R, Vijayan TA, Saravana Kumar S and Jayachandran M
Research Department of Physics, Sree Sevugan Annamali College, Devakottai, India

Sri Kumar SR
Department of Physics, Kalasalingam University, Krishnankoil, India

Ayeshamariam A
Department of Physics, Khadir Mohideen College, Adiramapattinam, India

Essa S
Department of Civil Engineering, Erbil Technical Engineering College, Erbil, Iraq

Asvath R and Suresh M
Department of Mechanical Engineering, Sri Krishna College of Engineering and Technology Coimbatore, India

Dou Z, Wang G, Zhang E, Ning F, Zhu Q, Jiang J and Zhang T
Hubei Collaborative Innovation Center for Advanced Organic Chemical Materials, Ministry of Education Key Laboratory for the Green Preparation and Application of Functional Materials and School of Material Science and Engineering, Hubei University, Wuhan, China

Fandem QA
Aramco Qatif, Eastern Region, Saudi Arabia

Lakhe MG, Bhand GR, Londhe PU, Rohom AB and Chaure NB
Electrochemical Laboratory, Department of Physics, Savitribai Phule Pune University, India

Chennaiah MB
Assistant Professor in ME Department, V.R.Siddhartha Engineering College, Vijayawada, India

Kumar PN
Professor in ME Department, N.B.K.R Institute of Science and Technology, Vidyanagar, India

Rao KP
Professor in ME Department, J.N.T.University College of Engineering, Ananthapuram, India

Nandan Kumar GM and Kingsly Jasper M
VIT University, Vellore, Tamil Nadu, India

R Patel and K Ghosh
Department of Physics, Astronomy and Materials Science and Center for Applied Science and Engineering, Missouri State University, Springfield, MO 65897, USA

A Bhaumik and A Haque
Department of Physics, Astronomy and Materials Science and Center for Applied Science and Engineering, Missouri State University, Springfield, MO 65897, USA Department of Materials Science and Engineering, North Carolina State University, Raleigh, NC 27695, USA

MFN Taufique
Department of Physics, Astronomy and Materials Science and Center for Applied Science and Engineering, Missouri State University, Springfield, MO 65897, USA School of Mechanical and Materials Engineering, Washington State University, Pullman, WA 99164-2920, USA

P Karnati
Department of Physics, Astronomy and Materials Science and Center for Applied Science and Engineering, Missouri State University, Springfield, MO 65897, USA Department of Materials Science and Engineering, Ohio State University, Columbus, OH, 43210, USA

M Nath
Department of Chemistry, Missouri University of Science and Technology, Rolla, MO 65409, USA

Walid K Hamoudi, Dayah N Raouf and Narges Zamil
Applied Sciences Department, University of Technology, Baghdad, Iraq

Nur Atikah Adnan, Kandasamy R and Mohammad R
Research Centre for Computational Fluid Dynamics, FSTPi, Universiti Tun Hussein Onn Malaysia, Malaysia, 86400, Batu Pahat, Johor, Malaysia

Martin A, Addiego F, Mertz G, Bardon J, Ruch D and Dubois P
Luxembourg Institute of Science and Technology, Materials Research and Technology, Luxembourg

Omoniyi FIS, Olubambi PA and Sadiku ER
Department of Chemical, Metallurgical and Materials Engineering, Tshwane University of Technology, Pretoria, South Africa

Mongelli GF
Department of Chemical Engineering, Case Western Reserve University, Cleveland, USA

Mohammad HM
Materials Engineering Department, College of Engineering, University of Basrah, Iraq

Ibrahim RH
Mechanical Engineering Department, College of Engineering, University of Basrah, Iraq

Khan A and Joshi S
Department of Physics, Motilal Vighyan Mahavidhyalaya, Bhopal, India

Ahmad MA
Department of Physics, Faculty of Science, University of Tabuk, Saudi Arabia

Lyashenko V
Laboratory "Transfer of Information Technologies in the Risk Reduction Systems", Kharkov National University of Radioelectronics, Kharkov, Ukraine

Ali Kalkanlı and Tayfun Durmaz
Department of Metallurgical and Materials Engineering, Middle East Technical University, Ankara, Turkey

Ayşe Kalemtaş
Department of Metallurgical and Materials Engineering, Bursa Technical University, Turkey

Gursoy Arslan
Materials Science and Engineering, Anadolu University, Eskişehir, Turkey

Prashanth S
Department of Mechanical Engineering, Gargi Memorial Institute of Technology, Bharathinagar, Mandya, India

Subbaya KM
Department of Industrial and Production Engineering, The National Institute of Engineering, Mysuru, India

Nithin K
Department of Chemistry, The National Institute of Engineering, Mysuru, India

Sachhidananda S
Department of Polymer Science and Technology, Sri Jayachamarajendra College of Engineering, JSS S&T University, Mysuru, India

Bingbing Fan, Jian Li and Hao Chen
School of Materials Science and Engineering, Zhengzhou University, Zhengzhou, Henan 450001, China

Fan Zhang
School of Materials Science and Engineering, Zhengzhou University, Zhengzhou, Henan 450001, China

Henan Information and Statistics Vocational College, Zhengzhou, Henan, 450002, China

Rui Zhang
School of Materials Science and Engineering, Zhengzhou University, Zhengzhou, Henan 450001, China

ZhengZhou Institute of Aeronautical Industry Management, Zhengzhou, Henan 450015, China

Kareem AA and Raheem Z
Department of Physics, College of Science, University of Baghdad, Iraq

Chandra RK, Majid M, Arya HK and Sonkar A
Department of Mechanical Engineering, SLIET, India

James CM Li
Materials Science Program, Department of Mechanical Engineering, University of Rochester, Rochester, New York, USA

Islam Bossunia MT, Sarwaruddin Chowdhury AM, Poddar P and Gulenoor F
Department of Applied Chemistry and Chemical Engineering, Faculty of Engineering, University of Dhaka, Bangladesh

Khan RA
Institute of Nuclear Science and Technology, Bangladesh Atomic Energy Commission, Bangladesh

Hasan MM
Institute of Leather Engineering and Technology, University of Dhaka, Bangladesh

Hossain MT
Bangladesh Jute Research Institute, Bangladesh

Islam MI Moustafa and Ihab A Saleh
Chemistry Department, Faculty of Science, Benha University, 13518, Egypt

Mohamed R Abdelhamid
Department of Medicine and Hepatology, Faculty of Medicine, Minia University, Egypt

Scharf C
Helmholtz Centre Dresden Rossendorf, Helmholtz Institute Freiberg for Resource Technology, Chemnitzer Strasse 40, 09599 Freiberg, Germany

Ditze A
MetuRec, Metallurgie und Recycling, Ingenieurbüro, An den Eschenbacher Teichen 16, 38678 Clausthal-Zellerfeld, Germany

Kaupp G
University of Oldenburg, Germany

Index